Composition by Acme Art, Inc.

Library of Congress Cataloging-in-Publication Data

Hofstadter, Douglas R., 1945–
    Fluid concepts and creative analogies: computer models of the fundamental mechanisms of thought / by Douglas Hofstadter and the Fluid Analogies Research Group.
      p. cm.
    Includes bibliographical references and index.
    ISBN 0–465–05154–5
    1. Cognitive science.   2. Analogy—Computer simulation.   3. Artificial intelligence.   I. Fluid Analogies Research Group.   II. Title.
BF311.H617   1994
153.4—dc20
                                 93–44294
                                   CIP

95 96 97 98   RRD   9 8 7 6 5 4 3 2 1

Fluid Concepts
&
Creative Analogies

*Computer Models
of the Fundamental Mechanisms
of Thought*

```
       *
   *       *
*      *       *
   *       *
       *
```

**by
Douglas Hofstadter
and the
Fluid Analogies Research Group**

BasicBooks
A Division of HarperCollins*Publishers*

To D. and M.

Table of Contents

List of Illustrations

Acknowledgments

Prior Sources for Chapters

Chapter 2 ("The Architecture of Jumbo" by D. Hofstadter) appeared in a highly condensed version in 1983, under the same title, on pages 161–170 of the *Proceedings of the International Machine Learning Workshop*, edited by R. Michalski, J. Carbonell, and T. Mitchell, published by the University of Illinois Press (Urbana, Illinois). Essentially the whole of Chapter 2 also appeared in 1985, in an Italian translation, under the title "L'architettura del 'Jumbo'", on pages 298–333 of *La Sfida della Complessità,* edited by G. Bocci and M. Ceruti, published by Feltrinelli (Milan).

Chapter 3 ("Numbo: A Study in Cognition and Recognition" by D. Defays) originally appeared in 1990, under the same title, on pages 217–243 of *The Journal for the Integrated Study of Artificial Intelligence, Cognitive Science, and Applied Epistemology,* vol. 7, no. 2.

Chapter 4 ("High-level Perception, Representation, and Analogy: A Critique of Artificial-intelligence Methodology" by D. Chalmers, R. French, and D. Hofstadter) appeared in a slightly longer version in 1992, under the same title, on pages 185–211 of *The Journal of Experimental and Theoretical Artificial Intelligence,* vol. 4, no. 3.

Chapters 5 and 6 ("The Copycat Project: A Model of Mental Fluidity and Analogy-making" and "Perspectives on Copycat: Comparisons with Recent Work" by D. Hofstadter and M. Mitchell) originally appeared together in 1993 as a single article, bearing the same title as Chapter 5, on pages 31–112 of *Advances in Connectionist and Neural Computation Theory, Vol. 2: Analogical Connections,* edited by K. Holyoak and J. Barnden, published by Ablex Corporation (Norwood, New Jersey). Reprinted with permission from Ablex Publishing Corporation.

Chapter 7 ("Prolegomena to Any Future Metacat" by D. Hofstadter) originally appeared in 1993 as the Afterword (pages 235–244) to M. Mitchell's book *Analogy-Making as Perception,* published by MIT Press (Cambridge, Massachusetts).

Parts of Chapter 8 ("Tabletop, BattleOp, Ob–Platte, Potelbat, Belpatto, Platobet" by D. Hofstadter and R. French) appeared in 1992, in a somewhat different and quite condensed form, in Chapter 3 of *Tabletop: An Emergent, Stochastic Computer Model of Analogy-making,* French's Ph.D. dissertation in Computer Science and Engineering at the University of Michigan.

The final six sections of the Epilogue ("On Computers, Creativity, Credit, Brain Mechanisms, and the Turing Test" by D. Hofstadter) appeared in 1993, in somewhat different form, as part of the article "Analogy-making, Fluid Concepts, and Brain Mechanisms", published in *Essays in Honour of Alan Turing, Vol. 2: Connectionism, Concepts, and Folk Psychology,* edited by P. Millican and A. Clark, and published by Oxford University Press (Oxford, England).

Figure Credits

Jumble cartoon, p. 98, reprinted by permission: Tribune Media Services.

Cartoon by Dana Fradon, p. 335, © 1992 The New Yorker Magazine, Inc.

"Adam and Eve", p. 469, reprinted by permission: W. H. Freeman.

Prologue:

The Why,
the When, the Where,
and the Who of This Book

A Short History of FARG and FARGonauts

This book attempts to present roughly a decade and a half of research in cognitive science carried out by a significant number of people. It all began in 1977, when I became an assistant professor of computer science at Indiana University and officially started doing research in artificial intelligence.

A word on the term "artificial intelligence"... In the 1970's, I enthusiastically embraced this provocative phrase (or its acronym, "AI") as a good way of describing my field of research and my own goals. For me and probably for a good many other people, the term conjured up an exciting image — that of questing after the deepest secrets of the human mind and expressing them as pure, abstract patterns. In the early 1980's, however, that term, as words are wont to do, gradually started changing connotations, and began to exude the flavor of commercial applications and expert systems, as opposed to basic scientific research about the nature of thinking and being conscious. Then, even worse, it slid down the slope that ends up in meaningless buzzwords and empty hype. As a result I came to feel much less comfortable saying or writing "AI". Luckily, a new term was just then coming into currency — "cognitive science" — and I started to favor that way of describing my research interests, since it clearly stresses the idea of fidelity to what actually goes on in the human mind/brain, as well as the pure-science nature of the endeavor. Nowadays, I seldom call myself an "artificial-intelligence researcher" any more, choosing instead to say that I am a cognitive scientist. But once in a blue moon, the term "AI" still manages to creep into my speech or writing.

My first AI research project, one in sequence extrapolation, led me to a series of related projects, and over the years, a number of graduate students joined me in developing them. In these early days — the late 1970's and early 1980's — Marsha Meredith and Gray Clossman were my closest co-workers, and both of them eventually completed Ph.D.'s under my supervision. In particular,

Marsha developed the Seek-Whence program, which perceived linear patterns and extrapolated them. Seek-Whence was the first large project representing our approach.

In 1983, I took a sabbatical year at MIT's famous Artificial Intelligence Lab, courtesy of Marvin Minsky, and there I was very fortunate to meet yet another "MM", named Melanie Mitchell, who subsequently did her Ph.D. with me on the Copycat project, which had sprung largely from Seek-Whence but was devoted to modeling creative analogical thinking. At MIT, I was also accompanied by David Rogers, who came as a postdoc and continued working with me for another few years, and graduate student Marek Lugowski.

About halfway through the year I spent at MIT, the University of Michigan offered me a very attractive position, and consequently I moved there in the fall of 1984. Peter Steiner, Dean of the School of Literature, Science, and the Arts, and Al Cain, Chair of the Psychology Department, made me feel very much at home at the U–M. Perhaps the strongest attraction for me of Michigan was the presence of John Holland, whose intellectual and human companionship helped make my four years at Michigan unforgettably rich.

The move to Michigan in the fall of 1984 marks, in my mind, the official beginning of the Fluid Analogies Research Group, or "FARG", as we — the FARGonauts (as we sometimes refer to ourselves) — usually refer to it. The silly-sounding acronym was more or less intended to be such.

As for the term "fluid", this has occasioned some puzzlement now and then, but I think that it exudes quite a clear image of flexibility, mutability, nonrigidity, adaptability, subtlety, pliancy, continuousness, smoothness, slipperiness, suppleness… By chance, as I was typing the preceding sentence, I noticed how rigid, unpliant, and unsupple my fingers were starting to get in this chilly Italian office. Given the context of my thoughts, I took this discomfort as an invitation to renew my acquaintance with the nature of fluidity. So I walked down the corridor, filled up a sink with nice hot water, and immersed my hands in it. As they grew warmer, I thought about what was so special about water and how it moves. A fluid responds to pressures by constantly changing shape in the most supple, flowing manner; it is neither rigid and brittle like a solid, nor volatile and insubstantial like a gas. Where do these special properties come from? From the invisible molecular substrate, of course.

As I pondered this, I recalled one of my favorite images and phrases from all of science — that of "flickering clusters". This poetic little phrase encapsulates a well-known theory of water according to which H_2O molecules continually make fleeting little associations, thanks to the very weak hydrogen bond that can form between the O of one and an H of another, if they happen to be passing close enough by each other (see Figure P–1). If the flickering-clusters model of water is correct (and when I last read about it, this was somewhat

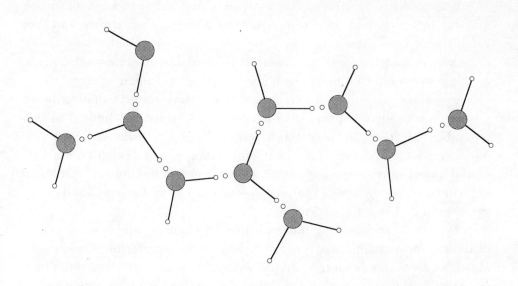

Figure P–1. A typical "flickering cluster" composed of water molecules fleetingly linked by hydrogen bonds. Each water molecule consists of a central oxygen atom (large shaded circle) linked to two hydrogen atoms, with a 107-degree angle between the two bonds. Either of the hydrogens is free to further bond, via a so-called "hydrogen bond", which is very weak and transitory, to an oxygen atom belonging to a different water molecule. In this diagram, a hydrogen bond is symbolized by an unattached dot, representing an electron having a hard time making up its mind which of the two atoms it really wishes to belong to. This period of indecisiveness defines the duration of the hydrogen bond. Since all the different hydrogen bonds are independent, they come undone at unrelated moments, so this cluster falls apart in an asynchronous manner, and even as pieces of it are decoupling, other pieces are forming new associations, so that new flickering clusters arise out of the remnants of old ones. All throughout any sample of water, such clusters are forming and unforming by the trillions every microsecond.

unclear), then all throughout every tiny droplet of water, trillions of complex, randomly-shaped clusters of H_2O molecules are forming and then falling apart every microsecond, all completely silently and invisibly. And thanks to this fantastically unstable, dynamic, stochastic substrate, the familiar and utterly stable-seeming properties of wateriness emerge.

   This image is ideal, I feel, for suggesting our philosophy, according to which the familiar and stable-seeming fluidlike properties of thought emerge as a statistical consequence of a myriad tiny, invisible, independent, subcognitive acts taking place in parallel. Concepts have this fluidity, and analogies are the quintessential manifestation of it. That is why we chose to call ourselves "FARG", and it is why this book's title is as it is.

Back to the story of FARG. I was joined in Ann Arbor by a small band of new students, including Bob French, Alejandro López, Greg Huber, and Roy Leban. Aside from these newcomers, there were Melanie Mitchell and David Rogers, who had followed me from MIT. Then David Moser, officially a graduate student in Chinese, joined us as a kind of "consultant-at-large", Gray Clossman came to finish up his Ph.D., and Dany Defays came from Belgium as a visiting professor for a year. Altogether, FARG, which was housed in the venerable old Perry Building, was a very lively group those days, and our discussions were extremely broad and stimulating. Our research efforts also moved ahead quite satisfyingly, with Melanie implementing Copycat, Bob implementing a new project in analogy-making called Tabletop, and Dany striking out on his own with Numbo.

In 1988, to my surprise, Indiana University made me an irresistible offer to return, and though I was very sad to leave some wonderful colleagues at Michigan, I was also delighted to rejoin many colleagues in Bloomington that fall. Mort Lowengrub, Dean of the College of Arts and Sciences, arranged for FARG to move into a delightful little house of its own just off the main part of campus. This house needed a name, so it was soon dubbed "Center for Research on Concepts and Cognition", but "CRCC" just didn't flow off the tongue as pleasingly as did the mellifluous "FARG", and so we still tend to refer to ourselves and our headquarters by the latter acronym.

Melanie and Bob and David Moser came with me on my move from Ann Arbor to Bloomington, and we were soon joined at IU by some new faces, including Dave Chalmers, Gary McGraw, and Jim Marshall. Helga Keller was enlisted almost from Day One as our Administrative Assistant, and with her energy, efficiency, and enthusiasm, she has made an enormous contribution to our group. Yan Yong, Liu Haoming, and, later, Wang Pei all came from Beijing, where they were involved in the Chinese translation of my book *Gödel, Escher, Bach* — a fascinating endeavor, to say the least. Also, during FARG's first few Bloomington years, Don Byrd, David Leake, Peter Suber, and Steve Larson joined us as, respectively, postdocs and visiting professors, bringing lively insights into music, humor, and philosophy, among other things. In the meanwhile, Melanie and Bob continued to work very hard on implementing their respective programs, and eventually, both of them completed their Michigan Ph.D.'s while working with me at Indiana.

At present, the research group consists of myself and a handful of graduate students, including Gary and Jim, both of whom are already seriously engaged in their doctoral work, and John Rehling, who is beginning his. As I write, I have just begun another sabbatical year, this time in northern Italy, at the remarkable Istituto per la Ricerca Scientifica e Tecnologica, set among hillside vineyards just outside Trento, and surrounded by beautiful mountains on every side. Thanks in large part to the generosity of IRST and its director, Luigi Stringa, FARG has effectively been temporarily transplanted across the Atlantic, and we expect to flourish here very well indeed.

The Double-strandedness of FARG's Intellectual Goals

From its very outset, the intellectual goals and activities of FARG have been characterized by two quite distinct strands, one of them concerned with developing detailed computer models of concepts and analogical thinking in carefully-designed, highly-restricted microdomains, and the other concerned with observing, classifying, and speculating about mental processes in their full, unrestricted glory. The latter has served as a constant stream of ideas and inspirations for the former. In particular, many of our most important intuitions about conceptual fluidity, analogical thinking, and the detailed mechanisms underlying them come from our own intense involvement in such diverse activities as the translation of poetry and wordplay, the investigation of how sexist language reveals the structure of concepts, the acts of collecting, classifying, and theorizing about speech errors and other natural-language phenomena, the careful observation of the discovery process in physics and mathematics, the study of counterfactual conditionals and "frame blends", the creation of artistic typefaces and other imaginative letter-based designs, the classification of jokes and analysis of their deep structure, and the activities of musical perception and composition.

For example, Bob French, Alejandro López, David Moser, Yan Yong, Liu Haoming, and Wang Pei were all deeply involved in published translations of my book *Gödel, Escher, Bach* — Bob into French, Alejandro into Spanish, and the remaining four into Chinese — and thanks largely to their involvement, these translations were superb. As a translation challenge, *GEB* is very difficult, being filled to the brim with wordplay and other highly language-dependent structural games, and countless discussions as well as several articles (French & Henry, 1988; Moser, 1991; Hofstadter, 1987c, 1995) came out of our efforts not only to characterize the nature of good translations but to describe the mechanisms responsible for discovering creative translations, which we claim are, not too surprisingly, very close to the mechanisms responsible for coming up with creative analogies.

As another example, while at Michigan, David Moser and I wrote an article (Hofstadter & Moser, 1989) that tried to describe our ideas on many types of speech and action errors, ranging from low-level spoonerisms to high-level conceptual errors that often cannot be distinguished from creative invention.

This article represented but the beginnings of our efforts to catalogue and theorize about our personal collections of errors and other linguistic phenomena, both of which reach back many years, even decades.

At this point, I cannot refrain from presenting one small example of the kind of error that intrigues us. While revising a draft of this Prologue, I decided to convert the phrase "once in a while" (near the end of the second paragraph) into the more vivid phrase "once in a blue moon". Thus I highlighted the word "while" and started typing its replacement. Before I knew it, what I saw before me on the screen was "once in a bloom", and noticing this stopped me dead in my tracks. (In a way, I wish I hadn't spotted it so soon, since I am now very curious what the next few keystrokes would have been.) It was a typical typo that I might have just gone back and fixed without any pondering, but something about this little glitch caught my attention. I thought, "How come that happened here and now?" Obviously the "oo" of "moon" had jumped the gun in the race from my brain to my fingers, but why the "oo" instead of the "m" or the "n"? Clearly because "oo" and "ue" encode the same vowel sound. But how were my *fingers* clever enough to know that? To be more precise, by what mechanisms did the phonetic sameness contribute to this anticipation error? This was interesting, but it seemed to me that there was more going on behind the scenes. After all, the "m" had actually jumped the gun as well — it had leapt ahead of the blank space and attached itself to the misspelling of "blue". Why did that happen? Well, I had just been writing about working in Ann Arbor and Bloomington, and the word "Bloomington" doubtless had a kind of lingering, fading presence in my mind. It seemed almost certain to me that this was a crucial contributing factor to this error. After all, I have typed "blue moon" many times before, and never made that error. Altogether, then, this subtle confluence of subterranean factors contributing to one tiny surface-level event — a glitch that most people would pay no attention at all to — represents the kind of phenomenon that David and I are profoundly interested in.

Similar examples could be given to illustrate the other types of studies that constitute FARG's more speculative strand, but these two will suffice to give the flavor. To me, this side of FARG represents a word that was considered but rejected as part of the title of our new center — namely, I had suggested "Center for Research on Concepts, Cognition, and Creativity", but this was voted down on the grounds that it was too long and that the word "creativity" is too vague, too trendy, and too "new-age". Be that as it may, there is no doubt that we are deeply concerned with studying creative acts and the mechanisms underlying them.

This book represents mainly the first strand of FARG — that is, our computer-modeling efforts. I hope to follow it up in the near future with one or more volumes devoted to the more theoretical and speculative side of FARG, which I consider equally important but not quite as urgent to get out soon.

This Book in a Nutshell

This book is structured essentially chronologically. Chapter 1, though written recently, is about the earliest days. It tells how the Seek-Whence project arose in the late 1970's from playful mathematical explorations I had long earlier engaged in — or more exactly, from my introspections about that activity. This chapter turns out to be an excellent way to motivate and perhaps even justify the entire approach to the mind on which FARG's research is based. Many of the principal questions about fluid concepts, analogy-making, and the art of generalization show up here, and one can even see glimmerings of the architectures of later projects in the sketchy ideas that were just being developed then — that is, in the years between 1977 and 1983 or so. (One can also see slightly earlier glimmerings in the sketch of an architecture to solve Bongard problems, presented in Chapter 19 of *Gödel, Escher, Bach*.)

From Seek-Whence the story turns to Jumbo, actually the first of all our computer projects that was brought to a reasonable level of realization, in 1982. Jumbo was concerned with the types of cognitive mechanisms — or, more properly speaking, *subcognitive* mechanisms — that allow people, especially experts, to figure out anagrams rapidly and nearly effortlessly in their heads. This chapter, written in 1983, has been left in its original form, with just the original references and no later ones. The third chapter deals with a cousin project to Jumbo, called Numbo, which was proposed and developed by Daniel Defays, a professor of psychology at the University of Liège in Belgium, during his one-year stay with us in Ann Arbor, in 1986–87. Numbo is a model of how people explore various arithmetical combinations of a few given numbers in trying to reach a given "target" number exactly. It is an elegant and charming piece of work. Defays' article, written in 1987, has likewise been left untouched.

Chapter 4 is quite different in mood, as it tackles more philosophical questions about how research in artificial intelligence is carried out. In a sense, the thrust of this three-author article — a collaborative effort by Dave Chalmers, Bob French, and myself — is that there is far too little respect for cognitive science and philosophy in AI. This article, drafted in 1989 and published a couple of years later, contains critiques of a number of AI projects, and toward its end, it sets the stage for a discussion of Copycat. (Actually, the final section of the original article, dealing specifically with Copycat, has been deleted here in the interest of reducing redundancy.)

Chapters 5, 6, and 7 describe in detail the accomplishments of the Copycat project, as well as some ambitious new goals for the future. Chapter 5, by Melanie Mitchell and myself, is by far the most detailed chapter of the book, and in some sense it can be considered the heart of the book, since Copycat represents the furthest that we have so far been able to carry out our research goals. This chapter describes the architecture in great detail, talks about its philosophical justification, presents results of many runs from two different perspectives (one from very close up, the other from far back), and at its end tries to explain the special virtues of the Copycat architecture as a model of the crux of cognition. The work described in Chapter 5 was begun in 1983 (the domain, its challenges, and a fairly specific description of the architecture were devised that year) and completed in late 1990 (the detailed working-out of the architecture and its computer implementation took that long). Chapter 6, also written by the two of us, compares Copycat with other approaches to analogy-making and tries to situate our approach relative to both old-fashioned AI and new-fangled connectionism. Chapter 7, written by me as the afterword to Melanie's definitive book on Copycat (Mitchell, 1993), points in some exciting new directions that I hope to see Copycat carried in, over the next few years.

Chapters 8 and 9 are devoted to Tabletop, joint work by Bob French and myself carried out in the period 1986–91. The first of these chapters is essentially a philosophical exploration of the nature of microdomains, starting out with the Tabletop domain and moving to others, and the complexity of the cognitive act of constructing analogies. The second of these chapters focuses specifically on the Tabletop computer model, pointing out its similarities to and differences from Copycat, and summarizing its performance on a large number of closely related analogy problems falling into three quite distinct families.

Chapter 10 describes incipient and future work by Gary McGraw and myself on what is almost certainly FARG's most ambitious computer model, the Letter Spirit project, which attempts to model creativity on a far larger scale than our previous projects have. As its name might suggest, Letter Spirit concerns the design, by a computer program, of artistically consistent alphabets called "gridfonts". The idea of getting a machine to figure out how to imbue all 26 lowercase letters of the roman alphabet with the same abstract essence — visual style, that is — is one of the most fascinating and deeply challenging tasks that we can imagine, permeated as it is by scads of creative mini-breakthroughs, to model any one of which would already be quite a challenge.

Finally, the Epilogue attempts to set our work in some overall perspective by discussing a few other computer models of creativity and making some general criticisms of such work and the way it is often portrayed to the general public. It then turns to a consideration of the philosophically provoking as well as rather amusing question of when it would be reasonable to assign credit for a new discovery to a computer program rather than to the program's authors. This leads to a discussion of the nature and the observability of mechanisms comprising a cognitive model, and the relationship of our views on these issues to the premises underlying the famous and controversial Turing Test.

Many of these chapters have appeared in some form or other in print elsewhere, although in the preparation of this book nearly every already-published article was revamped at least slightly, and usually a great deal. Because of their having been written independently, at least in their original form, these articles are somewhat redundant. In particular, I offer a deep bow of apology for the many, many times that terms like "codelet", "parallel terraced scan", and several others are defined. I might point out, however, that this redundancy confers upon the book the desirable quality that every chapter can be read pretty much on its own, without regard to anything else in the book. I hope that that makes the redundancy worth it.

As is obvious from the table of contents or the slightest amount of browsing, each chapter is preceded by a "preface". These prefaces were all written by me. The first purpose of such interstitial material is of course to make gradual and natural transitions between chapters, but the prefaces serve a second purpose as well — namely, they allow a much more informal level of discourse, and they provide the opportunity to introduce some historical and philosophical perspectives, as well as numerous anecdotes and personal impressions. I feel that the often lighter-in-tone prefaces are useful and important companion pieces to the generally more serious-in-tone chapters.

Some Vital Acknowledgments

I would particularly like to acknowledge the special catalytic role played by Bob Bolick in bringing this book into existence. In 1989, quite out of the blue, Bob wrote to me from England to suggest that Harvester–Wheatsheaf, the publisher he worked for, might publish a compilation of articles on my research. I thought the idea was excellent, and was very grateful for the suggestion. In the subsequent three years, he and I planned it together, and the whole thing started to come together very nicely. Then Bob was given a unique opportunity to direct a line of publication on his own in another publishing house, and so he left Harvester–Wheatsheaf. Of course, this book still bears many clear marks of his deep involvement with it, and I am very, very grateful to him for his key role in it.

After Bob left Harvester–Wheatsheaf, Farrell Burnett took over in helping me work on this book, and she has made some excellent suggestions that I have taken to heart. On the American side of the Atlantic, my usual publisher, Martin Kessler of Basic Books, was delighted to get involved as well, and his enthusiasm was much appreciated. Fortunately for Martin, he too was given a special opportunity to direct a line of books in his own name in another publishing house, and so, like Bob Bolick, Martin moved onward and upward just as this book was about to be finished. I remain very grateful to Martin for his interest in and support of this and my previous books. Kermit Hummel has stepped in to fill Martin's shoes at Basic, and is continuing Martin's tradition of graciousness and enthusiasm. In particular, one great thing Kermit did was to bring in Theo Lipfert and Vivian Selbo of Acme Art, who saved the book from what might otherwise have been typographical oblivion. Basic's Dana Slesinger and Mike Mueller have also played important mediating roles.

The funding for the projects described herein has come from numerous sources, including the National Science Foundation, Indiana University, and the University of Michigan. In particular, many thanks go to Deans Peter Steiner and Mort Lowengrub of Michigan and Indiana, respectively, for their enormously generous and unconstrained support of my research group in its incarnations at those excellent institutions. Somewhat earlier, Marvin Minsky's financial and intellectual support at the MIT AI Lab was also extremely helpful. In addition, four other individuals played very special roles in helping support our ideas. In the early days, Mitchell Kapor and Ellen Poss contributed both personal funding, through the Kapor Family Foundation, and corporate funding, through the Lotus Development Corporation. FARG simply couldn't have gotten started without it — it was crucial. And Larry Tesler and Barbara Bowen at Apple Computer also made sure that in lean times we were able to keep going, by channeling some of Apple's external research funds to our group. This support was truly invaluable and we remain most grateful for all of it.

Over the years, many friends and colleagues have been enthusiastic sounding-boards, helpful critics, and discussion partners. They are of course too numerous to list, but certain ones truly deserve special mention. In particular, I would like to thank (in no particular order) Paul Smolensky, Scott Buresh, Dan Dennett, Wayne Loofbourrow, Scott Kim, Bill Cavnar, Mark Weaver, Terry Jones, Gilles Fauconnier, Michael Conrad, Charles Brenner, Valentino Braitenberg, Benedetto Scimemi, Zamir Bavel, Pentti Kanerva, Ana Mosterín, Maggie Boden, Allen Wheelis, Jerry Fisher, Piet Hoenderdos, Achille Varzi, Francisco Claro, Dan Friedman, Mike Dunn, Henry Lieberman, Güven Güzeldere, Vahe Sarkissian, Eric Dietrich, Jay McClelland, Daniel Kahneman, Rich Shiffrin, Ryszard Michalski, Bob Axelrod, Dave Touretzky, and Dedre Gentner, all of whom have had significant positive influences on this book.

My family — my dear wife Carol, and our delightful and impish children Danny, now 5, and Monica, just 2 — has had to accept my weird nocturnal hours for a long time now, and they all can finally look forward to a more normal presence of "Babbo". They have been good sports, and now they deserve a break! They have also provided, intentionally as well as unintentionally, lots of ideas for this book.

The two people who have had the most important intellectual influence on me over the last ten years are unquestionably David Moser and Melanie Mitchell. It is to them that this book is dedicated, with much love and appreciation.

— D. R. H.
Povo, Italy
September, 1993

*   *   *
*   *
*

Post Scriptum.

On December 22, 1993, in Verona, Italy, my wonderful and beloved wife Carol Ann Brush Hofstadter died in a coma, ten days after having undergone emergency brain surgery. Up till a week before her operation we had never suspected there was anything wrong, and our little family had spent a joyous three months in Italy, traveling to many beautiful cities, enjoying magnificent scenery and superb food, and savoring the company of dear friends. The brutal shock of Carol's sudden death has forever shattered our happy little bubble of a family. Our children Danny and Monica have been irreparably impoverished. Each of them, in their own quiet way, mourns for their Mommy, little able to realize how much they truly have lost. As for me, I miss Carol terribly — her warmth, her wit, her depth, her dash, her magically radiant smile — but even more than I miss her, I mourn her own loss: up in smoke, the chance to watch her adored children grow up; down the drain, all our shared hopes and dreams. Words cannot express my grief at this tragedy.

I can only add that for years, Carol deeply wanted this book to come out; she would have been so happy to see this long-drawn-out project finally come to fruition. In the poignant afterglow of her life, her continuing inspiration has greatly helped me to keep on going and finish it.

Chapter 1

To Seek
Whence Cometh
a Sequence

DOUGLAS HOFSTADTER

Pattern-finding as the Core of Intelligence

In 1977, I began my new career as a professor of computer science, aiming to specialize in the field of artificial intelligence. My goals were modest, at least in number: first, to uncover the secrets of creativity, and second, to uncover the secrets of consciousness, by modeling both phenomena on a computer. Good goals. Not easy.

In an earlier incarnation as a math major, one thing I had become convinced of was that pattern-finding was close to the core, if not *the* core, of intelligence. In those long-gone days, I had taken great pleasure in devising curious problems and trying to solve them, and it was a routine occurrence that a problem I'd concocted would lead me to some completely unanticipated new number sequence. I would start calculating its terms, and it was truly exciting to have a sequence crop up that clearly exhibited a pattern, yet not a completely *clear* pattern; such cases acted as powerful lures, drawing me deeply into the exquisite quest for the sequence's secret essence.

The first few terms would give me a hint, and I would make a guess; then I would calculate a few more terms, often having my guess confirmed but sometimes getting thrown for a loop. Then more terms, a new guess, and either confirmation or a return to the drawing-board. This kind of oscillation among calculation, guesswork, and revision could go on for any length of time. But usually, after generating sufficiently many terms and shuffling ideas in my mind for sufficiently long, I would be able to uncover the rule — to lay bare the sequence's essence.

Sometimes this *dénouement* was a moment of great pleasure, other times a letdown. It all depended on how elegant and deeply hidden the rule was. The best type of sequence was of course one with such a clever, intricate rule that I myself *never* could have invented it — yet the paradox was that I was the very person who, by inventing the problem that gave rise to the sequence, *had* in fact created — albeit indirectly — the rule so clever and intricate that "I could never have invented it myself!"

Given this background, it is not entirely surprising that I chose, as my first domain in which to explore the nature of intelligence, the domain of integer sequences, their patterns, and their underlying rules. In particular, I decided to try to make a computer program that would *seek whence* a sequence came — that is, it would look at a finite pattern composed of integers and would try to discover its underlying rule, thus allowing the finite pattern to be extended *ad infinitum*. The program would act just as I had — beginning with a few terms, making a preliminary guess or two, then getting more terms, revising its guess, and so on.

To make more vivid for readers the kind of pattern-finding I used to engage in with such ardor, and to show the kinds of capability I hoped to impart to my hypothetical program, I will now go through an example — in fact, my canonical example of this type of activity, as it involves the first unanticipated sequence I ever came up with and decoded, something of which I was very proud at the time.

Triangles between Squares

To give the example, I need to define the so-called *triangular numbers* — 1, 3, 6, 10, 15, 21, 28, 36, 45, 55, 66, 78, 91, ... The rule behind the triangular numbers is that the nth element is the sum of the first n natural numbers. For instance, the fifth triangular number is 15, which is $1 + 2 + 3 + 4 + 5$. The triangular numbers go back to antiquity, and have quite a few attractive properties. It comes as something of a surprise to most people that the much more familiar squares — 1, 4, 9, 16, 25, 36, 49, 64, 81, 100, 121, ... — can be described in a similar manner — namely, the nth square is the sum of the first n odd numbers. For instance, the fifth square is 25, which is $1 + 3 + 5 + 7 + 9$. There are also pentagonal numbers, hexagonal numbers, and so on, but we shall not need them. (Readers might nonetheless be entertained by figuring out analogous series that yield them, and figuring out why these types of numbers all have polygon-related names.)

My sequence came out of wondering about the relationship between the triangular numbers and the squares — in particular, how they interleave along the number line. To explore this question of relative densities of the two sequences, I did the obvious — I simply wrote down squares and triangular numbers in order, keeping in mind which ones were which (I'll use two typefaces to make the distinction apparent):

$$\textit{1}, 1, 3, \textit{4}, 6, \textit{9}, 10, 15, \textit{16}, \ldots$$

Note that 1, being both a square and a triangular number, is included twice, wearing its "square" hat to the left of its "triangular" hat.

The two sequences seemed to mesh in a fairly nice way, with neither one hugely predominating over the other. The most natural thing to do, in my way of looking at things, was to *count* the triangles between the squares, as follows:

$$\textit{1}, 1, 3, \textit{4}, 6, \textit{9}, 10, 15, \textit{16}, \ldots$$
$$\quad\ 2 \quad\ \ 1 \quad\quad 2 \ \ldots$$

The lower sequence is what I was interested in; indeed, it is what launched my whole career in pattern-exploration. For this reason, and also because it raises so many of the issues I wish to raise in this article, I will step through this example very slowly and carefully — hopefully not *too* slowly and carefully.

What does one think when one sees a sequence that starts out "2, 1, 2"? What kinds of continuations seem plausible? What kinds seem implausible? How would one express a given hypothesis as a rule capable of giving the nth term for any n?

Of course, we are not in a random context. We know that these three terms have a mathematical origin, so our expectations are quite different from what they would be if those same numbers had come from a random-number table or some stock-market statistic. The mathematical origin of this sequence suggests that there will indeed be pattern or order, but as math itself is full of patterns of the most enormous diversity, the field is still wide open. Still, we have built-in prejudices in favor of simplicity and elegance, whether or not we can define those notions. So, given our expectation that we are likely to find a simple and elegant rule, what would we expect to happen next? On the flip side, what would we *not* expect to happen next? Why not?

Dot dot dot…

Given just "2, 1, 2" as the sole piece of evidence, most people would be comfortable with the idea that the behavior might be cyclic. In symbols:

$$2, 1, 2 \implies 2, 1, 2, 1, 2, 1, 2, 1, \ldots$$

The arrow is supposed to mean "suggests", and the dot-dot-dot at the end is supposed to mean something like "and so on", where the pattern presumably is obvious, at least to anyone of reasonable intelligence, by the time the dots are encountered. Still, it's somewhat ironic that, when the whole topic of discussion is the difference between *raw patterns* and *explicit rules,* I would dare to use a raw pattern as if it were no different from an explicit rule. It would seem that, in the context of this article, using three dots is a self-undermining act.

There is a way to justify the use of dot-dot-dots, though, making it not so ironic-seeming. We can simply establish a convention that a dot-dot-dot is allowed to express certain agreed-upon patterns of an extremely basic sort — either exact repetition of a fixed, explicit chunk of terms that has already been repeated a few times, as above, or some sort of counting operation, as in this very different but still plausible hypothesis about how our triangles-between-squares sequence might continue:

$$2, 1, 2 \Rightarrow 2, 1, 2, 2, 2, 3, 2, 4, 2, 5, 2, 6, \ldots$$

which would be understood as standing for a rule having a template $[2 \ n]$ with one unchanging element — the 2 on the left — and one "counting element", n, which steps through the natural numbers $1, 2, 3, \ldots$ (There's that dot-dot-dot again!) All I can say is, I will use the dot-dot-dot notation only when I think a pattern is so blatantly obvious (to a human, mind you, not to the program!) that the notation is completely innocuous — that is, unambiguous and non-self-undermining. I would never use it, for example, in the following way:

$$2, 1, 2, 2, \ldots$$

as if to imply that there was only one reasonable continuation for this four-term "opening gambit". That would be the height of absurdity. I would even be hesitant to use the three-dot notation in this way:

$$2, 1, 2, 2, 2, 3, 2, 4, \ldots$$

because the three adjacent 2's muddy the waters quite seriously. There is some room for doubt at this stage, even if not much. With another 2 and 5, however, things are considerably clearer, so that perhaps three dots would be warranted, and with another 2 and 6 after that, virtually all doubt about the pattern is removed from the mind of a human — at least from the mind of any "reasonable human".

My usage of three dots is thus intended to convey to "reasonable humans" the following meta-level piece of information: "The *obvious* extrapolation of this pattern is in fact the *correct* extrapolation." This is of course not a formal definition of the three-dot notation, because I haven't formalized what I mean by "obvious extrapolation"; however, it would be pretty easy to formally define a few simple kinds of obviousness, and then to use the three dots only in cases that that definition covered. But since this is not a technical logic paper, I will simply use the notation informally.

Introduction to Mathgod

With this remark, let us conclude our digression on the philosophically provocative usage of the three dots (much more could be said but this is not the place) and return to the sequence at hand. Let's take things easy, and take just one new element at a time, at least for a little while. This is, after all, a very delicate stage, where things might move in any number of ways, and it is crucial that we think carefully about what goes on in the mind of a human at highly ambiguous moments like this. So here's the next term:

$$1, 1, 3, 4, 6, 9, 10, 15, 16, 21, 25, \ldots$$
$$2 \quad\; 1 \quad\;\; 2 \quad\quad 1 \ldots$$

This knocks out the second hypothesis (the one with template [2 n]), but leaves the first one (with template [2 1]) well in the running. However, there are many — in fact, infinitely many — theoretically possible continuations for any sequence. In the present case, 7 or 777 could in principle be the next term, although if either of them were, it would hugely shock anyone in the slightest mathematical. How about 3? That wouldn't seem so outrageous. The pattern might plausibly go this way:

$$2, 1, 2, 1, 3, 1, 3, 1, 4, 1, 4, 1, \ldots$$

which might be expressed more clearly with some grouping symbols, as follows:

$$(2 \; 1 \; 2 \; 1) \quad (3 \; 1 \; 3 \; 1) \quad (4 \; 1 \; 4 \; 1) \ldots$$

However, it would seem fair to say that there is very little evidence for any such upwards motion as of yet.

Incidentally, let me introduce the term *packet* for a hypothesized group inside a sequence. A packet is a kind of perceptual overlay, a local bit of order that helps one to make sense of some specific region of the sequence. I will henceforth use parentheses to indicate packets. A *template,* by contrast, is a hypothesis about the structure of the entire sequence, and is thus in some sense a composite of infinitely many packets. I will always use square brackets to indicate templates.

A fanciful and amusing speculation about how things might go on would be this:

$$(2 \; 1 \; 2 \; 1) \quad (3 \; 1 \; 3 \; 1 \; 3 \; 1) \quad (4 \; 1 \; 4 \; 1 \; 4 \; 1 \; 4 \; 1) \ldots$$

featuring a variable integer having meanings on two levels at once — that is, influencing both the content and the form (specifically, the length) of the template. However, it is silly to make such byzantine hypotheses when nothing really suggests them. This is just idle speculation. Let us instead do what I myself did — go get more terms. The next triangle and square already furnish some valuable new information:

$$\textit{1},\, 1,\, 3,\, \textit{4},\, 6,\, \textit{9},\, 10,\, 15,\, \textit{16},\, 21,\, \textit{25},\, 28,\, \textit{36},\, \ldots$$
$$2 \qquad 1 \qquad 2 \qquad 1 \qquad 1 \ldots$$

Now we have something quite interesting about which to speculate. Perhaps the most obvious guess for future behavior would be this:

$$2, 1, 2, 1, 1 \Rightarrow (2\ 1)\quad (2\ 1\ 1)\quad (2\ 1\ 1\ 1)\quad (2\ 1\ 1\ 1\ 1) \ldots$$

Here we have a dot-dot-dot signifying a different sort of counting operation. Instead of a *numeral* (*i.e.*, a symbol standing for a number) that is hypothesized to change from one occurrence of the template to the next, we have a *number* — that is, an actual quantity or magnitude — that is changing. In this case, it's the number of 1's that is increasing by one each time. One could think of the packets as increasing in length each time, or else as consisting of two elements every time — a 2 followed by a group of 1's — in which case the group of 1's is what is changing in size.

Another way of writing down exactly the same prediction would be this:

$$2, 1, 2, 1, 1 \Rightarrow (2)\quad (1\ 2)\quad (1\ 1\ 2)\quad (1\ 1\ 1\ 2) \ldots$$

In a sense, this is just a trivial variant of the earlier way of expressing the hypothesis. It simply shifts the boundary of all the packets simultaneously, and makes the first packet a rather degenerate one, containing *zero* 1's. But psychologically, these two "equivalent" views are really quite different entities. Just listen to yourself as you read them aloud. They give rise to slightly different feelings in the head, and, depending on which of them you were thinking of at the time, slightly different ideas might pop to your mind if you were requested to produce "variations on a theme". (Toward the end of this article, I will discuss the variations-on-a-theme game in considerable detail; its importance cannot be overstated.) The ability to make this kind of *perceptual regrouping*, trivial though it may seem, is a very deep part of the acts of discovery and creativity. More on this later. Let us go on with "21211".

There is one other mildly plausible extrapolation that comes to my mind at this point, given just these five terms; it is the following one:

$$2, 1, 2, 1, 1 \Rightarrow (2)\quad (1\ 2)\quad (1\ 1\ 2\ 2)\quad (1\ 1\ 1\ 2\ 2\ 2) \ldots$$

The only thing I don't like about this is that it somehow starts out wrong. That is, the initial packet — "(2)" — doesn't fit the pattern. To be truly consistent with the later ones, the first packet would have to consist of *zero* 1's followed by *zero* 2's, whereas in fact it consists of zero 1's and *one* 2. So this guess seems a bit suspicious.

Why do I say this? Because "Mathgod" — the entity behind the scenes, the abstraction responsible for the patterns of mathematics — doesn't, generally speaking, like irregularities and inconsistencies. Mathgod likes perfect patterns, not imperfect ones. Why? Mathgod only knows. But anyone who has studied mathematics knows this intimately. It is true that some patterns in mathematics stumble a bit at their beginnings, and then regain their balance and proceed on to infinity in perfect style, like an Olympic skater who falls at the beginning of their routine but thereafter has a flawless run; we shall in fact look at just such a pattern later on. But all other things being equal, certainly a pattern without any blemishes at all is preferred by Mathgod to a slightly blemished one. And that's why this hypothesis is a little bit suspect.

A Strange Pattern Starts to Appear

Too much speculation! Let us see how the sequence actually continues. (By the way, 36, being both square and triangular, is included twice, and, as before, the square is placed to the left of the triangle.)

1, 1, 3, *4*, 6, *9*, 10, 15, *16*, 21, *25*, 28, *36*, 36, 45, *49*, 55, *64*, 66, 78, *81*, …
2   1   2   1   1   2   1   2 …

The two most recent hypotheses go down the drain. A rather boring hypothesis consistent with these first eight terms would be that the sequence simply consists of the repeating five-term template "[2 1 2 1 1]":

21211212 ⇒ 21211–21211–21211– …

A less boring idea is based on the fact that the eight terms so far seen form a symmetric pattern — "2121–1212". Given the human mind's predilection for symmetry, this might seem an elegant place for a packet to end. How would things go on from here? The simplest possibility would be repetition of this packet forever:

21211212 ⇒ (21-21 – 12-12) (21-21 – 12-12) …

Another possibility would be for the packets to preserve their symmetry while growing a little in complexity each time, perhaps in one of the following ways:

21211212 ⇒ (21-21 – 12-12) (21-21-21 – 12-12-12) (21-21-21-21 – 12-12-12-12) …
21211212 ⇒ (212 – 11 – 212) (212 – 111 – 212) (212 – 1111 – 212) …

Both of these are very pretty, but they also go way out on a limb. What evidence do we have for such elaborate theories? Hardly any; clearly, we need more data. Here is one more term, and it shoots all three of the latest guesses down:

1, 1, 3, *4*, 6, *9*, 10, 15, *16*, 21, *25*, 28, *36*, 36, 45, *49*, 55, *64*, 66, 78, *81*, 91, *100*, …
2   1   2   1   1   2   1   2   1 …

It's becoming very likely that the pattern involves just 1's and 2's, and that the 2's always occur alone, with the 1's sometimes alone and sometimes in pairs. But seeing the successive 1's as forming little groups may be the wrong way to look at it. Maybe there is a template whose boundary cuts *between* successive 1's, something like this:

$$212112121 \Rightarrow 2\text{--}121\text{--}12121\text{--}1212121\text{--}\ldots$$

This is quite an elegant idea — but it is wrong, as the next term, which is 2, reveals.

From here on out, I won't show any more triangles and squares. You can take my word for it, however, that the 2's and 1's that I will show do indeed come from that source. In fact, when I first explored this sequence, I painstakingly did hundreds of calculations of squares and triangles by hand, using many sheets of paper. At a certain intermediate stage, I had an initial segment that looked like this:

$$2121121212112121121212112121112$$

A surface glance suggested that it might well be periodic, but on closer look, this seemed dubious. Or if it *was* the case, then what was repeating was certainly not obvious. It definitely didn't simply alternate between "21" and "211".

By this time, I was hooked on this sequence, and sweated hard to get a lot more terms. (When my own mental resources were finally too taxed, I even resorted to using my father's mechanical desk calculator, which he used for doing his income taxes!) What I was producing didn't remind me of anything I had ever seen before, which meant I wouldn't be able to relate it to something else already known; instead, I just had to hope that it would have its own brand of self-contained logic.

Readers who are interested in this pattern are invited to try their own hand at finding the hidden rule. They have, of course, the advantage of having the all-important meta-level information that there *is* such a rule, and moreover that it is an *interesting* rule. Obviously, I had no such meta-level information, and that fact makes a big difference. (Of course, being a great believer in Mathgod, I had implicit faith that there *probably* was such a rule, and that it was probably an elegant rule, but that's not the same as being *sure*.) In any case, I will now proceed to reveal the answer, so don't read on if you want to work it out yourself.

Reenactment of a Discovery

Up to this point, I have discussed various hypotheses that some imaginary person or group of people might come up with, rather than the precise set of hypotheses that I myself considered (although I probably did entertain a few of them). By contrast, in what follows, I am going to be very faithful to my actual discovery process, not to some idealization thereof. This means that what I will show may seem quite awkward and stupid in spots; but that is the way discoveries are often made. After a discovery has been completed and its ideas well-digested, one quite understandably wishes to go back and clean it up, so that it appears elegant and pristine. This is a healthy desire, and doing so certainly makes the new ideas much easier and prettier to present to others. On the other hand, doing so also tends to make one forget, especially as the years pass, how many awkward notations one actually used, and how many futile pathways one tried out.

I have often observed that in retrospect, I tend to recall my own discoveries as having happened rapidly and easily, and as having been rather trivial observations. However, that impression is erroneous. It comes from the fact that in mentally looking back at the phenomenon some years later, I am doing so with exactly the right set of concepts and exactly the right amounts of emphasis in exactly the right places — which, of course, I couldn't possibly have done at the moment of discovery. That's the whole reason that I was excited by the discovery — it led me to a fresh new way of seeing things that I didn't have before!

Luckily, having always been fascinated by how the mind works, I have tended to keep careful records of my discovery processes — even back at age sixteen, which is when this particular exploration and discovery took place. When I go back and reread my records of some discovery, refreshing my mind on how it *actually* happened, the process never is nearly as clean as my memory thinks. There's clutter and confusion just about everywhere, and that's certainly true of this discovery, as we shall now see.

Once I had a large number of terms, it started to emerge quite clearly that the sequence did have a kind of oscillatory structure — namely, it would go back and forth between a "112" group and either a "12" or a "1212" group.

212 112 1212 112 12 112 1212 112 12 112 ...

What about the "212" at the beginning? I wasn't quite sure; sometimes I broke it into a "2" and a "12", and then just the "2" was anomalous. But I couldn't get it to go away altogether, to my annoyance.

In any case, the 112's being so predictable and regular, my focus turned entirely to the less clear pattern of 12's and 1212's. At first it looked as though there was an oscillation at this higher level as well — first a 1212, then a 12, then a 1212, then a 12 again, and so forth (ignoring the 212 at the beginning, of course). However, this hope was dashed after another ten or fifteen terms.

212 112 1212 112 12 112 1212 112 12 112 12 112 1212 112 ...

Darn! This was a very confusing stage.

Given the large amount of irregularity, I decided to pull the pattern of 12's and 1212's out in a very visual way by writing down just the *length* of each group instead of the group itself — thus, "12" would be represented by a label of "2", and "1212" by a "4". (That initial "212" was again a nuisance, with part of me feeling it deserved a label of "3" and another part of me arguing that it should be looked upon as a "12" preceded by an irrelevant hiccup, thus deserving a "2" label). Here's what that gave:

212 *112* 1212 *112* 12 *112* 1212 *112* 12 *112* 12 *112* 1212 *112* ...
(3?)   4   2   4   2   2   4 ...

I had of course hoped that this derived pattern would yield more readily to my eye — that regularities present in the upper sequence but hard to notice would become more obvious in the terser lower one. But it didn't seem to be helping.

Nonetheless, I pressed on, and derived more and more of the lower sequence — 15 terms or so of it, for which I needed about 100 terms of the upper one. (Imagine the number of triangles and squares I was banging out on my father's Friden calculator!)

$$3, 4, 2, 4, 2, 2, 4, 2, 4, 2, 4, 2, 2, 4, 2, ...$$

To my disappointment, there didn't seem to be much more clarity in this sequence than in the original sequence — it seemed equally irregular and chaotic.

I tried replacing the initial "3" by a "2":

$$2, 4, 2, 4, 2, 2, 4, 2, 4, 2, 4, 2, 2, 4, 2, ...$$

It was a pretty trivial change, but it at least made all the terms *even,* for what that was worth. And at some point, that suggested the slightly simplifying idea of dividing everything by two, which I did, getting a new sequence that started out as follows:

$$1, 2, 1, 2, 1, 1, 2, 1, 2, 1, 2, 1, 1, 2, 1, ...$$

From a purely mathematical point of view, this act of division was completely trivial. But from a psychological point of view, the distance it carried me was enormous! In particular, I was now suddenly struck by a resemblance I hadn't noticed before between this derived sequence and the original sequence. They both consisted of 1's and 2's exclusively, with 2's always occurring alone, and 1's occurring either alone or in pairs, and so on. In fact, I soon noticed that if you were willing to ignore the initial "1" of the derived sequence (which I was!), the two sequences looked identical — no wonder they had seemed equally irregular and chaotic!

This was quite astonishing to me, but it also set off a warning bell in my mind. I knew that in math and science it is easy to jump to wrong conclusions on meager evidence, and I didn't feel I had nearly enough evidence to justify such an unlikely-seeming idea. (Remember, I had never seen anything like this phenomenon before in my life.) So I went ahead and rather back-breakingly calculated 450 terms of the upper sequence, which furnished me with 53 terms of the lower sequence, every last one of which I found to be in agreement with the upper sequence. *Now* I was getting pretty sure of myself!

Having decoded the sequence, I could afford to go back and reflect on it all a little. In doing so, I realized that I could slightly increase the elegance and concision of my recipe for deriving the lower sequence from the upper one — all I needed to say was that I was *counting 21's between 211's*. This slight re-characterization of the process involved two simple mental shifts — from 12's and 112's to 21's and 211's, and to the word "between". These small shifts really cleaned things up, in that I no longer had to worry about any funny business at the very start. That is, since the leading 21 was not *between* a pair of 211's, it didn't need to be counted at all. Thus I had gotten rid of any faltering at the start; my rule was now unblemished. Hurrah!

By the way, when I recently looked very carefully at my records, I found that the actual story is even more involved than this, because as it happens (and as I had completely forgotten), I had accidentally started out by not including 1 as a triangular number. This meant that my original sequence had begun with "112112" instead of "212112", which really threw me for a loop, since everywhere else in it, successive 112's had either 12 or 1212 between them, but these two had *nothing* between them. Luckily, I caught my error fairly early and fixed it. This, too, removed a worrisome blemish.

A Magical Aperiodic Pattern

I had never seen anything like this type of pattern at all, and I found it extremely lovely and quite mysterious. Although there was something troublingly circular about this rule — after all, it defined a sequence in terms of itself! — it wasn't empty, as circular definitions generally are; it was meaning-ful. Thus, given the rule and the first couple of terms, I could generate more terms; then given those extra terms, I could generate yet more; and on and on. It felt like getting something out of nothing — magical, in a way. Who would have suspected that looking at how triangular numbers and squares interleave would lead to this kind of thing?

The proper way of looking at this sequence had seemingly been revealed: it involved *counting 21's between 211's*. But as this novel idea settled in my mind, it occurred to me that there was a simpler and perhaps more natural counting operation that I had ignored, perhaps because it was *too* simple — namely, counting 1's between 2's. I immediately tried this out:

2121121212112121121212112121121211212122112...
1 2 1 1 2 1 2 1 1 2 1 2 1 2 1 1 2...

To my disappointment, the lower sequence, although it certainly resembled the upper one in many ways, was definitely not identical to it. Frustrated, I hunted around for ways to *make* them agree. For instance, I noticed that if I dropped the first four terms of the lower sequence, the remaining sequence started out "2121121212112" — just the same as the top sequence! — but unfortunately, a discrepancy turned up soon thereafter, so my hopes that this simpler counting operation would turn the trick went down the drain.

    I didn't utterly give up on the idea of counting 1's between 2's, though, and at some point the idea occurred to me to re-apply that same operation to the lower sequence:

2121121212112121121212112121121211212122112...
1 2 1 1 2 1 2 1 1 2 1 2 1. 2 1 1 2...
2   1   2   1 1   2...

Aha! The bottom and top sequences agreed — at least for six terms! I checked it out considerably further, and found that the original sequence and its "second derivative", as I called it, agreed term by term as far as I looked.

    It didn't take me long to see that doing *two* levels of counting 1's between 2's comes to exactly the same thing as *directly* counting 21's between 211's in the original sequence. So now I had found two elegantly related ways of understanding my sequence's pattern.

    This *recursive* rule (as I would later learn was its proper description), if true, implied that the sequence could not be periodic, no matter how long the period. For if it *were* periodic — say, of period 100 — then any sequence derived from it by counting would also be periodic. Thus the sequence derived by counting 21's between 211's in the length-100 block would contain on the order of 12 terms, which would repeat over and over, assuming that the length-100 block did. This would pose no problem except for the fact that, if the recursive rule holds, the derived sequence is the *same* as the top sequence — thus implying that the top sequence's period must be of size 12, not size 100. The numbers 100 and 12 of course don't matter; the point is simply that whatever period-size one assumes, this argument proves it has to be shorter than that! The assumption of periodicity thus undermines itself, and is thereby ruled out. Coming to this realization gave me much pleasure, for it told me I had uncovered something complex and elusive, something much harder to pin down than mere repetition.

I soon wrote a computer program — my first serious program ever — that verified the hypothesized identicality of the derived sequence and the original sequence for several thousand terms. At that point, no doubt was left in my mind that I had unlocked the secret of the distribution of the triangular numbers among the squares. (A few months later, I also proved this result, after slowly working out the appropriate concepts with which to do it.)

This discovery launched a several-year period in my life during which I became truly obsessed with integer sequences. Over those years I invented hundreds of sequences, many of them having complexly tangled recursive properties that made this first one look almost trivial. And the variety of mathematical ideas that went into the sequences was also huge. But there was something beautiful about this "first love" that no other discovery ever quite equaled.

Pattern Extrapolation as Research Project and Class Assignment

I feel grateful to my teen-aged predecessor for having kept quite good records of his discovery process. To be sure, if I were exploring number-sequences today, I would keep an even more detailed record of every tiny step I took, right or wrong, but back in those days, I was interested mostly in the math — after all, I was aiming at being a mathematician. Although I was already interested in creativity even at that time, it certainly never occurred to me that the pathways of discovery I was taking would one day play as big a role in my life as the discoveries themselves!

In any case, we have now been through a typical process by which a sequence's structure is guessed at, revealed bit by bit, and finally understood fully. Such processes typically involve a good number of false guesses based on strange groupings and unwarranted extrapolations. It is this kind of exploratory mental process that, as a new professor just beginning a research project, I was so convinced contained the essence of intelligence, even of creativity, and that I therefore wanted to model on computer.

It took a long time, however, to really pinpoint what kinds of behavior it was that I was after. At first, I thought I wanted to model the entire activity of pattern search, where the patterns were mathematical ones made out of integers. As I said, my own glory days as a number-pattern seeker had involved sequences of a wide variety of types, and I thought that surely the vaster the domain, the better it would be.

Thus I set very high goals for my project. Typical kinds of sequences that I expected my program to figure out for itself were: the squares, the triangular numbers, the cubes, fourth powers, and so on; the powers of 2, powers of 3, and so on; the primes, the Fibonacci numbers (explained below), the factorials (n factorial, written "$n!$", is the product of the first n integers; thus $3! = 1 \times 2 \times 3 = 6$);

and all sorts of variants and combinations of these and many others. I also hoped that the program eventually would be able to figure out the recursive rule behind my triangles-between-squares sequence, but this was not my primary goal, because I thought it represented a somewhat higher level of sophistication. Later on, though, I completely reversed my opinion about which type of discovery was more fundamental.

During these early stages of planning my own research project, I was also teaching my first artificial-intelligence course, and I greatly enjoyed and re-spected my students. I could think of no better way to get them deeply involved in the challenges of designing a "thinking machine" than to share with them my own research ideas, and in fact to engage them in a competition. I declared that there would take place an official sequence-extrapolation contest at the end of the semester, with everyone, including me, required to submit their own program. Over a several-week span, I presented the issues and the kinds of strategies that they might consider, and of course I gave them many sample challenges, to illustrate the variety of the task.

Here is a list of many of the types of sequences that I discussed in class and that I said would resemble the sequences in the match. Of course, the students were well aware that the examples discussed in class represented but a tiny sampling of the huge space of sequences they were to aim at.

1, 2, 2, 3, 3, 3, 4, 4, 4, 4, 5, 5, 5, 5, 5, 6, 6, 6, 6, 6, 6, ...
> (n copies of n, for each n, including 0 or not including 0, depending on how you choose to look at it)

2, 3, 5, 7, 11, 13, 17, 19, 23, 29, 31, 37, 41, 43, 47, 53, 59, 61, ...
> (the prime numbers)

2, 3, 3, 5, 5, 5, 7, 7, 7, 7, 11, 11, 11, 11, 11, ...
> (n copies of p_n, the nth prime)

1, 1, 2, 2, 2, 3, 3, 3, 3, 3, 4, 4, 4, 4, 4, 4, 4, ...
> (p_n copies of n, for each n)

2, 1, 2, 2, 2, 2, 2, 2, 2, 2, 2, 2, 2, 2, 2, 2, 2, 2, ...
> (all 2's with one oddball thrown in)

2, 3, 5, 7, 9, 11, 13, 17, 19, 23, 29, 31, ...
> (the primes, with one oddball thrown in)

2, 3, 5, 7, 9, 11, 13, 15, 17, 19, 21, 23, 25, 27, 29, 31, ...
> (the odd numbers, with one non-odd ball thrown in)

1, 0, 0, 1, 0, 1, 0, 1, 1, 1, 0, 1, 0, 1, 1, 1, 0, 1, 0, 1, 1, 1, 0, 1, ...
> (zeroes in prime locations)

2, 1, 2, 4, 2, 9, 2, 16, 2, 25, 2, 36, ...
    (2's interleaved with the squares)

2, 1, 3, 4, 5, 9, 7, 16, 11, 25, 13, 36, 17, 49, ...
    (the primes interleaved with the squares)

1, 0, –6, 0, 120, 0, –5040, 0, 362880, 0, ...
    (zeroes interleaved with every other factorial, alternating in sign —
    reciprocals of the coefficients of the Taylor series for sin x)

2, 5, 11, 17, 23, 31, 41, 47, 59, ...
    (every second prime number)

3, 5, 11, 17, 31, 41, 59, 67, 83, ...
    ($p(p_n)$ — that is, the 2nd, 3rd, 5th, 7th, ... primes)

3, 4, 6, 8, 12, 14, 18, 20, 24, 30, 32, 38, 42, 44, 48, 54, ...
    ($p_n + 1$, for each n)

1, 4, 27, 256, 3125, 46656, 823543, 8388608, ...
    (n^n, for $n > 0$)

1, 1, 2, 3, 5, 8, 13, 21, 34, 55, 89, 144, 233, 377, 610, 987, ...
    (the Fibonacci numbers, which obey the recursive relationship
    $F_n = F_{n-1} + F_{n-2}$)

1, 2, 2, 3, 3, 4, 4, 4, 5, 5, 5, 6, 6, 6, 6, 7, 7, 7, 7, 8, 8, 8, 8, ...
    (a_n copies of n, where a_n is the nth element of the sequence itself —
    a recursively defined sequence)

2, 1, 1, 3, 4, 2, 5, 9, 2, 7, 16, 3, 11, 25, 3, 13, 36, 3, ...
    (interleaving the primes, the squares, and the first sequence in this
    list)

1, 2, 3, 5, 7, 8, 11, 13, 17, 19, 21, 23, 29, 31, 34, 37, 41, 43, 47, 53, 55, ...
    (merging the Fibonacci numbers and the primes — not by taking
    turns but in order of numerical size, and with no duplication)

2, 1, 2, 1, 1, 4, 1, 1, 6, 1, 1, 8, 1, 1, 10, ...
    (successive denominators in the simple continued-fraction expan-
    sion of Euler's constant e)

The frequent presence of the prime numbers may seem surprising. It is just
a bias on my part, because I find the semi-chaotic distribution of the primes to be
symbolic of the subtlety of mathematics. I certainly didn't expect my program to
be able to figure out the primes from scratch — that would have been far too
ambitious! Rather, I considered the presence of a *library* of certain standard

sequences to be imperative. Thus the sequence of primes (represented by its first 25 terms or so) would be found in the library, as would some other famous sequences that would be beyond the program's ability to recognize in any other way. (Although I didn't do this, I could have even thrown into my library the famous joke-sequence that goes like this: 14, 18, 23, 28, 34, 42, 50, 59, 66, 72, 79, 86, 96, 103, … The reason it's a joke is that there is no interesting mathematics here: these are just the subway stops on the Broadway IRT line in Manhattan. The next stop is 110th Street, also known as "Cathedral Parkway"; if you give the latter as the answer to an extrapolation puzzle, you'll get some laughs.)

Many of the sequences shown above spring directly or indirectly from well-known mathematical phenomena, such as continued fractions, Fourier series, and Taylor series. Even when they don't, they represent patterns of a sort that mathematicians have a natural sense for. To put it another way, a mathematician would not find it at all strange to see a dot-dot-dot after most of these sequences; the idea of "the natural extrapolation" would seem perfectly sensible.

This set of goals, though it provided a wonderful challenge for my AI class, turned out not to be a good target to aim my research project at — a fact that I only slowly came to realize, and that will hopefully become clear as we proceed.

Trying to Reduce a Sequence to Simpler Sequences

My students and I all plotted our own strategies at the same time, and we often shared ideas. Moreover, since all the terminals were located in a single basement room in Lindley Hall, we all got to watch each other's programs evolve over the course of a couple of months. It was great fun watching various programs flop on the easiest sequences, and then surprise us on occasion by succeeding on really tricky ones. Designing sequence-extrapolation programs for a class contest in which the professor himself was just another contestant was among the most stimulating and successful assignments I have ever given.

My personal entry to the contest was certainly not intended to be The Real Thing — that is, to be the full realization of my research goals. I thought of it more as a kind of warm-up exercise. Nonetheless, I took it extremely seriously and competitively.

When I first started planning this program-to-be, I had a definite strategy in mind for how to conduct a search for the underlying rule of a sequence. I knew that there were certain standard types of operation that yielded *derived sequences* — for instance, filtering out every other term, or counting the number of repeated terms — and that with an insightful (or very lucky) choice of such an operation, a derived sequence would turn out to have a structure *simpler* than that of the sequence itself. Thus it seemed that the trick to solving a sequence's structure was just to find the right operation(s) and the right derived sequence(s), and presto! — you'd have unlocked the target sequence's secret. A simple example is provided by the squares:

$$1, 4, 9, 16, 25, 36, 49, 64, 81, 100, \ldots$$

Though it's not the way most people would recognize the pattern here, one way of figuring it out is to take *first differences* — that is, differences between successive elements of the sequence. This extremely standard technique yields the following "child sequence":

$$3, 5, 7, 9, 11, 13, 15, 17, 19, \ldots$$

which is of course the sequence of odd numbers with its first element left out. It is certainly simpler than the squares, as hoped. If you already know or can figure out the rule for this sequence, then you can get the nth square from it by adding up the first $n - 1$ terms and then adding 1 (the first square) to that. For example, the fourth square is $1 + (3 + 5 + 7)$. Thus a rule for the parent sequence can be built up from the known rule for the child sequence.

If someone were so dense as to *not* recognize the odd numbers (and a computer might well be, especially given that the first one of them has been left out!), they could still resort to the same technique again — namely, taking the first differences of the child sequence (which is also called taking the *second* differences of the original sequence). What that gives is even simpler:

$$2, 2, 2, 2, 2, 2, 2, 2, \ldots$$

Even a computer should have no trouble guessing the rule here!

This grandchild sequence's rule — "2's everywhere" — allows us to construct a formula for the nth term in the child sequence: add up the first $n - 1$ terms of this sequence (thus getting $2n - 2$), and to that add 3 (the first term in the child sequence). One could put this recipe together with the previous one (which gives squares as sums of odd numbers); armed with these two recipes, even a dolt could build the nth square up mechanically, stultifying though doing all the additions would surely be.

There is, needless to say, something very peculiar about this technique of "solving" the squares — namely, it pays no attention to any individual element of the sequence on its own. Thus in this approach, the fact that 25 is 5 x 5 is totally irrelevant. From a human point of view, this seems very unnatural. Humans undeniably *do* notice such facts about individual numbers. To miss the idea that these numbers are *squares* certainly seems to miss the point!

Nonetheless, the strategy of applying various operators to a target sequence and trying to find "easier" sequences seemed very powerful to me, and I wanted my program to have a very big repertoire of such built-in operators, including taking first differences, taking "first ratios" between successive terms (analogous to first differences, except that it produces *two* child sequences — a sequence of quotients and a sequence of remainders), counting the number of occurrences of repeated elements, extracting every second term (or every third one, etc.), perhaps taking square roots, cube roots, or even *n*th roots of every element (which would provide an alternate route to solving the sequence of squares), and quite a few others.

I was somewhat familiar with earlier work on sequence extrapolation (*e.g.*, Simon & Kotovsky, 1963; Pivar & Finkelstein, 1964; Persson, 1966) and thought I had a good shot at doing better than they had done. At the time, I was fired up about the prospect of developing a program that could outdo me, its maker; attaining that goal seemed like it would constitute a wonderful proof of machine intelligence. Although part of me wanted my system to faithfully reflect how human minds work, another part of me was simply excited by the idea of pure performance power, no matter how it was gotten. And so I tried to pack as much clever machinery inside my code as I could, trying to develop the analogue to a smoothly-running fuel-injected combustion engine, a vehicle that could mentally outrace its designer with no trouble at all.

Strategies for Controlling a Search

In the development of such a program, one possible overall strategy — known as *breadth-first search* — would be to apply each known operator to the given sequence, thus producing, say, ten different child sequences. If any of those was recognized (meaning its rule was already known and stored in memory), then fine — the program, combining its knowledge of the *operator* that led from the target to the child sequence with its memorized *rule* for the child sequence, could construct a rule for the target sequence itself. The program could then come to a halt, having achieved its goal. (This is how we "solved" the squares in terms of the presumably known odd numbers, above.)

On the other hand, if nary a one of these ten child sequences was known, then the program would "recurse", in computer-science jargon — that is, it would treat each child sequence as a full-fledged target sequence in its own right. In other words, it would apply each of the ten known operators to each of the ten child sequences. The result would be a set of 100 "grandchild" sequences, all to be treated in the same way.

Outward and outward the search would go, pushing ever further away from the original challenge and involving ever more plentiful subsidiary challenges. But as soon as just *one* of the plentiful descendants was identified, the program could work its way back up the family tree. For example, starting from the rule from a known great-great-grandchild of the original sequence, it could make a rule for its parent (a great-grandchild), and from that great-grandchild's rule, a rule for a plain old grandchild, and so on, moving all the way back to the original sequence, which — just as in the case of the above-described two-level solution for the squares in terms of the grandchild sequence "2, 2, 2, ..." and the child sequence "3, 5, 7, ..." — would now be "understood" in terms of a chain of simpler sequences. (I put the word "understood" in quotes to underscore the psychologically strange "understanding" that may well be involved; think of the weird view you would have of squares if all you knew about them were that they are the outcome of starting with a row of 2's and doing bunches of additions.)

For all of its theoretical power, breadth-first search is a dangerous strategy, because it gives rise to a severe "combinatorial explosion" — 10 child sequences one ply down, 100 grandchild sequences two plies down, 1,000 great-grandchild sequences three plies down, and so forth — soon you're talking real computation!

Fortunately, there are more efficient strategies for exploring the potentially huge family tree of derived sequences. At the other end of the spectrum from the computationally costly breadth-first search strategy is the *depth-first* strategy, which makes just *one* child sequence, and from it makes just a single grandchild, and then from that one just one further descendant, and so forth. If and when some descendant down the tree is recognized, the program stops the search at once, and simply works its way back up the tree, building a rule for each intermediary sequence until it reaches the top-level one, as before.

On the other hand, if the program reaches a preset critical depth — say, eight levels — without having recognized anything, then instead of going further down, it backs up one level and tries other pathways. Thus suppose the program has descended all the way to the critical level, level eight, and not found anything known. Instead of going on to level nine, it moves back up to level *seven* and applies its *second-choice* operator there (yielding a new eighth-level sequence). If that fails to yield fruit, then the *third-choice* operator (again on the seventh level) is applied.

And of course, if it turns out that none of the ten operators *seven* levels down gives success, then a new choice is tried out at level *six*, and once again the full spectrum of operators at the seventh level is available. If this choice at level six fails to yield fruit, then a new choice is tried at level five, with all the levels below being re-explored again, and so on.

This pattern of searching through an abstract space is called *backtracking*, and although a depth-first strategy that backtracks in this way might in theory wind up executing a complete traversal of the family tree, it is set up in such a way as to minimize the chance of that dire fate. Its trick is that it explores the "children" in the family tree in a totally different order from the way breadth-first search does. In fact, the whole point of a depth-first approach is that by being very responsive to *cues* at each level, it can keep to a minimum the percentage of the family tree it actually ever generates and looks at.

For instance, suppose that at the top level, the probable relevance of Operator F (first differences, let's say) to Sequence 1 is "sniffed", and so that operator is picked first. It creates from Sequence 1 a child sequence, Sequence 2, which hopefully will be simpler than Sequence 1. Suppose that Sequence 2 actually *is* simpler than Sequence 1 but is nonetheless too complex to be recognized directly. Then it must itself become a parent sequence. This means that the "sniffer", whose purpose is to try to identify the most promising operator to apply, is run on Sequence 2. Say the sniffer comes up with the suggestion of Operator G (counting group lengths), and that this operator is applied to Sequence 2. This yields a grandchild sequence — Sequence 3. Perhaps Sequence 3 is known; in that case, we're done. If not, we again apply the sniffer, use the selected operator, and again produce a new derived sequence. At each new level, a quick but careful "sniff" is given, in an attempt to pinpoint the *most promising operator* to use to produce the next derived sequence.

Heuristics, or the Importance of Sniffing Before You Inhale Deeply

Let me give a simple example of why such "sniffing" is of the utmost importance. Consider the following sequence:

2, 10, 0, 3, 20, 1, 5, 30, 2, 7, 40, 3, 11, 50, 4, 13, 60, 5, 17, 70, 6, …

Although its pattern is not instantly transparent, it's not too hard to unravel, for a human. Something tips you off pretty quickly to the idea of looking at *every third element*. In other words, you break the sequence up into three independent child sequences (they are all "siblings"):

2, 3, 5, 7, 11, 13, 17, …

10, 20, 30, 40, 50, 60, 70, …

0, 1, 2, 3, 4, 5, 6, …

Each of these is a piece of cake — the primes, the multiples of 10, and the natural numbers; weaving them back together is also utterly trivial. So it really is quite a simple sequence, as long as one looks at it the right way (*i.e.*, by applying the right operator). However, imagine the disaster that taking first differences gives! Here it is:

8, –10, 3, 17, –19, 4, 25, –28, 5, 33, –37, 8, 39, –46, 9, 47, –55, 12, 53, –64, …

This is more complex, not simpler, than its parent, and is thus carrying us out to sea.

In fact, no operator other than "take-every-third-element" will simplify this sequence in any way. Not even taking every *second* element helps out. So it really is imperative that the "take-every-third-element" operator be applied to this sequence, and moreover, it ought to be the *first* operator tried out, if at all possible.

On the other hand, what an absurd strategy it would be if "take-every-third-element" were the operator that the program *routinely* applied first to every sequence it encountered! In fact, when you think about it, taking every third element is the kind of operator you'd want to keep on a shelf high in your mental closet almost all of the time, and only on very unusual occasions would you pull it out and apply it.

The same kind of argument can be made for virtually every type of operator. Some sequences seem to cry out for first differences; some for counting operations; some for extractions; some for ratios; some for roots; and so on. Efficiency demands that one be sensitive to these "sniffable" qualities of sequences (*i.e.,* qualities detectable from a fairly superficial scan). You certainly wouldn't want to waste your time taking first differences or cube roots on our old friend "2121121212112", for example, let alone look at every seventeenth element!

What all this boils down to is that you definitely don't want to apply *every* operator to every sequence, because it would be ridiculously wasteful of time and energy. (There is an even more compelling argument against applying *every* operator, which I could have mentioned before but skipped. It is simply the fact that there are *infinitely many* operators. This is because some operators, such as "take-every-nth-element", come in families, with a different operator for each value of n. Obviously, there's no way of applying *all* the members of such a family to any sequence.)

So trying all operators is out. Rather, given a sequence, what you want to do is this: scan it quickly for various easy-to-detect characteristics that suggest which operators to throw out (this can of course only be intelligent guesswork), and then take the operators that escaped the pruning and arrange them in a sensible order, again guided by information from the quick scan. Typical results of a quick scan would include such things as: the degree of presence of *repeated terms,* a quick-and-dirty measure of the sequence's *smoothness,* a rough estimate of its *rate of growth,* observations of *roughly periodic behavior,* and so on. This is the work of the "sniffer" alluded to above.

The trouble is, although throwing some operators away will, if done well, improve performance dramatically most of the time, it will once in a while be disastrous, causing a solution to be totally missed. Clever techniques for guessing — generally known as *heuristics* in artificial intelligence — cannot, by definition, do a perfect job.

This poses a dilemma: there are great advantages to reducing the search space, and yet doing so will necessarily throw out the baby with the bathwater, occasionally. However, this is nothing new; it is a problem endemic to search: when you are dealing with a world that is much bigger than you can fully explore, you have to make guesses, and guesses are risks, and risks sometimes don't pan out.

Glimpses of a Very Different Type of Architecture

Depth-first and breadth-first search, especially when augmented by well-designed and prudent pruning techniques, are both good ideas; they represent two important polar-opposite strategies for searching in huge spaces. However, as I was planning the design of my program, it seemed to me that both of them were far too rigid, and very unlike what people do. I felt that what people do — or at least what *I* do — is more like an initial very shallow breadth-first scan followed by a bit of depth-first in a local area highlighted by the breadth-first scan, then resurfacing for more of a broad overview, then plunging back in more deeply somewhere, re-emerging for a brief overview again, plunging back in perhaps somewhere else for a bit, and so on. In short, a constant interplay between episodes that tend toward the deep side and episodes that tend toward the broad side, with a constant willingness to jump out of any given mode and to try something different for a while. There is also no intense resistance to looking again at an area already looked at once or even more times, because sometimes one comes back a second or third time with fresh new eyes.

A visit to our Computer Science Department by Dave Slate, one of the programmers of Chess 4.6, at that time one of the world's top chess programs, helped confirm some of these fledgling intuitions of mine. In his colloquium, Slate described the strict full-width (*i.e.,* no pruning at all), depth-first search strategy employed by his enormously successful program, but after doing so, confided that his true feelings were totally against this type of brute-force approach. His description of how his ideal chess program would work reso-nated with my feelings about how an ideal sequence-perception program should work. It involved lots of small but intense depth-first forays, but with a far greater flexibility than he knew how to implement. Each little episode would tend to be focused on some specific region of the board (although of course implications would flow all over the board), and lots of knowledge of specific local configurations would be brought to bear for those brief periods. Not being a chess programmer myself, I could get only a vague impression from his musings, but talking with him about chess nonetheless helped my own intuitions to congeal.

Although my head was swimming with hunches about a highly sophisti-cated architecture, I certainly couldn't figure it all out before the arrival of the contest, so I stuck with a pretty straightforward depth-first search, but in compensation gave it the most sophisticated sequence-sniffing, operator-pruning-and-ordering techniques I could think up, since they were much easier to think about and to program. Tackling the sequence-extrapolation task in this manner, it must be pointed out, was much more like thinking about *math* than thinking about *thinking*.

When the time came for our tournament, I selected some 30 sequences out of a big pool that I had amassed during the semester, and tested each program on all of them. It was a little tense for me, as my own legitimacy as a professor of AI was somewhat on the line. It would be quite a crushing blow if my program were to come in last! Fortunately, it fared quite well, missing, as I recall, only about three of them. However, I was beaten out by one student, Bil Lewis, whose program figured out one sequence more than mine did; luckily, this pleased rather than humiliated me. In the class was Marsha Meredith, who later would become my graduate student and develop the much more sophisticated program called Seek-Whence, about which more later. Marsha's program did very well, coming in slightly behind mine.

General Intelligence versus Expert Knowledge

Once the pressure of the tournament was off, I came to feel quite disturbed in two ways by my own strategy. One thing that upset me was the realization that I had done my best to cram as much math sophistication as I could into my little program, and when I reflected on it, I wasn't interested in programming math sophistication, but in programming *intelligence*. Had the term existed in those days, I might have said that I was getting sucked into the *expert-systems* trap — the idea that the key to all of intelligence is just knowledge, knowledge, and ever more knowledge. This was an idea that, to be frank, repelled me.

Of course *some* domain knowledge is necessary to get off the ground, but I had very deep intuitions to the effect that intelligence has — and *has* to have — a powerful, general, and abstract knowledge-independent core. Perhaps this intuition was just prejudice, but it seemed to me it was grounded. I had known many people who were fountains of knowledge yet seemed to lack all insight; conversely, I knew of many examples in the history of science where blinding flashes of insight were produced by people who were relative novices to a field — often very young people whose knowledge was incredibly small, relative to the wizened experts. So this reliance on lots of expert knowledge instead of on a deeper, more basic and abstract type of intelligence was the first aspect of my own work that I was displeased with.

The second disturbing realization I had was that in making my assignment, I had somehow totally ignored one of the most central aspects of my own personal experience in sequence extrapolation: the fact that in my old sequence-crazed days, I would almost always work out just the first few terms of a sequence, and see what, if any, pattern I could find in them; then, if needed, I would calculate more new terms, try to find a pattern given this extra information, and so on. It was a very dynamic process, in which "experimentation" (*i.e.*, a data-gathering mode) oscillated with "theorizing" (*i.e.*, a data-structuring and -restructuring mode). Crucial to it was the fact of getting the sequence bit by bit, not all at once. By contrast, my class assignment had been to design programs that would accept one big batch of input terms, period. There was no way to ask for more terms, and of course no one in their right mind would have deliberately ignored terms that were given to them for free, so the entire assignment missed that all-important temporal dimension.

Mathgod Makes a Booboo

All this was brought out most clearly as I contemplated the last sequence on the list above — the one for the continued fraction of *e*. The tale is a little complicated, but I think it is worth telling, especially since it involves some beautiful mathematics. Here, then, is the origin of that sequence:

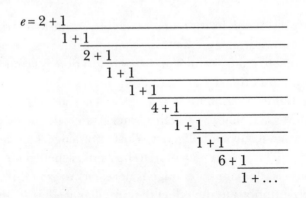

$$e = 2 + \cfrac{1}{1 + \cfrac{1}{2 + \cfrac{1}{1 + \cfrac{1}{1 + \cfrac{1}{4 + \cfrac{1}{1 + \cfrac{1}{1 + \cfrac{1}{6 + \cfrac{1}{1 + \cdots}}}}}}}}}$$

Leonhard Euler's famous irrational constant *e*, whose value is roughly 2.71828182845904523536..., can be expressed (as can any other real number) in a so-called *simple continued fraction*. The idea is that you break the number — call it *x* — into its *integral part* (2, in the case of *e*) and its *fractional part* (0.718..., in this case). You write down the integral part, which we'll call *i*, and then represent the fractional part (which by definition is less than 1) as $1/y$, where of course *y* will necessarily be greater than 1. Symbolically,

$$x = i + 1/y.$$

When *x* is *e*, *y* is about 1.3922...

Now you do for y exactly what you did for x — break it into its own integral and fractional parts, write down the integral part (call it j), and re-represent the fractional part as $1/z$, and so on. (The j-value for this particular y is of course 1, and its fractional part, when re-expressed as $1/z$, is $1/2.5496...$) Symbolically,

$$y = j + 1/z.$$

The two equations can be nested to yield this one:

$$x = i + \cfrac{1}{j + \cfrac{1}{z}}$$

Now, predictably, you do for z what you did for y, and so on and so forth. If you start with any rational number, this process will eventually terminate, but for irrationals like e, it just goes on and on forever, generating what is called the *simple continued fraction* for the number in question.

The natural way to symbolize the entire process is to construct an infinite-leveled fraction that descends forever, diagonally to the right. All the numerators are by construction equal to 1, whereas the denominators are the successive *integral parts* at all the intermediate stages along the way: $i, j, k, l, m, ...$ Thus the infinite sequence of denominators uniquely represents the real number that was its starting point.

The continued-fraction expansion of a real number is a much more fundamental entity than its decimal expansion, because the values of the denominators are not dependent on any base (such as 10, which is the base for the decimal expansion). It is therefore of great mathematical interest to investigate the continued fractions for important mathematical constants, such as π, $\sqrt{2}$, e, and so on. It turns out that the simple continued fraction for π is completely jumbly and chaotic, and the continued fraction for $\sqrt{2}$ is purely periodic, whereas that for e is a very elegant pattern, neither chaotic nor periodic. The sequence of denominators alone is as follows:

$$2, 1, 2, 1, 1, 4, 1, 1, 6, 1, 1, 8, 1, 1, 10, 1, 1, 12, 1, 1, 14, ...$$

The pattern starts to become pretty definite after 12 terms or so (*i.e.*, where the "8" occurs); it involves the template "[1 1 $2n$]", where n takes on the values 2, 3, 4, 5, ... The reason it takes so long to become clear is that there is a "glitch" at the beginning. For some reason, instead of making it go this way:

$$1, 1, 2, 1, 1, 4, 1, 1, 6, 1, 1, 8, 1, 1, 10, 1, 1, 12, 1, 1, 14, ...,$$

which would have made the template "[1 1 $2n$]" fit perfectly all the way from the start, Mathgod seems to have made a little booboo and put "2" rather than

"1" at the front. To coin a slightly blasphemous phrase, Mathgod works in mysterious ways.

Mathgod Redeemed

This glitch bothered me from the very first time I saw the continued fraction for e, because it was otherwise such a beautiful expression. It apparently also bothered my friend Bill Gosper, a super hacker (in the original sense of the term, meaning a deep computerist) and mathematician *sui generis* with a uniquely insightful understanding of continued fractions, and a deep believer in the wisdom of Mathgod. (See the interview with him in Albers, Alexanderson, & Reid, 1990.) Spurred by this strange anomaly, Gosper played around with the continued fraction for e in his mind and one day came up with a new way of writing it, which he told me about with considerable enthusiasm. It looked as follows:

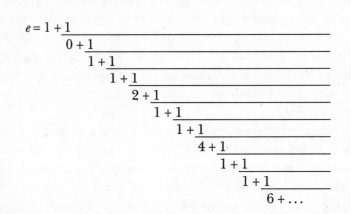

$$e = 1 + \cfrac{1}{0 + \cfrac{1}{1 + \cfrac{1}{1 + \cfrac{1}{2 + \cfrac{1}{1 + \cfrac{1}{1 + \cfrac{1}{4 + \cfrac{1}{1 + \cfrac{1}{1 + \cfrac{1}{6 + \cdots}}}}}}}}}}$$

Except for the way it starts out, it is the same as the old expression. Notice especially the "0" denominator. Now in an ordinary fraction, zero is not permitted as a denominator, and moreover, the algorithm for constructing simple continued fractions never produces a zero denominator either, so at first this "0" appears doubly shocking. And yet it is completely innocuous. There is no division by zero implied in this expression; what *appears* to be a zero denominator is actually added to a nonzero quantity to make the *true* denominator on that level.

So much for the legality of the new expression — but what does it gain you? Well, to tell the truth, at first I couldn't see any gain at all. Horizontally, the sequence now ran this way:

$$1, 0, 1, 1, 2, 1, 1, 4, 1, 1, 6, 1, 1, 8, 1, 1, 10, 1, 1, 12, 1, 1, 14, \ldots,$$

To my eye, it still seemed as blemished as ever. It merely had a *different* initial glitch ("1, 0" rather than "2, 1, 2") marring the perfection of the "[1 1 2n]"

template. How did Gosper see that as an improvement? A littler shorter, yes, but still a glitch.

I then noticed that I could shrink the glitch even more by switching over to a "[$2n$ 1 1]" template, with n starting out at 0. Now the glitch consisted in just a single number — the misfit "1" at the very front. Nice, but still a glitch!

But then something in my mind did a Necker-cube-like flip, and my perception was completely altered. All of a sudden, the boundary of the template had shifted again. Now it was "[1 $2n$ 1]", and it applied from the very start, with n running from 0 on up. The glitch had vanished entirely! Moreover, the new template had the esthetic bonus of *symmetry,* a quality that both previous templates had lacked. Gosper's reinterpretation of the scriptures proved, as I am sure he had hoped to do, that Mathgod had made no error after all. It was just that, as Einstein might have put it, "Raffiniert ist der Zahlengott, aber boshaft ist er nicht." ("Mathgod is subtle but not perverse.")

Esthetics-driven Perception

This may seem like a long shaggy-dog story, but it has a point. I learned about Gosper's new twist on the continued fraction for e at about the time I was deeply thinking about modeling sequence extrapolation, and it highlighted for me exactly those thought processes that I had been most fascinated by when I initially decided that sequence extrapolation was an ideal domain to tackle. Here was a beautiful pattern coming straight from the heart of classical mathematics, but it had a slight blemish. Despite its tininess, this blemish had been sufficiently disturbing, from the point of view of a particular individual with a powerful esthetic sense, to inspire a search for a way to reconceptualize the pattern's origins. Esthetics, not vast domain knowledge, had played a central role in this process.

And my own reaction to Gosper's new sequence was also fascinating to me. What I had perceived at first was this structure:

$$1 - 0 - (1 \ 1 \ 2) - (1 \ 1 \ 4) - (1 \ 1 \ 6) - (1 \ 1 \ 8) - ...$$

and then, by shifting the packets' boundaries, this structure:

$$1 - (0 \ 1 \ 1) - (2 \ 1 \ 1) - (4 \ 1 \ 1) - (6 \ 1 \ 1) - ...$$

But for some reason, and don't ask me why, it had been much harder for me to come up with this view of the new sequence:

$$(1 \ 0 \ 1) - (1 \ 2 \ 1) - (1 \ 4 \ 1) - (1 \ 6 \ 1) - (1 \ 8 \ 1) - ...$$

Actually, *do* ask me why, because I think I can tell you! The answer is partly *inertia* — after all, for years, I had perceived the old continued fraction for e in terms of a "[1 1 2*n*]" pattern — and partly a *default esthetic preference,* which caused me to strongly resist cleaving a pair of 1's down the middle. Specifically, the two identical numbers "wanted", in my mind, to be kept together, so it took considerable external pressure to knock me out of this default way of seeing things.

All of this was hugely reminiscent of the kinds of tiny, subtle perceptual shifts that had gradually brought about my triangles-between-squares breakthrough so many years before. This kind of insight involved in both cases seemed to have nothing to do with fancy search techniques and powerful armies of mathematical operators; rather, it had to do solely with *perception and reperception of structures in response to esthetic pressures.* More than anything, such universal but famously elusive essences as simplicity, consistency, symmetry, balance, and elegance seemed to be the driving forces behind an ability to make sense of patterns. All the rest was irrelevant complexity having to do with knowledge.

Elegance and Consistency Unravel a Sample Sequence

When I now looked back at all the messy sequences I had at first thought of as ideal targets for my ultimate sequence-extrapolation program, I could very clearly see this tension between domain expertise, on the one hand, and a domain-independent sense of elegance and consistency, on the other hand. To illustrate this idea, I will use one of the math-based sequences from my class, one of the simpler ones but complicated enough to make my point:

$$1, 4, 27, 256, 3125, 46656, 823543, 8388608, \ldots$$

What would be a typical route by which a mathematically competent human, but by no means a numbers fanatic or a calculating prodigy, would decipher this pattern? I will sketch below what to me seems as plausible a route as I can think of (illustrated in Figure I–1). However, I am not claiming that the exact details of this route make any difference. All I wish to do is highlight the interaction between narrow mathematical knowledge and a much more general esthetically-based pattern sensitivity. This type of interaction would be a feature of *any* pathway that led to the decoding of the pattern. But a concrete example is needed in order to bring out just what it is that I am talking about, so here we go.

Firstly, the three very big numbers on the right would tend to be seen as so threatening that they would probably be largely ignored, except insofar as they suggest that arithmetical operations that yield big numbers are involved. Thus we shall limit our analysis to what happens when just the initial fragment "1, 4, 27, 256, 3125" is seen. Another way of thinking about this is that perhaps just those first five terms are available at the outset, with later terms requiring computational effort, say, or money or time or whatever, hence being postponed until and unless they are needed.

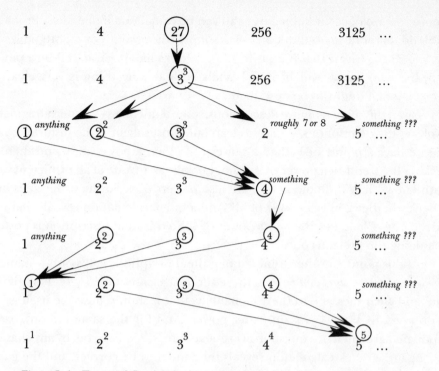

Figure I–1. Temporal flow of ideas in one person's way of discovering the pattern in a mathematical sequence.

The first two terms wouldn't attract much attention because the integers 1 and 4 are very common and have no single outstandingly salient idiosyncrasy. Likewise, the fourth and fifth terms wouldn't attract much attention because the integers 256 and 3125 are infrequent and unfamiliar (at least to this hypothetical individual). But the 27 would very likely jump out and seize the stage, because the integer 27 has one outstanding property, and only one — namely, being the cube of 3. Thus this description would be "tacked" to the 27. There is no guarantee, of course, that it is relevant — it might turn out to be a total red herring — but it is at least a suggestion, something to start the ball rolling. A side effect of this first local suggestion would be to trigger the larger-scale thematic suggestion of *describing numbers as powers* — perhaps cubes, perhaps not, but powers in any case. This, incidentally, jibes with the prior intuition that high-powered operations, so to speak, are involved.

The idea of powers slips on very comfortably, given the actual numbers in the sequence, although some aspects would at first be unclear. Thus, how to describe 1, since *every* power of 1 is 1? And how to describe 256? When the number 256 is considered in the context of "powers", a vague memory drifts up that 256 is some biggish power of 2, perhaps also the square of 16. (These things are often quite vague in people's minds, although the arithmetical facts themselves are of course objective, and can be calculated if need be.) Probably it will do temporarily to think of 256 as simply "the seventh or eighth power of 2". And 3125? Given that it ends in "25", it looks like it *might* well be a power of 5, but who knows which one? Luckily, at least one thing is very clear — namely, that 4 ought to be described as "2 squared".

Given this set of tentative descriptions, some tendencies — pattern-fragments involving the descriptions — start to leap out. Most salient is the disturbing fact that we have a power of 1, then a power of 2, then a power of 3, and then — bucking the trend — a power of 2 again. Luckily, as anyone at all proficient with math knows, half the powers of 2 are also powers of 4. So the order goes out: "Check whether 256 is a power of 4!" And a quick calculation reveals that it is indeed one — namely, the *fourth* power of 4. So this new description is posted, dislodging the previous one.

At this point, another trend among the descriptions leaps out — namely, *the identity of exponents and bases* in three successive cases: 2^2, 3^3, 4^4. This gives a loud and clear message to the 1 — namely, that it should describe itself as 1^1, as opposed to 1^0 or 1^2 or any other power. And for the same reason (base–exponent agreement), a question also goes out to 3125: "Are you, by any chance, 5^5?" Again, a quick calculation reveals the hunch to be correct, and the game is just about wrapped up at this point.

Number Savvy versus Pattern Sensitivity

In this process one sees an alternation between two very different types of activity. There is what might be called "number savvy", which involves such activities as:

- recognizing certain numbers because of salient memorized properties they have (*e.g.*, 27 is 3 cubed);
- making plausible guesses about properties of numbers (*e.g.*, 3125, as it ends in "5" — especially "25" — might be a power of 5);
- knowing relationships between different numbers' properties (*e.g.*, all powers of 4 are powers of 2);
- being able to make complicated calculations (*e.g.*, work out the value of 5^5);

and so forth. Then there is what might be called "pattern sensitivity", which involves such activities as:

- noticing samenesses (*e.g.*, this base equals this exponent);
- noticing simple relationships (*e.g.*, this number is the successor of that one);

- noticing analogies (*e.g.,* this little pattern-fragment looks like that one);
- imposing consistency (*e.g.,* let me alter this pattern-fragment so it looks more like that one);
- building abstractions (*e.g.,* this shared pattern-fragment can be summarized in a template);
- shifting boundaries (*e.g.,* this might better be grouped with this rather than with that);
- driving towards beauty (*e.g.,* let me alter this pattern-fragment because it would be more balanced this way);

and many others.

It is obvious how extremely different these two lists are. One of them has *everything* to do with the domain; the other has *nothing* to do with it. One seems very narrow and relatively unmysterious; the other seems very general and filled with mystery.

Another way of looking at this example is to see each number as a "packet", with the task being to find a pattern among the packets. More specifically, we have at an early stage a sequence of packets something like this:

$$(1 \ ?) \quad (2 \ 2) \quad (3 \ 3) \quad (2 \ ?) \quad (5? \ ?) \ldots$$

where the three stand-alone question marks represent undetermined items, and the question mark attached to the "5" represents uncertainty about whether the 5 itself is valid at all. The pattern-sensitive part of us, seeing very simple but appealing connections between some of the packets, wants the contents of some of them to change, if possible. Since the packets themselves are not primary givens but secondary perceptions or interpretations, they are indeed somewhat malleable under pressure. However, knowledge of *how* they can mutate resides in the math-savvy part of us, which must then be consulted. Thus when the pattern-sensitive part asks if the packet "(2 ?)" can slip to "(4 ?)", the math-savvy part checks it out and says "yes". Similarly, when the pattern-sensitive part asks a bit later if "(5? ?)" can slip to "(5 5)", the math-savvy part again checks it out and says "yes".

This characterization makes sequence extrapolation look like *pattern-play involving small packets whose ingredients are smallish integers,* and where the inter-relationships between packets involve no mathematical notions beyond *equality, successorship,* and *predecessorship.* This view of the essence of sequence extrapolation slowly gained strength in my mind, and eventually it became so strong that my original vision of what the project ought to be about was completely overturned.

A Droll Interlude

At this point, partly to give a break and partly to make a serious point, I would like to quote a very curious sequence-extrapolation puzzle given to me decades ago by a friend, who had heard it from a friend of his. I never was able to track down the origin of this little puzzle; perhaps by publishing it here, I finally will. The puzzle asks, of course, for the next number in a sequence, that sequence being this:

$$0, 1, 2, \ldots$$

Given that this triviality has been posed as a serious puzzle, you of course sense that there is a trick. Since you can think of no reasonable alternate answer to the obvious one, you ask for the answer. "All right," I say, "it is 720 factorial." "720 factorial??" you exclaim. "Yes," I reply, "720 factorial." "But how could that be?", you ask. And that is in fact the *real* puzzle. In other words, I now ask you: "What is the next number in this sequence?", where the sequence is:

$$0, 1, 2, 720!, \ldots$$

Number-savvy people will recognize 720 straight off as a factorial in its own right — namely, 6 factorial. (As far as I know, that is the only important identifying characteristic of 720, although of course it has infinitely many minor properties, one of them being that it is the difference between 3^6 and 3^2.) So this could be rewritten as:

$$0, 1, 2, 6!!, \ldots$$

where the double exclamation point means that the factorial operation is to be executed twice in succession.

Note that what is going on is an attempt to reduce a gigantic atomic entity (it has 1,747 digits) into a little packet involving smallish integers. The sequence could be rewritten this way:

$$0 \quad 1 \quad 2 \quad (6\ !!) \ldots$$

This in itself does not do much. It might give rise to a larger thematic suggestion, however — that we turn the other members of the sequence into packets as well:

$$(0\ ?) \quad (1\ ?) \quad (2\ ?) \quad (6\ !!) \ldots$$

Of course, the question marks and exclamation points have very different meanings!

A wild guess coming from the pattern-sensitive agent might be that, in imitation of the rightmost packet, double exclamation marks could be plugged in where all the question marks are, as follows:

$$(0 \ !!) \quad (1 \ !!) \quad (2 \ !!) \quad (6 \ !!) \ ...$$

The math-savvy agent has to check this out. At first, it looks very good, because 1 factorial is 1 (hence 1 double-factorial is also 1), and likewise, 2 double-factorial is 2. However, 0 factorial is, both by convention and by any reasonable logic, 1; hence 0 double-factorial is 1, not zero. How frustrating! The hypothesis, working out almost perfectly but snagged at the last moment on a technicality, has to go down the drain.

Still, some potentially important ideas have emerged. The fact that 1 and 2 are their own factorials seems extremely pertinent, even if it can't quite be pinned down just *why* it seems relevant. What if we try to salvage what we have by not taking the factorial of zero at all?

$$(0) \quad (1 \ !!) \quad (2 \ !!) \quad (6 \ !!) \ ...$$

Now the *values* are all correct, but there is no elegant pattern any more. On the other hand, there never *was* a completely elegant pattern, because nothing explained the "6". "It would have been nice if the '6' had been a '3'", muses the pattern-sensitive agent. "Aha!" says the math-part, "but 6 is 3 factorial. I should have thought of that before!" (The reason this simple fact might not have come to mind is that 6, being so small and so frequent an integer, has many interesting properties besides being a factorial, in contrast to 720. In fact, the other properties of 6 are so much more salient than its "factoriality" that they tend to obscure that property.) So we write:

$$(0) \quad (1 \ !!) \quad (2 \ !!) \quad (3! \ !!) \ ...$$

Of course, the three exclamation points in the rightmost packet should now all be grouped together, this way:

$$(0) \quad (1 \ !!) \quad (2 \ !!) \quad (3 \ !!!) \ ...$$

But now there is a hodgepodge of different numbers of factorial signs. Aside from the single packet having none, there are two packets with *two* of them, and in the last packet there are *three* of them.

Pattern-part suddenly perks up: "Look, moving leftwards from the right, first we have 3 exclamation points, then 2, then 2. It's almost 3–2–1. Couldn't we just substitute a *single* exclamation point for the double one in the second packet?" This would yield:

$$(0) \quad (1 \ !) \quad (2 \ !!) \quad (3 \ !!!) \ ...$$

The values are still all correct and it's quite elegant, but there still is a troubling anomaly. The packet on the left doesn't fit in with the pattern. Whereas the three others all feature exclamation points, this one doesn't. Could it be that we are simply dealing with yet another sequence that has an initial glitch? But surely, if the initial number *were* a glitch, the puzzle's inventor would simply have left it out of the puzzle. Nobody makes up puzzles with pieces that don't fit in — to do so would ruin the beauty of the puzzle! (Unless, of course, the puzzle is like one of those silly joke-riddles that end up denying one of their own premises ("I was lying about the bananas"), but that seems highly dubious in this context.)

Then out of nowhere comes the final epiphany: the "zeroth packet" has precisely the *right* number of exclamation points — zero! All of a sudden, a perfect, glitchless logic to the sequence has appeared: we have 0 with 0 factorial operations (which leaves it alone, as opposed to a single factorial, which would change it to 1); 1 with 1 factorial operation; 2 with 2 factorial operations; and 3 with 3 factorial operations. It's pretty obvious what the next element would be.

There is one curious aspect to the puzzle as originally stated, which was to extrapolate "0, 1, 2". We now know that the intended answer was 720 factorial, not 3, and we can even defend that weird answer. On the other hand, the case for the defense *tacitly assumes* the extrapolation from "0, 1, 2" to "0, 1, 2, 3" — in other words, it tacitly assumes an answer different from the answer it is attempting to uphold! Thus the 720-factorial answer to the original puzzle is self-undermining in a most amusing fashion, curiously reminiscent of the self-undermining of the hypothesis of periodicity of the triangles-between-squares sequence. (By the way, you only dig yourself into a deeper pit if you try to argue that the term following the 2 should be neither 3 nor 720 factorial, but 720 factorial followed by 720-factorial exclamation points — for that answer, in its own attempted self-justification, tacitly presumes the superiority of *two* rival answers to the same puzzle!)

Segmentation and Unification — Intertwined Facets of Pattern Sensitivity

There are both flip and serious sides to what we have just discussed. The flip side is obvious. The serious side (*i.e.*, the flip-side of the flip side) gets back to my earlier characterization of sequence extrapolation as an interaction in the mind between a math-savvy agent and a pattern-sensitive agent. Notice that the two sequences we analyzed recently are, when seen as *packet* sequences rather than *number* sequences, essentially identical. Both of them share the template $[n \; n]$; it's just that one sequence interprets the second element as the number of factors to multiply together and the other one interprets it as the number of factorial operations to carry out.

In some sense, then, the sequence we were dealing with in both cases was this:

$$1\ 1\ 2\ 2\ 3\ 3\ 4\ 4\ ...$$

(perhaps with a pair of leading zeros — that's a minor point). This looks like a pretty trivial sequence, and indeed it is, because the numbers are all explicit and on the table from the start, whereas in the two original sequences there were large integers (such as 3125 and 720) that had to be broken up in just the right mathematical way in order to yield packets made up of small numbers like these. Such breaking-up of big numbers into plausible packets made up of little numbers is a nontrivial task requiring considerable mathematical knowledge. On the other hand, once a large integer is broken up, its "components" are all sent to fill predetermined slots in a packet attached to that integer, so there is no ambiguity about the packets themselves — no conceivable confusion about whether a given small number belongs to *this* packet or to *that* one.

By contrast, in this unparenthesized small-integer sequence, the packets are *not* explicit, and therefore have to be discovered (or created, if you prefer). That is, one must move from the above structureless sequence to a coherent sequence of packets, like this one:

$$(1\ 1)\quad (2\ 2)\quad (3\ 3)\quad (4\ 4)\ ...$$

Not that there is a unique "correct" set of packets, given a sequence of small integers. For instance, an alternate and perfectly coherent way of putting the preceding sequence into a stream of packets would be this:

$$1\quad (1\ 2)\quad (2\ 3)\quad (3\ 4)\quad (4\ 5)\ ...$$

It has the disadvantage of a glitch in front, but that's a minor peccadillo. The little packets now represent a series of successorship-leaps, like salmon jumping up the steps of a fish-ladder on their way to spawn. The more serious drawback of this parsing is esthetic — namely, the fact that people are more inclined to go for packets built on sameness than ones built on successorship-leaps.

In contrast to these coherent views, here is an *incoherent* view of the same fragment:

$$1\quad ((1\ 2)\ 2)\quad (3\ 3)\quad (4\ 4\ 5)$$

There is no apparent logic to it — neither inside a given packet, nor from one packet to another. This lack of logic is, by the way, the reason I suppressed the dot-dot-dot from this line, but included it on the two above — each of those can be extrapolated without problem (*i.e.,* made into templates), whereas this hodgepodge of arbitrarily formed packets cannot.

The activity of building up a coherent stream of packets from an unpunc-
tuated, structureless sequence of small integers, and coming to understand their
interrelations, is thus a nontrivial task. In fact, it was precisely this type of activity
that I was engaging in when seeking the rule behind the triangles-between-
squares sequence, composed of just 1's and 2's, and when trying to see the point
of Bill Gosper's rewriting of the continued fraction for *e*.

In the case of Gosper's sequence, the trick was realizing that I could build
packets with default-violating boundaries — boundaries that ran against my
instinctive esthetic preferences — and that I might thereby come up with a
glitch-free pattern. Thus this involved what might be called *segmentation* — that
is, *figuring out where the boundaries of packets ought to lie.*

In the case of the triangles between squares, the trick was not so much
finding the boundaries of packets, although that was a bit of a problem, but
rather coming to the realization that the sizes of alternate packets gave rise to
a sequence identical to the original sequence. This, then, involved what might
be called *unification* — that is, *figuring out how the packets are related to one another.*

These two activities are of course not entirely independent — indeed, they
are so intertwined that it is sometimes even difficult to tease them apart. Nonetheless, at least conceptually, they complement each other. These two closely
related activities, then, emerged as the crux of the purely pattern-sensitive aspect
of sequence extrapolation. We will come back to this in a few pages.

The Austere Microworld of Seek-Whence

Considerations of this sort started leading me down a path that deviated
by quite a long ways from the mathematical-knowledge-intensive task I had
initially been so taken with. I felt that at long last I had come to the heart of
sequence extrapolation — the arena where "pure intelligence" was all that was
involved.

It took me a while to specify exactly how limited the task I was now
interested in should be. For example, while I could see that exponentiation
led quickly into the knowledge swamps I wanted to avoid, and whereas I was
more than willing to drop the idea that my program should know a lot about,
say, Fibonacci numbers or factorials, it nonetheless seemed to me that it should
be able to handle sequences whose definitions used any of the basic four
arithmetical operations — addition, subtraction, multiplication, division. But
then I found that by combining just a few such operations, I could easily create
sequences requiring great math-savvy to decode.

My first inclination, upon realizing this, was to drop just multiplication
and division. But then I discovered, to my surprise, that there were still plenty
of sequences based on nothing more complex than addition and subtraction

that would be completely unrecognizable to a program that possessed only the type of knowledge-free pattern-sensitivity I was now so keen on. (The Fibonacci numbers constitute a good example of mathematical complexity built out of just addition, but even without using much imagination one can invent far more complex sequences than that.) So, with a sigh of regret and a breath of relief, I dropped addition and subtraction as well, the final vestiges of the domain's arithmetical origins, and set off in a new direction.

But what was left? Had I not decimated the numerical domain? Well, I still had the integers themselves, starting with 0 (with subtraction gone, negative numbers were totally banished). I still had the notion of equality and the notion of immediate adjacency — that is, successorship and predecessorship. (Without that, the numbers themselves would lose every trace of numberhood!) In addition (as it were), I still had the possibility of clustering numbers together according to shared properties, and the idea of clustering packets together into larger packets — in other words, the idea of hierarchical perceptual structures. And finally, I still had the ability to use numbers to *count* — to count, for instance, the number of numbers in a packet, or the number of little packets in a larger packet, or even the number of hierarchical *levels* in a packet.

I had boiled down the task of sequence extrapolation into a very austere, stripped-down caricature of itself. Sometimes I wondered if I hadn't gone too far. But then a rich new vein of inspiration came to my attention: the patterns of music — specifically, *melodic lines* considered as sequences.

Good-bye, Math… Hello, Music!

Having played the piano for many years and composed a number of small pieces as well, I was intimately familiar with the basic building-blocks of melody in tonal music, and I knew that although my tiny domain was not nearly rich enough to capture the ineffable qualities that make truly deep melodies, the phenomena it dealt with were nonetheless an important part of the very core of melody-making. Thus in the stripped-down sequence world there were "scales", both ascending and descending; there were clusters (and clusters of clusters, etc.) analogous to short groups of notes, full measures, and phrases made of several measures; there were simple and complex rhythms, and even a crude notion of counterpoint (the interplay between two or more "voices" whose notes took turns rather than sounding simultaneously). Thus while I was bidding farewell to the patterned world of mathematics, I was ushering in the equally deeply patterned world of music.

Not to say that all of a sudden I decided to try to model genuine musical understanding — that would have been jumping from the frying pan into the fire. I was very aware that music is every bit as knowledge-intensive as mathematics is, if not more so, and that I would risk getting bogged down in another expert-systems trap if I tried to handle even a small amount of real musical understanding. Rather, I let myself be *influenced* by melodic patterns, just as many of my favorite sequences in the stripped-down domain had been inspired by various mathematical sources.

One common musical phenomenon is that a given pitch — middle C, say — will occur several times in fairly close spots in a given melodic line, yet in these diverse spots it will play a number of different functional roles. In one spot it may be heard as resolving a tension, in another spot as creating one, in yet another spot as a mere passing tone between two other notes, and so on. On occasion, a short passage will even consist of nothing but repetitions of a single note, in which some occurrences of the note are heard one way and others are heard another way, so that in effect there is a kind of "joke" or "pun" going on, whereby two distinct meanings are being carried by a single tone. Often such passages are part of longer "contrapuntal" or "polyphonic" passages in which two effective "voices" — both belonging to a single musical line — are taking turns and, because they are tracing out very different kinds of patterns, are clearly heard as separate entities. However, it can happen that these voices momentarily cross each other, or in some other manner wind up playing the same note, so that for a brief period it would be impossible to disentangle them from each other if the passage were heard in isolation, out of context.

This description may seem opaque without an example, so here is the sort of thing I mean. One of the "voices" in the following melodic fragment is just playing a long succession of repeated E's, while the other one is slowly moving up the scale, playing each note twice: A, then A, then B, then B, and so on. That they will soon collide is obvious. Here is a "raw" (*i.e.,* uninterpreted) representation of the fragment:

EAEAEBEBECECEDEDEEEEEFEF

The collision zone, toward the end, is a mass of undiscriminated E's in which, without the surrounding context, it would be impossible to tell that some of them play different roles from others. If, however, we do the simple trick of distinguishing the two voices by using upper case for one and lower case for the other (a little bit like playing the two voices on different instruments), things get considerably clearer:

EaEaEbEbEcEcEdEdEeEeEf Ef

And if we then pull the passage apart into its natural temporal chunks by constructing "note packets" at two distinct hierarchical levels, we get an even clearer representation of what a listener intuitively hears:

Ea–Ea Eb–Eb Ec–Ec Ed–Ed Ee–Ee Ef–Ef

This hierarchical build-up could be carried further, since the recent context might have told us that each pair of such chunks constitutes a measure:

(Ea–Ea Eb–Eb) (Ec–Ec Ed–Ed) (Ee–Ee Ef–Ef) ...

This entire passage could then be heard as a single structure, setting up quite clear expectations for coming phrases involving roughly the same number of notes. As those notes are actually heard, there will be an automatic attempt to fit them into a pattern resembling this one as much as possible, and the mind will feel the deviations from the pattern as interesting sources of tension.

Although examples like this might seem unusual and uncharacteristic of the majority of music, that is not really accurate. The deliberate game-playing with ambiguity here is perhaps more blatant than in many passages, but ambiguity and context-dependence are pervasive in music, and moreover are central to the way that music carries meaning. It seemed to me therefore that modeling the act of real-time perception of patterns containing ambiguous elements — elements that require surrounding context for their proper perception — would have special importance for the understanding of artistic and esthetic cognition.

The Theme Song of the Seek-Whence Project

I consider these ideas, which began to emerge in 1977 and had pretty much congealed by late 1978, to mark the genuine beginning of the Seek-Whence project. At about this point, I welcomed on board two graduate students — Gray Clossman and Marsha Meredith. Both of them eventually completed Ph.D.'s under my supervision, Gray doing original work on neural networks and Marsha implementing the Seek-Whence program, or at least the first incarnation thereof. The three of us, inspired both by music and by an evolving sense of the potential beauties and complexities of pure linear pattern-play, embarked on the rather pleasurable task of dreaming up interesting sequences of numbers in the austere new domain. Many of our sequences were attempts to model polyphony, ambiguity, context-dependent meanings of local elements, and related phenomena. Several will be shown below.

Eventually, out of this semi-systematic exploration of the Seek-Whence domain came the "theme song" of the project — a challenging target sequence for the program-to-be (whose architecture, incidentally, was evolving in synchrony with the domain). This sequence, which combined features of the musical example just given and the continued-fraction expansion for *e*, was not intended to be enormously hard — just hard enough to bring up the most central issues. Here are its first 16 terms:

$$2, 1, 2, 2, 2, 2, 2, 3, 2, 2, 4, 2, 2, 5, 2, 2, \ldots$$

It would of course not be presented in this all-at-once way to the program, but one term at a time, which makes the extrapolation task far harder. Obviously, the tricky part is handling the longish burst of 2's near the beginning. As one 2 after another arrives, there is a mounting sense of strain and discomfort (remember that you don't know that a 3 and a 4 are coming up quite soon). Once the 3 arrives, there seems to be some hope; then the arrival of the 4 really helps clear things up. Given those two clues, it is not so hard to spot the "different" 2 hiding amongst its identical siblings:

$$2, \mathit{1}, 2, 2, \mathit{2}, 2, 2, \mathit{3}, 2, 2, \mathit{4}, 2, 2, \mathit{5}, 2, 2, \ldots$$

This representation of the sequence suggests a primitive type of polyphony, consisting of one voice that sings nothing but "2" (or rather, "2 2") alternating with another voice that is singing a forever-ascending scale.

Having found this structure, one is nearly finished "seeking whence"; all that remains is to construct packets and, from them, a template. As with the continued fraction for e, one is at first inclined towards packets that keep the strongly appealing "2 2" chunks intact. But this leads to the same kind of initial glitch as we have seen before; as in Gosper's fraction, the only remedy is a systematic boundary-shift that repeatedly cleaves the sequence right between adjacent 2's, violating what seem like the most natural-seeming chunks, and yielding this:

$$(2 \ 1 \ 2) \ \ (2 \ 2 \ 2) \ \ (2 \ 3 \ 2) \ \ (2 \ 4 \ 2) \ \ (2 \ 5 \ 2) \ldots$$

In compensation for going against the grain and breaking the natural chunks, we are rewarded with a symmetric template [2 n 2], appealing to our sense of elegance, and of course we have also completely accounted for the initially confusing row of five 2's.

The Surprisingly Wide Gulf between Researchers' Goals

I must point out that, ironically, the most formulaic type of sequence-extrapolation algorithm, using a bunch of standard operators and a boring vanilla breadth-first search, would instantly and trivially solve this sequence if just ten — even eight! — terms were handed to the program as its input. All such a program would need to do is break it up into three neatly-interleaved subsequences (standard operator #17, say), and it would find that each of them was child's play. That would be it! But what would that grand success teach us about human perception, pattern recognition, theory formation, theory revision, and esthetics? Nothing — nothing at all.

This is not just an irony; it is also an important point, because it brings out the vastness of the gulf that can separate different research projects that on the surface seem to belong to the same field. Those people who are interested in *results* will begin with a standard technology, not even questioning it at all, and then build a big system that solves many complex problems and impresses a lot of people. Those people who are interested in the more abstract questions about the nature of intelligence and creativity will spend a lot of time seeking the essence of those phenomena and then trying to model that essence with maximal fidelity. The results may be harder to appreciate, because they are likely to be far less flashy. Today's wonderfully powerful chess programs, for instance, have not taught us anything about general intelligence — not even about the intelligence of a human chessplayer!

Well, I take it back. Computer chess programs *have* taught us something about how human chessplayers play — namely, how they do *not* play. And much the same can be said for the vast majority of artificial-intelligence programs.

Typical Sequences in the Seek-Whence Domain

Here is a short sampler of other sequences from the Seek-Whence domain, including several that are definitely on the hard side, just to hint at the richness of what might appear at first to be a nearly trivial domain:

2, 2, 2, 3, 3, 2, 2, 2, 3, 3, 2, 2, 2, 3, 3, ...
>(a "hemiolic bi-cycle" — *i.e.*, a trio of 2's alternating with a duo of 3's)

1, 0, 0, 1, 1, 1, 0, 0, 0, 0, 1, 1, 1, 1, 1, ...
>(*n* copies of either 1 or 0, alternately)

1, 0, 0, 1, 5, 1, 0, 0, 0, 5, 1, 1, 1, 1, 5, 0, 0, 0, 0, 5, 0, 1, 1, 1, 5, 1, 1, 1, ...
>(the same sequence as immediately above, but with every fifth term replaced by "5")

1, 2, 0, 2, 0, 2, 1, 3, 1, 3, 1, 2, 0, 2, 0, 2, 0, 3, 0, 3, 1, 2, 1, 2, 1, 2, 1, 3, 1, 3, ...
>(interleaving the first two sequences)

1, 2, 2, 3, 3, 4, 4, 5, 5, 6, 6, 7, 7, 8, ...
>(template is [n $n + 1$]; alternatively, template is the slightly more appealing [n n], but then the "1" at the start is a glitch)

1, 1, 2, 1, 2, 3, 1, 2, 3, 4, 1, 2, 3, 4, 5, 1, 2, 3, 4, 5, 6, ...
>(template is [1 2 ... $n-1$ n] — in other words, ascending scales of length n)

2, 1, 3, 1, 2, 4, 1, 2, 3, 5, 1, 2, 3, 4, 6, 1, 2, 3, 4, 5, 7, ...
(template is [1 2 ... $n-2$ n], with n starting at 2 — in other words, increasingly long ascending scales with the next-to-last note omitted every time)

1, 2, 1, 3, 2, 1, 4, 3, 2, 1, 5, 4, 3, 2, 1, 6, 5, 4, 3, 2, 1, ...
(template is [n $n-1$ ... 2 1] — in other words, descending scales of length n)

3, 2, 1, 0, 1, 2, 3, 2, 1, 0, 1, 2, 3, 2, 1, 0, 1, 2, 3, ...
("sawtooth sequence" — a simple periodic sequence with template [3 2 1 0 1 2])

0, 0, 0, 1, 1, 2, 4, 5, 5, 6, 6, 6, 7, 7, 8, 10, 11, 11, 12, 12, 12, 13, 13, 14, 16, 17, 17, ...
("sawtooth plateaus" — an endlessly ascending scale with repeated notes, the multiplicity of each pitch being dictated by the "sawtooth sequence"; the seemingly missing pitches — 3, 9, 15, etc. — are therefore *conceptually present* but their number of occurrences simply happens to equal zero)

1, 1, 2, 1, 1, 2, 3, 2, 1, 1, 2, 3, 4, 3, 2, 1, ...
(a "mountain chain", whose template is [1 2 ... n ... 2 1] — in other words, a "mountain" rising from 1 to height n and then redescending symmetrically to 1)

1, 1, 2, 3, 1, 2, 2, 3, 1, 2, 3, 3, 1, 1, 2, 3, 1, 2, 2, 3, ...
(the "marching doubler", in which the template [1 2 3] is modified by a "doubler" whose position inside the template shifts cyclically rightwards each time a new packet is made)

1, 1, 2, 3, 1, 2, 2, 3, 1, 2, 3, 3, 1, 2, 2, 3, 1, 1, 2, 3, 1, 2, 2, 3, ...
(the "bouncing doubler", like the preceding one except that the doubler bounces back and forth inside its [1 2 3] "cage")

1, 2, 2, 3, 3, 1, 1, 2, 3, 3, 1, 1, 2, 2, 3, 1, 2, 2, 3, 3, 1, 1, 2, 3, 3, 1, 1, 2, 2, 3, ...
(the "marching singler", a kind of "figure/ground reversal" performed on the marching doubler, in that the basic template is [1 1 2 2 3 3], involving doubled numbers, and the marcher removes rather than adds material)

(1-22-33 1-22-3 1-2-33) (11-2-3 11-2-33 1-2-33) (11-2-3 1-22-3 11-22-3) ...
(the "doubler/singler spliceroo", served to you here on a silver platter — that is, in pre-parsed form — because its pattern, though truly elegant, is too intricate to verbalize succinctly; figure it out yourself)

It is instructive to compare this list of sequence-extrapolation challenges with the original list — there's a world of difference. Readers who try their own hand at making up sequences that would fit into the Seek-Whence domain will soon develop an appreciation for what might be thought of as a miniature art form. One might say that these puzzles are the *haiku,* or perhaps the *bonsai,* of sequence-extrapolation puzzles.

The Deep Problem of Representing Rules Realistically

It is by no means a trivial problem to figure out how to represent the rules for some of these sequences. The simple, intuitive templates we have so often exhibited above fall far short of being able to capture many of these patterns, in which there is dynamic growth or even "moving parts". For example, consider the bouncing-doubler sequence. From a strictly formalistic point of view, it is a trivial sequence — it merely repeats this template *ad infinitum*:

$$[1\ 1\ 2\ 3\ 1\ 2\ 2\ 3\ 1\ 2\ 3\ 3\ 1\ 2\ 2\ 3]$$

But that is as mechanical and as lifeless a way of looking at it as could be imagined. What is fascinating and cognitively deep about this sequence is the *internal structure* of this short segment — the way in which the doubler bounces back and forth inside its three-unit "cage". If your theory of the sequence ignores this structure, you're missing the whole point of this sequence!

Another example is the hemiolic bi-cycle sequence. It too is periodic, in fact with a much shorter period, which makes it look even more trivial; psychologically, however, it is very subtle, in that it involves two objects, one playing *number* and the other playing *numeral,* which then turn around and switch roles. Any notation that failed to capture that subtle quality of role-swapping would completely miss the boat.

To be able to *generalize* a sequence — to think of it as a theme on which to make variations — one has to have a crystal-clear sense of *what it is about.* By this I mean that one has to understand what is central and must therefore be preserved in a variation, as opposed to what is peripheral and therefore "slippable". And since variations can be made from endlessly many different perspectives, one needs to be able to see slippability lurking in all sorts of hidden spots (this idea is explored in depth in Chapter 12 of Hofstadter, 1985). For instance, the *bouncing* doubler is, to human eyes, a pretty simple variation on the *marching* doubler, as is the marching *singler.* Yet nothing in a straightforward representation of the marching doubler would give the slightest hint of these dimensions of variability!

I spent literally years grappling with this problem of notation — how to represent the rules behind the sequences. I absolutely refused to yield to the easy temptation of representing sequences by computer programs, because I felt that what we build up in our heads to represent such a pattern bears no resemblance whatsoever to a standard computer program that would generate it. This intuition became much clearer as time went on. For example, the more carefully I thought about the challenge of making variations on a theme, the more deeply I realized that computer programs are extremely inflexible entities, in the sense that "humanly natural" variants of a pattern often correspond to exceedingly *un*natural transformations of a program that produces that pattern. Any computer person who thinks about how a program for the marching doubler could be turned into a program producing one or another of its various "cousin sequences" shown above should be able to see what I mean without trouble. Nothing in a program to generate the marching doubler *explicitly* suggests the idea of figure and ground, and yet a figure/ground switch makes a beautiful and entirely natural variation. Furthermore, lines in computer programs do not come labeled "I matter — leave me alone!" and "I'm slippable — tamper with me as you please!" How to represent the "untouchable essence" or the "core" of an idea, when the idea can be seen from unbelievably many different perspectives?

Over time, I gradually developed ways of "fluidifying" templates — allowing templates to have internal pieces that could move, or that could change under control of other pieces. Remaining faithful to my sense of what was psychologically accurate was an incredibly difficult challenge, but it wound up giving me enormous respect for the subtlety of what goes on in our heads when we do something so simple-seeming as recite a sequence of this sort out loud. By no means did I completely succeed in this task of capturing the fluidity of human representations in my Seek-Whence diagrams; however, the long struggle led me deep into uncharted territory.

I will not go into the Seek-Whence notation in this article, fascinating though it is. It played a central role in the development of Marsha Meredith's pioneering Seek-Whence program, and interested readers can find a good discussion of many aspects of it in her writings on her program (Meredith, 1986, 1991).

A Mountain-chain Sequence

We shall now take a closer look at the process of perceiving and extrapolating a sequence of this type. Rather than following it in an excruciatingly blow-by-blow manner, we will describe just the highlights. The subject of our scrutiny will be a sequence that is closely related to the "mountain chain" sequence exhibited above. Typical of the infinite number of mountains that compose it is this one: "1234544321" — short for "1, 2, 3, 4, 5, 4, 4, 3, 2, 1".

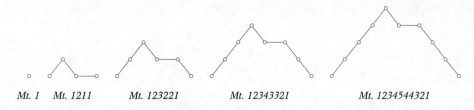

Mt. 1   Mt. 1211   Mt. 123221   Mt. 12343321   Mt. 1234544321

Figure I–2. Schematic representation of a "mountain-chain" sequence.

(Note: For the remainder of this discussion, I will compress segments of sequences by omitting commas and spaces, as just shown. Strictly speaking, this is ambiguous since it makes no distinction between such segments as "11, 2, 3" and "1, 12, 3" and "1, 1, 23", and so forth, which involve multi-digit numbers. However, we will simply ignore this, because in these next few paragraphs we will be concerned with just the very beginnings of our sequence, in which only single-digit numbers figure, so there will be no ambiguity.)

The mountain just shown is slightly asymmetric, with an "outcropping" near the top of its descending slope. All the mountains comprising the sequence share this same symmetry-marring "defect". Thus just before "1234544321" comes the littler mountain "12343321", and just before that comes the even littler "123221". What about the mountain before that? It has to ascend to 2, and then on the way down, the first number below the summit has to be doubled to give the outcropping. That gives "1211" — a somewhat degenerate case, but the reasoning behind it is clear. The mountain preceding this one is an even more degenerate case. Since its peak has to be 1, which in this sequence is playing the role of ground level, there can be no down-slope, *ergo* no number to double, *ergo* no outcropping, *ergo* no asymmetry. So this tiny first mountain is a kind of glitch, as it lacks the salient characteristic that all the others in the chain have. So here is our sequence, broken up into its natural substructures:

1   1211   123221   12343321   1234544321   123456554321 ...

We might even represent it pictorially as in Figure I–2, giving the mountains fanciful names inspired by the famous Himalayan peak known as "K-2".

When displayed this visual, graphical way, the sequence seems extremely simple and logical — practically trivial. To an extrapolator, however, this is how it would look:

1, 1, 2, 1, 1, 1, 2, 3, 2, 2, 1, 1, 2, 3, 4, 3, 3, 2, 1, 1, 2, 3, 4, 5, 4, 4, 3, 2, 1, 1, 2, 3, ...

The pattern is much more hidden. And it would be even more hidden if one were given just the first dozen terms:

1, 1, 2, 1, 1, 1, 2, 3, 2, 2, 1, 1, ...

This short stretch conveys a vague sense of ups and downs, but there is a great deal of ambiguity and confusion to overcome.

Building up Islands of Order

To start organizing these twelve terms perceptually, one needs to try to locate "islands of order" — short stretches that seem like they "make sense", or, put another way, stretches that have their own "internal logic". What I mean by this is exemplified by structures like these:

- *plateaus* such as "44", "111", etc., held together by "sameness glue";
- *up-runs* such as "12", "456", etc., held together by "successorship glue";
- *down-runs* such as "21", "654", etc., held together by "predecessorship glue";
- *palindromes* such as "5885", "71617", etc., characterized by mirror symmetry.

Thus we might see islands like these sprouting up:

11	2	111	232	2	11 …
plateau		*plateau*	*palindrome*		*plateau* …

or possibly these:

1	121	11	232	21	1 …
palindrome	*plateau*	*palindrome*	*down-run* …		

or, then again, these:

11	21	11	23	22	11 …
plateau	*down-run*	*plateau*	*up-run*	*plateau*	*plateau* …

These are just three of the vast number of different ways for islands of order to emerge in this short initial stretch of the sequence — and as it happens, each of these views is at least somewhat misleading, in the sense that each features one or more islands that mix parts of two different mountains. Such mixed-up islands have little or nothing to do with the actual rule behind the scenes. But *a priori*, how could one know that? You have to start your search for order somewhere, and there seems to be no alternative to doing things locally, to begin with.

What, then, is the best strategy for building up local islands of order? The presence of "perceptual glue" in plateaus and runs makes them jump out, to differing degrees, to the eye — either the real eye or an imaginary inner eye, more sensitive to abstractions than a physical eye is. Obviously, plateaus are the most salient type of island since sameness is such an elemental perceptual fact.

Runs come next, with palindromes being somewhat subtler, since recognition of a palindrome requires noticing long-distance relationships — and the longer the palindrome, the greater the leaps involved (*e.g.,* to see "2743472" as a palindrome, you have to jump across five intervening terms). One could spend a great deal of time studying the exact ratios of perceptual attractiveness to human subjects of different types of islands, but (to mix metaphors) that would be getting too involved with the trees instead of the forest. Let us simply accept the intuitively obvious qualitative facts, as stated above.

It might then seem reasonable to make a model that strictly followed this natural inclination, letting plateaus always emerge first, then runs, then palindromes. However, such a deterministic strategy would be very rigid and inefficient — reminiscent of a brute-force depth-first search. Worse yet, it would be fatal for all sorts of sequences. Recall, for example, the "theme song" of the Seek-Whence project:

$$2, 1, 2, 2, 2, 2, 2, 3, 2, 2, 4, 2, 2, 5, 2, 2, \ldots$$

A strategy that always built plateaus first would, given these terms, fall into the trap of building the perceptually immediate and appealing "22222" island and then being committed to that structure, which in this context is a hallucination, a decoy — a perceptual entity that, though perfectly real in a sense, nonetheless has nothing at all to do with the essence of this sequence. Indeed, the challenge of figuring out how to avoid falling for this alluring perceptual trap, or at least how not to get deeply mired in it, was the main reason for my intense interest in this sequence.

What one wants is to find types of structures that are *echoed* throughout the sequence, and hopefully at regular intervals. Thus it makes more sense to let different types of islands "bubble up" independently here and there in the sequence, and then see if there are correlations. The stronger the correlations, the more one will feel one is on the right track. Thus for the sake of efficient picking-up of ideas, one wants to encourage *diversity* in the types of islands being built up, rather than uniformity. On the other hand, too much diversity will simply turn the sequence into a jumble of random, uncorrelated islands of order, thus completely blocking the discovery of patterns, which, after all, involve uniformity, by definition. So there has to be a balance between the overly chaotic strategy of encouraging different kinds of islands to bubble up completely randomly and the overly rigid strategy of always trying one type first throughout, then another type, and so on.

This kind of subtle balance can be struck by employing *parallel processing with probabilistic biases*. The way this works is to let perceptual glue of various sorts bubble up in parallel in different regions of the sequence, with a *tendency* but not a rule for sameness glue to emerge the fastest, successorship and predecessorship glue a bit more slowly, and so forth. Each dab of glue then acts as a small local pressure towards building a particular type of island of order in a particular location. (Palindromes, lacking any local glue, would be handled in a related but more complex manner.) This way, natural perceptual biases can be respected but not slavishly so, and diverse ideas — "hunches", you might say — can arise independently and be explored simultaneously in different regions of the sequence.

A subtle point must be made here. Glue alone does not make an island come into existence; it merely serves as a hint or suggestion to build an island of a certain sort in a certain region. Islands, being larger and more global, are the next stage of perception beyond dabs of glue, and any actually-built island represents much more commitment to a particular theory of what is going on than mere dabs of glue do. However, even a fully-built island (such as the appealing but deceptive "22222" just mentioned) can be sacrificed, under pressure, for the greater good — destroyed, that is, releasing its constituents so that they can be perceptually reinterpreted and incorporated into different islands that hopefully will fit more coherently into the emerging global order. But how does global order emerge from little local hints, and how can it exert pressure on local regions to conform? That is the next topic.

Many Levels of Perception, All Going on in Parallel

As dabs of glue and islands of order are emerging out of the gloom, a *second tier* of exploratory processing can be going on as well — namely, perception of regularities among the islands themselves, leading to multi-level packets and ultimately to templates. However, this level of perception is considerably trickier because an island of order is a more complex entity than a mere number. (I say "mere" here not to convey any scorn for numbers, which I love and respect enormously, but because in this domain, numbers are so stripped-down as to have almost no properties; to be sure, numbers in their full mathematical glory are arbitrarily complex entities.)

Perception of an island in isolation is already a phenomenon with much hidden subtlety. For instance, the island "1111111" is certainly seen by humans as more than just a member of the category "plateau" — it is seen as a plateau consisting of 1's, and possibly but not necessarily as having length 7. Likewise, "23456" tends to be seen by people not just as an up-run, but as an up-run beginning on "2" and ending on "6". In addition, people may possibly notice that it has length 5, or, with considerably less likelihood, that it is centered on "4", or even that its penultimate element is "5". In general, an island of order is a little structure that can be characterized by a *name* and one or more *parameters,* with the parameters themselves having different degrees of interest to people, and therefore different probabilities of being perceived.

Searching on the second tier of abstraction therefore involves two intertwined activities: perceiving each island on its own, and perceiving relationships between different islands. Each activity of necessity affects the other. Thus, for instance, if the central number of some island happened to be noticed, that event would enhance the likelihood of noticing the central numbers of other islands. And once two or three central elements had been thus highlighted, it would be but a short step to looking for sameness or successorship or predecessorship connections between *them*. And if, say, a successorship connection was indeed found, then that would in turn reinforce the tendency to pay attention to central elements, as well as to successorship relations in general. And around and around it goes.

It is very important to understand that these two intertwined activities on the second tier of abstraction are also intertwined with the perceptual activities on the *first* tier of abstraction — the two tiers of perception are not serially separated. Many types of things are going on at once, and affecting each other: the placing of dabs of glue, the building of islands, the labeling of islands, the analogy-making between islands. To envision this kind of multi-ring circus, one has to imagine many little agents, each one searching in its own small area for appealing samenesses or other types of connections, none knowing about any of the others. If this sounds a little like ants in an ant colony, that's exactly the image I wish to suggest, with the overall colony-level activity being merely an emergent outcome of all the tiny activities inside the swarm.

Many different types of connections between islands can be made — for instance, there might be two different up-runs whose first term is "4", or there might be a few plateaus consisting of 0's, or several palindromes of length 3. Another type of higher-order regularity might be a reliable oscillation between two different types of islands of order, or between different sets of parameters (think of the "hemiolic" sequence, made of up "222"'s and "33"'s, that started out the list above).

To make matters very concrete, let us consider the following little stretch of a hypothetical sequence, which we will presume has already been cleanly perceived in terms of four local islands of order:

$$11 \quad 34 \quad 22 \quad 567 \ldots$$

One can't help but see connections between these islands — and the connections that leap out are far from arbitrary. Metaphorically speaking, different islands are "attracted" to one another to different degrees. Clearly, for instance, there is a strong natural affinity between the "11" and "22" islands, whereas there is little or none between the "11" and the "567". The degree of affinity of two islands comes from a number of features, among which the most obvious are: *being close to each other in the sequence, belonging to the same category,* and *sharing parameters.* The more such affinities there are between two islands, the more chance those islands should have of being explicitly linked up.

The Key Role of Analogies

The act of connecting up two different islands in one's mind — for example, linking the "11" with the "22", above — is a very simple instance of *analogy-making.* Making any such analogy is a subjective guess about the likely worthiness of a given pathway of exploration, and inevitably has numerous consequences. For example, if one "mates" those two islands in one's mind, then this act tends to push strongly for an attempt to mate the "34" and "567" islands as well. Another consequence of making an analogy between the "11" and the "22" would be to set up expectations that plateaus consisting of exactly *two copies* of various numbers will be found elsewhere in the sequence. Such a leap could of course be made on the basis of seeing just "11" or "22" in isolation, but it is the analogy between these two islands that really justifies it strongly. After all, "11" taken alone might well be interpreted as a plateau consisting of $n+1$ copies of n, with $n=1$, and "22" alone might well be interpreted as a plateau consisting of n copies of n, with $n=2$. Making simple "variabilizations" like this (*i.e.,* making plausible guesses as to how constants should be replaced by variables), whether based on individual structures or on analogies between structures, is a crucial ingredient of the pattern-perception process.

Analogies vary not only in their degrees of *salience* (*i.e.,* obviousness) but also in their degrees of *strength.* Thus the islands "34" and "567" are somewhat less obviously analogous than "11" and "22", but once the analogy is perceived, it is a very strong one. Somewhat weaker would be the analogy between "11" and "34", based solely on one affinity — namely, that they both have length 2 (plus the fact that they are neighbors in the sequence). Of course, if lengths were not being paid attention to, then this analogy would never even come to the fore.

The strength of a mini-analogy of this sort is a collective function of the various affinities on which it is based. To model this in a simple way, each affinity must be summarized by a number representing its strength, and there must also be a process that computes the strength of a particular mini-analogy as a function of the strengths of its component affinities. We will not worry here about the details of any of these computations, interesting and challenging though they are; my purpose is simply to point out how central they are to the activity of trying to figure out how a sequence is put together.

Thus in addition to the simple dabs of perceptual glue that bond adjacent numbers together to suggest islands of order, there is a more complex web of higher-level analogical connections, often leaping across islands, that is getting built up at the same time. And just as varieties of perceptual glue have different degrees of strength, so these analogical connections have different levels of strength. It would of course make sense that the stronger any inter-island analogy was felt to be, the more influence it would exert on the future course of the perceptual process.

And just what would it mean for some perception to "exert influence" on the perceptual process? The only reasonable idea would be for it to *enhance the likelihood of similar perceptions* to be made, and simultaneously to *weaken the commitment to dissimilar perceptual structures*. This means that probabilistic biases guiding the search for regularities should be altered on the basis of discoveries already made. Thus if at the outset the degree of interest in, say, palindromes is fairly low, but then one or two palindromes chance to be noticed, the degree of interest in looking for further palindromes should shoot up. Moreover, if two palindromes of, say, length 6 were found, then there should arise a bias favoring looking particularly hard for others of that length, and perhaps of lengths close to it. Obviously, having such a bias in from the start would be extremely counterproductive, since precious few sequences are chock-full of length-6 palindromes.

What we are describing, then, is a perceptual process that begins in a pure *bottom-up* manner but that is gradually invaded by increasing amounts of *top-down* influence. "Bottom-up" here describes perceptual acts that are made very locally and without any context-dependent expectations; "top-down" pertains to perceptual acts that attempt to bring in concepts, and to extend patterns, that have been noticed in the sequence (and are *ipso facto* presumed to be relevant to its underlying rule). Another term for "bottom-up" is thus "data-driven"; "top-down" corresponds to "theory-driven".

Once all of this has been laid out so explicitly, it seems a trivial thing to assert that *analogy-making lies at the heart of pattern perception and extrapolation.* What could be more obvious? And when this banality is put together with my earlier claim that *pattern-finding is the core of intelligence,* the implication is clear: *analogy-making lies at the heart of intelligence.* Yet these extremely simple ideas have seldom been stated in cognitive science, let alone explored in detail. Instead, analogy-making has usually been considered to be a specialized, isolated "tool" occasionally invoked in complex, intellectual problem-solving, and because it is seen as such an esoteric ingredient of thought — sort of a luxury add-on — it has been studied by relatively few specialists. It would not be exaggerated to say that all the research in this book came from the detailed working-out of these ideas in a few carefully-designed and very small domains.

The Mountain-chain Sequence — Solved

Let us now return to our "lopsided-mountain-chain" sequence and finish up our discussion of how its structure might be divined. It would be very hard, though certainly not impossible, to spot the hidden order from just the first twelve terms:

$$1, 1, 2, 1, 1, 1, 2, 3, 2, 2, 1, 1, \ldots$$

Let us therefore make things a little easier by throwing in another handful of terms:

$$1, 1, 2, 1, 1, 1, 2, 3, 2, 2, 1, 1, 2, 3, 4, 3 \ldots$$

The "4", being an attractive new stranger, draws one's immediate attention. Leading up to it on its left is the rather salient up-run "1234", which echoes the earlier up-run "123" (let us charitably assume that it too had been noticed). Now the mini-analogy between these two up-runs that both begin with "1", if made, will set up a strong desire to go back and recast earlier parts of the sequence in these same terms (*i.e.*, to look for related up-runs in reasonable spots). Let us symbolize what we have already noticed this way:

$$1 \ 1 \ 2 \ 1 \ 1 \ (1 \ 2 \ 3) \ 2 \ 2 \ 1 \ (1 \ 2 \ 3 \ 4) \ 3 \ldots$$

Clearly, the optimal up-run to find would be "(1 2)", and we would hope to find such an island somewhere to the left of the "(1 2 3)". No sooner said than done:

$$1 \ (1 \ 2) \ 1 \ 1 \ (1 \ 2 \ 3) \ 2 \ 2 \ 1 \ (1 \ 2 \ 3 \ 4) \ 3 \ldots$$

This is quite a promising finding.

Operating independently, bottom-up searchers for order might at roughly the same time come up with two length-2 plateaus "(1 1)" and "(2 2)", each one reinforcing the other's claim to validity:

$$1 \ (1 \ 2) \ (1 \ 1) \ (1 \ 2 \ 3) \ (2 \ 2) \ 1 \ (1 \ 2 \ 3 \ 4) \ 3 \ldots$$

Their obvious affinities suggest that we look at these two islands together, and when we do so we get a new mini-analogy that suggests the theme of *length-2 plateaus related to each other by successorship*. The emergence of this new theme gives rise to a top-down pressure to look for more such plateaus in other regions of the sequence. If the observation is made carefully enough, it will even suggest exactly *where* to look — namely, just to the right of the up-runs; if it is made more quickly and less carefully, it will merely set up a pressure to look *anywhere* for such plateaus.

In either case, the search for more such plateaus will turn up nothing. In the former case, this is because the desired plateau would involve a "3" that hasn't yet appeared, and which would hopefully be the next term. In the latter case, the search would be fruitless because almost all the numbers have already been "taken", in the sense that they are already involved in islands. The only way to discover more length-2 plateaus in this stretch would be to dismantle one or more islands, which would seem a rather desperate act. Such acts of desperation and dismantling certainly have their place, but this is not the moment. After all, we are just beginning our search and are in fact enjoying some good luck! So we are far from desperate. Let us thus simply accept the fact that we can't find any more plateaus in the stretch we have been given, and ask for a few more terms:

$$1 \ (1 \ 2) \ (1 \ 1) \ (1 \ 2 \ 3) \ (2 \ 2) \ 1 \ (1 \ 2 \ 3 \ 4) \ \textit{(3 3)} \ 2 \ 1 \ \textit{(1 2 3 4 5)} \ldots$$

Very nicely, we get reinforcements for *both* of the tendencies we have so far found: a new plateau of length 2 as well as a new up-run beginning with 1, each of them coming in exactly the expected place, to boot. Moreover, also fitting our prior patterns, all the up-runs seen so far form a neat ascending pattern of summit-heights, and all the plateaus seen so far ascend in a similar manner. We can therefore form a tentative but incomplete template:

$$[[1 \ 2 \ldots n] \ [n-1 \ n-1] \ ???]$$

The question marks represent uncertainty as to what is going on there. They naturally form a focus of attention, and lead us to looking explicitly at those regions, shown below in boldface parentheses:

$$1 \ \ ((1 \ 2) \ (1 \ 1) \ \textit{( )}) \ \ ((1 \ 2 \ 3) \ (2 \ 2) \ \textit{(1)}) \ \ ((1 \ 2 \ 3 \ 4) \ (3 \ 3) \ \textit{(2 1)}) \ \ ((1 \ 2 \ 3 \ 4 \ 5) \ldots$$

In the first packet, the site of the question marks is occupied by a "nullity", so to speak, which may or may not be disturbing. After all, at the outset of any sequence, we have to expect some degree of degeneracy, so it is quite possible that this is just one of those degenerate cases. Let us thus look further. In the second packet, the site of the question marks is occupied by just the number "1", and in the third packet, by the number-pair "2 1". If we now engage our analogy-making facility (this time it is being called upon in service of a theory — *i.e.,* in a top-down manner), we find an abstraction that all three of the boldface packets neatly fit: the notion of a down-run ending on 1, and whose length is given by the predecessor of the number in the plateau to its immediate left. (Even the null case fits this abstraction — after all, it could be seen as a down-run of length 0.) Thus we could formulate the whole template this way:

$$[[1 \ldots n] \ [n-1 \ n-1] \ [n-2 \ldots 1]]$$

Further terms in the sequence would then confirm this theory over and over again, so that the activity of seeking whence would seem to be over. Terrific!

Polishing the Solution

Actually, it's not quite over; there remains one last term to be explained — namely, the very *first* term! We had almost forgotten it, but now it comes back to haunt us. The template shown above would imply that the sequence starts out with the packet "((1) (0 0))", which is of course totally wrong. (You might be wondering about the third part of the template — the down-run. Why didn't I put in anything corresponding to it? Well, it is incoherent, in that it involves a subtraction that can't be done — namely, 1 – 2. For this reason, it simply vanishes into thin air. Poof!)

So the theory (as represented by this template) would have our sequence start out with "1 0 0", which is nonsense. What a pain! Does this mean we have to junk the whole theory? It would seem ridiculous to do so solely on the basis of one teeny-weeny glitch at the beginning. But it would seem almost as ridiculous to have a theory that was an ugly splice of *two* theories — namely, "1" just in case $n = 1$, and the big template for all other values of n.

Can't we instead somehow patch the theory? Well, we can try to reformulate the template by transferring one of the two trouble-giving $n - 1$'s from the plateau into the down-run, as follows:

$$[[1 \ldots n] \quad n - 1 \quad [n - 1 \ldots 1]]$$

In all nondegenerate cases (*i.e.*, for n greater than 1), this modified template yields exactly the same thing as the template from which it came, but for the case $n = 1$, it does not. Instead, it produces the packet "((1) 0)". (The third part of the template again poofs into thin air, but this time for a different reason — namely, the fact that for a down-run to begin on 0 and end in 1 is incoherent.) Thus we have made *some* progress, but we're still shy of a perfect, glitchless theory. Can we do even better?

Well, yes, but only if we dare to stretch the meaning of a "down-run" template. Consider, thus, the idea of transferring *both* of the $n - 1$'s into the down-run, as follows:

$$[[1 \ldots n] \quad [[n - 1 \quad n - 1] \, n - 2 \ldots 1]]$$

I would be willing to bet that you know what I mean by this notation, but I would also be willing to bet you are a human, not a machine. Down-runs are supposed to run between *numbers,* not between *structures.* Nothing heretofore suggested that the first item of a down-run could be a *pair.* Only a being with very fluid concepts could gracefully handle this type of informal, almost sloppy, notation-stretching.

In any case, this template acts just like the previous two for all values of n above 1, but for $n = 1$, it gives something different. The up-run yields "1", while the down-run self-destructs (because, as before, you can't run down from 0 to 1);

this leaves us with the solo "1", exactly as desired. So the glitch is finally gone, and we are *really* done! Note how tricky the process of reworking an already basically correct theory is, just so that it can account for a tiny exception at the very beginning. It might even be considered the subtlest part of the whole process.

Even though we are done, there is still room for improvement. An even more polished template would bring out the mountain's "essential symmetry", perhaps something like this:

$$[1 \ 2 \ldots n-1 \ n \ \boldsymbol{n-1} \ldots 2 \ 1]$$

This nearly-symmetric notation is meant to suggest that first you mentally build a perfectly symmetric mountain, rising from 1 to n and then falling back to 1, and then you go back and mess up the perfection *a posteriori,* by tampering with the item in boldface (in this case, replacing it by two copies of itself, although that precise action is not expressed above). This roughly approximates how a human might conceive of it.

There is *still* room for improvement! How can that be? Well, something very salient is still lacking in this notation — namely, *explicit* recognition of the symmetry (or near-symmetry) of the template. We *see* the symmetry of the notation, of course, but nothing in the notation itself *says* it in as many words — much less tells you that the symmetry is *critical.* As all this shows, figuring out how to capture what even the simplest linear pattern is to a human mind is an unbelievably subtle endeavor.

On Deciphering Shorter versus Longer Messages

Now that we've thought through the process in some detail, let's step back and consider the effect of receiving different-size batches of input at the outset. Suppose, for instance, that one were given a really big batch of terms to begin with, such as this:

1, 1, 2, 1, 1, 1, 2, 3, 2, 2, 1, 1, 2, 3, 4, 3, 3, 2, 1, 1, 2, 3, 4, 5, 4, 4, 3, 2, 1, 1, 2, 3.

Even the quickest of visual scans of this list reveals a sense of alternating valleys and peaks, with valleys that look like "1 1" and mountains that increase in both bulk and height. Especially salient would be the peaks of height 3, 4, and 5. These peaks would rapidly draw one's interest and as such would become strong foci of attention around which perceptual structures would rapidly grow, like raindrops forming around dust particles or suburbs spreading out around cities.

An asymmetric mountain like "1234544321" would leap out at the eye almost effortlessly, with the only trouble possibly coming at its left and right edges, where it's a little unclear where this mountain leaves off and its neighbor-mountains start. But any territorial battle over the outlying 1's would quickly be decided when "12343321" struck a truce with "1234544321", partitioning the "11" valley evenly between them and thereby making each mountain on its own a strong and coherent entity (which would not be the case otherwise).

This solution then easily propagates leftwards towards the start of the sequence, allowing "123221" to emerge without trouble; after that triumph, even "1211" and "1" come along without offering much resistance. In short, giving a strong boost to perceptual activity toward the right end of a long initial segment and then working leftwards makes things pretty much fall out naturally. Degeneracies and glitches near the left can be totally sidestepped at the outset, thus making the challenge seem almost trivial.

But truncate the initial segment somewhat — say to this:

$$1, 1, 2, 1, 1, 1, 2, 3, 2, 2, 1, 1, 2, 3, 4, 3, 3.$$

Now, even though there's still a lot of information here, things are much harder. Peaks and valleys still jump out, but the "11" and "111" valleys could easily be put together in one's mind with the "22" and "33" plateaus, thus setting one off on what would eventually turn out to be a false pathway. Even if one were not too distracted by all these plateaus, and somehow had the wise intuition to concentrate mostly on mountains, the biggest mountain's incompleteness would prevent a "prototypical mountain" from coming into existence and serving as an analogical mold on which to base other mountains. However, two smaller mountains with somewhat blurry edges could be formed earlier on in the sequence by letting islands spread out in parallel from the peaks of height 3 and height 2. In this way, three mountainous structures (the rightmost being incomplete) might grow outwards simultaneously, eventually abutting one another and jostling a bit over valley territory. With some effort, one could probably arrive at the whole pattern from just this sample. We need not spell out the details; suffice it to say that doing so would be a much more stringent test of any model of perception and insight.

And all the more so if the initial segment were just the first dozen terms:

$$1, 1, 2, 1, 1, 1, 2, 3, 2, 2, 1, 1.$$

What a welter of confusion! To arrive at the hidden pattern on the basis of just these terms would be a genuine triumph of intelligence.

With fewer terms yet — say these:

$$1, 1, 2, 1, 1, 1, 2, 3$$

— all one could make is a bunch of half-baked theories without much confidence that the theories were even on the right track, let alone right!

Trying to guess the full pattern from an initial segment is much like trying to predict the kind of adult a child will grow into. It will obviously make a huge difference if the child is an eighteen-month-old, a toddler, a five-year-old, a twelve-year-old, or an adolescent. In an eighteen-month-old, all is murk and confusion; in a toddler, there are numerous clues, but they are extremely subtle and ambiguous; in a twelve-year-old, the separate strands of the personality have started to become untangled and clear. Similarly, in a very short initial stretch, there are ambiguities galore and the challenge is enormous; in a very long one, there is so much perceptual clarity that the task is not much of a challenge. (Actually, of course, there is still huge complexity to the task, but for *humans,* it seems relatively unchallenging.)

In any case, the core of the process is analogy-making, but a somewhat peculiar variety of analogy-making, in that one isn't exactly sure of the *identity* or the *edges* of the structures between which analogies are being made. Of course this is all the more true of real-world analogy-making: no true-life situation comes with hard-and-fast boundaries; all situations blur out into a myriad other situations and facts, and an important part of the act of analogy-making is the determination of where one situation ends and others start. Despite this, most artificial-intelligence models of analogy-making involve situations that come prepackaged as tight little bundles of facts with perfectly crisp edges. Needless to say, this is most unrealistic.

Mathematicians' Deep Ambivalence towards Obvious Patterns

The Seek-Whence domain is not nature itself; it is an artificial microworld with artificially concocted patterns. Here there are no eternal verities to be discovered, no universal theorems to be proven; the only thing that determines which way you move is taste — a very abstract sort of taste, disconnected from real-world consequences. Despite this artificiality, a theoretical, mathematical mind should flourish in this kind of microworld, because mathematical thinking is permeated by esthetics.

Ironically, though, mathematical sophisticates often balk when asked to extend simple patterns. They will tell you that no finite pattern has a unique extension, that there are infinitely many defensible ways to extend any sequence; some will even go so far as to say that sequence extrapolation in a context-free vacuum like the Seek-Whence domain is an exercise in meaninglessness. They make these protestations because of years of exposure to tricky counterexamples, which have taught them to be suspicious of "obvious" patterns, no matter how simple they are (think of the bizarre extension of "0, 1, 2" that we discussed earlier); they do so also because of years of having had it drilled into them that an observed pattern, no matter how obvious, compelling, or beautiful it may seem, no matter how many mountains (or mountain-chains) of evidence it is supported by, is not a *real* pattern as long as it remains *unproven*. And of course in our little world, "proof" is not a meaningful concept.

Despite their protestations, though, mathematical sophisticates are often the very first to spot patterns, both simple and complex, in a Seek-Whence puzzle. And often they are the only ones to find the deepest patterns. Like everyone else, mathematicians have an unconscious esthetically-based sense of which kinds of extrapolations are more appealing or more plausible than others. Indeed, when working professionally, they make full and unashamed use of their esthetic intuitions about patterns, in the act of guessing which pathways would be best to explore in inventing new concepts, making hypotheses, devising proofs of theorems, and so on. In mathematics, pattern sensitivity acts like an abstract sense of smell, and informs all of the hardest decisions.

In fact, mathematics books and articles are filled with appeals to the "obvious" extension of a pattern. Readers are told that the proof of the general case is left to them, after they've seen a proof of one specific case. Or they are told that the other cases in a proof follow "by symmetry" or "by analogy", where the symmetry or analogy is presumed obvious. An infinite sequence (such as the integers running down the diagonal of a continued fraction) or infinite series (such as a Taylor series or Fourier series) is indicated by giving the first few terms, and then writing three dots to indicate that the rest is "obvious". An infinite diagram is presumed to be clearly communicated by showing a small typical piece and then using some kind of two-dimensional dot-dot-dot notation to suggest the rest. All over the place, one is asked (subliminally) to buy into other people's sense of pattern. In effect, an "objective" or "natural" esthetics of pattern is presumed by the way mathematicians communicate.

Expanding Conceptual Spheres in Mathematics

Progress in mathematics comes from repeated acts of generalization. If mathematics is anything, it is the art of choosing the most *elegant* generalization for some abstract pattern. Thus esthetics is central. What grabs or doesn't grab the mathematics community does so not simply because it is right or wrong, but because it appeals or does not appeal to the community's collective esthetic sense.

The most important factor that sets top-notch mathematicians apart from their peers is that after coming up with something new, rather than being concerned primarily with the question, "How can I prove this result?", they are concerned with questions like "How interesting is this idea? By fiddling around with it in various familiar ways, can I find ideas that are even more interesting?" This is pattern-play at a very high level, not so different from what a composer of music might do with a newly-discovered theme that seems full of potential.

Mathematicians are constantly inventing, creating, discovering (call it whatever you wish) new concepts by discovering patterns in known concepts. Over millennia, there has been an explosion of mathematical concepts as more and more patterns and types of patterns are encountered. The history of the concept "number", for instance, consists of a series of incredibly fertile generalizations, from natural numbers to fractions, negative numbers, irrational numbers, imaginary and complex numbers, vectors, quaternions, matrices, elements of groups and rings and fields, groups themselves, sets, transfinite ordinals and cardinals, functions, functors, morphisms, categories... Similar lists could be provided for the concept "arithmetical operation", the concept "point", the concept "line", the concept "space", the concept "distance", and so on. There has to be a tacitly shared sense of worthwhile pathways to follow in the development (via generalization) of a concept; otherwise, there would be no consensus, and mathematicians would constantly be asking one another, "Yes, but why did you invent *that* concept? It seems so arbitrary!"

Common Sense and Expanding Conceptual Spheres

Metaphorically speaking, the ever-growing family of concepts centered on some primary concept is an expanding sphere in a conceptual space shared by many individuals. This kind of communal expanding sphere doesn't exist just in mathematical thinking; in fact, it is an incredibly central aspect of everyday thought, and constitutes, to my mind, the essence of common sense.

Much like the mathematical concepts just cited, our ordinary concepts are also structured in a sphere-like manner, with the most primary examples forming the core and with less typical examples forming the outer layers. Such sphericity imbues any concept with an implicit sense of what its stronger and weaker instances are. But in addition to slowly building up richly layered spheres around *concepts* (a process that stretches out over years), we also quickly build spheres around *events* or *situations* that we experience or hear about (this can happen in a second or two, even a fraction of a second). Thus, surrounding every event on an unconscious level is what I have referred to elsewhere (Hofstadter, 1988a; Chapter 12 of Hofstadter, 1985) as a *commonsense halo,* or an *implicit counterfactual sphere,* so called because it consists of many related, usually counterfactual, variants of the event. Just as events themselves are quite fleeting, so are these spheres, which tend to fade rapidly as one's memory of the event itself fades. To make all of this a bit more concrete, here are a few examples, all based on emotionally disturbing events, since, quite understandably, wrenching situations tend to churn up much larger and longer-lasting counterfactual halos than bland ones do.

Several years ago, an interstate-highway bridge in Connecticut collapsed, sending several cars and trucks hurtling into the chasm below and killing a number of people. It soon came to light that for a few days before the accident, local residents had been complaining — unheededly — of "strange creaking noises" emitted by the bridge. As a consequence of this tragedy, the governor of Connecticut immediately ordered the inspection of all other interstate-highway bridges in the state. His choice for "how to generalize the given incident" was to replace a *specific* interstate-highway bridge by the general *category* of "interstate-highway bridge", leaving all else alone. This choice is certainly a natural, understandable, human instinct. But it sounds like a kind of vain attempt to undo, or make up for, something that simply can't be undone — a bit like shutting the barn door after the horse is out.

Why stop at just *interstate* bridges? Why even stop at *bridges*? A more sophisticated response might have been to set up inspection teams to pay attention to public constructions of any sort that people had been issuing warnings about. Admittedly, such a vague policy would be much harder to implement, given the kind of giant mechanical bureaucracy that any government is, but it certainly seems to capture the "essence" of the idea better. And while we're letting our sphere expand around a central essence while not losing touch with that essence, we might note that there was nothing particularly "Connecticutian" about this event. Thus when other states' governors heard of this event, shouldn't they have asked for immediate inspection of interstate-highway bridges (at the very least) in *their* states? It would seem only common-sensical. But where does it all end?

Here's another example. Recently, the young tennis star Monica Seles was stabbed while playing on a court in Germany. Immediately, people — especially other tennis stars — started worrying about the safety on the court of other female tennis players. There was of course *some* worry about males, but definitely less. And of course there was more worry about that particular court, and courts in Germany, than about other courts or other countries. Perhaps stars in a sport similar to tennis started worrying about their personal safety when in public. But do you think that professional bowlers or professional golfers started hiring bodyguards because of this incident? I seriously doubt it.

And why are we tacitly restricting our discussion to sports figures? What about singers? What about authors? Perhaps I, as a professor, should have started to worry that I might be stabbed when — when what? When I stepped onto a tennis court (especially in Germany)? Or when I stepped in front of a blackboard? Or what? Should I have worried more if I had been a female professor, or named "Monica", or a very accomplished amateur tennis player, or all three? Not very likely. Variations this far away from the core theme seem pretty far-fetched. It would seem that these ideas lie considerably beyond the sphere's blurry edges. But how does one intuitively know this?

One last example of the notion of "commonsense halo" involves the infamous Tylenol murders in Chicago over a decade ago, which many readers will recall. Some crazed individual managed to insert poisonous pills into bottles of the drug Tylenol, and the bottles, once distributed to stores, were purchased by random people. Of course, when the unsuspecting purchasers got around to using their pills, they died mysteriously and their deaths were highly publicized, presumably giving great glee to the tamperer. Acting quickly in response to this tragedy, the Food and Drug Administration imposed a set of new packaging regulations on the makers of drugs. Certain of the regulations had to be implemented posthaste by the drug manufacturers, others a bit less rapidly, and still others were given a fair amount of time. Although never enunciated publicly, the rationale behind this kind of staggering must have been essentially as follows. "We assume that criminal minds and average minds are basically comparable. Now to an average mind, some drugs are unconsciously conceived of as 'more similar' and others as 'less similar' to Tylenol. Therefore, to prevent similar crimes from taking place, we will enforce the new packaging regulations very quickly for very similar products, and less quickly for less similar products — roughly speaking, with a speed proportional to a drug's *perceived similarity* to Tylenol." In other words, the FDA was operating on the belief that there was a conceptual sphere with *predictable contents* and expanding at a *predictable rate,* which represented the generalizations that would most likely take place in the minds of sick individuals who might be inspired by the original crime.

Curiously, however, none of the FDA's new regulations extended beyond the domain of drugs. It is as if the FDA was operating on the assumption that the only conceivable generalization of the Tylenol murders was to other drugs (and most likely, to drugs similar to Tylenol) — not, say, to ketchup or spices (which come in bottles having screw-on tops), or to butter or cheese or meat (which are wrapped merely in paper), or to vegetables or fruits or nuts (which are found out in the open)...

In principle, there is no limit to the extension of the Tylenol-murders notion — yet the FDA, by making new regulations only for drugs, seemed to be hoping that maybe no one would think of these other possibilities. Is this hope sensible, or is it like an ostrich sticking its head in the sand? Did they simply recognize that it would be hopeless to try to enforce packaging on all types of foods, and hope that by not mentioning food in the new regulations, they would help to prevent the problem from spreading in that direction? It is interesting to note that the first "variations on the Tylenol theme" that occurred were indeed spikings of an extra-powerful variety of Anacin, a close cousin of Tylenol, and that the next variations were with other kinds of drugs (not foods). Eventually, however, the FDA was forced to extend its regulations outwards, to incorporate foods — at least some foods — as well.

How Far Can a Conceptual Sphere Stretch Before it Pops?

You may not have noticed it, but an expanding conceptual sphere is governing this very discussion of gruesome crimes. It seems perfectly natural to slide outwards from spiking Tylenol to spiking other drugs, then to spiking foods, but we stopped there, as if food *certainly* were the natural end of the line. But why couldn't the concept be further extended? What would amount to twisting the knob further? For instance, what about poisoning the water supply of a big city? Is any anonymous murder of random victims a generalization of this particular event? Is it likely that some demented mind would be inspired by the Tylenol murders to go out and place an old washing machine filled with heavy rocks on some railroad tracks, hoping to derail a train and kill a bunch of random people? (A friend of mine was once in a train that hit just such a "spiked" washing machine; fortunately, no one was hurt.) Based on the Tylenol murders, should the FDA have thought of this "variant" idea and immediately gotten in touch with the Department of Transportation and worked on some joint plan to prevent this kind of thing? As I said above, where would it all end?

And of course we have not generalized outwards from "murder". What about slashing the tires or unscrewing the license plates of random cars? And, while we're into making variations on a theme, why not effect a truly giant leap and switch from doing *bad* deeds to doing *good* deeds — for example, dropping a handful of $10,000 bills from a helicopter above a poor neighborhood, thus "spiking" the neighborhood with good fortune? This bizarre variation on the original theme is a little bit like using a major key to write a variation on a minor-key theme, or reversing figure and ground, as in the marching-singler and marching-doubler sequences.

My repeated usage of the term "variations on a theme" is of course a deliberate allusion to music. Most great classical composers have written sets of variations on a given theme: Bach's Goldberg Variations, Beethoven's Diabelli Variations, Mendelssohn's Variations Sérieuses, Chopin's Variations on Mozart's "Là Ci Darem la Mano", Brahms' Haydn Variations and Händel Variations, Tchaikovsky's Rococo Variations, Franck's Symphonic Variations, Rachmaninoff's Rhapsody on a Theme by Paganini, Copland's Piano Variations — these are just a few salient examples. In each variation of such a set, the idea is to move far away from the given theme on a surface level while at the same time remaining faithful to it on some deeper level. Often the variations grow wilder and wilder as the set progresses, yet one can still sense some intangible kind of link with the theme, though it may get stretched quite thin. The game of "variations on a theme" is, in a way, a big dare: to see how much one can get away with. What constitutes a theme's essence, and how far can that essence be stretched or tampered with before all touch is lost with the original idea?

Of course this question pertains to generalizations in any domain; thus someone might feel that the bills-from-the-sky variation on the Tylenol-murders theme pushes too far, for example. Though the sphere surrounding the core of any idea — musical, mathematical, or of any other sort — is usually capable of expanding far beyond one's initial expectations, there must be a vague kind of limit — a kind of "soft edge" roughly at or beyond which it ceases being realistic to claim that Y is a variation on X. But how to put one's finger on this blurry limit?

The "Me-too" Phenomenon

Generalization outwards from a conceptual center is an automatic, unconscious process that pervades thought — indeed, it *defines* thought. It's not as if there is just one rigid propositional or logical structure that captures what we understand when we read in the newspaper about a kidnapping or hear a throwaway remark about dieting. That is as far from the proper image of what thought is as one can get! Rather, all sorts of analogous events and related images from our own lives are activated to different degrees, and commingle and blur with aspects of the event itself to form a very complex, active, fluid structure, whose rules bear very little connection to those of any kind of formalizable logic. Thus, for instance, consider this verbatim transcript of an interchange between two fairly new acquaintances:

> *Carol:* I often forget my last name, still.
> *Peter:* How long have you two been married now — nine months?
> *Carol:* About.
> *Peter:* I have that trouble every year in January.

Should this seem unclear to you, by "that trouble" Peter meant that for the first few weeks of each new year, when he writes dates on checks and such, he tends to fill in the old year. But his phrase "that trouble" also referred, at the very same time, to Carol's tendency to sign with her maiden name. What he was referring to was thus a blend of the two situations, or something abstract midway between them, some kind of *shared essence*. But of course it was nonverbalized, totally implicit.

As this example shows, in the most mundane of conversations, we will casually refer to two different situations as if they were not only similar, but exactly the same thing. One of the most common of phrases used to signal such a throwaway analogy — or throwaway blend — is "Me, too", as in the following exchange (again a verbatim transcript) between Shelley and Tim, who had just had a drink together in a hotel lobby.

> *Shelley:*   I'm going to pay for my beer now.
> *Tim:*   Me, too.

What did Tim mean? That he too would pay for Shelley's beer? Clearly not. That he intended to pay for his own beer? Conceivably, unless you knew that he had had a Coke, not a beer, so that what he in fact meant was, "I'm going to pay for my Coke now."

These little throwaway remarks by Peter and Tim are completely unremarkable — yet for that very reason they are most remarkable, because they typify the astounding fluidity with which humans use language and, behind the scenes, concepts. This "me-too phenomenon", as I call it, is so commonplace that it is almost invisible to us. Yet in it lurk some of the deepest mysteries of cognition. Consider this typical me-too remark:

> *Marilyn:*   You remember Evelyn? She had a terrible accident last
> month. She was working on the roof of their house and
> fell off and broke her back. She was paralyzed very seriously
> — they're saying probably for life. Yet the first day she was
> in the hospital, she started playing her bassoon.
> *David:*   God, I don't know if *I* could do that.

As you might suspect, this particular David is not a bassoonist. On the other hand, he does play trumpet, but was music-making really what he had in mind when he spoke of himself "doing that"? Far more likely is that he was acknowledging Evelyn's fortitude, and wondering if he had as much. Or something roughly like that, something involving spunk and a will to go on despite adversity and other such admirable traits. Who knows — perhaps blended in there, there was even a trace of the self-doubting feeling, "If that happened to *me,* would *I* be able to sit up in my hospital bed and play my *trumpet*?" And any ordinary human listener basically grasps this whole blurry set of flavors effortlessly.

Here are two last examples of the me-too phenomenon, again taken from real conversations.

> *Computer scientist:*   I'm in artificial intelligence because it's a mixture
> of psychology, philosophy, linguistics, and com-
> puter science.
> *Architect:*   That's the reason I'm in architecture.

Uh-huh. Sure. Gotcha.

Ana: My parents are always calling me "Lucie" and my sister "Ana".

Bob: Oh, yeah — my parents used to do that, too, except it was with our dog and cat.

Your parents probably "did that", too, didn't they?

Hopefully, this handful of examples manages to convey the essence of the me-too phenomenon, allowing readers to become aware of it in their own speech and that of their associates. Beware of innocent phrases like "Oh, yeah, that's exactly what happened to *me*!", for they nearly always signal an impending fluid generalization outwards from a single event, conveyed through the most casual and spontaneous rhetorical gesture, a kind of verbal flip of the hand, behind whose nonchalance is hidden the entire mystery of the human mind.

Outward Generalization in the Seek-Whence Domain

But what, you might ask, does this all have to do with Seeking Whence — or Finding Whither? Strange though it might seem, in the pint-size challenges of the Seek-Whence domain no less than in the ocean-size challenges of real-life situations, generalization outwards from a single event or structure plays a key role. An extremely simple example is the "variabilization" that converts a packet like "(3 4)" into a template like $[n \ n+1]$, a type of act that is indispensable if one is to build theories from mere organizations of data. Another example, somewhat more difficult, is the conversion of an analogy, such as that between packets "123221" and "12343321", into a template like $[1 \ 2 \ldots n \ n-1 \ n-1 \ldots 2 \ 1]$.

But as has just been pointed out in several different ways, generalization in human thought is far, far richer than mere substitution of variables for constants. Generalization involves the ability to internally reconfigure an idea, by

- moving internal boundaries back and forth;
- swapping components or shifting substructures from one level to another;
- merging two substructures into one or breaking one substructure into two;
- lengthening or shortening a given component;
- adding new components or new levels of structure;
- replacing one concept by a closely related one;
- trying out the effect of reversals on various conceptual levels;

and so on. In Seek-Whence, such remedial actions are called for whenever a theory is cast in doubt, which can happen in a number of circumstances, such as these:

- when one's current theory clashes with a new term, and thus is proven to be wrong;
- when one's current theory, despite never making wrong predictions for *new* terms, is unable to account for terms at the sequence's very beginning;
- when one's current theory is judged esthetically unappealing.

Besides these, though, there is another very different reason for wishing to generalize theories in the Seek-Whence domain — namely, the fact that *coming up with interesting new sequences* is an important type of goal in itself, a goal that is of course on a completely different level from that of solving extrapolation puzzles. After you have sought whence a given sequence comes, a new and higher level of activity can occupy you — namely, *finding whither* your sequence leads you.

Nothing could make this clearer than if you try it on your own. I therefore urge you to take the following very simple sequence as your theme, and try making as many variations on it as you can. Here's the sequence:

$$1, 2, 2, 3, 3, 4, 4, 5, 5, 6, \ldots$$

See what kinds of techniques you use, and how many genuinely different directions you can think of to move outwards in, all while attempting to hold onto the spirit of the original at some deep level, of course.

Variations on a Theme by Chopin

In what follows, I shall present some of my own variations on this theme, beginning with rather tame ones whose credentials as variations on this theme will seem completely beyond doubt, and moving outwards to fairly wild ones, where one may legitimately begin to wonder whether the essence of the theme has been preserved. (Of course I like the wilder ones better!)

We have looked at a sequence very much like this before. The only difference was that it began with two 1's. We noted in passing that that sequence had two very distinct parsings, and an analogous fact holds for this sequence. We shall call them simply the A- and B-parsings:

A: 1 (2 2) (3 3) (4 4) (5 5) (6 …
B: (1 2) (2 3) (3 4) (4 5) (5 6) …

The "A" view is built on the more salient and perceptually more appealing plateaus, but has a glitch at the outset. The "B" view, on the other hand, has no glitch but is cast in terms of up-runs, which are a bit less salient and less appealing, perceptually. It's a trade-off: either you live with an unexplained (or at best semi-explained) first term, or you give up a stronger concept for a slightly weaker one and in return gain a uniform global explanation.

From a musical point of view, these two parsings are quite different. There is a Chopin piano prelude (Op. 28, No. 12, in G# minor) whose melody starts out in a long pattern exactly like the sequence given: it is an ascending chromatic scale in which each note but the first one is doubled. Now there are two ways such a pattern could be played: either you accent the even-numbered notes or you accent the odd-numbered notes. In the former case, you get what amounts to our A-parsing, and in the latter, our B-parsing. From the score, it is clear that Chopin himself intended the B-parsing, since the first note forms the downbeat of the prelude's first measure and is beamed together with the second note. If he had intended the A-parsing, he would have shifted the notes relative to the barlines as well as beamed them differently, making the first note an upbeat rather than a downbeat. If you play it in this reversed way, you get a curious (and quite mild) rhythmic variation on the prelude, but it certainly reduces its musical power considerably. (The far greater appeal of the original, incidentally, shows that my earlier claim that plateaus are "perceptually more appealing" has to be taken with a grain of salt. Often, what makes a piece of art appealing is precisely the fact that it violates some normal, easy way of doing things. And given that there are layers upon layers of structure and convention both in perception and in art, no simplistic rule of thumb about "perceptual appeal" is going to hold all across the board.)

Now let us finally look at some variations on our little theme. We begin with ones based on the A-parsing. Perhaps the most obvious knob to twist is the length of the packets. Thus if we jack the value up from 2 to 3, we get this sequence:

Variant A1: 1 (2 2 2) (3 3 3) (4 4 4) (5 5 5) ...

It is left as an exercise to the reader to imagine what would happen if the parameter took on other values.

Some people might feel that by leaving the initial "1" untouched, we have disrespected it. They might argue that if the number of occurrences of every other number was increased by 1, the leading "1" deserves similar treatment:

Variant A2: (1 1) (2 2 2) (3 3 3) (4 4 4) (5 5 5) ...

The counterargument would be that the "1" is not perceived as a plateau but simply as an isolated object; therefore, why should it be given the same treatment as a plateau?

A more extreme form of the view that led to the preceding variation might give rise to this one:

Variant A3: (2 2) (3 3 3) (4 4 4) (5 5 5) (6 6 6) ...

Here the systematic increase in length of the plateaus brings along on its coattails not just a parallel increase in the number of copies of the first term but also a jump of the very *value* of the first term. This vision sees everything in the sequence as very tightly tied together; of course, in the creator's mind, such tightness might not have been present. On the other hand, there can be no arguing that the effect is esthetically motivated and quite elegant.

The preceding variation suggests carrying the idea just a bit further, as follows:

Variant A4:   (1) (2 2) (3 3 3) (4 4 4) (5 5 5) (6 6 6) ...

Here the glitch at the outset is interpreted as a kind of "acceleration", a way of getting up to speed. Once the packets reach full size, then they stabilize. This variation has a particularly satisfying quality to it, because it gives the feeling that the supposed glitch in the original was not really a glitch, after all.

Along quite different lines, we can make a curious figure–ground reversal, somewhat like the transformation that converted the marching doubler into a singler:

Variant A5:   (1 1) 2 3 4 5 6 7 8 ...

Here, *glitch-like behavior* (in which a number appears solo) and *normal behavior* (in which a number appears twice in a row) have playfully exchanged roles. The trick involved in coming up with this variation — seeing the A-parsing in terms of figure and ground — is quite a creative act in itself.

Our next few variations will be based on the B-parsing, and will consequently have a very different flavor. Here, for starters, is a very obvious knob-twist on the up-runs; all it does is lengthen each up-run by 1, leaving their take-off points invariant.

Variant B1:   (1 2 3) (2 3 4) (3 4 5) (4 5 6) ...

Of course there is room for disagreement with this choice. In the original sequence, each new up-run took as its "head" the "tail" of the previous up-run. Thus "(1 2)" segue'd neatly into "(2 3)", and this into "(3 4)", and so on. This smooth joining of up-runs could well be perceived as absolutely central to the sequence's essence, in which case Variant B1 would seem a barbarism, and the following a far more reasonable variant:

Variant B2:   (1 2 3) (3 4 5) (5 6 7) (7 8 9) ...

The following variant on the B-parsing is extremely simple and obvious in one sense, yet in another sense it is quite weird and out of the blue:

Variant B3:   (2 1) (3 2) (4 3) (5 4) (6 5) ...

Each up-run has merely been turned around, to make a down-run. We can make a simple variant on this variant:

Variant B4:   (3 2 1) (4 3 2) (5 4 3) (6 5 4) (7 6 5) ...

Notice that here, each down-run's *middle* number becomes the *rightmost* number of the next down-run. We can make a very similar sequence that shares this property:

Variant B5:   (5 4 3 2 1)   (7 6 5 4 3)   (9 8 7 6 5)   (11 10 9 8 7) ...

But perhaps at this point, we are going too far afield.

The "A" and "B" parsings, while certainly the most obvious ones, are not the only reasonable ones. Consider this way of seeing (or hearing) the original sequence, for instance:

C: (1 (2 2) 3)   (3 (4 4) 5)   (5 (6 6) 7) ...

If this is how you perceive it, then one very simple and natural knob-twist would yield:

Variant C1:   (1 (2 2 2) 3)   (3 (4 4 4) 5)   (5 (6 6 6) 7) ...

A more elaborate set of interrelated knob-twists basically reflecting a similar vision would yield this:

Variant C2:   (1 (2 2 2) (3 3 3) 4)   (4 (5 5 5) (6 6 6) 7)   (7 (8 8 8) (9 9 9) 10) ...

More variations on the C-parsing can be found without too much trouble, but I will not go into them.

The next variation is a horse of another color:

Variant X: 2 1 2 1 1 2 1 1 2 1 1 1 2 1 1 1 2 1 1 1 1 2 1 1 1 1 2 1 1 1 1 1 2 ...

Déjà vu? It starts out identically to the old triangles-between-squares sequence. However, this is much simpler than that one — it simply converts each term of the theme into a group of 1's of the appropriate length, using 2's as separator elements. We can even make it represent, say, the B-parsing by turning half the 2's into 3's:

Variant Y: 2 1 2 1 1 3 1 1 3 1 1 1 2 1 1 1 2 1 1 1 1 3 1 1 1 1 3 1 1 1 1 1 2 ...

As a parting fillip, consider this last variant:

Variant Z: 2 1 2 1 1 2 1 1 3 1 1 1 3 1 1 1 3 1 1 1 1 2 1 1 1 1 2 1 1 1 1 1 2 ...

It is like hearing the original theme in a new rhythm: 1-2-2, 3-3-4, 4-5-5, 7-8-8, and so on.

The Blurry Edge of Essence

So... How do your variations on this theme compare with mine? Did I think of some cute ideas you didn't think of? Did you come up with some clever ideas that I missed? Undoubtedly so. Do you think that all of my variations managed to respect the theme sufficiently, or are some of them too fringy for your taste? And how about yours, now that you have a basis for comparison?

Which of all the variations pleases you the most? Which do you think would appeal the most to Chopin? Probably those that go quite far out on a limb yet clearly retain the flavor of the original are the best — but this still leaves entirely unanswered the question as to what it means to "retain the flavor of the original".

To see how delicate and intangible an issue that is, look at this variant suggested by someone whose set of variations was in most respects amazingly close to my own set:

Steve's variant: (1) (2 2) (3 3 3) (4 4 4 4) ...

When I saw this one, my reaction was, "This is really beyond the fringe. How could anyone call this a variation on the given theme?"

After the fact, Steve himself decided he didn't like it very much, either — it was just "too perfect" in comparison with the original theme. He had apparently been driven by a strong desire to get rid of the original sequence's glitch, and in a moment of weakness he allowed that goal to override the deeper goal of respecting the core of the theme. His trick for making the glitch go away was to have the group-lengths increase indefinitely. This is where, in my view, he lost touch with the theme's essence. In a clever bit of self-perception, Steve described his generation of this variation as a case of "two wrongs making something that was too right".

I agreed with this, but the fact that it was "too right" was not the crux of what bothered me. I felt that Steve's real sin was in allowing group-lengths to increase beyond bounds; to me, this was an utter betrayal of the theme. To Steve, on the other hand, doing so seemed a bit fringy but perhaps acceptable, depending on how it was done. To him, his variant didn't look so different from my Variant A4; to me, there was no analogy at all. In the end, we simply had to agree to disagree on this matter, even though in almost all other respects we saw eye to eye on the essence of the theme, and on the quality of diverse variations produced by ourselves and other people.

The act of composing a rich and diverse set of variations on a theme, such as we have just been through, is a creative and intellectually stimulating game. It is a purely intuitive, esthetically-driven process, seeking only to find ever new types of patterns, yet patterns that all retain some intangible "family resemblance".

One of my most ambitious goals in the Seek-Whence project, though it always remained just a tantalizing dream, was to impart to a computer the ability to play this game. To do this well of course requires a keen sense for when essence has been preserved, and when it has been lost. It also requires the ability to perceive a theme in all sorts of novel ways by bringing in unexpected concepts and "trying them on", to see how they fit. Lastly, it requires a sense of naturalness versus forcedness, and a sense of elegance versus clunkiness. Such senses and abilities, which, taken together, certainly deserve the label *intuition,* are so subtle and elusive that it is no wonder that many people are skeptical about ever giving computers intuition. I do not entirely share their skepticism, although I sympathize with their, well, their *intuition,* and I agree with their gut-level feeling that where we are today, in terms of computer models, is nowhere near the goal.

Coming Full Circle to Triangles between Squares

When I first discovered the recursive pattern of the triangles-between-squares sequence, I had no idea where it would lead me. But I was so thrilled by the discovery that I wanted to experience it again and again, and so, of course, I was led to trying one variation after another on my initial theme. For several years I remained on this ill-defined outward-bound quest in search of beauty and magic.

Basically, there were two main tacks I could take. One involved making variations on the numerical sequence "212112…" itself — for example, making another sequence having the same self-defining aspect but with the roles of 1 and 2 reversed, or making a self-defining sequence that contained more than just two different integers, or making more complex versions of the recursive rule, and so on, all the while paying no attention to the sequence's original *source.* The other tack was quite the reverse: I looked *only* at the sequence's source, paying no attention to the sequence itself — thus, for example, I tried counting squares between pentagonal numbers, squares between cubes, squares between double-squares, factorials between products of successive odd numbers, powers of 2 between powers of 3, and on and on. Once launched on this adventure, I was off and running.

These two tacks led me in very different directions, but each of them was in its own way extremely fruitful, and in the end, they wound up converging in all sorts of surprising ways. Eventually, however, as is typical in my experience, I found that my "generalization engine" was running out of steam, in the sense that the best ideas seemed to have been unearthed, and the newer ideas were getting a little too complicated to keep my interest at the same high level. Thus slowly, and without any clear cut-off point marking its end, my adventure wound down.

This long saga of mathematical exploration had a final payoff at a place and time in my life that I would never have expected: some ten years later, as I was doing my doctoral work in theoretical solid-state physics. I was trying to understand the elegant but tricky mathematics involved in describing the properties of electrons in a highly idealized two-dimensional crystal immersed in a uniform magnetic field. It turned out, to my immense surprise, that many of the most beautiful ideas I had discovered in my number-theoretical meanderings a decade earlier were precisely the tools needed to explain a mysterious mathematical structure that plays the starring role in this area of physics, and whose deeply recursive nature no one understood, or even suspected, at the time. Never in my wildest dreams had I expected to come back to number theory, and in a sense I didn't; rather, it came back to me — through the back door of solid-state physics. But it was a very happy reunion, in any case. (For technical details, see Hofstadter, 1976, and for a quick lay-level description, see Chapter 5 of Hofstadter, 1979.)

I mention all this because it is a quintessential, if microscopic, example of the totally unpredictable, snowballing character of the discovery process in mathematics, in scientific research — indeed in all creative intellectual activity, including design, writing, painting, musical composition, and so on. Not only was this a beautiful episode in my life, but it taught me unforgettable lessons about the creative process. Much of my work in cognitive science has been a quest to capture the essence of what I was doing in those wonderful heady days of unfettered mathematical exploration.

From Seek-Whence to Jumbo, Copycat, and Others

In this retrospective look at the origins of the Seek-Whence project, I have tried to give a broad feeling for the kinds of general cognitive issues that came to the fore, issues that gradually came to dominate my approach to modeling mental processes by means of computers. If I were asked to summarize the most central research ideas and goals that emerged from the years of work by myself, Marsha Meredith, and Gray Clossman on the Seek-Whence project — both its conceptual groundwork and its computational realization — I would have a very hard time, because there is certainly no single, pithy, catchy idea that on its own captures our philosophy. That would be very nice, but it just isn't the case. Rather, our philosophy is a complex of many tightly interrelated ideas. Some of the most important themes that crop up over and over again are the following:

(1) the *inseparability of perception and high-level cognition,* leading to the idea of a perceptual architecture being at the heart of cognition;

(2) the fruits of high-level perception being *easily reconfigurable multilevel cognitive representations* held loosely together by bonds of different types and different strengths;

(3) the idea of *subcognitive pressures* — namely, that the more "important" a concept or a representation is, the greater an influence it should be allowed to exert, in a probabilistic sense, on the direction of the processing;

(4) the *commingling of many pressures,* both context-dependent and context-independent, leading to a nondeterministic parallel architecture in which bottom-up and top-down processing co-exist gracefully;

(5) the *simultaneous feeling-out of many potential pathways* at differential rates governed by quickly-made estimates of degree of promise;

(6) *the centrality of the making of analogies and variations on a theme* in high-level cognition;

(7) the possession, by cognitive representations, of *deeper and shallower aspects,* with the former remaining relatively immune to contextual pressures, and the latter being more likely to yield under pressure (to "slip");

(8) the crucial role played by *the inner structure of concepts and conceptual neighborhoods* in all these goals, particularly context-dependent conceptual overlap and proximity, and context-independent conceptual depth.

Different combinations of these ideas culled from the Seek-Whence project have, over the years, given rise to various new projects. Thus, Jumbo and Numbo are models of fluidly regroupable hierarchical structures, and are concerned mainly with the issues on lines (1) through (5) in this list. Neither features a deep model of concepts. On the other hand, Copycat is a much more ambitious and integrated model of concepts, perception, and analogy-making (but not sequence extrapolation) in a one-dimensional domain that is a very close cousin to the Seek-Whence domain. It aims at modeling some of the key mechanisms responsible for insightful and creative thinking, independently of the domain in which those activities occur.

Tabletop and Letter Spirit carry the ideas on perception and analogy-making pioneered in Copycat into two very different two-dimensional domains, and push the architectural ideas in some important new directions. Tabletop, in particular, is concerned with analogy-making in a domain featuring "mushier" concepts, and typical situations in the Tabletop domain tend to be much less elegant and patterned than those in the Copycat domain. Letter Spirit, on the other hand, is deeply concerned with the spinning-out of variations on a theme, and represents an attempt to pinpoint the elusive nature of "essence" in a highly idealized artistic domain.

It is my firm belief that pattern perception, extrapolation, and generalization are the true crux of creativity, and that one can come to an understanding of these fundamental cognitive processes *only* by modeling them in the most carefully designed and restricted of microdomains. This article of faith has guided me and my research group over the past decade and a half, and the remainder of this book is dedicated to conveying the results of those investigations.

Preface 2:
The Unconscious Juggling of Mental Objects

The Joy of Anagrams

Jumbo is a curious and amusing program. It tries to make English-like words out of a set of letters by rearranging them and putting them into plausible orders. This activity of doing anagrams is one that I personally have enjoyed ever since I can remember. For me, there is something both delightful and tantalizing about doing anagrams. When I was a student, I used to solve Jumbles in the newspaper (see Figure II–1, page 98) on a regular basis, often timing myself to see how fast I could do them, and one time a friend and I worked out anagrams on the names of all 100 students who lived in his dormitory. These days, I often find myself idly doing anagrams on some word in my visual field without even having noticed when I started.

Because of my enjoyment of this activity and because of years of practice, I am quite a high-level practitioner of the art. Sometimes, solutions to Jumbles in the newspaper just pop into my mind in no time flat — blindingly fast. More often it takes five to ten seconds for a five- or six-letter challenge. And then there are those painful times when what looks like a very ordinary challenge stumps me for several minutes, and I find myself resorting to artificial techniques, such as writing out my latest couple of attempts backwards, or writing the letters out in some completely random order, to try to jerk myself out of the mental ruts I'm caught in.

I have always wondered just what it is that is going on in my mind when it carries out these tricks for me. I phrase it in that peculiar way deliberately, because it really feels quite magical when you look at six letters in some random order and a word just instantly jumps to mind. And it's even more interesting when it takes ten or twenty seconds, because then you get little glimmers of the shuffling taking place somewhere in there. But it's not *you* who's directing the show. You're just a passive observer, at least most of the time. Only when those artificial rut-breaking techniques are needed do *you* actually play a role in the process. But that's the boring part. The neat part is when you just sit back and watch.

Letters Go Up, Words Come Down

One metaphor I've often used to describe how it feels is this. You're given a set of letters and you throw them all up into the air at once, out of sight. When they come back down, somehow they're all stuck or "glommed" together in a way that is always pronounceable and usually very wordlike. You look at this "glom" and see if it indeed forms a word; if so, you're done. If not, toss the letters back up in the air and see how they come down the next time. Surprisingly, it feels as if they never come back in any previously tried order.

When the number of letters is just five or six, I seldom make errors, in the sense of leaving a letter out or adding a letter. In fact, I can't remember *ever* getting the number of letters wrong in tackling such a small-sized challenge. However, in a challenge with doubled letters such as "toonin" (try it yourself!), it is quite possible that I might come up with a word candidate having two "t"'s instead of two "n"'s, or two "i"'s instead of two "o"'s. As long as the letters are all distinct, though, I have found that I can pretty easily handle challenges of up to about ten letters. I don't at all mean that I easily *solve* ten-letter challenges — just that my mind can do this unconscious tossing-up and catching without losing any of the letters or sticking in wrong ones.

One fascinating phenomenon that I think would be very interesting to investigate through psychological experimentation is the initial period of accommodation to those ten letters. It's certainly not the case that, when I see ten randomly scrambled letters (with no duplicates), I instantaneously have them down cold and can start juggling them mentally. Six letters, yes, but ten, definitely not. With ten, I have to let them sink in for a few seconds, maybe a minute. During this time, I am of course playing with them but on a fairly conscious level, and I make lots of mistakes, in the sense of leaving letters out or sometimes even doubling them. (It's extremely rare that I introduce a letter type that wasn't present at all in the original set.) But there comes a point when things just start happening automatically; then I can "toss them up in the air", and they come back down very nicely on their own, with nothing left out or added. It feels almost as if, for each different letter, some kind of tangible, movable mental *object* were actually getting made, and once those objects are all really present "on stage", then the automatic level can take over.

Let me make this very concrete. Suppose the challenge is to unscramble the eight letters "ucilgars" (try it!). Once I've gotten these letters internalized, if at some point I happen to notice that "girl" is in there, then the remaining four letters immediately pop to mind in some random order — "csua", say. I don't have to consciously work out what's left after "girl" is removed (*e.g.,* mentally writing out the eight letters in a row, then mentally crossing four of them out and visualizing what remains); rather, what's left is just *there*; it simply presents itself to me without any effort on my part. It's definitely not visual — it's not that I *see* the remaining four letters on some mental screen. It's just that I sense their presence — "csua". And no sooner am I aware of the subchallenge "csua" than it is gone, the letters having instantaneously regrouped themselves into some stabler structure, such as "caus" or "suca". And around and around it goes.

On the other hand, before the letters are fully internalized, the manipulation is far more intellectual and conscious. For example, if the challenge is "abcelmnrsu" (this is pretty long, but you might try it anyway), it takes some time to get used to those ten letters. If I pull out "clamber", say, it's quite possible that instead of "uns" presenting itself to me, "buns" will — showing that I inadvertently doubled the "b". I may well have a hunch that something is wrong, but it will probably take me a couple of seconds to figure out just what it is. (Often I make worse errors than this in the initial build-up period.) And if the challenge itself contains a repeated letter, then manufacturing the appropriate duo of identical "mental objects" is considerably harder for me, although that too can be done. It just takes longer.

The Gulf between Virtual Mental Objects and their Physical Substrate

I have a hunch that this transitional build-up period involves a very significant and little-explored kind of interface activity between *long-term memory* (where exactly 26 distinct letter *types* constituting the Platonic alphabet reside, but certainly no duplication exists) and *working memory* (where any number of letter *tokens* can reside, including repeated instances of any particular Platonic letter). It's as if the transition period during which I "get good" at a ten-letter challenge is the time when these short-term tokens are actually getting built. Before they're all there, I'm somehow getting by with just the Platonic letter types in long-term memory — but since they're not designed for this kind of quick juggling, the shuffling process is slow and conscious. Then at some intermediate stage, *some* of the mental objects have been built in working memory, but I'm also depending on some of the long-term Platonic letters, which of course are not in working memory. That's a curious, blurry stage — most interesting but also most complicated.

Actually, this transitional stage is probably even subtler than what I have just said, in that the working-memory tokens most likely come into existence *gradually,* rather than jumping suddenly, in a black-and-white fashion, from the status of "not there" to that of "there". But how could such semi-existent or nascent tokens come to participate in the juggling process, alongside fully-made tokens, on the one hand, and long-term-memory types, on the other?

This, to my mind, is a deep and subtle mystery that goes to the heart of what mental objects are. After all, we are not talking about building physical structures that will literally move about inside the brain the way blood corpuscles do; we are talking about the manufacture of *virtual* objects, objects that float on neural hardware but that are certainly not easily describable in terms of neurons or networks of neurons. Such "objects" exist in a virtual space — working memory, that is — in which they are free to move around, mingle, associate, cluster, come apart, and so on.

A useful analogue is the image of a ball in a video game, which is neither a genuine physical object nor a group of fixed pixels, but rather something abstract that has its own persisting identity and its own types of behavior, and that floats on pixel hardware but is utterly different from pixels or groups thereof. There is a deep and fundamental level-distinction between such a virtual video object and the "stuff" on which it floats — and I would argue that this same kind of level-distinction holds between these mental letters and the neural hardware that they reside on. This level-distinction is a very deep breach that, it seems to me, few people in cognitive science have even noticed (an exception would be Daniel Dennett, 1991), and that certainly no one has yet grappled with adequately.

In any case, once *all* the working-memory tokens are fully made and installed, the Platonic types need play no further role and fade out of the picture. At this point, the fully-made tokens act like physical juggling balls, in the sense that when I play with them, they won't disappear or multiply. As I said earlier, when shuffling such tokens, I don't have to work at keeping track of them — they are just *there*.

This is is just about as far as my introspective musings about the blurry emergence of these "mental juggling balls" have carried me. However, I have many further intuitions about how it is that such juggling balls, when all of them are fully present in working memory, somehow glom together when they are "up there in the air" out of sight; indeed, that is what Jumbo is all about.

Why work so hard to model such a frivolous and atypical cognitive activity? I tried to answer this question in the article itself, but let me just add here that I think that such mental juggling is a very important, pervasive kind of mental activity that has nothing intrinsic to do with anagrams. Perhaps the slow letter-juggling that goes on in the heads of people who have almost never tried anagrams is not of much universality and therefore of little importance or interest, but I think that when the activity reaches expert level, where it is highly automatized and very rapid, it has something in common with the deep processes of reorganization and reinterpretation that take place in truly creative thought. Not to suggest that all good anagrammists are latent Einsteins, of course, but just that the activity itself, when done fluently, has a special and important quality. (Novick & Coté, 1992 describes a series of psychological experiments carried out on both novice and expert anagrammists; this work tends to confirm my intuition that parallel processing is central in experts, and that what novices do is of quite a different nature.)

Jumbo versus Brute Force

Developing the Jumbo program, during the years 1982–83, was a milestone for me personally, in that it represented the first time I was seriously implementing a set of ideas about the mind that I had been slowly developing over the previous five years. It was written in a thankfully now-defunct dialect of Lisp and ran on a Vax machine in our Computer Science Department. When given a six-letter challenge, for example, it ran quite fast, spewing out word candidates on the screen, one per line, almost as fast as it could print.

This might sound impressive, but of course raw speed was hardly the point. To make this clearer, let me mention that sometime in the 1980's I learned about the existence of superfast brute-force anagram programs that use unabridged dictionaries. Such programs take a far more difficult challenge — say someone's full name, consisting of a couple of dozen letters — and they find every possible way to anagrammize it using legal entries in the dictionary. There are often thousands of such solutions, yet it takes only a few seconds to do this mammoth task — infinitely faster than Jumbo. But such programs, based on highly mathematical rapid-search techniques, are the antithesis of cognitive models, and for that reason I have little interest in them, aside from genuinely admiring the clever hacks involved in programming them.

That Fateful Footnote

The influence of the Hearsay II project in speech understanding on my work cannot be overstated. I first learned of that project in 1976 or 1977, and was instantly struck by the idea of a parallel architecture in which bottom-up and top-down processing could coexist and influence each other. When I then read a collection of papers on the details of the architecture (Reddy *et al.*, 1976), I was fascinated by the way in which highly specific types of actions — KS's, as they were called (short for "knowledge sources") — were invoked by situations that came about in the central data-structure called the *blackboard*. One thing that particularly struck me was that for any KS to be invoked required a rather complex set of conditions on the blackboard to exist. Ideally, the KS would "just know" that its particular set of conditions was satisfied, and then and only then would it jerk into action. But of course that is pure fantasy; a KS isn't telepathic. It has to have "sensory organs" to inform it. In other words, there has to be something actively watching the blackboard in order to *detect* the existence of the appropriate conditions.

Thus computational tests called *preconditions* were attached to all KS's, and each precondition, when run, checked to see if the appropriate conditions for invoking its KS held on the blackboard. Now ideally, the different preconditions of all the different KS's would all be running in parallel with one another all the time, so that each KS could virtually instantly know when the proper circumstances had arisen for it to jump into action. But unfortunately, preconditions themselves were quite complex computations. In fact — and here I quote from a technical paper on the subject (Fennell & Lesser, 1975):

> Preconditions themselves have preconditions, call them "pre-preconditions." In HSII, knowledge-source preconditions.... may be arbitrarily complex. In order to avoid executing these precondition tests unnecessarily often, they in turn have pre-preconditions which are essentially monitors on relevant primitive data base events.... Whenever any of these primitive events occurs, those preconditions monitoring such events are awakened and allowed to test for full precondition satisfaction.

I found this an absolutely fascinating idea, appealing greatly to my esthetic sense. It immediately suggested to me a whole hierarchy of conditions, preconditions, pre-preconditions, pre-pre-preconditions, and so on — not an infinite regress, of course, but stopping after some finite number of levels. In this scheme, only the processes on the bottom-most level (*i.e.*, the level with the largest number of "pre"-fixes) would be kept constantly running in parallel, with higher-level processes being triggered when and only when appropriate events happened one level down. This would be a beautiful strategy for pinpointing exactly when to invoke a very precisely and subtly targeted type of action.

That short paragraph by Fennell and Lesser was perhaps the most memorable one I ever read in AI, and yet the curious thing is that it was not given a great deal of play in the article itself — in fact, it was merely a footnote, and a somewhat apologetic one at that. The notion that this strategy might be an important *principle* in the design of parallel systems was certainly not suggested, and as far as I could tell, it was never highlighted in any other publication on Hearsay II. For me, however, this footnote became enshrined as a central lesson about how to deal with parallel computations having different degrees of computational *specificity* and different degrees of computational *expensiveness*. When adapted to my own research goals and merged with my ideas of probabilistic processing, it became the strategy I came to call the *parallel terraced scan*.

The Parallel Terraced Scan Meets the Greek System

Perhaps the most important idea of Jumbo's architecture, and one that pervades the work in this book, was the parallel terraced scan. That idea is described in a formal and detailed manner in Chapter 2, but it seems worthwhile to include here one curious example of roughly the same idea, which I came across while reading the Indiana University student newspaper. It was a story detailing how a young woman went through "rush" — a standard ritual, in American college life, of trying to associate oneself with a "Greek" organization (a sorority or fraternity). I had never had any inkling of what an intricate selection process was involved, and when I read the article, I was strongly reminded of some aspects of courtship.

Basically, sorority rush at Indiana goes like this. The women who wish to join a sorority — and there are some 1700 of them each year, typically — go to open houses one weekend in November, spending a half hour each at all 22 different sororities. Then the residents of each sorority collectively decide which of the "rushees", as they are called, they would like to invite back. In early January, each rushee receives a list of the sororities inviting her back, and then on two successive evenings she goes to visit them. Some rushees receive up to 22 invitations, but they can only go to 16 parties, so they have to narrow down their choices. Each night they go to eight half-hour parties, during which time they size up the sororities and the sororities size them up. There follows, of course, a culling process, and each rushee is invited back to a smaller number of houses. At the same time, she herself is mentally culling out those places she likes the best. Invitations are issued for another round of parties — this time all taking place on a single evening called "eight-party night". That night, a rushee goes to at most eight 45-minute parties, and the same kind of mutual sizing-up takes place, only over a more extended period, and of course fewer rushees are at each house, so the contact time between members and would-be members is considerably higher. There follows further mutual culling, and then there is a "four-party night", which involves a rushee attending up to four parties of one hour in length. Once again, this is an intensification of the mutual scrutiny. Then there is more culling, and the culmination of it all is "preference night", which is a mere two-party night in which each party lasts an hour and a quarter. Finally, in February, bids are issued by the sororities, and each successful rushee makes her final choice. When you consider that the parties get gradually longer, and factor in the fact that fewer and fewer rushees are present on the various party-nights, you see that the intensity of the bidirectional scrutiny more than doubles at each successive stage of the process. It is really quite an arduous routine, and only some 800 rushees survive all the cuts and actually wind up pledging at some sorority. It's hard to imagine coming across a more vivid example of the notion of a parallel terraced scan!

To Read, or Toreador

Some time after Jumbo had been developed, I was interested in getting a student to revive the project and to push it further in a number of directions. As we talked about how Jumbo worked in some detail, I realized that Jumbo

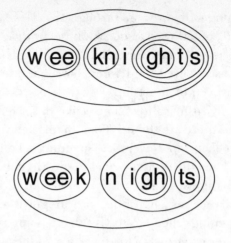

Figure II–0. When one reads a word, letters rapidly glom together in a hierarchical manner. Above are shown two reasonable ways of reading the string "weeknights", with the distance between the centers of any two chunked structures (letters or gloms) intended to suggest the strength of the bond connecting them (longer distances meaning weaker bonds, of course).

implicitly subsumed a very fundamental and everyday cognitive activity — namely, *word perception*. In particular, all you have to do is to *freeze all the letters in their original order,* and now the task for Jumbo is to figure out how they fit together.

We are totally unaware of the multi-level chunking that goes on effortlessly as we read, but just consider a word like "distract". It could be seen as "di–stract", as "dis–tract", or as "dist–ract". How do we put it together in a small fraction of a second? Or how do we perceive the word "weeknights" when it is flashed in isolation on a screen? (See Figure II–0.)

Consider the hierarchical complexity involved in perceiving the even simpler word "nights". The last four letters seem to form a single consonantal unit, but that unit then breaks down into two clear parts: "ght" followed by "s". Moreover, "ght" is not an elementary chunk, but consists of the chunk "gh" followed by "t". This is already complex enough, but there is a completely different way of perceiving "nights". Perhaps the "consonanticity" of the individual letters "g" and "h" tricks us into assuming that their fusion in "gh" must be consonantal as well; however, why could it not be that "igh" is just one big vowel, as in "high"? In that case, the vowel would break down into "i" followed by "gh". (Again, see Figure II–0.)

How does a fluent reader actually perceive "nights"? Such a question cannot be answered by introspection, because too many simultaneous things are going on at too fine-grained a level. I suspect that a fluent reader's perception of the word "nights" involves some kind of unconscious blend, wherein "igh" is seen to some extent as a vowel-like unit, and yet at the same time, "ghts" is seen to some extent as a consonantal unit. For such a blend to exist, the "gh" glom would have to have an indeterminate or probabilistic or very rapidly fluctuating status. But this would have to be studied carefully in the laboratory. To model such a cognitive phenomenon in a realistic manner would of course be extraordinarily subtle.

Ordinary reading is permeated by this kind of multi-level "glomming" process, and if one is aware of it, one can play games based on it. While driving through the town of Concord, Michigan, I noticed a sign that said "N. Concord Rd." Something about it grabbed my attention, and after a moment I realized that it could be seen this way: "NCO–NCO–RD–RD". A similar phenomenon is the nonstandard parsing of "hotshots" as "hots–hots", or the nonstandard breakup of "no nonsense" as "no–no–nse–nse". Ambiguities based on simple regrouping like this are legion, and it is rather remarkable that we so seldom find ourselves tripped up by them.

With Henry Velick, a student at Michigan, I worked for some time on developing a variant of Jumbo called "Toreador", whose purpose would simply be to take a *frozen sequence* of letters and, lacking any dictionary, to convert this sequence of isolated units into a hierarchically-chunked structure in the most plausible fashion, using the parallel and probabilistic glomming techniques at the core of Jumbo. Unfortunately, Henry left, and this project never came to fruition; it remains something to think about for the future.

To Learn, or Not to Learn

The original paper about Jumbo was written in the spring of 1983 for presentation at a workshop on machine learning that took place in May of that year in Monticello, Illinois. Unfortunately, the constraints for inclusion in the workshop's proceedings forced me to truncate that paper drastically, so that many ideas were only hinted at. I delivered the shortened paper and it was printed in the workshop's proceedings, but when those proceedings got turned into a book (Michalski, Carbonell, & Mitchell, 1983), the Jumbo paper was omitted, the editors having concluded that this work had little or nothing to do with learning. I do not agree with that judgment, although I can see why the connection eluded them. In any case, the version published here as Chapter 2 is the original uncut version, with some slight additions and modifications. The references have been left intact, and have not been updated to take into account more recent literature.

Chapter 2

The Architecture of Jumbo

DOUGLAS HOFSTADTER

From the guessing of newspaper enigmas
to the plotting of the policy of an empire
there is no other process than this.
— William James, in *Psychology* (1892)

Prologue: Not Mumbo-Jumbo

Jumbo is an artificial-intelligence research project with a concrete purpose and an abstract motivation. It is one of several projects, all focused on various facets of one shared long-term goal (Hofstadter, 1983b). In essence, that goal is to demonstrate the equivalence of cognition with "deep perception" — the non- modality-specific layers of perception that involve funneling into highly abstract and often nonverbalizable categories (Schank, 1980; Hofstadter, 1979, 1981, and 1982c). This thesis asserts that intelligence emerges out of the interactions of many thousands of parallel processes that take place within milliseconds and are inaccessible to introspection. It is captured in caricature form by the slogan "cognition equals recognition".

Explicitly or implicitly, this idea is denied by a large fraction of the work currently taking place in artificial intelligence. One of the most influential advocates of the antithetical viewpoint is Herbert Simon, who has claimed (Simon, 1981) that everything of interest in cognition takes place above the 100-millisecond level, which he characterizes as the time it takes for you to recognize your mother (or grandmother). According to this view, it would be a waste of time for cognitive scientists to try to analyze or reproduce the many microscopic parallel events in the brain that make up recognition or perception; all that they should care about is the macroscopically observable (and introspectively accessible) serial processing that we effortlessly sense in our minds, and that we label "thinking". Indeed, Simon once remarked in conversation (Simon,

Figure II–1. An example of the Jumble game, typical in most ways but atypical in a most curious way.

1982) that in his opinion, the billions of neurons involved in perception are no more relevant than the number of electrons in a diode. This statement certainly suggests that Simon believes that there is, metaphorically speaking, an "Ohm's Law" level of description of the brain's activity, allowing one to totally sidestep or ignore all biological substrates of thought.

This attitude is by no means unique to Simon. Indeed, the continuing prevalence of this type of belief makes it all the more necessary to point out, over and over again, that deep perception (Bongard, 1970; Hofstadter, 1982d) is an unsolved — in fact, almost untouched — problem of artificial intelligence. Deep perception (*i.e.*, the "recognition" mentioned in the slogan) is a far cry from low-level vision or hearing; it is the murky area where syntactic sensations interface with semantic categories. This syntactic–semantic transition zone, I feel, is the core mystery of all of intelligence. Before we can make machines that can learn, we must understand how concepts are structured and can be addressed and compared. Therefore Jumbo is a prelude to a more ambitious project in learning.

Jumbo and Jumbles

Jumbo's concrete purpose is to imitate human skills used in the well-known newspaper anagram game "Jumble" (Arnold & Lee, 1982). In this game, illustrated in Figure II–1, a human player seeks an English word made out of a small set of given letters. The idea of the Jumbo system is for a computer to construct English-like *word candidates* ("pseudo-words") from a given set of letters.

Jumbo possesses no dictionary of English; it makes its constructions purely by referring to stored knowledge defining how, in English, consonant and vowel clusters are formed out of letters, syllables out of clusters, and words out of syllables. Thus Jumbo is a building program; it begins with isolated atomic units (letters), and out of them gradually constructs *gloms* — conceptual molecules or chunks. There are gloms at various levels: clusters, syllables, and words.

Why no dictionary? Because that is irrelevant to the mental processes I am attempting to model. Humans who are good at Jumbles find that pseudo-words flash into their mind rapidly and effortlessly, then metamorphose fluidly and spontaneously into other pseudo-words. In skilled "Jumblists", the process is completely unconscious. They do not guide it; they simply watch it happening. The fact that most of the word candidates that people's minds invent are not genuine English words but have the right ring to them is the charm of the game. I am interested in *this* aspect of the process: the spontaneous and unconscious composition of coherent wholes out of scattered parts, using only a fairly small knowledge base (the local structure of English-like words).

I am *not* interested in the secondary aspect, that of checking whether a manufactured pseudo-word is actually in the English vocabulary or not. A big dictionary could of course be added to the program's knowledge base, but it would be utterly irrelevant to Jumbo's serious purpose, which is to be a model of the mental processes of assembly and transformation — not an expert system for playing an insignificant word game. The hope is that these abstract processes (and an architecture allowing their smooth integration) are notions exportable to more complex domains.

The Significance of Jumbo's Task Domain

What is significant about coming up with some English-like rearrangements of a set of letters? Well, doing Jumbles, though in itself trivial, exemplifies an extremely important facet of human intelligence — the way in which we mentally juggle many little pieces and tentatively combine them into various bigger pieces in an attempt to come up with something novel, meaningful, and strong. (A trivial example is that I just rearranged the words "strong, meaningful, and novel" into the reverse order, because I felt it would be slightly more effective. A more substantial example is that ten years after completing this article, I swapped this whole section — minus this sentence, of course! — with the one called "Two Basic Analogies on which Jumbo Rests", in an attempt to improve the logical flow.) Multi-level cleaving, splicing, regrouping, reordering, and rearranging — such operations permeate the process of creation, whether it is composition of music, art, or literature, or the invention or discovery of new ideas in science. As such, these operations play a key role in the engendering of the most important and innovative ideas.

Furthermore, the perceptual process, too, is essentially one of constructing larger units out of smaller ones, with temporary structures at various levels and permanent mental categories trying to accommodate to each other. Thus the activity of playing with letters to try to make a coherent structure out of them is closely related to perception. Let me spell this out a bit more carefully.

In a temporal sense such as hearing (listening to language or music in particular), there is an intrinsic linear order to the constituents on all levels, but the boundary lines between constituents are not provided. They must be found by a process of trial and error. Good guesses at the most elemental level (the shortest time unit) may result in correct larger structures getting manufactured at higher levels. By guessing well at each level, the system may succeed the first time around in putting together the top-level structure it is aiming for. On the other hand, the entire process must be organized so as to be able to handle failure or partial failure at any level as easily as it handles instant success. This is a very complex matter.

In a nontemporal sense such as static vision (perception of an unchanging scene, such as a drawing or photo), there is no intrinsic scanning order, and decisions about where boundary lines are at various levels become even more complex. Still, the process is essentially one of constructing larger perceptual units out of smaller ones.

In any type of perception, much back-and-forth motion must occur — that is, an intimate mixture of construction, destruction, regrouping, and rearrangement of tentative structures. Any architecture for a system to carry out this type of process is the result of many subtle decisions about how independent processes should interact, how structures should be put together or broken apart, what kinds of things form stable structures, what easy ways are of making new possible structures when old ones are seen to be inadequate, and so on (Lea, 1980; Hanson & Riseman, 1978; Waterman & Hayes-Roth, 1978).

The Hearsay II speech-understanding system (Reddy *et al.*, 1976; Erman *et al.*, 1980) is a sophisticated example of such an architecture; indeed, it has had a great influence on my ideas. Jumbo is an attempt to construct another such architecture, and to focus clearly on the theoretical aspects of the problem, rather than building a large-scale system with a highly pragmatic goal, as Hearsay had. This theoretical focus is one reason why Jumbo's domain is so small: idealization makes things easier to analyze.

Two Basic Analogies on which Jumbo Rests

The strategy of Jumbo is based on two analogies. One is to the way that complex molecules are constructed inside a living cell (Lehninger, 1975); the other is to the way that bonds of human friendship or romance are formed in a chaotic world. I will describe the biological analogy first.

In a cell, the lowest level of molecule is that of atoms, which come in many varieties. These correspond to letters. Above that, there are very small molecules such as water (H_2O), hydroxide (HO), carbon dioxide (CO_2), and so on, which are bound together by covalent bonds — the strongest sort of chemical bonds. These might be compared to very tight consonant clusters (such as "th", "ng", or "ck"). The next level up in a cell is that of, for instance, amino acids, whose constituents, which include both atoms and such tightly-bound small molecules as H_2O, HO, and CO_2, are bound together slightly less tightly. Structures at the amino-acid level could be compared to higher-level clusters, such as "thr" (made out of "th" and "r"), or "ngth" or "cks". Then there are linear chains of amino acids — polypeptide chains (which often are complete proteins), bound to- gether by "peptide bonds", which are yet weaker. These could be likened to syllables (which often are complete words). And then, in the same way as many proteins consist of an agglomeration of several polypeptide chains, many words are multisyllabic. Of course, if such a structure is put under external stress, its natural breaking-points will be between the highest-level constituents rather than inside them, but under unusual circumstances it might break anywhere. This idea of flexible multi-level structures held together by bonds of many different strengths and having natural breaking-points is very central in Jumbo, as we shall see later.

The assembly of molecules in a cell takes place throughout its *cytoplasm* (the region surrounding the cell's nucleus), not at any central factory. For each type of molecule, there is a standard chemical pathway by which it is assembled. The assembly pathway for a given molecule type may involve dozens of steps that build it piecemeal. All over the cytoplasm, that same pathway is followed by separate, independent, parallel assembly processes. These physically sepa- rated processes are not in any way in phase or synchronized with one another; each one proceeds entirely on its own, oblivious to the others. A similar process whereby letters "glom together" asynchronously to form higher-level structures takes place in Jumbo. Before describing this process, however, I will briefly sketch out the "friendship" analogy — or perhaps the "romance" analogy would be a better term, since the building of romances is the main driving image.

In society, the basic atomic units are individual persons. Then there are two-person units (couples). Above that there are friendships among couples, and larger social groupings. This metaphor, however, has less to do with the multi-leveledness of society than with the time-dependent manner of formation of bonds at any fixed level. We shall use the romantic bond as our main example. The basic idea is that it takes some amount of experimentation ("flirtation" and "dalliance", to use evocative terms) for two people to decide if they wish to be romantically involved.

The very first stage is of course chance-dependent: do the two individuals come near enough to each other to notice each other (or at least for one to notice the other)? If not, there is no hope of a bond forming. If they do come close enough, the initial noticing, if at all positive, may generate the germ of a romance: a "spark" propelling the partners into a desire for further exploration. If they survive this next stage of flirtation with each other, then I call what results a "flash": a mutual infatuation not yet resulting in a bond. The next stage is that of casual dating ("dalliance"), allowing them to check out their compatibility on a more serious level. If they survive this stage, they then go on to form a genuine bond — a "romance". There is still a further stage beyond this, of course, called "marriage".

Romantic relationships often break up. Some breakups result from internal tensions in the relationship; others are instigated by the arrival of a seemingly more desirable partner — an externally provoked breakup. If an attractive potential partner turns up, many factors have to be seriously weighed against each other. Sometimes the romance or marriage will survive, sometimes it will break up. It all depends on subtle factors involving the internal "happiness" of the marriage and the degree of excitement seemingly promised by the potential partner, as well as the general overarching mood of society regarding divorce. If society strongly disapproves of divorce, then it is very hard for either marital discord or a new partner to provoke the disbanding of a marital bond, whereas if society condones divorce, then many divorces will take place, some for the most trivial of reasons.

In the cell, too, breakups of molecules on various levels occur constantly. Large molecules break into smaller constituents, and from these pieces new molecules are then assembled. This is the very crux of life, at the molecular level. Assembly and disassembly are not truly spontaneous but are mediated by agents that are complex molecules in themselves — enzymes. And amusingly, enzymes are actually proteins — an intriguing loop — but for our purposes we can ignore the fact that what is doing the building and what is getting built are both the same type of object. For our analogy, the essence of the situation is the fact that for each type of operation, there is a specific type of agent responsible for carrying it out. The same holds for disassembly: each kind of molecular breakup must be carried out by a specific kind of enzyme.

Jumbo and Parallelism

Jumbo is a parallel system, yet its parallelism, based on the cell, diverges considerably from that of standard models of parallelism. In a cell, many metabolic activities take place simultaneously at different spatial locations. Each activity, whether anabolic (*i.e.,* bond-forming) or catabolic (*i.e.,* bond-breaking), is carried out by an enzyme. The typical anabolic enzyme's action involves the bonding-together of two molecules. How does the enzyme find those molecules? It has "crevices" called "active sites" that match only molecules of the right type. It wanders around the cytoplasm, and when it bumps into a structure that fits either of its active sites, it latches onto it; then the two wander together until the enzyme's second active site gets similarly filled at random. At that point, the enzyme is triggered into action. It performs its joining operation and releases its chemical product. This product is then potential material for a new enzyme to act upon, by grafting something else onto it. Thus larger pieces get built from smaller ones in something of a random way. Yet despite the randomness, specific building tasks defined by elaborate chemical pathways (*e.g.,* the Krebs cycle, also known as the "citric acid cycle") get carried out efficiently and reliably.

Jumbo's parallelism is modeled on the distributed parallelism of the cell. A salient difference is that Jumbo runs on a sequential computer, so its parallelism must be virtual rather than actual. However, Jumbo could be adapted to run on a distributed parallel computing architecture of the right sort.

Sparks and Affinities

At the outset, all the letters are "lonely" and clamoring for attention. What chooses who gets to flirt or bond with whom? In essence, it is random, just as in the cell or the real social world. To visualize Jumbo, imagine the letters floating around in a three-dimensional fluid — Jumbo's own "cytoplasm". If two letters come close enough to each other, they can "see" each other and possibly "spark".

What determines whether a spark is successful? At the outset, when all letters are unattached, it is pure *affinity* (or "chemistry", as the intriguingly accurate popular metaphor would have it). For example, "s" and "h" (in that order) are strongly attracted to each other, but in the opposite order, they have no mutual attraction at all.

Incidentally, do not be misled by the fact that "h" and "s" can occur in the order "hs" inside standard English words (*e.g.,* "withstand", "booths"); this is a red herring. Red or not, it is an important herring to analyze. What matters is not mere *adjacency* but *chunkedness*. Consider, for example, the word "withstand". Obviously, the "h" is part of "with", the "s" part of "stand"; the two letters have no more connection than two people who by chance are sitting back to back in adjacent booths in a restaurant. In the word "booths", the neighbor-status of "h" and "s" results not from their forming a chunk on their own, but rather, from a chunk made up of the *cluster* "th" and the *letter* "s" — quite another matter. Likewise, in words like "gashouse" or "mishap", the "sh" adjacency, though superficially looking like an "sh" chunk, is entirely coincidental and meaningless.

Affinity between letters (and gloms) is no more all-or-none than affinity between people. Some letters are strongly attracted to each other ("n" and "g"), some only mildly attracted ("d" and "w", as in "dwell"), and some not at all ("j" and "x"). There are many degrees of natural affinity.

Jumbo's knowledge of affinities resides in a permanent, static data structure that is clunkily but picturesquely dubbed the "chunkabet". This table was put together by me, simply by drawing on my own introspective sense of how inclined I personally feel to put various letters together in particular orders when I do Jumbles. Here are two excerpts from the chunkabet, to give its flavor:

sc:	*initial 2*
sh:	*initial 8, final 8*
sk:	*initial 4, final 4*
sl:	*initial 5*
sm:	*initial 5, final 2*
sn:	*initial 2*
sp:	*initial 4, final 2*
s-ph:	*initial 2*
sq:	*initial 3*
ss:	*final 5*
st:	*initial 8, final 4*
str:	*initial 3*
sw:	*initial 3*
oa:	*initial 2, middle 4*
oi:	*middle 4*
oo:	*middle 5, final 2*
ou:	*initial 2, middle 3*
ow:	*middle 3, final 3*
oy:	*final 3*

The meaning of the entry "*sp: initial 4, final 2*" is that "sp", considered as a potential *initial* consonant cluster (one that can serve as the beginning of a syllable, as in "spit"), is being assigned an attractiveness rating of 4, and as a potential *final* consonant cluster (one that can serve as the ending of a syllable, as in "clasp") is being assigned an attractiveness rating of 2. Similarly, the meaning of "*ou: initial 2, middle 3*" is that "ou", considered as a potential *initial* vowel cluster (one that can open a syllable, as in "our" or "out"), is being assigned an attractiveness rating of 2, and as a potential *middle* vowel cluster (one that can sit between consonant clusters, as in "flour" or "shout") is being assigned an attractiveness rating of 3.

Note that a few clusters involve three letters (there are even a couple of four-letter ones elsewhere in the table). Consider the entry "*s-ph: initial 2*". This obviously stands for the possibility of syllables like "sphere" and "sphinx". The hyphen indicates the conceptual breakdown of "sph" — namely, into "s" and "ph"; the opposite grouping — namely, into "sp" and "h" — is impossible, even though both pieces individually are possible. On the other hand, the entry "*str: initial 3*" has no hyphen, which means that it can be considered either as a union of the chunk "st" and the letter "r", or as a union of the letter "s" and the chunk "tr".

I repeat that these numbers in the chunkabet were assigned entirely intuitively by me. I felt no compunction to resort to objective frequency analyses or to do psychological experiments to determine these values. There are two reasons for this. One is that, as I have stressed earlier, the purpose of Jumbo is not to be an expert system for doing anagrams but simply a model of fluid transformation processes in the abstract, so getting the numbers "exactly right" (whatever that would mean) doesn't matter. The other reason is that perfectly accurate frequency tables are certainly not what good Jumblists have in their heads. Rather, they have *subjective* affinities, and so a set of rough subjective estimates from me was perfectly faithful to psychological truth. Obviously, if I had assigned these numbers entirely at random, the program's behavior would have looked completely incomprehensible — as if a Martian were doing Jumbles. That would have made it unjudgeable by people, and hence useless. Accordingly, I set the numbers as well as I could, so that the program's behavior would look reasonable, but I didn't worry about it beyond that.

Codelets and the Coderack

Since affinities come in a wide range of levels, some sparks are strong, some are weak, some simply fizzle at the outset. But what exactly is the computational meaning of "sparking"?

A spark is a short-lived simple data-structure telling who is sparking with whom, and in what order. Manufactured at the same time as the spark and associated with it is a small piece of code called a *codelet,* which is placed in a queue-like structure called the *Coderack,* where codelets sit while waiting to be selected to run. (The term "Coderack" is meant to suggest the image of a coat rack upon which coats are constantly being placed, and from which coats are also constantly being removed, at any spot along it.) This particular codelet, when selected to run, will look at the spark, evaluate its viability, and then suggest whether it is worthwhile going on with further exploration in this tentative "romance" between the given pair of letters. If this flirtation fails, then both letters will go on their merry ways, each of them free to spark with other partners instead. If the flirtation seems promising, though, then the codelet will create a "flash" — the next stage of a romance.

Not only sparking is implemented via codelets; *all* processing in Jumbo is. Thus, there are all sorts of codelets on the Coderack at any given time. Which codelet on the Coderack shall run first? Associated with each codelet is a number — its *urgency*. The choice of which codelet to run next is a random but weighted choice, the urgency of a codelet being proportional to its probability of being picked from the Coderack to run next. Thus a codelet with urgency 10 is five times as likely to be picked to run as a codelet with urgency 2. But there is no guarantee that it will be run first. It is all probabilistic, so that only in a statistical sense will what "should" happen more often *actually* happen more often.

If the Coderack happened to contain 100 codelets of urgency 1 and a single codelet of urgency 10, then although the high-urgency codelet would be the most likely one to run next, its chances would still only be 1 out of 11. Ten times more likely would be that one of the low-urgency codelets would be chosen, although *a priori*, no particular one of them is at all likely to be the lucky one.

Whenever a codelet is run, it is removed from the Coderack. The only traces it leaves behind are: (1) changes that it has caused in the cytoplasm; and (2) follow-up codelets it has placed on the Coderack. The self-propagating nature of codelets enables lengthy processes to be carried out in small disjoint steps, each step setting up its own possible continuation. It resembles the way that long chains of chemical reactions get carried out in independent small steps in the cell.

The Concept of a Terraced Scan

Why a spark, then a flash, then perhaps a dalliance, and so on? Why not check out the whole darn thing at once? The answer is: real-time pressures. It often takes considerable effort to determine how well two things or two people fit together. If this potential union were the only possibility in the world, then time wouldn't press. But if there are many possible options to sort through and choose from, time does indeed press. One needs to circumvent what at first might seem an awful necessity: that of exploring all potential matchups to an equal degree before making up one's mind about which one works best.

On entering a bookstore, do you read the first book you come across from cover to cover, then the next one, and so on? Of course not. There is a profound need for protecting oneself from this sort of absurdity. People develop ways of quickly eliminating books of little interest to them and homing in on the good possibilities. This is the idea of the *terraced scan*: a parallel investigation of many possibilities to different levels of depth, quickly throwing out bad ones and homing in rapidly and accurately on good ones. (The term is mine, but much of the idea was present in an implicit form in Hearsay II.)

The terraced scan moves by stages: first one performs very quick superficial tests, proceeding further only if those tests are passed. Each new stage involves more elaborate and computationally more expensive tests, which, if passed, can lead to a further stage — and so on. Furthermore, "passing a test" is not an all-or-nothing affair; each test produces a score, indicating how promising this line of investigation appears at that stage. One uses this score to determine the urgency of follow-up codelets (if indeed the score is high enough to warrant any). This provides the desired layeredness to the evaluation of the quality of a potential matchup.

If a system has true (hardware) parallelism, it can perform quick tests on many items in parallel, slower tests on a smaller number of items in parallel, and so on. On the other hand, if the hardware is serial, the various tests making up the parallel explorations must instead be *interleaved,* so that many possibilities can be simultaneously probed, with some being in the earliest stages of testing, and others at various stages further along in the scanning process. This is the Jumbo strategy.

In a system with parallel hardware, if a process is gauged to be of importance, it should be assigned much computing power, so that it will finish sooner than less important processes being carried out simultaneously. Then, while a process is running, if it shows itself to be very promising, it can request more computing power, whereas if it shows itself to be of little promise, it can relinquish power. In Jumbo's virtually parallel architecture, this dynamic adjustment of speeds is approximated by breaking each process up into small pieces (codelets) and assigning urgencies of varying degrees to the component codelets. As was stated above, each stage sets the urgency of its follow-up stage, so that the overall speed of a long (multi-codelet) process is regulated from moment to moment in accordance with the progress it is judged to be making.

The way Jumbo works, then, is that it examines many possibilities in simulated parallel, letting the more promising ones move ahead faster than the less promising ones. The hope is that accurate quick tests will screen out obviously bad matches, thus allowing a smaller number of slower, more deeply probing tests to screen out matches that are bad for subtler reasons, and so on. There could be screening tests on arbitrarily many scales. So it is with romances, it seems, where the tests range from that tenth-of-a-second glimpse that seems to promise so much to the many-year courtship period filled with anxieties and doubts.

A possible objection to the terraced scan is that it might on occasion defeat its own purpose: something very valuable might get eliminated in a cursory test. A great scientific paper might be thrown out because of a misspelled word in its title. You might pass over a great book in a bookstore because of the ugly typeface on its cover or because of unpleasant connotations its publisher has for you. There are untold numbers of scenarios to illustrate this kind of objection. However, in most cases the terraced-scan principle will hold up, if it is realized in a careful manner — in other words, if the tests are not too simple-minded and trivial.

Luckily for us, much of life has the feature that quick elimination is not only possible, but sensible and reliable. You can be pretty confident of a decision to throw out a scientific paper purporting to prove that the earth is flat. You can be reasonably certain you don't want to buy a book on a topic you usually find deadly dull or in a language you don't read. On a romantic plane, most people can eliminate all but a tiny number of potential mates on the basis of a small set of quick but telling tests, based on such quickly perceptible aspects as gender, age, appearance, style of dress, mannerisms, and so forth.

Were life not this way, we would feel far less in control of our own destinies. At every moment, we would have the sense that a million rich potentialities lay all about us, yet completely undetectable to us. There would be no superficial cues that would tip us off. It would be a most tantalizing feeling. Evolution, however, saw to it that we were constructed in such a manner that quick filters *do* work. Of all that is out there to potentially explore, only a small percentage attracts us, allowing us to discount most claims to our attention. We pay little attention to most ads, most books, most people, most music, most radio and television shows, most countries — in short, to most things in the world. We cannot possibly explore everything in depth, and luckily, we do not need to in order to do well in life.

Romances among the Letters

So it is with Jumbo. Letters begin by sparking with one another. Each successful spark engenders a flash. Each successful flash engenders a codelet that, if and when run, will bond the given letters into a new structure: a glom. Gloms, like letters, also spark together, flash together, and bond together into larger and larger gloms.

While two given items are sparking, two others may be flashing. And others may already have been bonded together to make a glom. Although any particular "romance" (exploration of mutual compatibility) has to go through a fixed sequence of increasingly elaborate testing-stages, many such explorations may be happening in parallel, and of course need not be in phase. Flash-evaluators, spark-evaluators, and other types of codelets commingle and coexist in the Coderack without conflict.

Furthermore, nothing prevents a given item from sparking with several other items at once. Thus parallel exploration of more than one "mate" can occur. Up to a point, a given item can pursue two romances at once, just as a person can pursue two romantic possibilities at once, not by dating two people literally simultaneously, of course, but by interleaving the dates and letting the rival romances develop in parallel. There comes a point, however, where if things are to go any further in a romance, commitment is necessary. This means bonding and especially glomming.

Bonds, Chains, Gloms, and Membranes

The idea of bonding and glomming is to create compound entities that have stability and that are difficult to penetrate or break apart. Bonding and glomming are two levels of this process. The distinction is subtle but vital. A bond officially couples two neighboring structures together with some degree of strength. Once bonded, two items form a short *chain*, both ends of which are open to further bonding. The chain can thus grow, and can come to include any number of items in a linear order. The bonds holding a chain together can be of different strengths and can represent unrelated types of affinities. Although a chain is bigger than a letter and is thus in some sense a higher-level entity, a chain is not a truly independent new object having its own distinctive properties. It is not a new object with an *identity*. Making such new entities is what glomming is about.

The items forming a linear chain of bonds are still seen as separate objects — there is no membrane shielding the constituents of the chain from external view. Thus, bonded items inside a chain are just as available to spark and flash with passers-by as are totally free items.

Glomming is the next stage, whereby a suitable set of bonded items is enveloped by a kind of official "membrane", making of their union a higher-level structure unto itself, and reducing the susceptibility of inner items from sparking and flashing externally. On the other hand, although its *parts* tend not to spark and flash, the glom *itself* can now easily do these things, at its own level. Understanding this notion of membranes and the level-distinction that they bring about is absolutely crucial.

Consider, for example, the three letters "t", "h", "e". The letters "t" and "h", taken in that order, have a strong natural affinity in English, so they are very attracted to each other. It is very likely that they will bond together to form the short chain "t–h". Then, since consonants generally like to bond with vowels and vice versa, the "h" at the tail of the little chain might further bond with the free "e", resulting in the three-unit chain "t–h–e". However, this chain would not in any sense be the same as the *two-unit* chain "th–e". For that to arise, the higher-level glom "th" would first have to be made, and then a bond would have to be built between that glom *as a whole* and the "e". (Such a bond is totally different from the earlier bond between the mere letters "h" and "e".) For such a buildup to take place would require a crucial extra step whereby the "t–h" chain was upgraded into the consonant-cluster glom "th". This wraps the "t" and "h" up inside a membrane and establishes a higher-level structure having its own properties.

Once this glom exists in the cytoplasm, then of course it is able to bond, with its propensities to do so being determined by its own natural affinities, which are quite independent of those of its constituent letters. The cluster "th", for example, is strongly attracted to "r" as a right-neighbor (this attraction is dictated by the chunkabet), whereas "h" itself has no interest in "r" as a right-neighbor. It would thus be impossible for the word "three" to be built up if only bond-bound chains existed, and no gloms.

Whereas a bond always involves precisely two items, a glom can fuse more than two items together, provided their union has a "natural" structure. For example, a *syllable* glom (*e.g.,* "broach") can pull together up to three subunits: an initial consonant cluster ("br"), a vowel cluster ("oa"), and a final consonant cluster ("ch"). Typical gloms include consonant and vowel clusters, syllables, and words.

To any freshly-made glom in the cytoplasm is assigned a new *node.* This serves to imbue the glom with its own unique identity, in the following way. Attached to the node, and therefore accessible through it, are all the properties, static and dynamic, of the glom, such as its *constituents,* its *type* (is it a syllable, word, etc.?), its *happiness* (is it complete or not? how happy are its parts?), and so on. (The notion of the "happiness" of a glom will be explicated in detail below.)

To the lowest approximation, being encapsulated inside a glom renders an item inaccessible to the outside world, much as being married — at least in theory — prevents a person from having new romances. In that sense, a new glom effectively replaces its pieces in the cytoplasm. Thus as soon as a new glom has been manufactured, sparks or flashes or bonding-codelets involving its constituents become obsolete. However, the creation of a new glom does not set off a massive search in which all codelets referring to parts of the new glom are purged. Those codelets are allowed to stick around, and when run they will usually just fizzle harmlessly. Such fizzling is like learning that someone you find attractive is involved; usually you just give up with a sigh.

It is possible, though, for a piece inside a happy glom to spark so strongly with an outsider that the glom is threatened. Depending on the happiness of the glom, the challenger may be defeated or it may emerge victorious, which will result in the dissolution of the bonds making the old glom, and the establishment of a new glom. Thus in Jumbo just as with people, being "committed" (glommed) acts as a damper, but not a total inhibitor, of exploration of rival romances. What *really* inhibits such exploration is happiness, not mere commitment.

A Single Reality with Many Parallel Counterfactual Musings

As the parallel structure-building — the glomming-together — progresses, an ever-smaller number of ever-larger gloms results. (Imagine a cell in which the overall goal of all the enzymes were to put together, out of a given set of atoms, the most viable "chemical sculpture" they could, with "viable" being a complex notion, having to do with satisfying at one and the same time as many esthetic criteria as possible.) Thus Jumbo aims at converting the contents of the cytoplasm from an initial array of unconnected and unrelated letters into a single final structure — a coherent and hierarchically structured assembly of parts, a candidate for "wordhood".

Hearsay II likewise built hierarchical structures out of an initially formless chaos, but with one salient difference: it constructed *multiple* top-level structures — rival interpretations of what the speaker was saying — at once. An analogous vision program would construct several top-level interpretations for a scene, all in parallel.

This type of parallelism strikes me as highly implausible. We do not experience a cognitive superposition of states, seeing (or hearing) something two or more ways at once. Examples of visual ambiguity — the flipping Necker cube, the famous vase–face and rabbit–duck pictures, and many others — support this intuition. Whereas we can rapidly flip back and forth between the two interpretations of such a picture, our cognitive systems simply do not allow us to see it both ways at once.

This suggests that parallelism, while vital for the assembly of a single coherent structure out of many parts, becomes less and less relevant at progressively higher levels of activity. This idea is mirrored in the way Jumbo works. As structures coalesce, the system gradually makes a transition from being distributed and parallel to being localized and serial; at the end, there is only one top-level glom left in the cytoplasm.

A certain kind of parallelism is thus nonexistent in Jumbo — namely, the parallelism that would correspond to consciously entertaining two rival and very different full-fledged thoughts at once, or consciously perceiving something in two contradictory ways at the same instant. At any time, Jumbo has but one overall interpretation (the current contents of its cytoplasm). On the other hand, parallel *dipping-into* or *scouting-out* of alternative realities (*i.e.,* slightly different alternative states of the cytoplasm) is permissible in Jumbo; in fact, it is of the essence. Jumbo is free to dip into, or muse about, many possible counterfactual worlds, as long as they are "close" to the current world, not radical variants thereof. In and of themselves, such musings do not affect the cytoplasm, which is why they are called "musings" — they are like inconsequential little daydreams.

However, as in real life, an occasional daydream, or act of musing, can be so tempting as to invite the taking of genuine action. Thus an act of musing in Jumbo may on occasion offer so much promise that it actually gets carried out. This no longer constitutes mere musing, as it actually changes the state of the cytoplasm, destroying its previous state (rather than letting the old state remain around in parallel with the new state). Unlike Hearsay II, then, Jumbo is committed to a single reality, rather than multiple simultaneous realities.

Both of the basic metaphors support this idea. From the cellular point of view, it's obvious that one amino acid cannot simultaneously be a piece of two different proteins; from the romance point of view, one person — at least a normal person — can't split into pieces so as to be involved in two truly committed romances at once. Of course, mild dipping-into or slight scouting-out of alternate romances (*i.e.,* light flirting, possibly even with several people at once) is commonplace even among committed marital partners.

Let us thus draw a distinction between two types of codelet. A *musing codelet* contemplates a future possibility without actually creating it; an *action codelet* realizes a contemplated possibility by carrying out a change in the cytoplasm. It could thus be said that musing codelets perform "read-only" operations on the cytoplasm (*e.g.,* evaluating sparks or flashes), while action codelets perform "write" operations in the cytoplasm (*e.g.,* making, breaking, strengthening, or weakening a bond).

In summary, in the early stages of glom-building, when Jumbo's cytoplasm is composed of many scattered pieces, nonoverlapping actions in it can be viewed as occurring simultaneously (whether the simultaneity is virtual or real); however, as gloms merge and become larger but fewer in number, parallelism of action inside the cytoplasm gradually gives way to seriality, although the parallel contemplation of possible actions to take — the terraced scan — continues unabated.

Unhappy Gloms, Squeaky Wheels

At any stage, Jumbo's cytoplasm contains gloms having various degrees of "happiness" — a numerical measure applicable to all gloms. There are two components to happiness. The *internal* component tries to summarize how well the glom, as a specific instance of a general concept (*e.g.,* "vowel cluster", "syllable", "word"), realizes that Platonic ideal. How strong are the bonds holding its components together? Is this glom tightly fused and hard to break apart, or does it have internal structural weaknesses? The *external* component tries to summarize the current situation of this glom *vis-à-vis* the cytoplasm. Is it part of a larger glom (thus "secure"), or is it unattached (thus "lonely")?

Consider a syllable-level glom. It must ask itself, "How good a syllable am I?" The answer will be a number taking into account the degree of completeness of the glom, the way its pieces fill the roles they are supposed to play, the affinities of its parts for each other (*i.e.*, the strengths of the bonds between them), and — recursively — the internal happinesses of its own parts.

Take the syllable-candidate "thic", for instance. This glom, seen as a potential syllable, is complete: it has a filler for the three main roles required for syllablehood — initial consonant cluster, vowel group, and final consonant cluster. (Such roles are analogous to the roles of "wife" and "husband" in a marriage, although not all three of them have to be present in every syllable, "ing" being a good example of an internally happy syllable lacking an initial consonant cluster.) However, being a *complete* syllable is not enough to make "thic" an internally *happy* syllable, because "c", although excellent when considered as a *consonant,* is only mediocre when considered as a *final* consonant. By contrast, "thick" is an internally very happy syllable, since "ck" is a very strong final consonant cluster.

Internal unhappiness makes a glom "squeak". For instance, the incomplete syllable "thi" wishes to be made complete, so it squeaks quite loudly. The complete but not-too-happy syllable "thic" will also squeak, but not quite so loudly. Such squeaks are like the "personals" ads in newspapers, where people seeking mates advertise their qualities and desires. (What squeaking really is will be explained in a moment.)

In real life, George may ask himself, "How good a husband would I make?" without having any particular wife in mind. He may feel he fills the abstract "husband" role ideally without there being any specific marriage into which he fits. An analogous notion holds for gloms. Take the short syllable "in". Internally, it is quite happy, but that is not enough to make it totally happy. Until it has been incorporated into a larger structure, it will be unhappy in a different way. This brings us to the external component of happiness.

The external component of any glom's happiness is a function of how well the glom fits into the current state of the cytoplasm. Initially, all letters are unhappy but not desperate, because they are all in the same boat — unmatedness. As some get hitched up and others don't, however, a new "emotion" enters the picture: jealousy. If, given the four starting letters "a", "b", "o", and "t", the latter three have glommed together into the strong syllable "bot", leaving the letter "a" out in the cold, the "a" will be very unhappy. The longer a letter goes unmated, the unhappier it becomes, and especially as more and more of its peers get involved and it remains exceptional, the more loudly it will squeak to express its unhappiness. Of course, this goes not just for unmated letters, but also for unmated higher-level gloms.

In Jumbo, the "squeaking" of a cytoplasmic object is implemented by increasing the likelihood of that object's sparking with other items in the cytoplasm — that is, by giving a very high urgency to a codelet that will place the item on a queue of potential "sparkees". As in real life, it's the squeakiest wheel that gets the oil — at least probabilistically speaking.

"Happiness" is thus a measure that a glom computes, taking into account both its self-estimated internal stability, and its self-perceived "normalcy" relative to the external society. Jumbo's overall goal could be succinctly stated as: producing one happy top-level glom. En route to this goal, it is guided by the happinesses of the intermediary structures it makes.

Looking for Alternative Solutions to a Given Jumble

It would be nice if the randomly-driven building processes that take place from the ground up always gave rise to an excellent candidate for wordhood. However, they often don't, and so there must be ways of fiddling around a bit with what has been constructed in an attempt to come up with something better without going to too much extra trouble. You certainly do not want to build up a whole structure, and then, on finding it in some manner unsatisfactory, to have to simply tear it completely apart into its basic-level constituents and start again from scratch. That would be hugely inefficient. Clearly a lot of care has gone into the buildup of the constituents on various lower levels, and it would be reckless to disband the whole thing without trying some reworking first.

Even if the word candidate produced on the first go-round were stunningly English-looking (*e.g.*, "glinced" or "knoodler" or "wrovening"), a human user might well wish to see other possibilities made out of the same set of letters. The user's criterion for rejecting Jumbo's suggested pseudo-word could be anything — perhaps they don't like its sound, perhaps they are hoping to find a *genuine* English word, or perhaps they merely want to explore the space of possibilities. Whatever the reason for seeking alternative solutions, it would be ridiculous to restart Jumbo from scratch several times in a row on the same set of letters, given that it has already built up a lot of high-quality structure. Far smarter and more efficient would be to fiddle around in different ways with what has been made and to try to reach other appealing possibilities from that starting point.

Totally Rational Choice versus Rationally Biased Coin-flipping

A critical part of Jumbo's architecture is therefore concerned with how to proceed *after* a word candidate has been built. At this point, one is trying to undo some of the decisions that have been made. Traditionally in artificial intelligence, reaching a stage where one wishes to go back and undo some prior decision is a critical moment. The standard strategy for how to proceed is called "intelligent backtracking" (Bobrow & Raphael, 1974), in which you look at the last decision made, undo it, and then either take a different branch in the tree at that point, or else backtrack further. The latter amounts to undoing one decision after another, moving gradually back towards the very earliest decisions made, until you hit a situation that looks promising enough, and then starting afresh from that point.

Taking a different branch in the tree of course requires you to choose — either at random or by reason. The latter is obviously preferable, but the world does not always allow enough time or knowledge to reason everything out. Even the best-calculated branch in the tree may lead to no success, in which case you would have been better off choosing at random, for you wouldn't have invested so much time attempting to foresee the unforeseeable. Thus the advantage of choosing at random is that you don't waste precious time worrying over matters that you can't possibly anticipate — you simply make a snap educated guess, plunge forward, and see what happens.

This is what Jumbo does. However, this strategy should not be confused with blindness. Although the "choices" at branch-points are made at random, there are powerful biases built in. Not everything has an equal chance; indeed, smart ideas will tend to be assigned high urgencies, and dumb ones low urgencies. There is no guarantee of this, of course, but then again, there is not even any guarantee that high-urgency codelets will get run first, since after all, randomness is randomness. However, in the long haul, things will most likely work out favorably. This, at any rate, is the hope on which Jumbo is based.

There is another aspect to the hope placed in randomness: to a program that exploits randomness, *all pathways are open,* even if most have very low probabilities; conversely, to a program whose choices are always made by consulting a fixed deterministic strategy, many pathways are *a priori* completely closed off. This means that many creative ideas will simply never get discovered by a program that relies totally on "intelligence". In many circumstances, the most interesting routes will be more likely to be discovered by accidental exploration than if the "best" route at each junction is invariably chosen.

Furthermore, having an extremely simple control structure allows the system to do things that a very complex control structure would get thoroughly tangled up in. Complex control structures are highly brittle, whereas probabilistic control structures are rubbery and don't care what happens in what order.

Transformations in Jumbo

So, what are the possible remedies to choose from, when Jumbo comes up with an unhappy pseudo-word? There are two basic types of transformation: "entropy-preserving" and "entropy-increasing". By "entropy", I mean "perceived disorder". Thus, a cytoplasm containing no bonded letters has maximal entropy; a cytoplasm containing just one happy pseudo-word has minimal entropy. Entropy-preserving transformations preserve the number of gloms and thus do not increase disorder. By contrast, entropy-increasing transformations destroy gloms, bringing the cytoplasm closer to its initial state of total disorder.

Of the two varieties of entropy-preserving transformation, the milder one is *regrouping*: readjusting internal boundaries so as to yield new substructures without rearranging anything. Some amusing examples of regrouping are the shifts from "no–where" to "now–here", from "super–bowl" to "superb–owl", from "week–nights" to "wee–knights", and from "man–slaughter" to "mans–laughter". The other, more drastic, type of remedy is *rearrangement*: taking pieces of a glom and shuffling them around to make new gloms on the same level (this is somewhat related to shuffling-processes in genetic algorithms, as discussed by Holland, 1975). A far more radical remedy for an unhappy cytoplasm is the entropy-increasing transformation of *disbanding*: pulling a glom's top-level bonds apart and perhaps even doing so to some of the thus-revealed lower-level structures. The ultimate in radical remedies would be to smash the unhappy glom to smithereens, and to start again from scratch.

Entropy-preserving Transformations

Suppose the unhappy or rejected pseudo-word is "pangloss", perceived as the two syllables "pang" and "loss". First consider regrouping. One possibility is to move the "g" from the first syllable to the second, thus making "pan–gloss". On the other hand, it is not possible to detach the entire final consonant cluster "ng" from the first syllable and to reattach it to the second syllable's initial consonant cluster, yielding "pa–ngloss", because "ngl" is not an acceptable initial consonant cluster. "Pa–ngloss" is the kind of structure that would never enter the mind of a human working on a Jumble puzzle — so it should never enter Jumbo's mind, either.

A more complex type of regrouping can occur when two syllables coalesce into one, or one breaks into two, like drops of liquid. A simple example can be given using four letters found inside "pangloss". Suppose the two syllables "ap" and "so" have been bonded together in this order: "so–ap". Now "soap" might be bisyllabic (like "react" or "boa"), but its preferred perception is as a single syllable. Therefore Jumbo should be able to carry out this type of regrouping fluently. One type of trigger event that could suggest such an idea would be the noticing of newly juxtaposed vowels whose combination is known to be a standard vowel group. The converse process — "fission" of one-syllable words like "coop" and "soap" into two-syllable words "co–op" and "so–ap" — should be as natural for Jumbo as the "fusion" process. Note that the word "coax" is ambiguous as to its inner structure: it can be either "co–ax" (as in "coax cable", short for "coaxial cable"), or the standard verb "c–oa–x". Nonetheless, in a human mind, it is always perceived in one of these ways — never both ways at once. (There is a seeming exception — namely, the fact that by drawing attention to the very fact of its ambiguity, we *are* in some sense seeing the word "coax" both ways at once. However, this is an illusion based on confusing use with mention. When we *mention* the word "coax", we can see it as two things in one; however, we do not *use* the word "coax" simultaneously both ways.)

The next entropy-preserving alternative, rearrangement, offers many possibilities. One would be to perform the operation known as *spoonerism*: interchanging the initial consonant clusters of two syllables. Operating on "pang–loss", this produces a new two-syllable candidate, "lang–poss". The term comes from the name of an Anglican minister, the Reverend Spooner, but it is tempting to ignore that dignified etymology and to invent parallel culinary terms: *forkerism* (the interchange of final consonant clusters) and *kniferism* (the interchange of vowel groups). Operating on "pang–loss", those would produce "pass–long" and "pong–lass" respectively. Spoonerism need not involve entire clusters; sometimes pieces of clusters can be exchanged, as in the transformation of "pan–gloss" to "gan–ploss".

Rearrangement can involve the *exchange* of syllables ("pang–loss" to "loss–pang"). It can also involve the internal *reversal* of a syllable either at the cluster level ("stan" to "nast") or at the letter level ("stan" to "nats"). Reversal at either level, unfortunately, poses tricky problems. Obviously, not all syllables can be reversed at the cluster level, since some consonant clusters cannot function initially, while others cannot function finally. For example, cluster-level reversal of "knots" leads to the nonsensical "tsokn". No human Jumblist would ever entertain this sequence, even fleetingly.

Reversal at the letter level is equally knotty: "thump" leads to the nonsensical "pmuht", and "head" leads to the impossible "daeh". These would simply not spring to the mind of a good Jumblist. Thus one cannot just let Jumbo barge ahead with either type of reversal; musing codelets must prepare the way by checking out the desirability of contemplated actions. This emphasizes once again that, although Jumbo is permeated by randomness, it is also permeated by dynamic tests imbuing it with caution and foresight.

Another type of rearrangement is where one syllable wants to trade its *final* consonant cluster for another one's *initial* consonant cluster. For instance, in "plag–noss", "plag" might want to trade its final "g" for the initial "n" of "noss", to yield "plan–goss". However, the feasibility of such a trade depends on whether the given clusters can function in the desired new positions. A natural nomenclature for such trades suggests itself: *sporkerism* (in which "plag–noss" would become "plan–goss"), and its complementary operation would of course constitute a *foonerism*.

A musing codelet's proposal of a trade resembles an offer by one baseball team to trade its pitcher for another team's left-fielder. Such offers are usually made out of recognition of weaknesses, and the feeling that both teams can benefit from the switch. The weaker a team is in a certain position, the more disposed it is to give away something in order to fill that role well. Likewise in Jumbo: the more valuable a trade is perceived to be by the gloms involved, the more urgent will be the corresponding action codelet manufactured by the musing codelet. Thus the principle of the terraced scan is at work at this stage of the game no less than in the earlier, glomming-together stages.

Fluid Data-Structures: A Principal Aim of Jumbo

The combination of regrouping and rearrangement provides a surprisingly powerful tool. It preserves a great deal of structure at all times, yet achieves a significant reshuffling of high-level pieces. Here is a random walk through pseudo-word space, beginning at "pang–loss":

pang–loss	*(start)*
pong–lass	*kniferism*
long–pass	*spoonerism*
pass–long	*switch syllables*
pas–slong	*regroup "s"*
sap–slong	*reverse syllable*
slap–song	*spoonerism*
slang–sop	*forkerism*
sop–slang	*switch syllables*
slop–sang	*spoonerism*
slos–pang	*sporkerism*
los–spang	*spoonerism*
loss–pang	*regroup "s"*
pang–loss	*switch syllables*
pan–gloss	*regroup "g"*

You can obviously continue with this type of meandering in word-candidate space for an indefinite length of time.

Here is another little example of such entropy-preserving reshuffling:

now–here	*(start)*
no–where	*regroup "w"*
on–where	*letter-level reversal of first syllable*
whon–ere	*spoonerism (degenerate form)*
ere–whon	*syllable-level reversal of full word*

The operations just demonstrated bring letters together in unanticipable ways, and thus encourage serendipitous interactions. Purely random associations of letters would do this as well, but in a far less efficient manner.

One might suppose this type of restructuration process to be idiosyncratic to the domain, an obscure type of transformation having little relevance to general intelligence. Such an impression would, however, be quite wrong. Recall that Jumbo's deep aim is to aid in demonstrating the thesis "cognition equals recognition". Now any study of recognition reveals the incredibly fluid boundaries of percepts and concepts. There are thousands of ways of drawing the letter "A" (see Hofstadter, 1982a), yet we recognize any of them at a glance. For any mental category (concrete ones like "dog" or "game", and abstract ones like "weakness" or "metaphor"), we immediately recognize thousands of instances, despite vast differences.

To recognize something — that is, to match an item with the proper Platonic abstraction — there must be room for flexing both the Platonic abstraction and the mental representation of the thing itself, to make them meet. What kinds of distortion are sensible? Readjustment of internal boundaries, reassessment of connection strengths, and rearrangement of parts are at the core of the recognition process. Thought's fluidity emanates from the nonrigidity — the fluidity — of representational structures. They have an ability to adjust, to alter themselves effortlessly, to assume many conformations. For fluidity to emerge at a high level in a computer model of thought, the underlying data structures must be permeated by an ability to slip back and forth at the drop of a hat (an idea discussed in the context of Bongard problems in Chapter 19 of Hofstadter, 1979).

It is important that this capacity reside inherently inside the structure itself: it must be ready and willing to be internally reconformed if you just "tap" it the right way. It is as if the structure itself contained numerous natural "hingepoints" where it will flex. This is to be contrasted with a passive data structure that can reconform, but only if manipulated from the outside by a complex and "intelligent" process. The term "change of representation", a goal of some AI projects, usually refers to the conversion by a so-called "reasoning engine" of one representation into another. With Jumbo, by contrast, reconformability is inherent in the structures themselves, with only the slightest push from the outside.

This inherent reconformability is a result of the fact that at every level, the structures in Jumbo are bound together by bonds of different strengths, and more generally, have different levels of internal and external happiness. This means that they are anything but homogeneous. Rather, they tend to break apart or flex in certain places far more easily than in others. Moreover, there are natural ways that, once taken apart, they want to come back together again.

The analogy to complex hierarchically-structured organic molecules in the cell helps make this clear, since those molecules have thousands of bonds of many different types, and therefore break easily at certain points and greatly resist breakage at other points. There are also many cellular molecules that have natural hingepoints — points where they easily flex rather than break. One type is called *allosteric enzymes*. These are enzymes that have two highly stable but utterly different configurations, like the ambiguous structures "nowhere", "superbowl", "weeknights", "coax", and so forth. The presence of a chemical catalyst makes such an enzyme snap back and forth between its two stable modes.

In summary, perhaps the major part of the intelligence of Jumbo as a whole system resides collectively in the hingepoints of its data-structures. The import of that key idea transcends the confines of this little domain.

It is my belief that a nearly constant background activity of creative minds is a playful twisting-around of mental structures of all sorts — turning them inside-out or backwards, regrouping parts, erasing or inserting new levels of structure, moving things back and forth between levels, and so on. In short, taking full advantage of the fluidity of their mental representations, and simply being sensitive to the crazy, quirky, unanticipated side effects that crop up when this happens.

If, for example, you chance upon the word "astronomer" and idly start doing anagrams on it, do you anticipate that anything special will happen? Obviously not — there is no way to anticipate what might lurk therein — but it is a delicious feeling if you happen to discover "moon starer" hiding therein. (See Bergerson, 1973 for many stunning examples of this sort.) This kind of discovery happens only to those who take delight in constantly browsing among their mental representations, idly mixing them in new ways and turning them over, like gaily colored toys of all shapes and sizes...

Entropy-increasing Transformations

If Jumbo deems its single top-level glom unsatisfactory, its first attempted remedy will be to create regrouping and rearrangement codelets and to give them high urgencies. Thus, a bit of shuffling-about as illustrated above will take place. During this, Jumbo will monitor the overall happiness of the cytoplasm.

If, after a while, that happiness has not reached a satisfactory level, Jumbo will alter its priorities and pour disbanding-codelets into the Coderack.

An example of disbanding would be: smash the word candidate into its syllables, then smash those syllables further into their cluster-level constituents. In the case of "pang–loss", this yields: "p", "a", "ng", "l", "o", and "ss". Now one may hope that some fresh gloms will bubble up. If not, one can disband further!

Disbanding is the ultimate recourse, as no operation is more drastic than reduction to the elemental smithereens. Actually, humans sometimes *do* do more drastic things: they jump out of the system entirely by changing, consciously or unconsciously, one or more of the given letters. However, jumping out of the system definitely goes beyond the aims of this model!

Action codelets for disbanding are called "dissolvers". A disbanding operation is not undertaken lightly, since it risks destroying a happy glom. Therefore, a musing codelet must check the idea out first, and if it approves of the disbanding, it creates a dissolver — but if there is any reason to think that some gentler remedy might make the cytoplasm happier, it will assign the dissolver a low urgency.

There are two sorts of dissolvers: targeted and nontargeted. A dissolver of the former type is aimed at a specific glom that has given trouble for a while. A dissolver of the latter type is a beast whose mission is to find *any* glom and tear it into its immediate constituents.

Why would nontargeted dissolvers ever be useful? Well, if you have been troubled by a glom that has been unfruitful no matter how you shuffled it about, one way to restart quickly would be to load up the Coderack with a bunch of nontargeted dissolvers. In a flash, they, not unlike a horde of hungry piranhas, will tear that pesky glom from limb to limb, leaving you to start out again with a totally disordered cytoplasm.

There is a problem here. Suppose you put ten nontargeted dissolvers in the Coderack, and after just five have run, the cytoplasm has been reduced to raw letters. Five nontargeted dissolvers remain in the Coderack. What's to prevent them from later being called upon to run, and at that time totally ruining the new emerging order?

Temperature and Self-watching

One way to protect emerging order is to have a global variable called the *temperature* of the cytoplasm, which is low when the top-level gloms are happy, and high when they are unhappy. The happiest state is when there is just one single happy glom. Since we want to preserve such a state, this state is assigned the "freezing" temperature. When the cytoplasm is freezing, dissolving is prevented.

When the cytoplasm is boiling, nontargeted dissolvers have permission to tear apart any glom, whereas when it is merely warm, they are inhibited (*i.e.,* have a low probability of carrying out their mission) in proportion to the happiness of the gloms they are attacking. The cytoplasmic temperature is thus Jumbo's equivalent to the "mood" of society regarding divorce, mentioned earlier. High temperature encourages split-ups, low temperature discourages them.

Cytoplasmic temperature is a very general concept. It provides quick global feedback about the system, helping to dynamically govern the probabilities of various kinds of activity. The temperature regulates the degree of chaos in the system, something like the rods in a nuclear reactor. When things start to get too wild, you lower the temperature a bit, and when things seem to be going nowhere, you raise it and pump in some craziness.

Paul Smolensky has suggested (Smolensky, 1983a) that temperature could be used to modulate the interpretation of urgencies. In particular, he suggests that as temperature rises, the selection probabilities of codelets in the Coderack should tend to become equal, and that conversely, as temperature falls, the selection probabilities should favor high-urgency codelets more and more, until at a freezing temperature, urgencies would turn into absolute rankings, making codelet choice totally deterministic. At very high temperatures, all processes would tend to run at about the same speed, whereas at low temperatures, only the processes made up of the very highest-urgency codelets would run at all, and processes made up of low-urgency codelets would flow like molasses in January (an apt metaphor, since it is the coldness that slows the flow).

Temperature will be useful only if kept updated. If not assessed frequently enough, its outdatedness could be harmful. This would imply that temperature-measurement codelets must vie for recognition in the Coderack along with everybody else. They must be accorded a high enough urgency that they will run fairly often, yet not such high urgency that the system spends all its time taking its own temperature!

This brings up the tricky question of how the system keeps itself balanced, with so many complex things happening at once, all mingling in the same Coderack. In any self-organizing system, there is serious danger that the system will get caught in a self-propagating runaway reaction, like a nuclear reactor undergoing a chain reaction and melting down. You might think that such deterministic behavior would not occur when there is so much randomness in the system. But this is not necessarily so. If some codelet somehow got assigned a huge urgency, and then, after running, replaced itself by a copy of itself (again with the same huge urgency), the system would suddenly find itself trapped in a quite deterministic and quite fatal loop. There are many other ways in which fatally loopy behavior could arise as well. For instance, a top-level word candidate might be deemed inadequate and broken into two syllables, which are then immediately rejoined to form the same word again. This cleaving-and-resplicing cycle could be repeated over and over again. Worse yet, the two syllables might be spliced back together sometimes in one order ("pan–gloss") and sometimes in the other order ("gloss–pan"), thus giving you loop-like but not purely cyclic behavior — an even harder type of loop to detect.

How can these kinds of loop-like behavior be obviated? One way is for the system to watch its own cytoplasm and Coderack (Hofstadter, 1982b), and to make sure that if any suspicious loopiness is detected, preventive actions are taken. Such simple remedies as raising or lowering the temperature may suffice, as would changing the values of other important global variables.

But how does self-monitoring really take place? This is, as yet, unexplored territory in Jumbo. The general idea is that there should be a way of *summarizing* the contents of the cytoplasm and Coderack in simple terms, so that no excruciatingly detailed self-scrutiny is needed. For the Coderack, a good start might involve a "census" telling what kinds of codelets are present, and for each type, giving its percentage of the total urgency. That way, any single codelet that was overwhelmingly dominating the scene would be quickly detected (unless, of course, even the self-watching codelets had been locked out as well!). For the cytoplasm, listing the top-level gloms' appearances (*i.e.*, pure sequences of letters without any indications of bonds or membranes) might be a good beginning. That way, periodic and quasi-periodic assembly and disassembly of structures would be easily detectable. But again, the tricky problem of self-reference arises: cytoplasm-watchers and Coderack-watchers must themselves be codelets, vying with all other codelets for time!

These twisty ideas are not yet implemented. Clearly they will be significant in determining how Jumbo tunes itself and thus controls itself dynamically. Such feedback mechanisms offer the potential for great flexibility and power, but they are also very tricky. They are reminiscent, in ways, of human consciousness, an alluring mystery towards which AI is slowly moving (Hofstadter, 1979, 1982e; Hofstadter & Dennett, 1981; Smith, 1982). Mechanisms for self-watching are therefore something to look forward to in the developing architecture of Jumbo.

A Self-sensitive, Self-driven System

In Jumbo, codelets of all types commingle in the Coderack. There may be some for sparking, some for spark-evaluation, some for flash-evaluation, some for disbanding, some for regrouping, some for rearranging, some for temperature measurement, and so on. If a disbanding codelet is run, for instance, it will, as a parting gesture, insert into the Coderack a number of new codelets for generating sparks, so as to get the process of assembly restarted, now that unmated parts are present in the cytoplasm once again. And of course, spark-making codelets will, after running, replace themselves by flash-making codelets, which will in turn replace themselves by bond-making codelets, and so on.

Thus the process drives itself. That is, there is no top-level "driver" making decisions about what to do next. The flow is a result of many weights and many random numbers, not a result of careful consideration and reflection at each one of thousands of microscopic branch-points.

All this stress on random numbers probably makes Jumbo sound stupider than it is. In life, you often must make a decision randomly in order not to get bogged down in an infinite regress of exploring consequences and meta-consequences, reasons and meta-reasons, and so on *ad infinitum*. At some level, because of time pressure, reflection must give way to reflex. But this need not be disastrous — in fact, it can be advantageous — provided your reflexes are well-tailored ones. That is the purpose of the urgencies attached to codelets, and of the measures of happiness.

Together, urgencies and happinesses guide Jumbo's processing. The overall behavior of the system is governed entirely by these quantities, so it matters greatly how they are assigned. Clearly, the flow of control, to the extent that it is determinate, is determined by urgency values. What determines those values? Some micro-actions, such as spark-generating codelets, have fixed urgencies. Others, such as codelets for spark-evaluating, flash-evaluating, and bonding, have variable urgencies. The variation comes from the perceived mutual suitedness of the gloms involved. This in turn is determined by the *intrinsic affinities* of the structures involved — a fixed feature of English — and by the *happinesses* of those structures. Thus happinesses and affinities determine the urgency values for bonding- and disbanding-codelets.

In this way, happinesses and urgencies are deeply intertwined, and they combine to yield a total outcome that is the overall, observable behavior of the system. Jumbo thus emerges as an exquisitely self-sensitive collection of processes: a coherent whole composed of many independent interwoven strands. (See the "Prelude...Ant Fugue" and comments on it in Hofstadter, 1979 or Hofstadter & Dennett, 1981. See also Hinton & Anderson, 1981, especially Chapter 1; Rumelhart & Norman, 1982; and Feldman & Ballard, 1982.)

Epilogue: Jumbo's Epiphenomenal Intelligence

Jumbo's intelligence, if indeed it has any, clearly has not been directly programmed; rather, it emerges as a statistical consequence of the way that many small program-fragments interact with each other. It is like a chess program that has a subtle tendency, but one that is crystal-clear to sufficiently keen chess observers, of "liking to get its queen out early" — a tendency taking its programmers completely by surprise, as they never knowingly or explicitly put any such strategic concept into their program. Daniel Dennett has referred to this as an "innocently emergent" quality (Dennett, 1978); I have called it an "epiphenomenon" (Hofstadter, 1979, 1982d, 1982e).

One could therefore bring in a third basic analogy relevant to Jumbo's architecture: that of statistical mechanics, which explains how macroscopic order (the large-scale deterministic laws of thermodynamics) emerges naturally from the statistics of microscopic disorder. The reliable emergence of macro-laws from micro-chaos could even be summarized in a metaphorical equation:

$$thermodynamics = statistical\ mechanics.$$

The philosophy of Jumbo comes from an analogous vision. This philosophy goes against the grain of traditional AI work, which seeks to find explicit rules (not emergent or statistical ones) governing the flow of thoughts.

The traditional holy grail of AI has always been to *describe thoughts at their own level* without having to resort to describing any biological (*i.e.,* cellular) underpinnings of them. This could be likened, on the one hand, to wishing to describe how the heart works and hoping not to be forced to take into account the fact that hearts are made out of muscle cells of certain sorts. Of course, it turns out that such a description can be made quite well. Everyone knows the *pump* metaphor for the heart; it is undeniably very successful in explaining what hearts are all about. Indeed, it could be argued that it is so powerful that it is not even a metaphor; the heart simply *is* a pump, quite literally. If the human mind and brain are like the heart, that's good news for traditional AI.

On the other hand, AI's holy grail could just as plausibly be likened to wishing to find laws for cloud motion and hoping to get away with treating clouds as stable, solid, sharp-edged objects — that is, hoping not to be forced to take into account the fact that clouds are tenuous, amorphous, boundaryless puffs of fluff made out of molecules rushing all about every which way. The hope of finding pure laws of cloud behavior is an admirable dream for a meteorologist, but it is not *a priori* obvious that reliable laws of "nubodynamics" — laws governing clouds at their own level — exist. Nor, to return to AI, is it *a priori* obvious that reliable high-level laws of "thinkodynamics" — laws governing thoughts at their own level — exist.

I call traditional AI's holy grail the "Boolean Dream" after George Boole, who in the 1850's set forth an elegant calculus called "The Laws of Thought" (Boole, 1855), which today we know as Boolean algebra, with its commutative, associative, and distributive laws, and so forth. Despite its esthetic appeal, we have come to realize how minuscule a role Boole's formal calculus plays in everyday thought, even though the mathematical study of axiomatic deduction could justly be said to be entirely grounded in Boolean algebra (nowadays usually called the "propositional calculus"). Rigorous reasoning, as a few decades of research have finally revealed for all to see, is but a minor and atypical twig on the tree of human thought.

One might still ask, is the Boolean Dream realizable? Note that this is not to ask whether the laws of Boolean algebra itself, or even the laws of any form of deductive logic, constitute the laws of thought. It is simply to ask whether there exist *any* formal laws governing or describing thought at its own level — laws that *in spirit* seem to embody Boole's vision, even if they are very different in detail from his wonderfully symmetric logical calculus.

Most AI people seem to hope and believe that the answer to this question is yes; some even feel that a claim to the contary is a mystical, antimaterialistic, or at least antiscientific claim. I see things differently. My view is that AI and the Boolean Dream are closely related but not synonymous. Thus the perishing of the Boolean Dream would not be a setback to AI, but simply a recognition that some high-level mental phenomena *are*, in fact, innocently emergent. The goal of AI would then shift from being one of trying to capture the "laws of thought" in an elegant mathematical formalism to one of trying to understand the laws governing subcognitive events (let us call such events "mentalics") that collectively go to make up thinkodynamics — the dynamics of thinking.

To sum matters up, then, we could say that the shifted burden of AI should eventually become the clarification of the meaning of the following corny but hopefully catchy "equation":

$$thinkodynamics \ = \ statistical \ mentalics.$$

Perhaps even the venerable Dr. Boole himself would, if he dreamed of this, be moved to exclaim, "O, a noble dream!"

Preface 3:

Arithmetical Play
and Nondeterminism

"Le Compte Est Bon"

In the summer of 1985 I received a letter from Daniel Defays, a mathematician and professor of psychology at the University of Liège in Belgium. He explained that he wished to spend a sabbatical year learning about artificial intelligence, and that he was particularly interested in the kinds of approaches I was taking. He even had a preliminary idea about a project he might like to work on, which I found very interesting. Sensing a good deal of promise in what he proposed, I invited him to join my research group for a year.

Late in the summer of 1986, Dany and his family came to Ann Arbor, where I was at the time, and he joined FARG. From the outset, it was clear that his heart's desire was to carry out the project suggested in his letter, so over a series of lunches, the two of us worked out the basic architecture for a program to play the popular television game *Le compte est bon*. This game, as it says in the article, involves five randomly-chosen smallish integers — "bricks" — and one randomly-chosen larger one — the target. The idea is to use additions, subtractions, and multiplications to make the target out of the bricks. In some ways, this activity resembles doing Jumbles — you are taking small pieces and putting them together into chunks, and then bigger chunks, and so on, trying to make a top-level chunk that matches some conditions. On the other hand, the kinds of combinations one can make and the type of goal one is seeking are very different from those in Jumbo.

After a month or two of planning, Dany was chomping at the bit to start implementing things, so he began his computer project in earnest. In addition to developing this program, which we soon dubbed "Numbo", for obvious reasons, he was sitting in on several courses in AI and cognitive science. He also spent much time with his family, so it seemed he was having a very busy year.

At the school year's end, not only had Dany hung in there with all his courses *and* managed to complete the entire Numbo project (true, not every possible loose end had been cleared up, but he had built a very impressive working program), but also he revealed to me that at the same time, he had been quietly working on a book on AI and had finished its first draft! He gave it to me, and I read it from start to finish with great enthusiasm. (Little wonder, given that it described in glowing terms the ongoing projects in FARG, and included an excellent description of his own program.) I was amazed by how well Dany had used his year off, and we were all very sad to see him and his family leave in late summer 1987. Not long thereafter, his book (Defays, 1988) was published under the title *L'esprit en friche* — literally "Mind Lying Fallow", but perhaps "Seeds of the Mental" would be a better translation.

Although there has been no subsequent FARG work that has grown directly out of Numbo, this project retains for me a great deal of charm and elegance. It supports my thesis that making mental recombinations of primitive objects is not limited to Jumbles but has more general interest, although admittedly this activity is still a variety of game-playing. Nonetheless, I think readers will agree that there is a liveliness of mind involved in playing this game that goes along with a general creative attitude, and that Numbo helps clarify the importance of the idea of mental juggling of little pieces in an attempt to build coherent wholes.

The Human Mind as a Stochastic Processor

A few years after Numbo was completed, I received a very interesting manuscript in the mail from Ann Dowker, a psychologist at Oxford University (Dowker *et al.*, 1995). She and her colleagues were studying computational estimation — how people mentally estimate the magnitudes represented by arithmetical expressions such as "76 x 89" or "546/33.5". Her experiments on human subjects, including many mathematicians, revealed that there are a large number of standard techniques that people bring to bear on such tasks, including rounding one or two numbers up or down, replacing numbers by more familiar numbers, factoring numbers, carrying out regroupings of multi-factor products, using fractions or powers, and exploiting various simple algebraic identities such as $(a + b)(a - b) = a^2 - b^2$.

One of the most interesting aspects of Dowker's work was that she tested people twice, with a six-month interval between tests, to see how consistent they were. It wasn't so much consistency of *answers* that interested her, but consistency of *methods*. That is, her central concern was whether a given person would tackle a given problem using the same strategy on both occasions. She found that the better estimators used different strategies on their two tests quite often — and certainly much more often than did the poorer estimators. That is to say, skill and consistency seemed to be negatively correlated. Her explanation for this finding was that skilled people by definition have many strategies at hand,

and thus can choose from a wider palette than less-skilled people. In other words, expertise and flexibility are really two faces of the same coin. As Dowker put it:

> Research mathematicians, unlike people who are simply skilled at solving mathematical problems, are paradoxically experts at dealing with situations where they are not experts. Metaphorically they have not only learned specific arithmetical routes, but have an effective cognitive map of the territory, which makes it possible for them to take unfamiliar routes without risking losing their way here in a serious and irreversible sense. In this respect they contrast markedly with people with limited "number sense", for whom it may be necessary to keep to known arithmetical paths if they are not to become lost. Though this study concerned mathematicians, this willingness to venture along unfamiliar paths is in fact likely to be a general characteristic of scientifically, artistically, or otherwise creative people.

The question remained, "Why would someone use a different method on the same problem when given it on different occasions? What does this say about nondeterminism in thought processes?" Dowker is inconclusive on this score, although she quotes two researchers who have studied creativity and who have come to opposite conclusions about the role played by randomness or nondeterminism. Psychologist Philip Johnson-Laird tends to support the idea of some kind of randomness in the brain, whereas philosopher Margaret Boden is dubious about the need for randomness. Clearly, however, Dowker herself is predisposed to believe in some kind of cognitive nondeterminism, especially in creative thinking, and she cites a number of computer models of thought that use nondeterminism in various fashions, including Copycat and Numbo.

In summary, Ann Dowker's experimental conclusions about how skilled humans use numbers and numerical operations in flexible, probabilistically-governed ways seem to provide an excellent *a posteriori* justification for many of the intuitive choices that went into the creation of Daniel Defays' Numbo program.

Chapter 3

Numbo:
A Study in Cognition
and Recognition

DANIEL DEFAYS

Introduction

Hobbes once said "Reasoning is reckoning", and today, some 300 years later, a growing group of artificial-intelligence researchers is rallying around a similar shibboleth. By contrast, "Cognition equals recognition" could be said to be the motto underlying the approaches to AI of a different group, including Douglas Hofstadter, David Rumelhart, Roger Schank, and others. Even if the borderline between perception (*i.e.,* the activation of appropriate semantic categories by syntactic sensations) and cognition (*e.g.,* reasoning, problem-solving, and generalization) is far from well-defined, the two processes are worth comparing. Perception is generally considered to be parallel, unconscious, and goal-independent. Reasoning, on the other hand, is usually conceived of as serial, conscious, and goal-driven.

My aim in the Numbo project has been to help clarify the relations between perception and cognition, by developing a computer program that plays a simple number game known in French as *Le compte est bon* (literally, "the total is correct"). In this paper, I will call the game "Numble", because of its similarity to the anagram game "Jumble", which appears in many American newspapers. (Jumbles are scrambled sets of letters, usually five or six in number, that constitute an English word; the game is to rearrange them and find that word.) The program I have developed to play this game is called "Numbo"; it belongs to a family of programs (Jumbo, Seek-Whence, Copycat) that attempt to simulate the human ability to discover patterns and to structure concepts fluidly, independently of any problem-solving context (Hofstadter, 1983a; Meredith, 1986; Hofstadter, Mitchell, & French, 1987).

To be sure, to tackle the broad issue of cognition versus recognition, one needs a multidisciplinary approach. The role of computer scientists is to test certain basic architectural principles, by exploring whether a particular system allows a computer to behave in a humanlike way. In the Numbo project, I was interested in reproducing a particular aspect of human mental behavior: the ability to fluidly group, take apart, and restructure the components of ideas, and to use these structures to achieve a goal.

This paper has four sections. The first is a description of the game and a discussion of its relevance to the matters mentioned above. The second proposes an architecture for a system capable of exhibiting some of the skills humans use when they play the game. The third section analyzes a sample run, to give the reader a concrete idea of how the system works. The fourth section contains a brief comparison with other work and a discussion of the advantages and limitations of the approach.

The Game of Numble

The goal in a game of Numble is to construct a given number, henceforth called the *target,* from a given set of five other numbers, called the *bricks.* The target is chosen at random from the integers between 1 and 150, and the bricks are randomly and independently chosen between 1 and 25. Three basic arithmetical operations — addition, subtraction, and multiplication — are available. Any combination of bricks and operations may be used, but a given brick can be used at most once. For the sake of simplicity, all problems discussed herein will have solutions. Here is a sample problem, which readers are urged to tackle:

Puzzle #1: Target: **114**
Bricks: **11 20 7 1 6**

Two solutions to this problem are: **20 x 6 − 7 + 1** and **(20 − 1) x 6**.

The game of Numble is interesting to cognitive scientists for the following reasons, among others:

- It is clearly representative of a large class of problems. There is a well-defined goal, various operations and operands are allowed, and standard techniques for searching a problem space are applicable.
- The human mental processes used in seeking solutions seem to demand the following important abilities:
  - construction of larger units out of smaller ones, with creation and destruction of temporary structures at various levels;
  - rearrangement and dismantling of these structures;

    *   interaction of *a priori* knowledge, familiar concepts, and
salient features of the input.

An example will illustrate this last point concerning interacting types of
knowledge. Suppose we are given the following puzzle:

            Puzzle #2: Target: **87**
                      Bricks: **8 3 9 10 7**

Almost everyone rapidly comes up with one of the following solutions: **8 x 10 + 7**
or **9 x 10 – 3**. If asked to think aloud when solving this problem, subjects generally
report something like this:

    "I see a 10 and an 8. The target can be decomposed as 80 + 7. I still
    need a 7. Oh, there is a 7! Therefore, the solution is 8 x 10 + 7."

Clearly, certain bricks are more salient than others, and certain routes
more inviting than others. The solution **9 x 7 + 8 x 3**, for instance, is never given.
The "syntactic" fact that the target's leftmost digit is itself a brick, namely **8**, is
often reported as being a key factor in motivating the solution **8 x 10 + 7**.
Moreover, any human who gives the response **9 x 10 – 3** is obviously motivated
by the fact that 90 is close to 87. Syntactic features (*e.g.*, the presence of a 10, or
the fact that a particular brick is visually part of the target) and *a priori* knowledge (*e.g.*, 87 is close to 90) interact to catalyze a rapid solution.

At least as interesting as the distribution of responses people give is the
set of solutions that are never — or hardly ever — noticed by humans. In puzzle
1, for instance, the shortest solution, namely *(20 – 1) x 6*, is virtually never
reported, and in puzzle 2, two further solutions are possible: **7 x 10 + 8 + 9**
and *(8 + 3) x 7 + 10*. Such facts about what solutions people fail to find clearly
provide clues about the mental processes involved.

The game of Numble allows one to study many other interesting cognitive
phenomena. A few examples will serve to illustrate this point.

            Puzzle #3: Target: **31**
                      Bricks: **3 5 24 3 14**

When most people tackle this problem, they try out one path after another, but
no path turns out to yield a solution. An interesting question is therefore: What
is going on during the search process? If a given subject is re-confronted with
this problem after a good deal of time, so that the routes already explored have
presumably long since been forgotten, it is most unlikely that those same
pathways will be explored again, and certainly not in the same order.

In a situation of this sort, where knowledge that could guide a systematic
search is lacking, some type of *random* search seems to take place. Note, however,

that by this I do not mean that all paths are equally likely to be chosen; I simply mean that the structure of the search will be different each time. The amount of knowledge used by a subject will determine how systematic the search looks — that is, how easily describable it is in terms of nested subgoals (*e.g.*, the subgoal of 90 in puzzle 2), or how erratic it is (as in puzzle 3). One can summarize these observations as follows: *The amount of knowledge available determines the nature of the strategy followed.*

Let us consider some more examples.

Puzzle #4: Target: **25**
             Bricks: **8   5   5   11   2**

Puzzle #5: Target: **102**
             Bricks: **6   17   2   4   1**

Both of these puzzles are simple, but the type of knowledge used to solve them seems to be different. In puzzle 4, no computation is needed: no sooner has one spotted the two **5**'s than the solution jumps to mind. In puzzle 5, on the other hand, one might well try to multiply 6 and 17 — but not because one knows that their product is 102; one tries it simply because one "sees" that the result will be somewhere *near* the target. The type of knowledge involved here concerns *approximate size*. We are not certain of the result, so we carry out a computation, and only by accident does it turn out to yield the solution. This illustrates another central fact about human thought processes that Numble brings out clearly: *Very different types of knowledge are called upon in Numble problems, and uncertainty affects the strategies used.*

Let us look at one more example:

Puzzle #6: Target: **146**
             Bricks: **12   2   5   7   18**

In this puzzle, the target instantly reminds some subjects (though few) of 144, the square of 12. Given that **12** is already a brick, one creates a subgoal of finding (or making) another 12. It is quickly made as **5 + 7**, yielding the solution **12 x (5 + 7) + 2**. The goal suggested a strategy in a top-down way. Of course, if one is not reminded of 144, the bricks will be played with much more randomly. The idea of making 12 will not be an explicit subgoal. The lesson to be drawn from this example is: *Top-down and bottom-up strategies not only coexist, but are deeply intertwined in Numble.*

The domain thus seems quite rich, and a simulation of a human player should be instructive. On the other hand, it would be folly to wish to *perfectly* simulate the mechanisms underlying human play. One must inevitably utilize *ad hoc* solutions in certain aspects of the simulation. For instance, the undeniable fact that humans sometimes make errors in carrying out arithmetical operations was deemed irrelevant to the Numbo project. Of course, this somewhat complicates any assessment of the validity of the simulation. I will return to this matter in the concluding section.

The Architecture of Numbo

The architecture of Numbo was inspired by that of the Copycat project (Hofstadter, Mitchell, & French, 1987), which in turn derives largely from that of Jumbo (Hofstadter, 1983a). It consists of three major components: a *spreading-activation network* that encodes the permanent knowledge needed to solve the task, a *working memory* where all the problem-solving activity takes place, and a set of so-called *codelets,* which are small operators that encode the needed procedural knowledge. Codelets can do any of the following:

- act on the contents of working memory to create, access, or modify various data structures;
- create other codelets and place them on what is called the *Coderack,* where they stay until chosen to run;
- increase or decrease activation of specific nodes in the network.

As in Jumbo and Copycat, all processing is carried out by codelets, which are probabilistically selected from the Coderack. Each codelet, when placed on the Coderack, is assigned an *urgency,* and its likelihood of being chosen is proportional to its urgency. Consequently, given two "competing" codelets (*i.e.,* codelets simultaneously waiting on the Coderack, "competing" in the manner of two people waiting for a taxi), no guarantee can be made as to which one will get run first.

The Permanent network

It was stated above that people playing Numble use a variety of types of knowledge. In particular, we make use of *rote small-number arithmetic* (*e.g.,* $6 + 1 = 7$), we make use of *knowledge of approximate sizes of numbers* (*e.g.,* multiplying 20 by 6 should bring us into the vicinity of 114), and we make use of *procedural arithmetical knowledge* (*e.g.,* we can multiply 6 by 19). In the course of solving a Numble problem, people effortlessly call on all three types of knowledge, blending them smoothly without any trouble. One standard way in AI of simulating this human ability would be to encode all the different pieces of knowledge in production rules, which would then be invoked by an appropriate control module, as, for instance, in the Soar system (Laird, Rosenbloom, & Newell, 1987). The approach taken in Numbo is quite different, however.

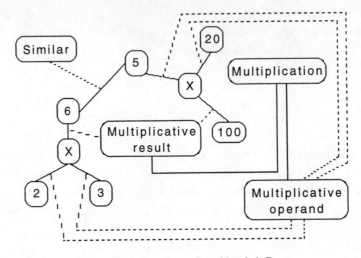

Figure III–1. An extract from Numbo's Pnet.

In Numbo, a network is used to encode the rote declarative knowledge needed in the game. This network, called the *Pnet,* encodes additive and multiplicative decompositions of three basic types of number: (1) small integers; (2) "landmark" numbers, such as 10, 20, 30, ..., 100, and 150, which are something like regular markings on a "mental ruler"; and (3) salient numbers (*i.e.,* numbers with which an individual just happens to be familiar, such as 128 for a computer scientist). In addition, the Pnet encodes a small amount of rote knowledge about the basic concepts and operations of arithmetic. Figure III–1 shows a small portion of the Pnet. In its current version, the Pnet contains about 100 different nodes.

For humans, facts about *small integers,* such as "6 = 2 x 3", are not found by calculation; rather, they are pieces of memorized declarative knowledge. Thus, it seems reasonable to store them in the Pnet. This particular multiplicative decomposition of 6 forms a little "biped" in Figure III–1, whose feet are the two factors 2 and 3, whose tummy is the times-sign "x", and whose head is the product, 6. (Such bipeds play a key role in Numbo, and will be discussed extensively later.) The solid lines connecting up the nodes of this biped are *labeled links,* whose labels are shown by the dotted lines leading off to Pnet nodes such as "multiplicative operand" and "multiplicative result".

Landmark numbers are a key ingredient in Numbo's way of simulating human intuition concerning numerical size. To each landmark are attached one or more standard decompositions. For instance, to the landmark "100" are attached the multiplicative decompositions "10 x 10" and "5 x 20". On its surface, the decomposition "100 = 5 x 20" appears to be strictly about the integers 100, 5, and 20, but in reality it is about numbers *near* 100, 5, and 20. It tells Numbo that if it is trying to build a number near 100, and if there are bricks close to 5 and 20, it might try out their product. Precisely how such "suggestions" are engendered will be explained later.

Salient numbers (not shown in Figure III–1) and their decompositions form the third ingredient of a typical human's knowledge about numbers. It seems obvious that in our minds, certain large numbers and particular decompositions of them are memorized. Thus it seems sensible to store "144 = 12 x 12" in the Pnet in the same way as "3 x 3 = 9" is stored. Of course, this type of knowledge varies highly from individual to individual, and such variation is undoubtedly responsible, at least in part, for the diversity of responses given by different people to the same problem.

Nodes representing integers are, as we have seen, not the only nodes in the Pnet. Given a target of **114**, an average person thinks of trying multiplication even before starting any calculation, almost as if the concept "multiplication" had "lit up" somewhere in the mind. To simulate this feature of the human approach, nodes labeled "addition", "subtraction", "multiplication", and "similar" are also contained in the Pnet. We will come to their function shortly.

It is admittedly not obvious just what should and what should not be put into the Pnet. For instance, should facts like "100 + 7 = 107", "20 + 20 = 40", "8 x 111 = 888", and "3 x 21 = 63" be included? I think not. Such arithmetical facts, even if they seem nearly immediate to us, seem nonetheless to have some manufactured flavor — that is to say, to be compounded from facts that are more basic. For instance, "20 + 20 = 40" is basically "2 + 2 = 4", modified by some trivial (*i.e.*, very syntactic) changes to both sides. Much the same could be said about the other examples. I believe that only *very* elementary, immediate things should be represented in the Pnet. In a certain sense, we do not "see" that 22 x 3 = 66 in quite as primordial a way as we "see" that 2 x 3 = 6.

At any moment of processing, each node has a certain degree of *activation*, which reflects Numbo's current "interest" in that node. Activation spreads from a node to its nearest neighbors, but (much as in Copycat's Slipnet) not in a uniform, isotropic manner. Depending on certain intrinsic factors as well as on the context, certain links will be more prone to transmitting activation than others. I call a link's capacity to transmit activation its *weight*.

As was mentioned earlier, every link has a *label* indicating the type of relationship it encodes. For instance, in Figure III–1, consider the biped encoding the fact "5 x 20 = 100". In it, the link joining the "5" with the "x" is labeled "multiplicative operand", as is the link joining "20" with that same "x". On the other hand, the link joining "100" with that "x" is labeled "multiplicative result". Notice that these labels come about by attaching the *link* in question to the appropriate Pnet *node*, via a dotted line. Dotted lines are thus, in a sense, "meta-links". Each different variety of link is represented by a different Pnet node.

There is a close connection between labels and weights of links: the more highly activated a given label is, the greater is the weight of each link that it labels. Thus in Figure III–1, if the Pnet node "multiplicative result" is "hot" (*i.e.*, highly activated), then the link connecting the node "100" with its neighbor-node "x" will be capable of transmitting a lot of activation. The overall pattern of activation of Pnet nodes thus changes continually during the processing, and in fact partly guides the processing.

Here is a simple example of how Pnet activations influence the flow of processing. If the target is close to 100, the nodes "100", "multiplication", and "multiplicative result" will all be activated, and therefore activation will spread to the particular "x" node forming the tummy of the biped that links "100" with both "5" and "20", thereby in effect suggesting a possible multiplicative decomposition of the target into "something like 5 times 20".

Association (simulated by spreading activation) is clearly the key notion here — but total reliance on blindly spreading activation can give rise to uncontrolled, chaotic behaviors of the network. Therefore, more focused modes of activating nodes must exist. For example, if *5* is a brick in the puzzle being worked on, some activation will periodically be pumped into the "5" node. Similarly, if *146* is the puzzle's target, nearby landmark-number nodes (*e.g.,* "140" and "150") and salient-number nodes (*e.g.,* "144") will receive periodic bursts of activation. (Note that Numbo has no "146" node, reflecting a belief that 146 is too rare a number for an average person to maintain a permanent dossier on it.)

Among the many advantages of this associative manner of storing rote knowledge are these:

- it allows different types of knowledge to be stored in a uniform way;
- it allows a given target to simultaneously evoke many different ideas and strategies, thus in effect surrounding each node by a "halo" of connotations;
- it allows great flexibility in controlling the processing, by making it possible to change the focus of attention in a very general way;
- it is task-independent;
- network structures have proven useful in many previous AI contexts.

The cytoplasm

All building and dismantling of temporary, puzzle-specific structures takes place in what is called (as in Jumbo) the *cytoplasm*. This term can be thought of as a synonym for "working memory" or "blackboard"; however, its biological connotations, reflecting some of the key intuitions behind this type of architecture, make it seem preferable. The basic image is that of a living cell, in which multitudes of enzymes (the role of which is played by codelets) are continually at work, inspecting various structures, modifying others, creating or destroying yet others, and so on. The area of a biological cell in which such activity takes place is the cytoplasm. The image is therefore one of distributed, parallel computations, rather than one of serial computation (despite the fact that codelets run one at a time).

One way to think about the effective parallelism in Numbo's cytoplasm is this: since there can be any number of totally disjoint structures inside Numbo's cytoplasm, the order in which actions are carried out on them does not matter — hence several mutually unaffecting actions can be thought of as taking place simultaneously but remotely. The cytoplasmic structures are like so many molecules being manipulated by a host of independent enzymes — over here, two are being connected up, over there some others are being pulled apart, somewhere else another one is being modified internally, and so on. The term "cytoplasm" is intended to invoke an image of swarms of enzymes carrying out many such small actions in parallel.

The entities resident in the cytoplasm are small "network fragments" created by codelets acting under the influence of the Pnet. These structures can in fact be thought of as temporary copies of specific parts of the Pnet, which can then be augmented with new nodes and links, all of which are produced by the actions of codelets. In contrast to the Pnet, *the cytoplasm has nodes and links that come and go*; they are characterized by various parameters (type, status, attractiveness, etc.), and they interact with each other via codelets.

At the very outset of a run, a node representing the target is placed in the cytoplasm. Then, in random order, the bricks are read in, and for each brick, a new node is added. The brick-nodes then begin to interact with one another and with the target-node.

What types of interaction can take place? The most important types are: (1) *grouping* of bricks into *blocks*; (2) *dismantling* of blocks into more elemental pieces; (3) production of *new targets,* called *secondary targets*; and (4) *bonding* of bricks, blocks, and targets. All these operations are carried out by codelets.

Figure III–2 shows a portion of the cytoplasm during a fairly typical Numbo run, at a critical moment in the solution of puzzle 2.

As was mentioned earlier, network representations have many advantages. The fact that Numbo's cytoplasm and Pnet are made up of the same type of components should also be stressed. Indeed, this sharing of structure proves interesting for several reasons, including these:

- The Pnet can act very simply on the cytoplasm by "downloading" pieces of its structure. Thus, permanent knowledge does not require restructuring before being used.

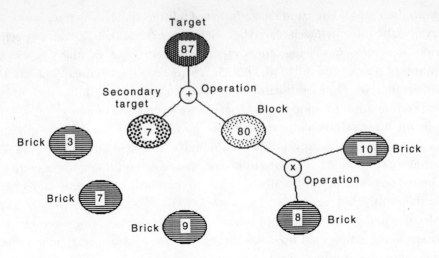

Figure III–2. A possible configuration of Numbo's cytoplasm.

- The difference between declarative and procedural knowledge of arithmetic vanishes, once an operation has been performed. Indeed, the result of an operation is stored in exactly the same way as it would have been, had it been a declarative piece of knowledge to begin with.
- Learning can be envisioned as an "upward" transfer of material from the cytoplasm to the Pnet.

Incidentally, confusion is possible since both the Pnet and the cytoplasm contain nodes (and links). When needed for clarity, a prefix of "P" or "cyto-" will be tacked on (*e.g.,* "Pnode" or "cyto-node"), in order to indicate where a given node is located.

The three most important parameters characterizing any cyto-node are its *type, status,* and *attractiveness.* Here are some of their most important characteristics:

- *Type* can assume any of five different values: "target", "secondary target", "brick", "block", or "operation". This makes it possible to define levels, or at least compartments, in the cytoplasm. The notion of "subgoal" is also partially captured through these distinctions.
- *Status* has two possible values: "taken" and "free". In Figure III–2, bricks *10* and *8* are taken, because they are involved in an operation. On the other hand, bricks *3* and *7* are still free.
- *Attractiveness* is a numerical value attached to all cyto-nodes; it influences the likelihood that the node will be used. (Most of the mechanisms used in Numbo are probabilistic, but not stupidly

random in the sense of "everything is equally likely". Therefore there have to be factors, such as the attractiveness of nodes, that work together to increase or decrease the likelihood of certain pathways being followed.)

The notion of attractiveness is of central importance. For a node representing a number (*i.e.*, a target, brick, or block), its attractiveness is mainly a function of its numerical value. For instance, a multiple of 5 — or better yet, a multiple of 10 — will be judged *a priori* more interesting than other numbers, and such numbers therefore tend to be used more often. (This type of bias is unquestionably observed in human players.) But the attractiveness of a node is a function of other factors as well. Any newly formed block, for instance, is automatically considered highly attractive. However, as it ages without being used, its attractiveness will gradually decline. Once a node is involved in a group (*i.e.*, it is "taken"), its level of attractiveness is temporarily frozen.

Certain codelets can modify the attractiveness of nodes. For example, there is a type of codelet that inspects various nodes and notices syntactic (*i.e.*, simple and surface-level) similarities among them. If such a codelet notices that the brick *11* shares some digits with the target *114*, it can create (and place on the Coderack) another codelet to increase the attractiveness of the *11* brick.

The cytoplasm as a whole is characterized by a *temperature* (this notion is similar to that described in Hinton & Sejnowski, 1983; Hofstadter, 1983a; Kirkpatrick, Gelatt, & Vecchi, 1983; Smolensky, 1983b). The closer the system feels it is to a solution (how it has a sense for this will be explained shortly), the more it lowers the temperature. This reduces the chances of drastically changing any of the structures in the cytoplasm. By contrast, when the temperature is high, codelets are produced that dismantle blocks and secondary targets; this is the system's way of "backtracking", although it must be pointed out that it is very different from the standard type of backtracking, which involves retracing a pathway to reach a prior state, and then starting from there down a new branch of exploration. Rather, this kind of "backtracking" involves giving up gains more or less randomly, and starting anew in some less-ordered state that has quite probably never been encountered before.

The temperature is computed as a function of the overall state of the cytoplasm, and takes into account such things as: attractiveness of the nodes, the number of free nodes, the number of secondary targets, etc. Essentially, the temperature will be low when the state of affairs appears promising. Characteristics typical of a promising state would be these: there are several secondary targets, most nodes are highly attractive, and some nodes are still free. On the other hand, if some node's attractiveness is very low, or if the number of free nodes is too small, then the state looks less promising and the computed value of the temperature will be higher. As a consequence of the high temperature, "dismantler" codelets will be loaded onto the Coderack. The structures to be dismantled are chosen randomly but not equiprobably: the lower a node's attractiveness is, the more likely it is to be chosen for dismantling. The idea behind this is, of course, that unpromising nodes should be eliminated first.

Codelets

All operations are carried out by small pieces of code called, as was stated earlier, *codelets*. Numbo's codelets are similar to those in Jumbo, Seek-Whence, and Copycat. They are characterized by a "mission" and an urgency. Once a codelet is on the Coderack, its chance of being the next selected to run is, at any time, proportional to its urgency (specifically, it is the ratio of the codelet's urgency to the sum of all the urgencies of all the codelets in the Coderack). Codelets can be produced by the Pnet, the cytoplasm, or other codelets.

Codelets can be classified in various ways. Let us first break them down according to *locus of action*. Three types can be distinguished: (1) codelets that modify the Pnet; (2) codelets that modify the cytoplasm; and (3) codelets that perform some sort of test and subsequently either post new codelets on the Coderack or modify already-posted codelets (*e.g.*, by changing their urgency).

Another way of classifying codelets is according to their *type of activity*. Codelets that *create* structures should be distinguished from codelets that *destroy* them. A trivial example of a constructive codelet is the "create-node" codelet, which, whenever a brick or target is read or a new block is manufactured, creates a corresponding node in the cytoplasm. A typical example of a destructive codelet is the "kill-secondary-nodes" codelet, which destroys a previously-built block or secondary target, and frees up the constituent nodes.

Insofar as possible, I have tried to make the actions of codelets be simple and "dumb". For example, a codelet can notice that a target and a brick share digits (this is a trivial syntactic check — a visually obvious fact for people), but no codelet can check whether or not the target is a multiple of the brick (that would be too semantic, too "smart"). Typical codelet actions are: creating a cyto-node, sending activation to a Pnode, comparing a given cyto-node with a given target, and so on. Codelets act at various levels of abstraction (*e.g.*, comparison of digits, creation of nodes, etc.), with no overseeing agent. This implies, among other things, that *different paths can be explored in parallel*. In more traditional AI approaches, once a subgoal (*i.e.*, a secondary target) is set up, all the system's resources are devoted to pursuing it; in Numbo, by contrast, competition with other possible goals continues. For instance, a subgoal can be dropped before being completely explored if a more promising path is discovered.

Not only are codelets supposed to be dumb, they are also supposed to be "myopic" (*i.e.*, to have no global overview, either spatially or temporally). Thus, for example, the cytoplasm cannot be systematically scanned by a codelet to check if a given number is present. Nor can codelets "plan" or "conspire"; the only sense in which codelets can cooperate is that a given codelet's offspring will presumably carry out an operation that furthers the direction in which the parent codelet was headed.

Another simple consequence of Numbo's probabilistic architecture is that it has virtually no chance of getting trapped in an endless loop. The probability of exploring the same wrong avenue over and over again simply goes exponentially to zero as the number of repetitions increases. No sophisticated backtracking mechanisms are thus necessary.

In summary, whatever "intelligence" Numbo manifests is a simply a by-product of the competitive interplay of many myopic and dumb processes operating in parallel at several levels of abstraction.

A Sample Run of Numbo

Having described Numbo's architecture, I now proceed to show how the Pnet and cytoplasm interact, via codelets, to solve the first puzzle.

> *Puzzle #1:*   Target: ***114***
>   Bricks: ***11   20   7   1   6***

This puzzle is interesting because, as was mentioned above, the simplest response — *(20 – 1) x 6* — is essentially never given by people. How does Numbo fare on this problem? Figure III–3 shows an annotated trace of an actual run.

The last three lines are a summary, provided by Numbo, of how it has managed to construct the target out of the given bricks. Let us now examine in more detail the process by which Numbo found this solution.

(1) The target is read. The Pnet landmark closest to the target — namely, the Pnode "100" — is activated; it will serve as a sort of beacon or focus. In addition, the Pnodes "multiplication", "subtraction", and "addition" are activated. The system knows that with large targets, the most promising avenues involve multiplication, and since the target, *114*, is much bigger than the biggest brick, the "multiplication" node gets more activation than do the two others.

(2) One brick, *11*, is read. A "syntactic-comparison" codelet happens to get run, which compares this brick with the target and notices that they start with the same two digits. This observation gives rise to a new codelet, which, when run, increases the attractiveness of the first brick.

Target-114 created	Numbo reads the target and creates a cyto-node representing it.
Brick-11 created *Brick-20 created* *Brick-1 created* *Brick-6 created* *Brick-7 created*	Numbo reads the various bricks in a random order, and creates cyto-nodes representing them.
11-times-7 created *Block-77 created*	Numbo tries multiplying 11 by 7, and creates a new block in the cytoplasm.
20-times-6 created *Block-120 created*	Another multiplicative block, this one suggested by the Pnet, is created.
120-minus-114 created *Target-6 created*	By subtracting, Numbo notices that the newly-created block is close to the primary target. Numbo thus creates a secondary target.
Block-77 killed *11-times-7 killed*	Block-77 has not been used; it is thus dismantled. Nodes Brick-11 and Brick-7, which had both been used in supporting the defunct Block-77, are now freed up.
7-minus-1 created *Block-6 created*	Under pressures from the Pnet, a new block equal to the secondary target is created from Brick-7 and Brick-1.

Done! Operation "times" applied to Brick-20 and Brick-6 to make Block-120.
Operation "minus" applied to Brick-7 and Brick-1 to make Block-6.
Operation "minus" applied to Block-120 and Block-6 to make Block-114.

Figure III–3. Trace of a run of Numbo.

(3) The remaining bricks are read. Notice that, because of the prob-abilistic way codelets are chosen, the bricks are not read in left-to-right order. For instance, in this run, the last brick, *6*, is read before the middle brick, *7*. As with the first brick, codelets whose mission is to syntactically compare each of these bricks with the target are loaded onto the Coderack, and may possibly be run.

(4) The Coderack's population always includes some fairly low-urgency codelets that try out random mathematical operations. By chance, such a codelet is run. It selects two bricks at random (but of course it is biased towards attractive ones), and also an arithmetical operation (likewise biased towards highly activated Pnodes). Recall that brick *11* has been judged quite attractive, and that "multiplication" has been highly activated. Thus *11* gets chosen to be multiplied by some other brick, which, again by chance, turns out to be *7*; as a result, block *77* is created, and, together with its underpinning structure *11 x 7*, it is placed in the cytoplasm.

(5) In the Pnet, activation will now spread from activated nodes to their neighbors. Notice that the head and feet of the biped "100 = 20 x 5" are all activated (landmark node "100" was acti-vated by its proximity to the target *114*, landmark node "20" was activated by being equal to the brick *20*, and small-integer node "5" was activated by its proximity to the brick *6*). Consequently, activation from three independent sources will converge on the node "x" ("times") — the biped's tummy. This arithmetic-operation node is therefore lurched all of a sudden into a state of high activation. Such "tummy" nodes, when highly activated, are very important, because they represent the potential for creating blocks equal to, or near, a target.

To try to realize this potential, a "seek-reasonable-facsimile" codelet is loaded on the Coderack, which — if and when it gets run — will try to build a new block by searching for a "reasonable facsimile" of 20 (*i.e.,* something numerically close to 20) and a "reasonable facsimile" of 5, and multiplying them. The search for reasonable facsimiles is, like most searching in Numbo, probabil-istic. In this case, the "seek-reasonable-facsimile" codelet might come up with either *20 x 6* or *20 x 7*, although the former would be more likely because 6 is a better facsimile of 5 than 7 is. Of course, nothing guarantees that this codelet will ever run, since other codelets — possibly even ones of the same type — are in competition with it. In fact, in this case, a rival "seek-reasonable-

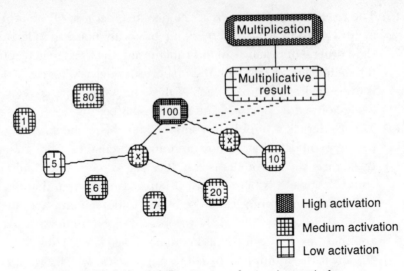

Figure III–4. Numbo's Pnet at an early stage in a typical run.

facsimile" codelet — this one hoping to get close to target **114** by finding a product resembling "10 x 10"—is also on the Coderack. (However, this latter codelet, if run, would get nowhere, because the cytoplasm, at present, simply does not contain two different nodes sufficiently close to 10.)

The running of the first "seek-reasonable-facsimile" codelet ("20 x 5") results in the loading of a "test-if-possible-and-desirable" codelet, which, once run, will examine the feasibility of the proposed operation: the component bricks (or blocks) must be *free,* and the proposed new block must be judged *worthwhile.* To determine whether a proposed block is worthwhile, the codelet examines the Pnet. If it finds that the activation of the corresponding Pnode (*i.e.,* the arithmetically closest Pnode) is high enough, a high-urgency codelet to create the proposed brick will be loaded on the Coderack, and because of its high urgency, it will probably soon get run. During all these considerations, of course, other rival avenues are being explored, and may even block this avenue, by taking needed blocks; this is the price of parallelism in Numbo.

In this run, as luck would have it, the block **120** does get created ("20 x 6"). Figure III–4 shows a portion of the Pnet just prior to block **120**'s creation. Only the most active Pnodes are included. (A node's density of cross-hatching symbolizes its degree of activation.)

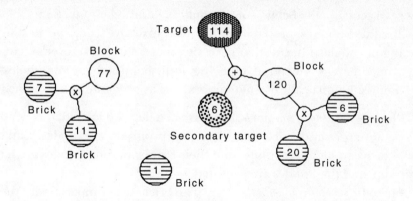

Figure III–5. Numbo's cytoplasm at an intermediate stage of the same run.

(6) The newly-created block *120* is now compared (by another codelet, of course) to the target. As they are quite close, the creation of a *secondary target* now enters the picture as a possibility. To this end, a target-making codelet having fairly high urgency is posted. When it is run, it subtracts 114 from 120, gets 6, and therefore inserts a new secondary-target node into the cytoplasm. This fresh cyto-node, block *6*, sends a new wave of activation into the Pnode "6", just as reading in a new brick would. Figure III–5 shows the cytoplasm at that moment.

(7) Since block *77* has not been used, its attractiveness has been decreasing, which fact has in turn been helping to raise the temperature. Recall that a rising temperature implies, among other things, that "dismantler" codelets will start to invade the Coderack; as this happens, the probability grows that any block — but especially unattractive ones — will be "attacked" and broken up. In this run, block *77* itself becomes a victim at this point.

(8) The simultaneous activation of Pnodes "1", "7", and "6" (coming from bricks *1* and *7*, and secondary target *6*, respectively) provokes the creation of a new "seek-reasonable-facsimile" codelet, which, after various routine tests by different codelets, results in the creation of the new block *6* ("7 – 1"). This block is compared with the current target. Since they are equal, a codelet checks to see if the overall goal has been reached. It has indeed, so the solution is printed and the run comes to a conclusion.

Numbo's processing of any problem can be decomposed into four rough "phases", corresponding to four different types of operations. These phases are presented sequentially below, but it would be a mistake to take this sequence too literally. Numbo's probabilistic control structure (*i.e.,* the posting and selecting of codelets) means that in reality, these phases can overlap considerably, sometimes to the point of being indistinguishable. Nonetheless, the following is a heuristically useful temporal analysis of Numbo's approach.

Phase I: *Reading of the puzzle.* Primitive codelets are loaded; they read the target and bricks, and install appropriate cyto-nodes. Links between these ctyo-nodes and their corresponding Pnodes are set up, and the Pnodes are activated.

Phase II: *Comparison with the target.* The bricks are compared with the target. The codelets that do this are capable of detecting equality, numerical proximity, or shared digits. Depending on the results of these comparisons, potential decompositions may be suggested.

Phase III: *Search for useful associations suggested by the Pnet.* As activation spreads in the Pnet, some arithmetical-operation Pnodes (tummy nodes inside bipeds) may get sufficiently excited to cause "seek-reasonable-facsimile" codelets to be placed on the Coderack. These codelets hunt in the cytoplasm for two free numbers close to two of the numbers forming the biped in question, so as to make a new block close to the biped's third number.

Phase IV: *Background-mode associations between bricks.* There is a background activity in Numbo that becomes important only when pressures from the Pnet are not too strong (*i.e.,* when the system does not have any strong motivation to perform a particular operation). Codelets randomly choose two free bricks (statistically favoring attractive ones) and an arithmetical operation (favoring highly-activated nodes). The operation is carried out, a new block is suggested, and then it is examined by a "test-if-possible-and-desirable" codelet. If the resulting number seems promising (*i.e.,* it is close to some target either syntactically or numerically), the construction of the block is recommended.

Discussion

Comparison with other computer models

Enough of the system has now been presented to allow some connections to be drawn between this project and others. The basic mechanisms of Numbo derive largely from earlier work by Hofstadter and his associates (Jumbo, Seek-Whence, and Copycat). But in many aspects, Numbo is undeniably reminiscent of other projects as well. For instance, a variant of "means–end analysis" seems to be used to reduce the differences between the bricks and the target. In particular, in Phase II, the proximity of a brick to a target suggests taking their difference, thereby creating a secondary target — in essence a subgoal. One might be tempted to infer that Numbo is simply using methods developed over twenty years ago in the General Problem Solver project (Ernst & Newell, 1969).

A careful comparison, however, reveals many substantial differences between Numbo and GPS. For example, GPS uses the heuristic-search paradigm directly: a problem is stated in terms of objects and operators. The process is then guided by a central executive, and all the entities are represented according to their role in a specific task: there is no permanent, task-independent representation of knowledge (in other words, in GPS there is no counterpart to the Pnet). A record is kept of all objects already considered, and all goals already tried. Subgoaling is the name of the game, and all processing is goal-directed. This means that "arbitrary" associations of bricks that have not been motivated by subgoals are impossible. Once a subgoal has been selected, it is carefully explored, but no others are explored during that time.

These methodological choices are in sharp contrast to those in Numbo. One would be at pains to cast Numbo's method of proceeding in terms of search in a problem space; after all, no central control is involved, and rote knowledge is represented in a task-free way. A goal is just one of many features of a situation. No systematic exploration takes place in Numbo. Taken together, competition between codelets and the element of randomness allow the system to jump from one idea to another, sometimes in a chaotic-seeming (but rather humanlike) way.

Another system whose overlap with Numbo cannot be overlooked is Anderson's ACT* (Anderson, 1983). An obvious mapping of the two architectures seems possible:

Declarative memory   ⇔   Pnet
Production memory   ⇔   Coderack

ACT*'s declarative memory is a spreading-activation network, and in that sense resembles Numbo's Pnet. But ACT*'s nodes are more diverse in nature (they include such things as temporal strings, spatial images, and abstract propositions). ACT*'s production memory could be likened to Numbo's permanent repertoire of all possible *codelet types* (as opposed to the Coderack, which is a constantly-changing inventory of *codelet tokens,* and as such is a rather poor analogue of production memory). But in Numbo, the structure of a task is less constrained than in ACT*: it is not required to be a condition/action production rule.

Perhaps the biggest difference between Numbo and ACT*, however, is in the overall control of processing. In ACT*, production rules are selected by a sophisticated pattern-matching procedure. In Numbo, the determination of the next task to be executed is less symbolic, and is a two-phase operation. A preliminary sort of task selection occurs when a given task (a codelet) is loaded onto the Coderack, and a second sort occurs when that codelet is actually chosen from the Coderack to be run. The reason for this strategy in Numbo is to bring about a type of *probabilistically-biased parallel exploration of possibilities,* by means of the interleaving of small tasks that form part of larger avenues of exploration. This idea, dubbed the "parallel terraced scan", has no counterpart in ACT*.

Notice also that patterns are detected in a more random way by Numbo: it is sometimes possible to miss an obvious solution, due to the probabilistic nature of the selection of the codelets and to the random way they scan the configuration. Finally, ACT*'s learning mechanisms have no counterparts in Numbo.

Further parallels with other systems can be drawn. The basic mechanisms used in the Pnet have a connectionist flavor. Research by Rumelhart, McClelland, and their colleagues is, like Numbo, based on a bottom-up, statistically-emergent approach to cognition (see McClelland, Rumelhart, & Hinton, 1986).

One might also compare Numbo with Lenat's AM project (Lenat, 1979, 1983a). AM's agenda of tasks is reminiscent of our notion of Coderack. But again, nothing like a stochastic selection of tasks biased according to priorities is used in AM, and the reason is that Lenat, like Anderson, is not striving after any kind of parallel exploration of pathways. Furthermore, in Lenat's work, much use is made of high-level control mechanisms (heuristics and meta-heuristics). Numbo is fundamentally different in that sense. It relies almost exclusively on very elementary activities (tentative associations of bricks, preliminary explorations of paths, suggestion of potential subgoals, and so on).

A more thorough discussion of the architecture used in Copycat, Jumbo, and Seek-Whence (and therefore in Numbo) is found in Hofstadter, Mitchell, & French, 1987.

Comparison with human performance

One of the aims of the Numbo project was to contribute to the understanding of the mental fluidity exhibited by human subjects in playing Numble. The Numbo program gives the impression of playing the game in a human style. Here are some of its similarities with human performances:

(1) Obvious solutions are found at once.
(2) Ideas are not necessarily systematically explored, and in fact are often abandoned before having been fully examined.
(3) The combinations created are not always strongly goal-driven, so that unmotivated-appearing avenues are occasionally embarked upon.

(4) Solutions are often found that involve the chaining of several arithmetical operations in a seemingly logical way.

Despite these similarities, a rigorous analysis of human performance in Numble will not be found in this paper. I do not consider such concerns to be irrelevant, of course. During the year when I was implementing the Numbo program, I tested several subjects on quite a few different problems, asked them to think aloud, and tape-recorded their commentary. Then I carefully examined their protocols, and based several aspects of my architecture on what I found. However, I feel that a considerable amount of additional research would be needed before anything substantial could be said about Numbo in comparison with humans.

A *strict* comparison of the performance of human subjects with the performance of Numbo is not possible for at least three reasons:

(1) Numbo's knowledge base is impoverished; major aspects of a typical adult's arithmetical background are lacking in the Pnet.
(2) Certain characteristics of the way humans tend to approach a given problem have been deliberately ignored. For instance, Numbo pays no attention to the linear order of the bricks in the initial presentation of a puzzle, whereas human subjects seem often to be influenced, mostly on a subconscious level, by that order.
(3) In the design of Numbo's architecture, *ad hoc* solutions were of necessity given to some important and difficult problems (*e.g.,* how to measure the degree and the nature of the similarity of two given numbers, how to set the temperature, and so on).

Another factor that would certainly complicate any comparison between Numbo and people is the highly questionable nature of human protocols. I was repeatedly told by solvers that it is difficult, if not impossible, to keep track of everything going on when one is solving a problem. An *a posteriori* reconstruction seems to be necessary.

A rough comparison of such a protocol with a trace given by Numbo is, of course, always possible, provided one keeps the preceding warnings in mind. Two protocols are shown in Figure III–6: one is a trace of a Numbo run, while the other is a human's protocol (not necessarily in that order). A couple of notational conventions: when a subgoal is defined, the number is followed by a question mark, essentially meaning, "Can I see how to get this number?". Also, "*no*" means that a given approach is being abandoned. The puzzle on which these are based is number 3:

Puzzle #3: Target: *31*
Bricks: *3 5 24 3 14*

Protocol 1	*Protocol 2*
31?	31?
3 x 3 = 9; *no*	24 + 5 = 29
24 + 3 = 27	(24 + 5) + 3 = 32
4?	1? *no*
(24 + 3) + 5 = 32	3 x 5 = 15
1? *no*	14 + (3 x 5) = 29
3 x 5 = 15; *no*	(14 + (3 x 5)) + 3 = 32
24	1? *no*
7?	3 x 3 = 9; *no*
15 + 4 = 19; *no*	3 + 5 = 8; *no*
24 + 5 = 29;	(3 + 5) x 3 = 24; *no*
2? *no*	24 − 14 = 10
24	(3 + 5) + 3 = 11
7? *no*	((3 + 5) + 3) + (24 − 14) = 21; *no*
3 + 5 = 8	
24 + (3 + 5) = 32	*etc., etc., etc.*
1? *no*	
14 − 3 = 11	
3 x 11 = 33	
2?	
etc., etc., etc.	

Figure III–6. Comparative protocols of a human and Numbo, tackling the same challenge.

Protocol 1 is the trace given by Numbo. Notice that neither Numbo nor the human subject solved the problem (even though it has four different solutions!). The vacillations of Numbo seem to have a human flavor. But obviously, this alone does not demonstrate the validity of Numbo as a model of the human mind.

Conclusions

My aim was to build a system capable of recognizing obvious regroupings and solutions, as we are, and also capable of chaining operations together so as to reach a goal in a humanlike way. To show some of the ways in which I succeeded, I will now discuss Numbo's performance on a few puzzles, some very simple and some more complex. Consider first this very simple puzzle:

Puzzle #7: Target: *6*
Bricks: *3 3 17 11 22*

Numbo immediately comes up with the solution *3 + 3* (and never *17 − 11*, let alone *17 + 11 − 22*). In other words, it quickly spots what is obvious to people, and does not waste time going down pathways that people would not try. Similarly, if Numbo is given this simple puzzle:

> *Puzzle #8:*   Target: *11*
>   Bricks: *2   5   1   25   23*

then it will immediately answer *2 x 5 + 1*. Here, the system instantly sniffs the promise of multiplying *2* by *5*, and does not waste time exploring other routes, even though there are some that work, such as *23 − 2 x (5 + 1)*.

The preceding examples show how Numbo quickly arrives at obvious solutions. The next one shows its ability to solve more complicated problems:

> *Puzzle #9:*   Target: *116*
>   Bricks: *20   2   16   14   6*

Numbo finds the rather subtle solution *6 x 20 − 2 − (16 − 14)*.

What are the advantages and the drawbacks of the proposed architecture? Three main comments are in order:

(1) The fluidity exhibited by Numbo is largely due to its probabilistic codelet-based architecture (based on the image of distributed enzymatic activity in a biological cell), featuring parallel exploration of different paths and constant competition between alternative solutions, both of which are powerful mechanisms. If the system is exploring a barren path, a better suggestion will soon arise and will make it possible to begin examining a potentially better path *at the same time*. In addition, seemingly unmotivated combinations will be tried out from time to time. This may open new routes that will be explored to the extent they seem promising, even if pressures for more obvious paths exist. Take, for instance, the following example.

> *Puzzle #10:*   Target: *127*
>   Bricks: *6   4   22   5   7*

Once in a while, Numbo starts out by multiplying *6* and *5*. Once the block *30* has been created, the solution *4 x 30 + 7* is not far away, and is, in fact, discovered by Numbo, although more obvious routes of approach (such as via *6 x 22* and *5 x 22*) exist.

(2) The knowledge encoded in the Pnet is critical. Numbo currently has problems with some fairly simple examples, such as this one:

Puzzle #11:   Target: **41**
Bricks: **5   16   22   25   1**

because it does not know that 40 = 20 + 20, and hence doesn't recognize the plausibility of trying reasonable facsimiles, such as **16 + 25** or **22 + 25**. This lack of knowledge induces a nearly random search: the combinations suggested by the Pnet are poor, and the system simply has to stumble across a solution through the random combinations it conducts as a background activity.

(3) Because the Pnet is fairly dense, a low level of activation is nearly always present in each node. This "background noise" makes it difficult, at times, to fully benefit from the network architecture. The Pnet is powerful in evoking direct connotations, but seems relatively weak in suggesting long chains of associations.

The Numbo project has shown that, with an appropriate architecture, a system can behave, at least in a limited domain, in a very fluid, humanlike way, combining the ability to spontaneously perceive chunks, the ability to manufacture groups, and the ability to achieve goals through the chaining of different operations.

Of course, Numbo's capabilities could be greatly improved. One very difficult but crucial open question concerns how to combine pieces of knowledge in the Pnet into new ones (*e.g.*, given the facts "144 = 12 x 12" and "12 = 3 x 4", how to come up with the new fact "144 = 9 x 16"?). Another open question is how the outcome of one task can be applied to another task. I believe that rather than attempting to extend the system to new domains to probe its generality, one should undertake a careful study of the basic mechanisms. The current incarnation of Numbo is thus but a first step down that long road.

Preface 4:

The Ineradicable Eliza Effect
and Its Dangers

Programs with Apparent Real-world Prowess

It's probably fair to say that the following paper, first drafted in mid-1989, was inspired by a mounting degree of exasperation and worry in our research group. The problem was this. By that time, Copycat had been implemented to a large degree, and was really starting to come into its own (see Chapter 5 for details). Tabletop, too, was starting to exist as a working program (see Chapters 8 and 9). There was a general feeling of excitement in the group on this account, in that our research ideas were showing fruit and were revealing the fascination and complexity of our carefully-designed microdomains. In the meantime, however, a great deal of uncritical publicity was being given to a number of AI programs that gave the appearance of creating very complex real-world analogies or making sophisticated scientific discoveries, rivaling in insight such pioneers as Galileo, Kepler, and Ohm. Such favorable publicity could not help but make our achievements in microdomains look rather microscopic, by comparison. And yet, we felt that any such conclusion about our work would be superficial and unwarranted. It's obvious why this was cause for concern on our part.

Typical of the reports on such AI work was an article by Mitchell Waldrop in the prestigious journal *Science* (Waldrop, 1987), which described in flattering terms the analogy-making achievements of SME, the Structure Mapping Engine (Falkenhainer, Forbus, & Gentner, 1990), a computer program whose theoretical basis is the "structure-mapping theory" of psychologist Dedre Gentner (Gentner, 1983). After a very brief presentation of that theory, Waldrop's article went through an example, showing how SME makes an analogy between heat flow through a metal bar and water flow through a pipe, inferring on its own that heat flow is caused by a *temperature* differential, much as water flow comes about as a result of a *pressure* differential. Having gone through this example, Waldrop then wrote:

> To date, the Structure Mapping Engine has successfully been applied to
> more than 40 different examples; these range from an analogy between the

solar system and the Rutherford model of the atom to analogies between
fables that feature different characters in similar situations. It is also serving
as one module in . . . . a model of scientific discovery.

There is an insidious problem in writing about such a computer achievement, however. When someone writes or reads "the program makes an analogy
between heat flow through a metal bar and water flow through a pipe", there is
a tacit acceptance that the computer is really dealing with the *idea* of heat flow,
the *idea* of water flow, the *concepts* of heat, water, metal bar, pipe, and so on.
Otherwise, what would it mean to say that it had "made an analogy"? Surely, the
minimal prerequisite for us to feel comfortable in asserting that a computer
made an analogy involving, say, water flow, is that the computer must *know what
water is* — that it is a liquid, that it is wet and colorless, that it is affected by gravity,
that when it flows from one place to another it is no longer in the first place,
that it sometimes breaks up into little drops, that it assumes the shape of the
container it is in, that it is not animate, that objects can be placed in it, that
wood floats on it, that it can hold heat, lose heat, gain heat, and so on *ad
infinitum*. If the program does not know things like this, then on what basis is it
valid to say "the program made an analogy between *water flow* and such-and-so
(whatever it might be)"?

Needless to say, it turns out that the program in question knows none of
these facts. Indeed, it has no concepts, no permanent knowledge about *anything
at all*. For each separate analogy it makes (it is hard to avoid using that phrase,
even though it is too charitable), it is simply handed a short list of "assertions"
such as "Liquid(water)", "Greater(Pressure(beaker), Pressure(vial))", and so
on. But behind these assertions lies nothing else. There is no representation
anywhere of what it *means* to be a liquid, or of what "greater than" means, or of
what beakers and vials are, etc. In fact, the words in the assertions could all be
shuffled in any random order, as long as the permutation kept *identical* words
in corresponding places. Thus, it would make no difference to the program if,
instead of being told "Greater(Pressure(beaker), Pressure(vial))", it were told
"Beaker(Greater(pressure), Greater(vial))", or any number of other scram-
blings. Decoding such a jumble into English yields utter nonsense. One would
get something like this: "The greater of pressure is beaker than the greater of
vial." But the computer doesn't care at all that this makes no sense, because it
is not reaching back into a storehouse of knowledge to relate the words in these
assertions to anything else. The terms are just empty tokens that have the *form*
of English words.

Despite the image suggested by the words, the computer is not in any sense
dealing with the idea of water or water flow or heat or heat flow, or *any* of the
ideas mentioned in the discussion. As a consequence of this lack of conceptual

background, the computer is not really *making an analogy*. At best, it is *constructing a correspondence* between two sparse and meaningless data structures. Calling this "making an analogy between heat flow and water flow" simply because some of the alphanumeric strings inside those data structures have the same spelling as the English words "heat", "water", and so on is an extremely loose and overly charitable way of characterizing what has happened.

The Slippery Slope into the Eliza Effect

Nonetheless, it is incredibly easy to slide into using this type of characterization, especially when a nicely drawn picture of both physical situations is provided for human consumption by the program's creators (see Figure VI–1, page 276), showing a glass beaker and a glass vial filled with water and connected by a little curved pipe, as well as a coffee cup filled with dark steaming coffee into which is plunged a metal rod on the far end of which is perched a dripping ice cube. There is an irresistible tendency to conflate the rich imagery evoked by the drawings with the computer data-structures printed just below them (Figure VI–2, page 277). For us humans, after all, the two representations feel very similar in content, and so one unwittingly falls into saying and writing "The computer made an analogy between this situation and that situation." How else would one say it?

Once this is done by a writer, and of course it is inadvertent rather than deliberate distortion, a host of implications follow in the minds of many if not most readers, such as these: computers — at least some of them — understand water and coffee and so on; computers understand the physical world; computers make analogies; computers reason abstractly; computers make scientific discoveries; computers are insightful cohabiters of the world with us.

This type of illusion is generally known as the "Eliza effect", which could be defined as the susceptibility of people to read far more understanding than is warranted into strings of symbols — especially words — strung together by computers. A trivial example of this effect might be someone thinking that an automatic teller machine really was *grateful* for receiving a deposit slip, simply because it printed out "THANK YOU" on its little screen. Of course, such a misunderstanding is very unlikely, because almost everyone can figure out that a fixed two-word phrase can be canned and made to appear at the proper moment just as mechanically as a grocery-store door can be made to open when someone approaches. We don't confuse what electric eyes do with genuine vision. But when things get only slightly more complicated, people get far more confused — and very rapidly, too.

The Eliza effect borrowed its name from the ELIZA program, written by Joseph Weizenbaum in the mid-1960's. That infamous program's purpose was to act like a nondirective Rogerian psychotherapist, responding to the typed lamentations of patients with very bland questions that echoed their own words back to them, most of the time simply urging them to continue typing along the same lines ("Please go on"), and occasionally suggesting a change of topics. The most superficial of syntactic tricks convinced some people who interacted with ELIZA that the program actually understood everything that they were saying, sympathized with them, even *empathized* with them.

Since that time, mountains of prose have been written about the Eliza effect (see, for example, Boden, 1977; Weizenbaum, 1976; and McDermott, 1976), and yet the susceptibility remains. Like a tenacious virus that constantly mutates, the Eliza effect seems to crop up over and over again in AI in ever-fresh disguises, and in subtler and subtler forms.

Please understand that what I am saying is not meant as a criticism of the developers of SME, or even of Waldrop. It is meant much more as a critique of the whole mentality swirling around the complex intellectual endeavor called "AI" — a surprisingly unguarded mentality in which anthropomorphic characterizations of what computers do are accepted far too easily, both outside and within the field.

"Just This Once"

Consider, for example, the fact that on July 2, 1993, the *New York Times* featured a front-page article with the attention-grabbing headline "Potboiler Springs From Computer's Loins". It told of a Silicon Valley programmer named Scott French, who a decade earlier had made a bet with some friends that he could program a computer to churn out a trashy novel in the style of Jacqueline Susann, author of the giant bestseller *Valley of the Dolls*. Since then, French had toiled away for some eight years, and finally, the week the story appeared in the *Times,* a novel entitled *Just This Once* (French, 1993) was being published. Underneath the title of the novel on the book's cover, it says: "A novel written by a computer programmed to think like the world's best-selling author, as told to Scott French".

According to the article, written by Steve Lohr, French programmed a Macintosh that he nicknamed "Hal" using, as Lohr put it, "so-called artificial intelligence, an advanced form of programming that tries to emulate human thought". Here is Lohr's description of the program's functioning:

> The writing of a scene would amount to a dialogue between Mr. French and his software. The computer would ask questions, he would answer them, and then the machine would spit out the story, a couple of sentences at a time. He would then change a word here and there, correct a misspelling,

whatever. Then, based on what went before, the computer would ask more questions that Mr. French would answer, and so on.

"It doesn't write whole paragraphs at a time," Mr. French said. "You can't get up, walk away, come back and find a completed chapter. It's not that advanced."

Much of the tone and plotting was based on thousands of rules that Mr. French programmed into the computer, formulas he derived by carefully analyzing two of Ms. Susann's books, *Valley of the Dolls* and *Once is Not Enough.*

When two key female characters were to meet, for example, the computer would ask Mr. French about the "cattiness factor" that should be used in the scene. Mr. French would be presented with choices 1 through 10. If he keyed in 8 — high cattiness — the computer reached into its memory to craft a sentence that was likely to employ words like "screaming" or "shrieking".

The claims made by the publisher on behalf of Scott French and his program, and quite uncritically reported by the *New York Times,* are typical of this lax mentality. Even Marvin Minsky, one of AI's most important and creative pioneers, was quoted in Lohr's article as saying, "It sounds great.... It sounds like he's taken the computer generation of language further than other people." I do not know why Minsky, who had just finished co-authoring a novel himself, was so willing to play along with the claim that a computer program had written a novel, as if he really believed it. Certainly he knew how unbelievably much understanding and imagery and human experience goes into writing a novel, even the most mediocre one.

I myself was infuriated that this "AI breakthrough" was not only printed by the *Times,* but on its front page, to boot! The evening the story appeared, I spent several hours trying to dispel my rage by composing a letter to the editor of the *New York Times,* which I sent off the next day. Unfortunately, they didn't run it, but that editorial oversight won't prevent it from seeing the light of day! Here, then, is my little outburst.

Dear Editor,

Within the space of scarcely over a week, the *New York Times* has given front-page coverage to two remarkable intellectual achievements: the solution of one of mathematics' most outstanding problems, after several centuries of frustrated attempts, and the creation of a novel by a machine, after thousands of years of people thinking of machines as the epitome of lifelessness. Both of these achievements were the fruits of seven to eight years' intense labor on projects deemed hopeless by the vast majority of professionals in their fields, both resulted in documents of two to three hundred pages in length that will soon be published, and both were hailed in the *New York Times* by experts as great developments going beyond anything done before.

What a wonderful parallel, one might think, and what an amazing era we live in! The problem is, however, that these two achievements are in reality at opposite poles in terms of genuine scientific importance. The proving of Fermat's Last Theorem (assuming that the proof does indeed hold up) is truly monumental, and it was altogether fitting that the *Times* celebrated it in a front-page article.

On the other hand, the claim that a novel has been "written by a computer" is extraordinarily distorted and misleading. It's most unclear from the article what Scott French's computer was actually able to do, but the impression was clearly given that the computer was handling sophisticated concepts such as "jealousy", "sex", "competition", and so forth, not to mention more everyday ones like "woman", "throat", and "jump". All in all, a naïve reader would easily get the impression that the computer was manipulating the same repertoire of concepts as is in the mind of a best-selling human author such as Jacqueline Susann.

However, in truth, no program in the world understands even *one concept* at the same level of complexity as an ordinary person does. This does not mean that computers do not excel at certain tasks, such as spell-checking and chess-playing, where manipulation of genuine concepts is not required. However, understanding how people feel, how they interact, how they talk, what motivates them, and so on — these are the kinds of things that are completely out of reach of today's technology and only at the fringes of today's theories and speculations in cognitive science.

Although the article implied that the work was collaborative in somewhat the same way as two humans collaborate, there can be no doubt that this collaboration was totally asymmetric, with French's understanding of the physical world and the human world being the basis for all the real decisions, and the computer merely providing a set of constraints and suggestions of possible pathways to follow.

It may have seemed to some readers of the article that French chose to mimic the potboiler style because there was some hope of getting a computer to succeed in that limited task whereas writing a truly great novel would have been out of the question. But this is a total deception. The mind of even the trashiest of human authors is filled to the brim with life experiences adding up to the most unanalyzable depths of mental complexity and subtlety. The difference between Vikram Seth, say, and Jacqueline Susann is microscopic in comparison with the difference between the mind of Jacqueline Susann and the micromind of any computer program in existence today.

It is mildly amusing that someone has played the game of computers and prose enough to have reached the point where a publisher has decided to risk some money in an elaborate publicity stunt, but it is hardly front-page news meriting the same level of respect as the proof of Fermat's Last Theorem. Indeed, it is an outrage, in my opinion, that these two topics were given similar treatments by the *Times*. French's work is not "the cutting edge of literary artificial intelligence", as it was touted in the article, but the cutting edge of empty syntactic play.

We have a long way to go in the scientific study of how human minds work before we will come up with a computer that can produce even a single good joke, let alone a novel.

Sincerely,
Douglas R. Hofstadter

Can Computers Understand Metaphors by Shakespeare and Plato?

Scott French's "computer novel" is an easy target, in a way, in the sense that even most outsiders will probably view claims of computer novel-writing with some skepticism. The hard targets are the cases where even cognitive-science professionals seem unable or unwilling to distinguish between what some program has done and what people do, provided there is some minimal degree of surface-level resemblance. Claims that a program has read and understood a newspaper article about various economic experts' prognoses about interest rates (Riesbeck & Schank, 1989), for instance, or that a program has rediscovered Kepler's laws of planetary motion (Langley *et al.*, 1987), or that a program has come up with a new type of play in football (Riesbeck & Schank, 1989), typify what I am talking about. When one really looks closely at what has been done, the achievement typically shrinks and shrinks until one sees that there was very little knowledge of the real-world concepts purported to have been manipulated, and that moreover, exactly the proper concepts (in an unimaginably diluted form) were supplied, and few others.

The distinguished British philosopher Margaret Boden, in her provocative and largely praiseworthy book *The Creative Mind: Myths and Mechanisms* (Boden, 1991) falls prey to the Eliza effect on several occasions, unfortunately marring her book's accuracy and muddying the waters that she is working so hard to clarify. In one chapter on artistic and literary achievements by computers, for instance, she attributes almost uncanny powers of understanding to what is actually a quite simple program called "ACME", developed by psychologist Keith Holyoak and philosopher Paul Thagard (Holyoak & Thagard, 1989, also described and discussed in Chapter 6 of this book). To set the stage for this drama, Boden begins by quoting these poetic lines from *Macbeth*:

> Sleep that knits up the ravelled sleeve of care,
> The death of each day's life, sore labour's bath,
> Balm of hurt minds, great nature's second course,
> Chief nourisher in life's feast.

Then she goes on to describe a complex analogy in Plato, in which Socrates describes himself as a "midwife of ideas". Here is how she puts it:

> Socrates. . . . is too old to have new philosophical ideas, but he can help his pupils to have them. He can ease Theaetetus' labor-pains (his obsession with a seemingly insoluble philosophical problem). He can encourage the birth of true ideas, and cause the false to miscarry. He can even matchmake successfully, introducing non-pregnant (non-puzzled) youngsters to wise adults who will start them thinking. His skill, he says, is greater than the real midwife's, since distinguishing truth from nonsense in philosophy is more difficult than telling whether a newborn baby is viable.

As anyone can see, this is an exceedingly rich brew of intertwined ideas, the understanding of which would require knowledge of the aging process, the goals and nature of philosophizing, many facts about human reproduction and mating customs, the profession of teaching, the contrast between youth and wisdom, and on and on. Indeed, it is quite clear that Boden's aim is precisely to impress her readers with the depth and complexity of this literary metaphor. Having done so, she now launches into the drama:

> Sleep and knitting, philosophy of midwifery. Could computers even appreciate such verbal fancies, never mind come up with them?
>
> Well, they already have. A computer model of analogical thinking called "ACME" ("M" is for Mapping) has interpreted Socrates' remark appropriately, showing *just how* a midwife's role in aiding the birth of a new baby resembles a philosopher's elicitation of ideas in his pupil's mind — and how it does not. . . .
>
> So far as I know, this program has not been tried out on Macbeth's speech about sleep. But I'd be prepared to bet that it could make something of it. For ACME uses highly abstract procedures for recognizing (and assessing the strength of) analogies *in general.*

The italics are Boden's.

It is instructive to supplement this glowing characterization of ACME with a display showing exactly how ACME's input looked (taken from Holyoak & Thagard, 1989). Here is the sum total of what ACME "knew" about midwifery (all input was provided in predicate-logic notation, with predicates such as "midwife" on the left and their subjects and objects coming thereafter):

m1: (midwife (obj–midwife))
m2: (mother (obj–mother))
m3: (father (obj–father))
m4: (child (obj–child))
m5: (matches (obj–midwife obj–mother obj–father))
m6: (conceives (obj–mother obj–child))
m7: (cause (m5 m6))
m8: (in-labor-with (obj–mother obj–child))
m10: (helps (obj–midwife obj–mother))
m11: (give-birth-to (obj–mother obj–child))
m12: (cause (m10 m11))

Line m2, for instance, is supposed to encode the statement "obj–mother is a mother", and line m6 the statement "obj–mother conceives obj–child". Terms like "obj–midwife", "obj–child", and so on are simply dummy names, something like algebraic variables that have no specific numerical value. The unpronounceable prefix "obj" presumably stands for "object", and was used simply to distinguish nouns from predicates.

And now, here is what ACME was given about Socrates (it is not, as Boden describes it, "Socrates' remark", no matter how you slice it):

s1: (philosopher (Socrates))
s2: (student (obj–student))
s3: (intellectual-partner (obj–partner))
s4: (idea (obj–idea))
s5: (introduce (Socrates obj–student obj–partner))
s6: (formulates (obj–student obj–idea))
s7: (cause (s5 s6))
s8: (thinks-about (obj–student obj–idea))
s9: (tests-truth (obj–student obj–idea))
s10: (helps (Socrates obj–student))
s11: (knows-truth-or-falsity (obj–student obj–idea))
s12: (cause (s10 s11))

This looks pretty sparse, but unfortunately, it is much more — or rather, much less — than just sparse. One has to bear in mind that the English words are completely vacuous. Not only is there an utter lack of imagery behind them, but there is not even any kind of attempt at a dictionary definition. (In her book, Boden says there is, but she was confusing ACME with ARCS, another program by the same researchers.) A compound word like "knows-truth-or-falsity", so transparently evocative to readers of English, might as well be, for all the computer could care, "xjs-beuglh?" or "doesn't-give-a-damn-about" or the digit "8" or any other alphanumeric string.

The computer's task was to figure out how well the set of lines m1 through m12, considered as meaningless formal patterns, corresponded to the set of lines s1 through s12. Note that there is a discrepancy in number of lines, as there is no line m9 to map onto line s9. This discrepancy was the only thing that kept the two sets of lines, considered purely as meaningless patterns, from being absolutely interchangeable with each other. In seeking a mapping between them, the computer paid no attention to any *ideas* encoded by these lines, because to it, there were none. It dealt just with meaningless patterns.

To make this absolutely crystal-clear, let us take lines m1 through m12 and replace the English words in them by capital letters, an act to which the program is completely insensitive. Of course, we have to do this replacement systematically. Thus if line m1 becomes, say, "(A (B))", then wherever "obj–midwife" formerly occurred, we will of course have to substitute "B". (Note that "A" never occurs anywhere again, nor does it occur anywhere else in the program's memory.) So here's how the first "situation", the one supposedly defining the "midwife" notion, looks (with line numbers left out):

(A (B)), (C (D)), (E (F)), (G (H)), (I (B D F)), (J (D H)), (K (L M)), (N (D H)), (P (B D)), (Q (D H)), (K (R S))

Now what about the second "situation", the one pertaining to Socrates? We'll again replace words by letters, but this time — just for fun — we'll use lowercase letters. Here we go!

(a (b)), (c (d)), (e (f)), (g (h)), (i (b d f)), (j (d h)), (k (l m)), (n (d h)), (o (d h)), (p (b d)), (q (d h)), (k (r s))

By creating a huge horde of mappings (in fact, every possible mapping, under certain simple constraints), including the mapping that equates "A" with "a", "B" with "b", and so on, and letting them all vie for dominance, ACME would eventually spot the structural resemblance between these two displays, and this act of letter–letter alignment would constitute the supposed "understanding", by a computer, of a subtle literary metaphor about Socrates and midwifery. Such string–string alignments discovered by brute-force methods are the "highly abstract procedures" that Boden touts as being able to deal with "analogies in general".

Here is another set of assertions — I made it up myself — that is isomorphic to the Socrates set of assertions. (I even retained the words "cause" and "helps", just to make the isomorphism a bit stronger.)

q1:   (neglectful-husband (Sluggo))
q2:   (lonely-and-sex-starved-wife (Jane-Doe))
q3:   (macho-ladykiller (Buck-Stag))
q4:   (poor-innocent-little-fetus (Bambi))
q5:   (takes-out-to-local-bar (Sluggo Jane-Doe Buck-Stag))
q6:   (somehow-or-other-conceives (Jane-Doe Bambi))
q7:   (cause (q5 q6))
q8:   (unwillingly-gives-birth-to (Jane-Doe Bambi))
q9:   (wraps-in-burlap-sack-and-throws-off-high-bridge (Jane-Doe Bambi))
q10: (helps (Sluggo Jane-Doe))
q11: (neatly-solves-the-problem-of (Jane-Doe Bambi))
q12: (cause (q10 q11))

Here the computer learns about a fellow named Sluggo taking his wife Jane and his good buddy Buck to a bar, where things take their natural course and Jane winds up pregnant by Buck. She has the baby but doesn't want it, and so, aided by her husband, she drowns the baby in a river, thus "neatly solving the problem" of Bambi.

Because the two situations are deeply similar — or, perhaps, because the sets of expressions chosen to encode them are totally isomorphic — ACME would manage to ferret out the compelling literary analogy between Socrates and Sluggo, thereby demonstrating in yet another way its prowess in understanding our complex world.

The Dubious Claim of Cross-domain Analogy-making

Boden is not the only one to crow about the real-world analogy-making prowess of ACME. As might be expected, its creators give a rundown of its many successes in their article. Among the complex real-world analogies they cite it as having made are a political one (involving alleged terrorists in Nicaragua, Hungary, and Israel), numerous scientific analogies (including the water-flow and heat-flow case), a number of "jealous animal stories", and some that link vastly different domains of knowledge. In fact, giving their program the ability to make *cross-domain analogies* is clearly one of the achievements of which Holyoak and Thagard are most proud. They devote a full paragraph to a critique of programs that can only do "intradomain analogies", which they characterize as "virtually trivial". (In passing, they say that Copycat — described in detail in this book's Chapter 5 — does "intradomain correspondences".) Here is how their critique of other projects concludes:

> If analogy is restricted to a single domain. . . . mapping becomes a very simple special case of the more complex processes used by ACME and SME for cross-domain analogies.

But when Holyoak and Thagard's situations are stripped bare of the English words and phrases that make them seem so pregnant with meaning, mapping "(A (B))" onto "(a (b))" hardly seems like the kind of deep and intellectually impressive leap as is conjured up by a grand term like "cross-domain analogy-making".

Such close scrutiny, carried out regrettably seldom by AI researchers, forces one to confront in great depth questions about where and when meaning is present — questions about how and when meanings are truly carried by symbols. For instance, is it accurate to say that ACME constructs "scientific analogies", "political analogies", or "analogies between fables"? What would make an analogy really be *about* science, as opposed to, say, politics? Would it not have to be the content of the words — the meanings that they carry? If the words are as empty as isolated letters, how can the analogy be "about" anything?

In one "political" analogy problem given to ACME, the typographical expression "(aim-to-overthrow (Contras Sandinistas))" was used. Referring to this expression, Holyoak and Thagard write:

> The "Contras" structure contains only the minimal information that the Contras aim to overthrow the government of Nicaragua, leaving open the question of whether the U.S. should support them, and whether they should be viewed as terrorists or freedom fighters.

But when one keeps in mind that the alphanumeric string "aim-to-overthrow" might just as well have been "holds-one-teaspoonful" or "bling-blang-blotch", it seems dubious that the above-quoted expression contains information that the Contras, or anyone else, ever aimed to do anything at all. In fact, even the use of the word "information" seems an exaggeration. The program has been informed of nothing — it has merely been handed a string of letters and punctuation marks. As a result of this, no ideas will be created, no knowledge will be consulted, no imagery will be formed. Yet simply because highly evocative English words are embedded in the string, it is hard to resist this easy slide down a very slippery epistemological slope.

Holyoak and Thagard were eager to demonstrate the psychological realism of their product, so they carried out a number of experiments on it designed to bring out its humanlike ability to be thrown off by irrelevancies. In one experiment, for instance, they modified the input to the midwifery metaphor, throwing in a dozen or so "decoy" expressions, including these three:

(drink (Socrates   obj–hemlock))
(matches (obj–soc-midwife   obj–soc-wife   Socrates))
(give-birth-to (obj–soc-wife   obj–soc-child))

They then assert without any qualification that this set of expressions "contains the information that Socrates drinks hemlock juice.... the information that Socrates himself was matched to his wife by a midwife; and that Socrates' wife had a child with the help of this midwife." But given that their program doesn't know one whit about Socrates or marriage or children or drinking, how could it conceivably been given all that "information"?

They go on to say that in this decoy-filled run, "Socrates himself maps to the father." What in the world can they mean? Undeniably, the alphanumeric string "Socrates" is brought into association with the alphanumeric string "obj–father" by the program, but how does this mapping of strings have anything to do with Socrates himself? Of course, this complaint might be seen by many as mere nitpicking. Obviously — so the argument would run — Holyoak and Thagard were simply avoiding a whole series of complex and awkward turns of phrase by speaking *as if* the terms in the expressions handed to ACME really referred to real-world entities and relationships — and surely this is a useful and harmless move to make. Surely nobody would misunderstand it.

Ah, but there's the rub. People, even sophisticates like Boden and Waldrop, make misunderstandings of this sort all the time. Do even Holyoak and Thagard themselves understand exactly how empty their symbols are? Let us assume they do. But in that case, what I don't understand at all is how they could feel comfortable in claiming that ACME carries out "cross-domain analogy-making", when in fact ACME deals with no domains at all — just strings.

Copycat: A Kid Doing a Somersault

The operational term here is, I suppose, "hype" — and yet it is, as I say, inadvertent. Clearly, *all* AI researchers, myself included, want to brag about their programs' achievements; on the other hand, we all know that we can't get away with out-and-out anthropomorphism. What generally results is some kind of intermediate level of description, in which a bit of caution is used but much is left ambiguous, so that readers are still free to draw conclusions that often will amount to some kind of Eliza effect — benefiting the researchers, needless to say. It may well be that in this book, precisely this kind of thing takes place in our discussions of our own work, but there is one difference: our domains are deliberately so stripped-down that the claims made *cannot* be very grandiose.

In fact, that's exactly the problem. It's what was bothering us back in 1989, and is still bothering us today — perhaps even more so. Whereas many research groups appear to be tackling domains of such complexity that even human experts are cowed — thermodynamics, international terrorism, atomic physics, computer-system configuration, VLSI chip design, economic forecasting, and on and on — we in our group are dealing with such microscopic domains that our programs' achievements appear to be nearly trivial. When there's a little kid trying somersaults out for the first time next to a flashy gymnast doing flawless flips on a balance beam, who's going to pay any attention to the kid? Our projects in microdomains come across, at least on the surface, as the equivalent of the little kid doing a somersault. Even the name "Copycat" was deliberately chosen to downplay the program's level of expertise — to empha-size its childlikeness.

When highly respectable magazines and newspapers and professional journals and books give complete credence to the claims that scientific discovery is being modeled, that metaphorical language is being routinely handled, that deep analogies are being made in — or between! — highly sophisticated domains, that cooking and coaching and designing and daydreaming and economics and engineering and so forth and so on all fall comfortably within the capabilities of today's artificial-intelligence programs, why should anyone ever pay any attention to a program that is trying to figure out connections between little strings of letters like *abc* and *xyz*?

Thus we were discouraged. We were worried that it would be hard to convince our colleagues, not to mention the wider public, that what we were doing had any merit. This worry eventually catalyzed a desire to write something about all this, and what resulted was the following article (actually, what was originally published was a somewhat longer version that includes a few pages of discussion of Copycat, but that section was mostly cut out here in order to avoid overlap with Chapter 5). In some ways, this article is a fundamental statement of FARG's philosophy, but perhaps calling it a "manifesto" would go a little far. Nonetheless, it is a kind of rallying cry for our style of work.

My co-authors — Dave Chalmers and Bob French, who at the time were graduate students working with me, and have since both gotten their Ph.D.'s and done excellent work bringing them considerable recognition — had both studied philosophy as well as cognitive science, and thus they brought a certain philosophical perspective and verbal style to our joint paper. Both had been influenced by Immanuel Kant's ideas, and I know that Bob had at one time been particularly struck — I'm not sure whether positively or negatively — by Kant's pompously-titled "Prolegomena to Any Future Metaphysics".

Chapter 4

High-level Perception, Representation, and Analogy:

A Critique of Artificial-intelligence Methodology

DAVID CHALMERS, ROBERT FRENCH,
and DOUGLAS HOFSTADTER

The Problem of Perception

One of the deepest problems in cognitive science is that of understanding how people make sense of the vast amount of raw data constantly bombarding them from their environment. The essence of human perception lies in the ability of the mind to hew order from this chaos, whether this means simply detecting movement in the visual field, recognizing sadness in a tone of voice, perceiving a threat on a chessboard, or coming to understand the Iran–Contra affair in terms of Watergate.

It has long been recognized that perception goes on at many levels. Immanuel Kant divided the perceptual work of the mind into two parts: the faculty of Sensibility, whose job it is to pick up raw sensory information, and the faculty of Understanding, which is devoted to organizing these data into a coherent, meaningful experience of the world. Kant found the faculty of Sensibility rather uninteresting, but he devoted much effort to the faculty of Understanding. He went so far as to propose a detailed model of the higher-level perceptual processes involved, dividing the faculty into twelve Categories of Understanding.

Today Kant's model seems somewhat baroque, but his fundamental insight remains valid. Perceptual processes form a spectrum, which for convenience we can divide into two components. Corresponding roughly to Kant's faculty of Sensibility, we have low-level perception, which involves the early processing of information from the various sensory modalities. High-level perception, on the other hand, involves taking a more global view of this information, extracting *meaning* from the raw material by accessing concepts, and making sense of situations at a conceptual level. This ranges from the recognition of objects to the grasping of abstract relations, and on to the understanding of entire situations as coherent wholes.

Low-level perception is far from uninteresting, but it is high-level perception that is most relevant to the central problems of cognition. The study of high-level perception leads us directly to the problem of mental *representation*. Representations are the fruits of perception. In order for raw data to be shaped into a coherent whole, they must go through a process of filtering and organization, yielding a structured representation that can be used by the mind for any number of purposes. A primary question about representations, currently the subject of much debate, concerns their precise structure. Of equal importance is the question of how these representations might be *formed* in the first place, via a process of perception, starting from raw data. The process of representation-formation raises many important questions: How are representations influenced by context? How can our perceptions of a situation radically reshape themselves when necessary? Where in the process of perception are concepts accessed? Where does meaning enter, and where and how does understanding emerge?

The main thesis of this paper is that high-level perception is deeply interwoven with other cognitive processes, and that researchers in artificial intelligence must therefore integrate perceptual processing into their modeling of cognition. Much work in artificial intelligence has attempted to model conceptual processes independently of perceptual processes, but we will argue that this approach cannot lead to a satisfactory understanding of the human mind. In support of this claim, we will examine some existing models of scientific discovery and analogical thought, and will argue that the exclusion of perceptual processes from these models leads to serious limitations. The intimate link between analogical thought and high-level perception will be investigated in detail, and points the way to alternative architectures (which are discussed in the next several chapters of this book).

Low-level and high-level perception

The lowest level of perception occurs with the reception of raw sensory information by various sense organs. Light impinges on the retina, sound waves cause the eardrum to vibrate, and so on. Other processes further along the information-processing chain may also be usefully designated as low-level. In the case of vision, for instance, after information has passed up the optic nerve, much basic processing occurs in the lateral geniculate nuclei and the primary visual cortex, as well as the superior colliculus. Included here is the processing of brightness contrasts, of light boundaries, and of edges and corners in the visual field, and perhaps also location processing.

Low-level perception is given short shrift in this paper, as it is quite removed from the more cognitive questions of representation and meaning. Nonetheless, it is an important subject of study, and a complete theory of perception will necessarily include low-level perception as a fundamental component.

The transition from low-level to high-level perception is of course quite blurry, but we may delineate it roughly as follows. High-level perception begins at that level of processing where concepts begin to play an important role. Processes of high-level perception may be subdivided again into a spectrum from the concrete to the abstract. At the most concrete end of the spectrum, we have *object recognition,* exemplified by the ability to recognize an apple on a table, or to pick out a farmer in a wheat field. Then there is the ability to grasp *relations.* This allows us to determine the relationship between a blimp and the ground ("above"), or a swimmer and a swimming pool ("in"). As one moves further up the spectrum towards more abstract relations ("George Bush is *in* the Republican Party"), the issues become distant from particular sensory modalities. The most abstract kind of perception is the processing of entire complex *situations,* such as a love affair or a war.

One of the most important properties of high-level perception is that it is extremely flexible. A given set of input data may be perceived in a number of different ways, depending on the context and the state of the perceiver. Due to this flexibility, it is a mistake to regard perception as a process that associates a fixed representation with a particular situation. Both contextual factors and top-down cognitive influences make the process far less rigid than this. Some of the sources of this flexibility in perception are as follows.

Perception may be influenced by belief. Numerous experiments by the "New Look" theorists in psychology in the 1950's (*e.g.,* Bruner, 1957) showed that our expectations play an important role in determining what we perceive even at quite a low level. At a higher level, that of complete situations, such influence is ubiquitous. Take, for instance, the situation in which a husband walks in to find his wife sitting on the couch with a male stranger. If he has a prior belief that his wife has been unfaithful, he is likely to perceive the situation one way; if he believes that an insurance salesman was due to visit that day, he will probably perceive the situation quite differently.

Perception may be influenced by goals. If we are trying to hike on a trail, we are likely to perceive a fallen log as an obstacle to be avoided. If we are trying to build a fire, we may perceive the same log as useful fuel for the fire. Another example: reading a given text may yield very different perceptions, depending on whether we are reading it for content or proofreading it.

Perception may be influenced by external context. Even in relatively low-level perception, it is well known that the surrounding context can significantly affect our perception of visual images. For example, an ambiguous figure halfway between an "A" and an "H" is perceived one way in the context of "C__T", and another in the context of "T__E". At a higher level, if we encounter somebody dressed in tuxedo and bowtie, our perception of them may differ depending on whether we encounter them at a formal ball or at the beach.

Perceptions of a situation can be radically reshaped where necessary. In Maier's well-known two-string experiment (Maier, 1931), subjects are provided with a chair and a pair of pliers, and are told to tie together two strings hanging from the ceiling. The two strings are too far apart to be grasped simultaneously. Subjects have great difficulty initially, but after a number of minutes some of them hit upon the solution of tying the pliers to one of the strings, and swinging the string like a pendulum. Initially, the subjects perceive the pliers first and foremost as a special tool; if the weight of the pliers is perceived at all, it is very much in the background. To solve this problem, subjects have to radically alter the emphasis of their perception of the pair of pliers. Its function as a tool is set aside, and its weightiness is brought into the foreground as the key feature in this situation.

The distinguishing mark of high-level perception is that it is semantic: it involves drawing *meaning* out of situations. The more semantic the processing involved, the greater the role played by *concepts* in this processing, and thus the greater the scope for top-down influences. The most abstract of all types of perception, the understanding of complete situations, is also the most flexible.

Recently both Pylyshyn (1980) and Fodor (1983) have argued against the existence of top-down influences in perception, claiming that perceptual processes are "cognitively impenetrable" or "informationally encapsulated". These arguments are highly controversial, but in any case they apply mostly to relatively low-level sensory perception. Few would dispute that at the higher, conceptual level of perception, top-down and contextual influences play a large role.

Artificial Intelligence and the Problem of Representation

The end product of the process of perception, when a set of raw data has been organized into a coherent and structured whole, is a *representation*. Representations have been the object of much study and debate within the field of artificial intelligence, and much is made of the "representation problem". This problem has traditionally been phrased as "What is the correct structure for mental representations?", and many possibilities have been suggested, ranging from predicate calculus through frames and scripts to semantic networks and more. We may divide representations into two kinds: long-term knowledge representations that are stored passively somewhere in the system, and short-term representations that are active at a given moment in a particular mental or computational process. (This distinction corresponds to the distinction between long-term memory and working memory.) In this discussion, we will mostly be concerned with short-term, active representations, as it is these that are the direct product of perception.

The question of the structure of representations is certainly an important one, but there is another, related problem that has not received nearly as much attention. This is that of understanding how such a representation could be arrived at, starting from environmental data. Even if it were possible to discover an optimal type of representational structure, this would leave unresolved two important problems, namely:

> *The problem of relevance:* How is it decided which subsets of the vast amounts of data from the environment get used in various parts of the representational structure? Naturally, much of the information content at the lowest level will be quite irrelevant at the highest representational level. To determine which parts of the data are relevant to a given representation, a complex filtering process is required.

> *The problem of organization:* How are these data put into the correct form for the representation? Even if we have determined precisely which data are relevant, and we have determined the desired framework for the representation — a frame-based representation, for instance — we still face the problem of organizing the data into the representational form in a useful way. The data do not come pre-packaged as slots and fillers, and organizing them into a coherent structure is likely to be a highly nontrivial task.

These questions, taken together, amount in essence to the problem of high-level perception, translated into the framework of artificial intelligence.

The traditional approach in artificial intelligence has been to start by selecting not only a preferred type of high-level representational structure, but also the data assumed to be relevant to the problem at hand. These data are organized by a human programmer who appropriately fits them into the chosen representational structure. Usually, researchers use their prior knowledge of the nature of the problem to hand-code a representation of the data into a near-optimal form. Only after all this hand-coding is completed is the representation allowed to be manipulated by the machine. The problem of representation-formation, and thus the problem of high-level perception, is ignored.

(These comments do not, of course, apply to work in machine vision, speech processing, and other perceptual endeavors. However, work in these fields usually stops short of modeling processes at the conceptual level and is thus not directly relevant to our critique of high-level cognitive modeling.)

The formation of appropriate representations lies at the heart of human high-level cognitive abilities. It might even be said that the problem of high-level perception forms the central task facing the artificial-intelligence community: the task of understanding how to draw *meaning* out of the world. It might not be stretching the point to say that there is a "meaning barrier", which has rarely been crossed by work in AI. On one side of the barrier, some models in low-level perception have been capable of building primitive representations of the environment, but these are not yet sufficiently complex to be called "meaningful". On the other side of the barrier, much research in high-level cognitive modeling has *started* with representations at the conceptual level, such as propositions in predicate logic or nodes in a semantic network, where any meaning that is present is already built in. There has been very little work that bridges the gap between the two.

Objectivism and traditional AI

Once AI takes the problem of representation-formation seriously, the next stage will be to deal with the evident flexibility of human high-level perceptual processes. As we have seen, objects and situations can be apprehended in many different ways, depending on context and top-down influences. We must find a way of ensuring that AI representations have a corresponding degree of flexibility. William James, in the late nineteenth century, recognized this aspect of cognitive representations (James, 1890, pages 222–224):

> There is no property ABSOLUTELY essential to one thing. The same property which figures as the essence of a thing on one occasion becomes a very inessential feature upon another. Now that I am writing, it is essential that I conceive my paper as a surface for inscription.... But if I wished to light a fire, and no other materials were by, the essential way of conceiving the paper would be as a combustible material.... The essence of a thing is that one of its properties which is so *important for my interests* that in comparison with it I may neglect the rest.... The properties which are important vary from man to man and from hour to hour.... many objects of daily use — as paper, ink, butter, overcoat — have properties of such constant unwavering importance, and have such stereotyped names, that we end by believing that to conceive them in those ways is to conceive them in the only true way. Those are no truer ways of conceiving them than any others; there are only more frequently serviceable ways to us.

James is saying, effectively, that we have different representations of an object or situation at different times. The representational process adapts to fit the pressures of a given context.

Despite the work of philosopher–psychologists such as James, the early days of artificial intelligence were characterized by an objectivist view of perception, and of the representation of objects, situations, and categories. As the linguist George Lakoff has characterized it, "On the objectivist view, reality comes complete with a unique correct, complete structure in terms of entities, properties and relations. This structure exists, independent of any human understanding." (Lakoff, 1987, page 159) While this objectivist position has been unfashionable for decades in philosophical circles (especially after Wittgenstein's work demonstrating the inappropriateness of a rigid correspondence between language and reality), most early work in AI implicitly accepted this set of assumptions.

The Physical Symbol System Hypothesis (Newell & Simon, 1976), upon which most of the traditional AI enterprise has been built, posits that thinking occurs through the manipulation of symbolic representations, which are composed of atomic symbolic primitives. Such symbolic representations are by their nature somewhat rigid, black-and-white entities, and it is difficult for their representational content to shift subtly in response to changes in context. The result, in practice — irrespective of whether this was intended by the original proponents of this framework — is a structuring of reality that tends to be as fixed and absolute as that of the objectivist position outlined above.

By the mid-seventies, a small number of AI researchers began to argue that in order to progress, the field would have to part ways with its commitment to such a rigid representational framework. One of the strongest early proponents of this view was David Marr, who noted (Marr, 1977, page 44) that

> the perception of an event or object must include the simultaneous computation of several different descriptions of it, that capture diverse aspects of the use, purpose or circumstances of the event or object.

Recently, significant steps have been taken toward representational flexibility with the advent of sophisticated connectionist models whose distributed representations are highly context-dependent (Rumelhart & McClelland, 1986). In these models, there are no representational primitives in internal processing. Instead, each representation is a vector in a multidimensional space, whose position is not anchored but can adjust flexibly to changes in environmental stimuli. Consequently, members of a category are not all represented by identical symbolic structures; rather, individual objects will be represented in subtly different ways depending upon the context in which they are presented. In networks with recurrent connections (Elman, 1990), representations are even sensitive to the current internal state of the model. Other recent work taking a flexible approach to representation includes the classifier-system models of Holland and his colleagues (Holland *et al.*, 1986), where genetically-inspired methods are used to create a set of "classifiers" that can respond to diverse aspects of various situations.

In these models, a flexible perceptual process has been integrated with an equally flexible dependence of action upon representational content, yielding models that respond to diverse situations with a robustness that is difficult to match with traditional methods. Nonetheless, the models are still somewhat primitive, and the representations they develop are not nearly as complex as the hand-coded, hierarchically-structured representations found in traditional models; still, it seems to be a step in the right direction. It remains to be seen whether work in more traditional AI paradigms will respond to this challenge by moving toward more flexible and robust representational forms.

On the possibility of a representation module

It might be granted that given the difficulty of the problem of high-level perception, AI researchers could be forgiven for starting with their representations in a made-to-order form. They might plausibly claim that the difficult problem of representation-formation is better left until later. But it must be realized that behind this approach lies a tacit assumption: that it is possible to model high-level cognitive processes independently of perceptual processes. Under this assumption, the representations that are currently, for the most part, tailored by human hands would eventually be built up by a separate lower-level facility — a "representation module" whose job it would be to funnel data into representations. Such a module would act as a "front end" to the models of the cognitive processes currently being studied, supplying them with the appropriately-tailored representations.

We are deeply skeptical, however, about the feasibility of such a separation of perception from the rest of cognition. A representation module that, given any situation, produced the single "correct" representation for it would have great difficulty emulating the flexibility that characterizes human perception. For such flexibility to arise, the representational processes would have to be sensitive to the needs of all the various cognitive processes in which they might be used. It seems most unlikely that a single representation would suffice for all purposes. As we have seen, for the accurate modeling of cognition it is necessary that the representation of a given situation be able to vary with various contextual and top-down influences. This, however, is directly contrary to the "representation module" philosophy, wherein representations are produced quite separately from later cognitive processes, and then supplied to a "task-processing" module.

To separate representation-building from higher-level cognitive tasks is, we believe, impossible. In order to provide the kind of flexibility that is apparent in cognition, any fully cognitive model will probably require a continual inter-action between the process of representation-building and the manipulation of those representations. If this proves to be the case, then the current approach of using hand-coded representations not only is postponing an important issue but will, in the long run, lead up a dead-end street.

We will consider this issue in greater depth when we discuss current research in the modeling of analogical thought. For now, we will discuss in some detail one well-known AI program for which great claims have been made. We argue that these claims represent a lack of appreciation of the importance of high-level perception.

BACON: A case study

A particularly clear case of a program in which the problem of repre-sentation is bypassed is BACON, a well-known program that has been advertised as an accurate model of scientific discovery (Langley *et al.,* 1987). The authors of BACON claim that their system is "capable of representing information at multiple levels of description, which enables it to discover complex laws involv-ing many terms". BACON was able to "discover", among other things, Boyle's law of ideal gases, Kepler's third law of planetary motion, Galileo's law of uniform acceleration, and Ohm's law of electrical resistance.

Such claims clearly demand close scrutiny. We will look in particular at the program's "discovery" of Kepler's third law of planetary motion. Upon examination, it seems that the success of the program relies almost entirely on its being given data that have already been represented in near-optimal form, using after-the-fact knowledge available to the programmers.

When BACON performed its derivation of Kepler's third law, the program was given only data about the planets' average distances from the sun and their periods. These are *precisely the data required to derive the law.* The program is certainly not "starting with essentially the same initial conditions as the human discoverers", as one of the authors of BACON has claimed (Simon, 1989, page 375). The authors' claim that BACON used "original data" certainly does not mean that it used *all* of the data available to Kepler at the time of his discovery, the vast majority of which were irrelevant, misleading, distracting, or even wrong.

This pre-selection of data may at first seem quite reasonable: after all, what could be more important to an astronomer–mathematician than planetary distances and periods? But here our after-the-fact knowledge is misleading us. Consider for a moment the times in which Kepler lived. It was the turn of the seventeenth century, and Copernicus' *De Revolutionibus Orbium Cœlestium* was still new and far from universally accepted. Further, at that time there was no notion of the forces that produced planetary motion; the sun, in particular, was known to produce light but was not thought to influence the motion of the planets. In that prescientific world, even the notion of using mathematical equations to express regularities in nature was rare. And Kepler believed — in fact, his early fame rested on the discovery of this surprising coincidence — that the planets' distances from the sun were dictated by the fact that the five regular polyhedra could be fit between the five "spheres" of planetary motion around the sun, a fact that constituted seductive but ultimately misleading data.

Within this context, it is hardly surprising that it took Kepler thirteen years to realize that conic sections and not Platonic solids, that algebra and not geometry, that ellipses and not Aristotelian "perfect" circles, that the planets' distances from the sun and not the polyhedra in which they fit, were the *relevant* factors in unlocking the regularities of planetary motion. In making his discoveries, Kepler had to reject a host of conceptual frameworks that might, for all he knew, have applied to planetary motion, such as religious symbolism, superstition, Christian cosmology, and teleology. In order to discover his laws, he had to make all of these creative leaps. BACON, of course, had to do nothing of the sort. The program was given precisely the set of variables it needed from the outset (even if the values of some of these variables were sometimes less than ideal), and was moreover supplied with precisely the right biases to induce the algebraic form of the laws, it being taken completely for granted that mathematical laws of a type now recognized by physicists as standard were the desired outcome.

It is difficult to believe that Kepler would have taken thirteen years to make his discovery if his working data had consisted entirely of a list where each entry said "Planet X: mean distance from sun y, period z". If he had further been told "Find a polynomial equation relating these entities", then it might have taken him a few hours.

Addressing the question of why Kepler took thirteen years to do what BACON managed within minutes, Langley *et al.* (1987) point to "sleeping time, and time for ordinary daily chores", and other factors such as the time taken in setting up experiments, and the slow hardware of the human nervous system (!). In an interesting juxtaposition to this, researchers in a recent study (Qin & Simon, 1990) found that university students starting with the data that BACON was given could make essentially the same "discoveries" within an hour-long period. Somewhat strangely, the authors (including one of the authors of BACON) take this finding to support the plausibility of BACON as an accurate model of scientific discovery. It seems more reasonable to regard it as a demonstration of the vast difference in difficulty between the task faced by BACON and that faced by Kepler, and thus as a *reductio ad absurdum* of the BACON methodology.

So many varieties of data were available to Kepler, and the available data had so many different ways of being interpreted, that it is difficult not to conclude that in presenting their program with data in such a neat form, the authors of BACON are inadvertently guilty of 20–20 hindsight. BACON, in short, works only in a world of hand-picked, prestructured data, a world completely devoid of the problems faced by Kepler or Galileo or Ohm when they made their original discoveries. Similar comments could be made about STAHL, GLAUBER, and other models of scientific discovery by the authors of BACON. In all of these models, the crucial role played by high-level perception in scientific discovery, through the filtering and organization of environmental stimuli, is ignored.

It is interesting to note that the notion of a "paradigm shift", which is central to much scientific discovery (Kuhn, 1970), is often regarded as the process of *viewing the world* in a radically different way. That is, scientists' frameworks for representing available world knowledge are broken down, and their high-level perceptual abilities are used to organize the available data quite differently, building a novel representation of the data. Such a new representation can be used to draw different and important conclusions in a way that was difficult or impossible with the old representation. In this model of scientific discovery, unlike the model presented in BACON, the process of high-level perception is central.

The case of BACON is by no means isolated — it is typical of much work in AI, which often fails to appreciate the importance of the representation-building stage. We will see this in more depth in the next section, in which we take a look at the modeling of analogy.

Models of Analogical Thought

Analogical thought is dependent on high-level perception in a very direct way. When people make analogies, they are perceiving some aspects of the structures of two situations — the *essences* of those situations, in some sense — as identical. These structures, of course, are a product of the process of high-level perception.

The quality of an analogy between two situations depends almost entirely on one's perception of the situations. If Ronald Reagan were to evaluate the validity of an analogy between the U.S. role in Nicaragua and the Soviet Union's role in Afghanistan, he would undoubtedly see it as a poor one. Others might consider the analogy excellent. The difference would come from different perceptions, and thus representations, of the situations themselves. Reagan's internal representation of the Nicaraguan situation is certainly quite different from Daniel Ortega's.

Analogical thought further provides one of the clearest illustrations of the flexible nature of our perceptual abilities. Making an analogy requires highlighting various different aspects of a situation, and the aspects that are highlighted are often not the most obvious features. The perception of a situation can change radically, depending on the analogy we are making.

Let us consider two analogies involving DNA. The first is an analogy between DNA and a zipper. When we are presented with this analogy, the image of DNA that comes to mind is that of two strands of paired nucleotides (which can come apart like a zipper for the purposes of replication). The second analogy involves comparing DNA to the source code (*i.e.,* non-executable high-level code) of a computer program. What comes to mind now is the fact that information in the DNA gets "compiled" (via processes of transcription and translation) into enzymes, which correspond to machine code (*i.e.,* executable code). In the latter analogy, the perception of DNA is radically different — it is represented essentially as an information-bearing entity whose physical aspects, so important to the first analogy, are of virtually no consequence.

In cases such as these, it seems that no single, rigid representation can capture what is going on in our heads. It is true that we probably have a single rich representation of DNA sitting passively in long-term memory. However, in the contexts of different analogical mappings, very different facets of this large representational structure are selected out as being relevant, by the pressures of the particular context. Irrespective of the *passive* content of the long-term representation of DNA, the *active* content that is processed at a given time is determined by a flexible representational process.

Furthermore, not only is analogy-making dependent on high-level perception, but the reverse holds true as well: perception is often dependent on analogy-making itself. The high-level perception of one situation in terms of another is ubiquitous in human thought. If we perceive Nicaragua as "another Vietnam", for example, the making of the analogy helps to flesh out our representation of Nicaragua. Analogical thought provides a powerful mechanism for the enrichment of a representation of a given situation. This is well understood by good educators and writers, who know that there is nothing like an analogy to provide a better mental picture of a given situation. Analogies affect our perception all the time: in a love affair, for instance, it is difficult to stop parallels with past romances from modulating one's perception of the current situation. In the large or the small, such analogical perception — the grasping of one situation in terms of another — is so common that we tend to forget that what is going on is, in fact, analogy. Analogy and perception are tightly bound together.

It is useful to divide analogical thought into two basic components. First, there is the process of *situation-perception,* which involves taking the data involved with a given situation, and filtering and organizing them in various ways to provide an appropriate representation for a given context. Second, there is the process of *mapping*. This involves taking the representations of two situations and finding appropriate correspondences between components of one representation with components of the other to produce the match-up that we call an analogy. It is by no means apparent that these processes are cleanly separable; they seem to interact in a deep way. Given the fact that perception underlies analogy, one might be tempted to divide the process of analogy-making sequentially: first situation-perception, then mapping. But we have seen that analogy also plays a large role in perception; thus mapping may be deeply involved in the situation-perception stage, and such a clean division of the processes involved could be misleading. Later, we will consider just how deeply intertwined these two processes are.

Both the situation-perception and mapping processes are essential to analogy-making, but of the two the former is more fundamental, for the simple reason that the mapping process requires representations to work on, and representations are the product of high-level perception. The perceptual processes that produce these representations may in turn deeply involve analogical mapping; but each mapping process requires a perceptual process to precede it, whereas it is not the case that each perceptual process necessarily depends upon mapping. Therefore the perceptual process is conceptually prior, although perception and mapping processes are often temporally interwoven. If the appropriate representations are already formed, the mapping process can often be quite straightforward. In our view, the most central and challenging part of analogy-making is the perceptual process: the shaping of situations into representations appropriate to a given context.

The mapping process, in contrast, is an important object of study especially because of the immediate and natural use it provides for the products of perception. Perception produces a particular structure for the representation of a situation, and the mapping process emphasizes certain aspects of this structure. Through the study of analogy-making, we obtain a direct window onto high-level perceptual processes. The study of which situations people view as analogous can tell us much about how people represent those situations. Along the same lines, the computational modeling of analogy provides an ideal testing-ground for theories of high-level perception. Considering all this, one can see that the investigation of analogical thought has a huge role to play in the understanding of high-level perception.

Current models of analogical thought

In light of these considerations, it is somewhat disheartening to note that almost all current work in the computational modeling of analogy bypasses the process of perception altogether. The dominant approach involves starting with fixed, preordained representations, and launching a mapping process to find appropriate correspondences between representations. The mapping process not only takes center stage; it is the only actor. Perceptual processes are simply ignored; the problem of representation-building is not even an issue. The tacit assumption of such research is that correct representations have (somehow) already been built.

Perhaps the best-known computational model of analogy-making is SME, the Structure Mapping Engine (Falkenhainer, Forbus, & Gentner, 1990), based upon the structure-mapping theory of Dedre Gentner (1983). We will examine this model within the context of our earlier remarks. Other models of analogy-making, such as those of Burstein (1986), Carbonell (1983), Holyoak & Thagard (1989), Kedar-Cabelli (1988a), and Winston (1982), while differing in many respects from the above work, all share the property that the problem of representation-building is bypassed.

Let us consider one of the standard examples from this research, in which the SME program is said to discover an analogy between an atom and the solar system. Here, the program is given representations of the two situations, as shown in Figure IV–1. Starting with these representations, SME examines many possible correspondences between elements of the first representation and elements of the second. These correspondences are evaluated according to how well they preserve the high-level structure apparent in the representations. The correspondence with the highest score is selected as the best analogical mapping between the two situations.

A brief examination of Figure IV–1 shows that the discovery of the similar structure in these representations is not a difficult task. The representations have been set up in such a way that the common structure is immediately apparent. Even for a computer program, the extraction of such common structure is relatively straightforward.

We are in broad sympathy with Gentner's notion that the mappings in an analogy should preserve high-level structure (although there is room to debate over the details of the mapping process). But when the program's discovery of the correspondences between the two situations is a direct result of its being explicitly given the appropriate structures to work with, its victory in finding the analogy becomes somewhat hollow. Since the representations are tailored (perhaps unconsciously) to the problem at hand, it is hardly surprising that the correct structural correspondences are not difficult to find. A few pieces of irrelevant information are sometimes thrown in as decoys, but this makes the task of the mapping process only slightly more complicated. The point is that if appropriate representations come presupplied, the hard part of the analogy-making task has already been accomplished.

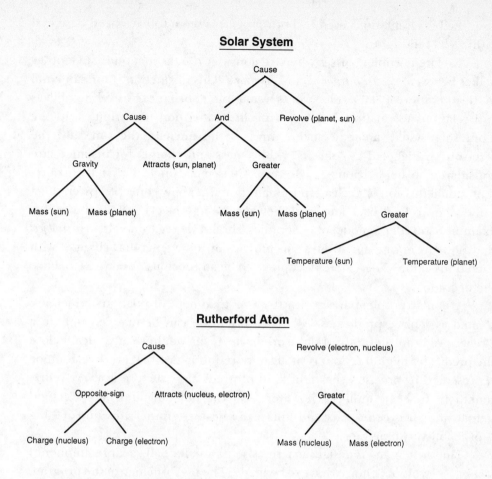

Figure IV–1. Predicate-calculus representations of two situations between which the SME program constructs a mapping.

Imagine what it would take to devise a representation of the solar system or an atom independent of any context provided by a particular problem. There are so many data available: one might, for instance, include information about the moons revolving around the planets, about the opposite electric charges on the proton and the electron, about relative velocities, about proximities to other bodies, about the number of moons, about the composition of the sun or the composition of the nucleus, about the fact that the planets lie in one plane and that each planet rotates on its axis, and so on. It comes as no surprise, in view of the analogy sought, that the only relations present in the representations that SME uses for these situations are the following: "attracts", "revolves around", "gravity", "opposite-sign", and "greater" (as well as the fundamental relation "cause"). These, for the most part, are precisely the relations that are relevant factors in this analogy. The criticisms of BACON discussed earlier apply here also: the representations used by both programs seem to have been designed with 20–20 hindsight.

A related problem arises when we consider the distinction that Gentner makes between *objects, attributes,* and *relations*. This distinction is fundamental to the operation of SME, which works by mapping objects exclusively to objects and relations to relations, while paying little attention to attributes. In the atom/solar-system analogy, such things as the nucleus, the sun, and the electrons are labeled as "objects", while mass and charge, for instance, are considered to be "attributes". However, it is most unclear that this representational division is so clean in human thought. Many concepts, psychologically, seem to float back and forth between being objects and attributes, for example. Consider a model of economics: should we regard wealth as an *object* that flows from one agent, or as an *attribute* of the agents that changes with each transaction? There does not appear to be any obvious *a priori* way to make the decision.

A similar problem arises with the SME treatment of relations, which are treated as *n*-place predicates. A 3-place predicate can be mapped only to a 3-place predicate, never to a 4-place predicate, no matter how semantically close the predicates might be. So it is vitally important for SME that every relation be represented by precisely the right kind of predicate structure in every representation. It seems unlikely, however, that the human mind makes a rigid demarcation between 3-place and 4-place predicates — rather, this kind of thing is probably very blurry.

Thus, when one is designing a representation for SME, a large number of somewhat arbitrary choices have to be made. The performance of the program is highly sensitive to each of these choices. In each of the published examples of analogies made by SME, these representations were designed in just the right way for the analogy to be made. It is difficult to avoid the conclusion that at least to a certain extent, the representations given to SME were constructed with those specific analogies in mind. This is again reminiscent of BACON.

In defense of SME, it must be said that there is much of interest about the mapping process itself; and unlike the creators of BACON, the creators of SME have made no great claims for their program's "insight". It seems a shame, however, that they have paid so little attention to the question of just how SME's representations could have been formed. Much of what is interesting in analogy-making involves extracting structural commonalities from two *situations,* finding some "essence" that both share. In SME, this problem of high-level perception is swept under the rug, by starting with preformed representations of the situations. The essence of the situations has been drawn out in advance in the formation of these representations, leaving only the relatively easy task of discovering the correct mapping. It is not that the work done by SME is necessarily *wrong*: it is simply not tackling what are, in our opinion, the really difficult issues in analogy-making.[1]

Such criticisms apply equally to most other work in the modeling of analogy. It is interesting to note that one of the earliest computational models of analogy, Evans' ANALOGY(Evans, 1968), attempted to build its own representations, even if it did so in a fairly rigid manner. Curiously, however, almost all major analogy-making programs since then have ignored the problem of representation-building. The work of Kedar-Cabelli (1988a) takes a limited step in this direction by employing a notion of "purpose" to direct the selection of relevant information, but still starts with all representations pre-built. Other researchers, such as Burstein (1986), Carbonell (1983), and Winston (1982), have models that differ in significant respects from the work outlined above, but none of these addresses the question of perception.

The ACME program of Holyoak and Thagard (1989) uses a kind of connectionist network to satisfy a set of "soft constraints" in the mapping process, thus determining the best analogical correspondences. Nevertheless, their approach seems to have remained immune to the connectionist notion of context-dependent, flexible representations. The representations used by ACME are preordained, frozen structures of predicate logic; the problem of high-level perception is bypassed. Despite the flexibility provided by a connectionist network, the program has no ability to change its representations under pressure. This constitutes a serious impediment to the attempts of Holyoak and Thagard to capture the flexibility of human analogical thought.

The necessity of fusing high-level perception with more abstract cognitive processing

The fact that most current work on analogical thought has ignored the problem of representation-formation is not necessarily a damning charge: researchers in the field might well defend themselves by saying that this process is far too difficult to study at the moment. In the meantime, they might argue, it is reasonable to assume that the work of high-level perception could be done by a separate "representation module", which takes raw situations and converts them into structured representations. Just how this module might work, they could say, is not their concern. Their research is restricted to the mapping process, which takes these representations as input. The problem of representation, they might claim, is a completely separate issue.

---

1. Disclaimer: Since this article was written, Ken Forbus, one of the authors of SME, has worked on modules that build representations in "qualitative physics". Some work has also been done on using these representations as input to SME. Despite this conceptual advance in the architecture, the two processes are still not intertwined and interdependent, as we claim they are in human thought.

This approach would be less ambitious than trying to model the entire perception/mapping cycle, but lack of ambition is certainly no reason to condemn a project *a priori*. In cognitive science and elsewhere, scientists usually study what seems within their grasp, leaving problems that seem too difficult for later. If this were all there was to the story, our previous remarks might be read as pointing out the limited scope of the present approaches to analogy, but at the same time applauding their success in making progress on a small part of the problem. There is, however, more to the story than this.

By ignoring the problem of perception in this fashion, artificial-intelligence researchers are making a deep implicit assumption — namely, that the processes of perception and of mapping are temporally separable. As we have already said, we believe that this assumption will not hold up. We see two compelling arguments against such a separation of perception from mapping. The first argument is simpler, but the second has a broader scope.

The first argument stems from the observation, made earlier, that much perception is dependent on processes of analogy. People are constantly inter-preting new situations in terms of old ones. Whenever they do this, they are using the analogical process to build up richer representations of various situations. When the controversial book *The Satanic Verses* was attacked by Iranian Moslems and its author threatened with death, most Americans were quick to condemn the actions of the Iranians. Interestingly, though, some senior figures in Christian churches in America had a quite different reaction. Seeing an analogy between this book and the controversial film *The Last Temptation of Christ,* which had been attacked in Christian circles as blasphe-mous, these figures were hesitant about condemning the Iranian action. Their perception of the situation was significantly altered by such a salient analogy.

Similarly, seeing Nicaragua as analogous to Vietnam might throw a par-ticular perspective on the situation there, while seeing the Nicaraguan rebels as "the moral equivalent of the Founding Fathers" (as Reagan once characterized them) is likely to give quite a different picture of the situation. Or consider rival analogies that might be used to explain the role of Saddam Hussein, the Iraqi leader who invaded Kuwait, to someone who knows little about the situation. If one were unsympathetic, one might describe him as analogous to Hitler, producing in the listener a perception of an evil, aggressive figure. On the other hand, if one were sympathetic, one might describe him as being like Robin Hood. This could produce in the listener a perception of a relatively generous figure, redistributing the superfluous wealth of the Kuwaitis to the rest of the Arab population.

Not only, then, is perception an integral part of analogy-making, but analogy-making is also an integral part of perception. From this, we conclude that it is impossible to split analogy-making into "first perception, then mapping". The mapping process will often be needed as an important part of the process of perception. The only solution is to give up on any clean temporal division between the two processes, and instead to recognize that they interact deeply.

The modular approach to the modeling of analogy stems, we believe, from a perception of analogical thought as something quite separate from the rest of cognition. One gets the impression from the work of most researchers that analogy-making is conceived of as a special tool in reasoning or problem-solving, a heavy weapon wheeled out now and then to deal with especially tough problems. Our view, by contrast, is that analogy-making is going on constantly in the background of the mind, helping to shape our perceptions of everyday situations. In our view, analogy is not separate from perception: analogy-making itself is a perceptual process.

For the time being, however, suppose we accept this view of mapping as a "task" in which representations, the products of the perceptual process, are used. Even in this view, the temporal separation of perception from mapping is, we believe, a misguided effort, as the following argument will demonstrate. This second argument, unlike the previous one, has a scope much broader than just the field of analogy-making. Such an argument could be brought to bear on almost any area within artificial intelligence, demonstrating the necessity for "task-oriented" processes to be tightly integrated with high-level perception.

Consider the implications of the separation of perception from the mapping process, by the use of an isolated representation module. Such a module would have to supply a single "correct" representation for any given situation, independent of the context or the task for which it is being used. Our earlier discussion of the flexibility of human representations should already suggest that this notion should be treated with great suspicion. The great adaptability of high-level perception suggests that no module that produced a single context-independent representation could ever model the complexity of the process.

To justify this claim, let us return to the DNA example. To allow the full system to understand the analogy between DNA and a zipper, the representation module would have to produce a representation of DNA that highlights its physical, base-paired structure. On the other hand, to understand the analogy between DNA and source code, a representation highlighting DNA's information-carrying properties would have to be constructed. Such representations would clearly be quite different from each other.

The only solution would be for the representation module to always provide a representation all-encompassing enough to take in *every possible aspect* of a situation. For DNA, for example, we might postulate a single representation incorporating information about its physical, double-helical structure, about the way in which its information is used to build up cells, about its properties of replication and mutation, and much more. Such a representation, were it possible to build, would no doubt be very large. But its very size would make it far too large for immediate use in processing by the higher-level task-oriented processes for which it was intended — in this case, the mapping module. The mapping processes used in most current computer models of analogy-making, such as SME, all use very small representations that have the relevant information selected and ready for immediate use. For these programs to take as input very large representations that include all available information would require a radical change in their design.

The problem is simply that a vast oversupply of information would be available in such a representation. To determine precisely which pieces of that information were relevant would require a complex process of filtering and organizing the available data from the representation. *This process would in fact be tantamount to high-level perception all over again.* This, it would seem, would defeat the purpose of separating the perceptual processes into a specialized module.

Let us consider what might be going on in a human mind when it makes an analogy. Presumably people have somewhere in long-term memory a representation of all their knowledge about, say, DNA. But when a person makes a particular analogy involving DNA, only certain information about DNA is used. This information is brought from long-term memory and probably used to form a temporary active representation in working memory. This second representation will be much less complex, and consequently much easier for the mapping process to manipulate. It seems likely that this smaller representation is what corresponds to the specialized representations we saw used by SME above. It is in a sense a projection of the larger representation from long-term memory — with only the relevant aspects being projected. It seems psychologically implausible that when a person makes an analogy, their working memory is holding all the information from an all-encompassing representation of a situation. Instead, it seems that people hold in working memory only a certain amount of relevant information with the rest remaining latent in long-term storage.

But the process of forming the appropriate representation in working memory is undoubtedly not simple. Organizing a representation in working memory would be another specific example of the action of the high-level perceptual processes — filtering and organization — responsible for the formation of representations in general. And most importantly, this process would necessarily interact with the details of the task at hand. For an all-encompassing representation in long-term memory to be transformed into a usable representation in working memory, the nature of the task at hand — in the case of analogy, a particular attempted mapping — must play a pivotal causal role.

The lesson to be learned from all this is that separating perception from the "higher" tasks for which it is to be used is almost certainly a misguided approach. The fact that representations have to be adapted to particular contexts and particular tasks means that an interplay between the task and the perceptual process is unavoidable, and therefore that any "modular" approach to analogy-making will ultimately fail. It is therefore essential to investigate how the perceptual and mapping processes can be integrated.

One might thus envisage a system in which representations are gradually built up as the various pressures evoked by a given context manifest themselves. In such a system, not only would the mapping be determined by perceptual processes, but the perceptual processes would in turn be influenced by the mapping process. Representations would be built up gradually by means of this continual interaction between perception and mapping. If a particular representation seemed appropriate for a given mapping, then that representation would continue to be developed while the mapping continued to be fleshed out. If the representation seemed less promising, then alternative directions would be explored by the perceptual process. It would be of the essence that the processes of perception and mapping be *interleaved* at all stages. Gradually, an appropriate analogy would emerge, based on structured representations that dovetail with the final mapping.

In fact, two such systems have been built in our research group (they are described in the next few chapters of this book). Such systems are very different from the traditional approach, which assumes the representation-building process to have been completed, and which concentrates on the mapping process in isolation. But in order to be able to deal with the great flexibility of human perception and representation, analogy researchers must integrate high-level perceptual processes into their work. We believe that the use of hand-coded, rigid representations will in the long run prove to be a dead end, and that flexible, context-dependent, easily adaptable representations will be recognized as an essential part of any accurate model of cognition.

Finally, we should note that the problems we have outlined here are by no means unique to the modeling of analogical thought. The hand-coding of representations is endemic in traditional AI. Any program that uses pre-built representations for a particular task could be subject to such a "representation module" argument similar to that given above. For most purposes in cognitive science, an integration of task-oriented processes with those of perception and representation will be necessary.

The Utility of Small Domains

A model of high-level perception is clearly desirable, but a major obstacle lies in the way. For any model of high-level perception to get off the ground, it must be firmly founded on a base of low-level perception. But the sheer amount of information available in the real world makes the problem of low-level perception an exceedingly complex one, and success in this area has understandably been quite limited. Low-level perception poses so many problems that for now, the modeling of full-fledged high-level perception of the real world is a distant goal. The gap between the lowest level of perception (cells on the retina, pixels on the screen, waveforms of sound) and the highest level (conceptual processes operating on complex structured representations) is at present too wide to bridge.

This does not mean, however, that one must admit defeat. There is another route to the goal. The real world may be too complex, but if one *restricts the domain,* some understanding may be within our grasp. If, instead of using the real world, one carefully creates a simpler, artificial world in which to study high-level perception, the problems become more tractable. In the absence of large amounts of pixel-by-pixel information, one is led much more quickly to the problems of high-level perception, which can then be studied in their own right.

Such restricted domains, or microdomains, can be the source of much insight. Scientists in all fields throughout history have chosen or crafted idealized domains to study particular phenomena. When researchers attempt to take on the full complexity of the real world without first having some grounding in simpler domains, it often proves to be a misguided enterprise. Unfortunately, microdomains have fallen out of favor in artificial intelligence. The "real world" modeling that has replaced them, while ambitious, has often led to misleading claims (as in the case of BACON), or to limited models (as we saw with models of analogy). Furthermore, while "real world" representations have impressive labels — such as "atom" or "solar system" — attached to them, these labels conceal the fact that the representations are nothing but simple structures in predicate logic or a similar framework. Programs like BACON and SME are really working in stripped-down domains of certain highly idealized logical forms — their domains merely *appear* to have the complexity of the real world, thanks to the English words attached to these forms.

While microdomains may superficially seem less impressive than "real world" domains, the fact that they are explicitly idealized worlds allows the issues under study to be thrown into clear relief — something that generally speaking is not possible in a full-scale real-world problem. Once we have some understanding of the way cognitive processes work in a restricted domain, we will have made genuine progress towards understanding the same phenomena in the unrestricted real world.

In our research group, we have built two systems along the lines sketched above. One of them, the Copycat program (see Chapter 5 as well as Mitchell, 1993), works in a domain of alphabetical letter-strings. This domain is simple enough that the problems of low-level perception are avoided, but complex enough that the main issues in high-level perception arise and can be studied.

Copycat is capable of building up its own representations of situations in this domain, and does so in a flexible, context-dependent manner. Along the way, many of the central problems of high-level perception are dealt with, using mechanisms that have a much broader range of application than just this particular domain. Such a model may well serve as the basis for a later, more general model of high-level perception.

Copycat's highly parallel and nondeterministic architecture builds its own representations and finds appropriate analogies by means of the continual interaction of perceptual structuring-agents with an associative concept network. It is this interaction between perceptual structures and the concept network that helps the model capture part of the flexibility of human thought. The Copycat program is a model of both high-level perception and analogical thought, and it uses the integrated approach to situation perception and mapping that we have been advocating.

The architecture of Copycat could be said to fall somewhere along the spectrum stretching between the connectionist and symbolic approaches to artificial intelligence, sharing some of the advantages of each. On the one hand, like connectionist models, Copycat consists of many local, bottom-up, parallel processes from whose collective action higher-level understanding emerges. On the other hand, it shares with symbolic models the ability to deal with complex hierarchically-structured representations.

Copycat illustrates possible mechanisms for dealing with five important problems in perception and analogy. These are:

- the gradual building-up of representations;
- the role of top-down and contextual influences;
- the integration of perception and mapping;
- the exploration of many possible paths toward a representation;
- the radical restructuring of perceptions, when necessary.

The successful implementation of Copycat amounts to a feasibility proof that these ideas actually can be made to work.

The architectural core of Copycat, incidentally, is applicable much more widely than to just the particular domain in which Copycat functions. For instance, substantially the same architecture has been used in the Tabletop program to deal with the problem of perceiving structure and making analogies involving configurations of silverware and crockery on a tabletop — a micro-domain with a more "real world" feel (see Chapters 8 and 9, as well as French, 1995). Additionally, Letter Spirit, a related architecture involving perception of the shapes and styles of visual letterforms, as well as generation of new letterforms sharing an abstract style suggested by other letterforms, has been proposed and partially implemented (see Chapter 10).

There is certainly nothing in Copycat or these other projects that corresponds to the messy low-level perception that goes on in the visual and auditory systems. It might well be argued by a skeptic, however, that just as we have insisted herein that high-level perception exerts a strong influence on and is intertwined with later cognitive processing, so low-level perception is equally intertwined with high-level perception. This point is well taken. We acknowledge that in the end, a complete model of high-level perception will have to take all levels of low-level perception into account as well, but we believe that for now, the complexity of this task is such that key features of the high-level perceptual processes must be studied in isolation from their low-level base.

The Tabletop program takes a few steps further than Copycat does towards lower-level perception, in that it must make analogies between visual structures in a two-dimensional world, although this world is still highly idealized. Letter Spirit goes even further toward low-level vision, but still stops far short of rock bottom.

We are familiar with a couple of AI projects that attempt to combine perceptual and cognitive processes. It is interesting to note that in this work, microdomains are almost always used. Chapman's "Sonja" program (Chapman, 1991), for instance, functions in the world of a video game. Starting from simple graphical information, it develops representations of the situation around it and takes appropriate action. As in Tabletop, the input to Sonja's perceptual processes is a little more complex than in Copycat, so that these processes can justifiably be claimed to be a model of "intermediate vision" (more closely tied to the visual modality than Copycat's high-level mechanisms, but still abstracting away from the messy low-level details), although the representations developed are less sophisticated than Copycat's. Along similar lines, Shrager (1990) has investigated the central role of perceptual processes in scientific thought, and has developed a program that builds up representations in the domain of understanding the operation of a laser, starting from idealized two-dimensional inputs.

Conclusion

It may sometimes be tempting to regard perception as not truly "cognitive", something that can be walled off from higher processes, allowing researchers to study such processes without getting their hands dirtied by the complexity of perceptual processes. But this is almost certainly a mistake. Cognition is infused with perception. This has been recognized in psychology for decades, and in philosophy for longer, but artificial-intelligence research has been slow to pay attention.

Two hundred years ago, Kant provocatively suggested an intimate connection between concepts and perception. "Concepts without percepts", he wrote, "are empty; percepts without concepts are blind." In this paper we have tried to demonstrate just how true this statement is, and just how dependent on each other conceptual and perceptual processes are in helping people make sense of their world.

"Concepts without percepts are empty." Research in artificial intelligence has often tried to model concepts while ignoring perception. But as we have seen, high-level perceptual processes lie at the heart of human cognitive abilities. Cognition cannot succeed without processes that build up appropriate representations. Whether one is studying analogy-making, scientific discovery, or some other area of cognition, it is a mistake to try to skim off conceptual processes from the perceptual substrate on which they rest, and with which they are tightly intermeshed.

"Percepts without concepts are blind." Our perception of any given situation is guided by constant top-down influence from the conceptual level. Without this conceptual influence, the representations that result from such perception will be rigid, inflexible, and unable to adapt to the problems provided by many different contexts. The flexibility of human perception derives from constant interaction with the conceptual level. We hope that the models of concept-based perception that we have developed go some way towards drawing these levels together.

Recognizing the centrality of perceptual processes makes artificial intelligence more difficult, but it also makes it more interesting. Integrating perceptual processes into a cognitive model leads to flexible representations, and flexible representations lead to flexible actions. This is a fact that has only recently begun to permeate artificial intelligence, through such models as connectionist networks, classifier systems, and the Copycat/Tabletop architecture and its generalizations. Future advances in the understanding of cognition and of perception are likely to go hand in hand, for the two types of process are inextricably intertwined.

Preface 5:

Conceptual Halos and Slippability

Analogy Puzzles in the Seek-Whence Domain

The Copycat project sprang out of difficulties that arose in Seek-Whence. It had become clear to me already by around 1980 that the ability to spot abstract resemblances — in other words, analogies — between short segments of a sequence lies at the heart of the ability to perceive regular patterns and to formulate rules describing those patterns. So at about that time, I began extracting analogy puzzles from the Seek-Whence domain, such as the following quite simple one:

What in "12344321" corresponds to "4" in "1234554321"?

For ease of reference, I will call the longer, taller, and latter of these mountains "structure A" and the shorter, shorter, and former of them "structure B". (Please note that this is the reverse of the order you might expect.)

This puzzle is not very challenging. Most people would effortlessly answer "3", seeing it as *playing the same role* in B as the "4" plays in A, and without even seeing any alternative answer. Still, someone[1] could pedantically say "4", arguing that it *has the same magnitude,* or, if you prefer, that it is *the same number.* Thus to the very limited extent that this puzzle evokes a conflict of pressures, it pits a rather dull and machine-like desire for preserving what might be called "object sameness" against a more lively and human-seeming desire for preserving a sense of *role played.* I italicize the phrase to make you think about it for a moment. Is this really the right way to think about the "4" — that is, does "4" really play a *single* role inside structure A? To help stimulate thoughts on this, I offer the following "cousin puzzle":

What in "123475574321" corresponds to "4" in "1234554321"?

The new mountain will be called "structure C". Obviously, what gives this mountain interest is that it makes a sudden brief leap upwards and then has a

---

1. *E.g.,* Richard Feynman. See Chapter 24 of Hofstadter, 1985.

dip, so that the apex is not located at the structure's center, as was the case for mountains A and B. Structure C is more like a sharp volcano with a sunken crater at its top.

The effect of this different shape is to break what seemed like just *one* role played by the "4" in A (and essentially also played by the "3" in B) into a *set* of roles. That is, in the light of structure C, we now see that the "4" in A was really serving in (at least) the following four non-identical capacities:

(1) physical neighbor of the centermost number-pair;
(2) next-to-largest number in the structure;
(3) numerical predecessor of the centermost number-pair;
(4) numerical predecessor of the largest number in the structure.

It just happened that the way structure A was, these different roles all coincided and looked like *one* role. Another way of saying this is that structure C *induced* a breakup of what seemed like one monolithic role into four separate facets. But this splitting happened *a posteriori*. That is, if there hadn't been a structure like C to force this new perception on us, we would have blithely continued to see the "4" in A as playing one single role. Would that notion have been an illusion, or would it have been correct?

The Nancy Reagan of England

In tandem with the foregoing microdomain puzzle, I always enjoyed presenting this real-world puzzle:

Who in England corresponds to Nancy Reagan in the United States?

which I more often phrased this way:

Who is the First Lady of England?

Keep in mind, of course, that this was all happening around 1980, when Margaret Thatcher was prime minister.

There are basically two types of confounding factors involved in this puzzle. One of them is that England has no president, but instead has *two* figures each of whom plays a role somewhat reminiscent of the presidential role — the monarch and the prime minister. The other confounding factor is that each of these people was, in 1980, a woman, whereas Ronald Reagan was, of course, a man. These confounding factors introduced a number of unexpected pressures into the situation.

There is, *a priori*, considerable pressure to map Nancy Reagan onto a woman. This pressure would be especially strong if the puzzle explicitly used the phrase "First Lady", but even if it did not, there would still be quite a bit of implicit pressure for a same-sex mapping, resulting from the fact that "First Lady" was by far the most salient epithet for Nancy Reagan, and would thus certainly be implicitly present even though not explicitly so. This pressure would push for the counterpart of Nancy Reagan to be either Queen Elizabeth or Margaret Thatcher, simply on the basis that a First Lady is the most prominent woman, more or less, in a country.

But of course there was a very strong counterpressure against either of these choices, coming from the obvious fact that these women were themselves *at the center* or *at the peak,* whereas Nancy Reagan was not quite there — she was *one step away.* So, depending on how one interprets "center" or "peak", one would then feel pressure to select either Prince Philip or Denis Thatcher, Margaret Thatcher's husband. (Is the role of Queen more like a *center* or a *peak?*) But either of these choices violates the desire to have a female as the answer. By the way, this droll question came to my mind when I chanced upon a newspaper article about Denis Thatcher, in which he was characterized as the "First Lady of England".

The analogy between this real-world analogy puzzle and the numerical analogy puzzle involving structures A and C is pretty obvious — both involve an *a posteriori* breakup of what at first appears to be just *one* role into a set of conflicting roles, giving rise to pressures pushing against each other, and for different answers. Moreover, both puzzles involve a figure that is "one step away from the center/peak".

Let us go back to the four facets into which the role of "4" was broken by our exposure to structure C.

If facet (1) — "physical neighbor of the centermost number-pair" — were taken as the most important, then the answer to the puzzle would turn out to be "7".

If facet (2) — "next-to-largest number in the structure" — were taken as the dominant one, then "5" would be the answer.

If facet (3) — "numerical predecessor of the centermost number-pair" — were dominant, then the answer would be "4".

Finally, if facet (4) — "numerical predecessor of the largest number in the structure" — were felt to be the most faithful to the precedent, then "6" — a number nowhere present in the structure itself! — would be the answer.

These four different answers — 7, 5, 4, and 6 — correspond, in no particular order, to the four "First Ladies of England in 1980": Prince Philip, Denis Thatcher, Queen Elizabeth, and Margaret Thatcher. Which answer maps onto which person depends on how you look at things, of course. (For further discussion of the First Lady riddle as well as of a host of analogy problems in the Seek-Whence and Copycat domains, see Chapter 24 of Hofstadter, 1985.)

I soon found that these analogy problems involving the stripped-down numbers of the Seek-Whence domain proliferated beyond anything I had ever expected. Soon I had invented so many of them I couldn't keep track of them, and they came in all levels of complexity, with intricate combinations of pressures pushing for or against all sorts of strange conceptual slippages. By "conceptual slippage", I mean the *context-induced dislodging of one concept by a closely related one*, inside the mental representation of some situation. But what would make one concept be dislodged by another one? How exactly would such a mental event happen, and when? It all has to do with conceptual overlaps, which are an automatic side-effect of the fact that concepts are like *halos*. Let us look at this idea, absolutely central to the Copycat architecture.

Words, Concepts, and Halos

Earlier, I asked whether the breakup into four different roles of what had previously seemed like just one role revealed that the oneness was an illusion. This is a subtle and deep matter, and can be usefully approached by considering an analogous question involving language. Consider a common word like "hard". When you think about what it means, it probably feels to you like "one idea" — you might say "not soft", for example. It seems quite monolithic. But I have just opened my thesaurus to its index, and under "hard" I am reading the following list of flavors that "hard" can take on: *addictive, adverse, alcoholic, bitter, callous, difficult, hard to understand, heartless, impenitent, industrious, inelastic, obdurate, painful, pitiless, rigid, severe, solid, strict, strong, substantial, tough, wicked*. If you think about each of these, you will see that "hard" actually subdivides into many ideas. But you might still feel that the overarching concept is still a unity — one idea.

However, now consider going into another language — say German, which is relatively close to English. It turns out that "hard" splits up into many different words in German, depending on what it is modifying. Thus, for a substance, it is *hart*; for a problem or test, *schwierig*; for a diligent worker, *fleißig*; for rough times or a severe blow, *schwer*; for life in general, *mühsam*; for rain, *heftig* or *stark*; for a frost, *streng*; for drinks, *alkoholisch*; for thinking hard, *gut* or *scharf*; for hitting hard, *kräftig*; for a trying person, *anstrengend*; and on and on. A German speaker would not see nearly as much of a unity behind all these concepts as an English speaker would. And do not think that each of these German words is narrower in meaning than "hard", in the sense of being a subset of the concept. Quite to the contrary, each of them on its own is extremely broad. For instance, among the many translations of *schwer* are these: "heavy", "solid", "powerful", "strong", "rich", "hard", "severe", "serious", "grave", "difficult", "tough", "ponderous", and "clumsy". To a German speaker, this broad constellation of flavors adds up to what feels like "one idea".

Are we English speakers thus deluded, when we think to ourselves that the concept denoted by "hard" is a unified, monolithic notion? Or are German speakers deluded when they analogously think to themselves that the concept denoted by *schwer* is a unified, monolithic notion? I would say that in the context of English, the set of flavors of "hard" seems to form a coherent, cohesive unity, but that seen from a German perspective, this unity seems to be a delusion. Note the word "seems" on both sides. Context and culture determine the boundary lines of concepts.

Perhaps a simpler example is provided by the following comparison among English, Indonesian, Chinese, German, and Italian.

In English, we standardly ask a new acquaintance, "Do you have any brothers or sisters?"

In Indonesian, people standardly ask if someone has *kakak*'s and *adik*'s. It's exactly the same question, yet not at all the same question, since *kakak* means "older sibling" and *adik* "younger sibling", irrespective of sex.

In Mandarin Chinese, people standardly ask if someone has *xiōng-dì-jiě-mèi*. Once again, it's exactly the same and not at all the same. This time there are four distinct questions being asked, since *xiōng* means "elder brother", *dì* "younger brother", *jiě* "older sister", and *mèi* "younger sister".

Our next case, German, is simpler. People ask *Haben Sie Geschwister?*, which just means "Do you have siblings?" Very simple. But beware of this seeming simplicity. *Geschwister* looks and sounds so similar to *Schwester* ("sister") that it certainly carries a strong undercurrent of "sisters", even if it is not at all consciously experienced as such.

Our final case, Italian, is sort of the flip side of the German. If an Italian asks me, *Lei ha fratelli?* — "Do you have brothers?" — I will be entirely correct to answer, *Sì, due sorelle* — "Yes, two sisters." Thus in this context, *fratelli*, which normally means only "brothers", floats somewhere between "brothers" and "brothers and sisters".

So, leaving specific languages aside, how many concepts are actually engaged in the human mind when the sibling-number question is asked — one, two, or four? Is that a meaningful question? Suppose it were the case that in some language, *plubibwa* meant "humorous sibling" and *vazil* meant "serious sibling", and one of the standard questions new acquaintances asked each other was *Exement-ci plubibwa flo vazil?* Would this mean it had been a delusion to think that there were merely *four* concepts hiding inside the concept "sibling"? Would we finally have attained enlightenment, realizing that the ultimate fact of the matter is that there are *eight* concepts hiding in there (ranging, of course, from "humorous younger sister" to "serious older brother")? Obviously not. There simply is no exact "number of concepts" to be counted here; it depends on context and culture. What there is in any mind, however, regardless of culture, is a blurry overlap of many, many specific instances of siblings, with strong clustering-patterns, tending to make certain potential subconcepts more natural-seeming, and others less natural-seeming.

These are not, by any stretch of the imagination, special cases. An analogous shattering of what had seemed a smooth, monolithic unity into many unexpected meaning-shards happens to every single word one knows when it bashes into any foreign language, even one as close as, say, French or Italian is to English. Thus, for example, such common notions as are expressed in English by "hit", "throw", "window", "box", "bag", "day", "top", "fast", and "in fact", all of which I once naïvely thought were structureless, basic, primary concepts that would be universals across languages, turn out to disintegrate, each and every one of them, into five or ten or even more varieties of hitting, throwing, and so on when they bang into Italian — and of course vice versa for Italian words of equal fundamentality when they bang into English. What seemed indivisible splits and splinters into microstructure.

After the fact, I can understand the "logic" behind the Italian breakup of "in fact" into *in effetti, di fatto, infatti, in realtà, anzi, però, tant'è vero che, per esempio, a dire il vero*, and so forth, but I never would have thought of it myself. It just didn't seem that complex a concept. Little did I know! And this shattering into pieces happens, I repeat, with *every single word one knows,* until finally one reaches that rarefied region of semantic space containing only highly specialized, technical terms such as "photosynthesis", which pretty much does go across intact, the Italians mercifully having decided that it is not the case that there is one kind of photosynthesis for flowering plants, another kind for deciduous trees, yet another kind for evergreens, and so on. But for common words, it is *always* the case — and in fact, the higher-frequency the word is, the more complex and startling its split-up is.

This is why it is so hilarious — or perhaps "pathetic" is what I really mean — to see the proliferation of pocket bilingual dictionaries and so-called "translation computers" filled to the brim with nothing but one-line English-to-French entries like "hit: *frapper*", "throw: *jeter*", "picture: *image*", "in: *dans*", and so on. The obverse side of the coin is the great pleasure it gives me to browse through my well-worn copy of Roget's International Thesaurus and to savor the haze and halo of so many sublimely interpenetrating concept-clouds.

From Conceptual-halo Slips to Conceptual Slippages

The overlapping and clustering of concepts in our brains, giving rise to a "semantic halo" surrounding every content word, is revealed surprisingly often by slips of the tongue, of which *word blends* are a common variety. Here are a few that provide evidence for such overlapping halos in the brain. Notice how often these errors are clearly direct results of pressures invoked by some particular context. A friend said to me, "No one could get tickets on that flane", obviously blending "flight" and "plane". A linguist heard himself say, "Don't shell so loud" — a clear blend of "shout" and "yell". I blurted out, "I'll chake a look", blending "check it out" and "take a look". My wife said, "Doug went to graduate stude with Pranab", blending "was a graduate student" and "went to graduate school". Instances of blending like this are legion, revealing fleeting moments of indecision where competing pressures could not be fully resolved under the constraints of real-time speech.

Somewhat stranger are full *substitution errors* such as the following few. I said to my little son, "Close your cookie jars, please", intending to refer to two *toy chests* in his bedroom. No bashful blending here — rather, the brazen spilling-out of a completely wrong term in its entirety. Another time, I said, "The bathroom door won't close — the faucet doesn't work right", when what I meant was, "The doorknob doesn't work right". Clearly, both "faucet" and "doorknob" are salient denizens of the region of semantic space that deals with twisty handles or controls, but in the context of bathrooms, "faucet" becomes especially prominent, and in this case it beat out "doorknob" in the utterance sweepstakes.

A friend said, "I'm going to go to the begin — uhh, the entrance of the store." I remarked, "We moved to Princeton when I was just two o'clock", meaning, obviously, "just two years old". My wife asked, "Should I put the garbage — uhh, the dirty clothes — in the back of the car?"

And one last example of this genre, perhaps my favorite... A grocery-store checkout clerk asked me, "Plastic bag all right?", to which I replied, "Prefer a wood one... uhh, a... a *paper* one, please." Contributing toward this slip might have been the following factors: paper is made from wood pulp; grocery bags are brownish, somewhat like wood and unlike standard paper; they are also considerably "woodier" in texture than ordinary paper is; and plastic and wood are both common materials out of which many household items are made, whereas paper is not.

Substitution errors like these reveal aspects of the subterranean landscape — the hidden network of overlapping, blurred-together concepts. They show us that under many circumstances, we confuse one concept with another, and this helps give a picture of what is going on when we make an analogy between different situations. The same properties of our conceptual networks as are responsible for our proneness to these conceptual-halo slips make us willing to tolerate or "forgive" a certain degree of conceptual mismatch between situations, depending on the context; we are congenitally constructed to do so — it is good for us, evolutionarily speaking. My term "conceptual slippage" is in fact no more and no less than a shorthand for this notion of "context-dependent tolerance of conceptual mismatch". (For more types of errors and detailed discussion of these kinds of issues, see Aitchison, 1994; Cutler, 1982; Dell & Reich, 1980; Fromkin, 1980; Hofstadter & Moser, 1989; and Norman, 1981.)

Copycat Is Conceived

I was beginning to become aware of the remarkable pervasiveness of these, and other kinds, of conceptual-halo effects in cognition at the time I was working on the Seek-Whence architecture, and this had a deep influence on my ideas. But the more clearly I recognized the key role played by analogy-making, the more I began to wonder whether the Seek-Whence project wasn't reaching perhaps a bit too far too soon. Gradually I became persuaded that it was imperative for me first to confront analogy-making head-on in a fully autono- mous project, and that I might eventually return to Seek-Whence at some later point, armed with the proper methods with which to attack the issues at its core.

So in the spring of 1983, I started focusing most of my energy toward the devising of an architecture for discovering analogies in a microdomain very much like that of Seek-Whence; in this microdomain, *conceptual slippage* was to be the key new ingredient. But for some reason — perhaps just a desire for slight variety — I slid over from using *numbers* to using *letters*. What startled and disappointed me over the next few years was how many people related better to these letter puzzles than to my earlier number puzzles, even if two puzzles were precise translations of each other. People would say to me, "I never liked math, so I don't like number puzzles, but these letter puzzles are fun!" I never could figure this out.

Anyway, here is a letter puzzle that shows some of the issues I was thinking about:

I change *efg* into *efw*. Can you "do the same thing" to *ghi*?

What I liked about this puzzle was that it had two answers with very different sorts of appeal. One answer is *whi*, gotten simply by replacing the instance of *g* by a *w*. Sure, that's "doing the same thing", but it seems kind of crude. The other answer is *ghw*, based on replacing the rightmost letter by *w*. This too "does the same thing", but in a very different sense — perhaps a slightly more refined sense, perhaps not. (By the way, any answer based on the idea of the "alphabetic distance" between *g* and *w* is strictly out of bounds. That is just not in the spirit of the domain.)

A closely related analogy puzzle is this one:

I change *efg* into *wfg*. Can you "do the same thing" to *ghi*?

Again the very same two strings are the most plausible answers: *whi* and *ghw*. However, the reasons behind them have changed. The former is easy to see:

simply replace the leftmost letter by *w*. The second answer is subtler, and involves seeing a kind of symmetry between *efg* and *ghi*. Specifically, *efg* is a string with *g* at its *right* end, *ghi* is a string with *g* at its *left* end. This vision can be parlayed into the idea that the concept "left" in *ghi* plays the same role as "right" plays in *efg*, which suggests that the two strings are "the same", if we allow this *left/right* conceptual slippage. From this it is easy to get to the answer *ghw*.

There's something ironic here. In the first problem, the answer *whi*, based on equating two instances of *g*, seemed quite crude, and yet in the second problem, the answer *ghw*, which also was based on equating two instances of *g*, seems quite sophisticated. How can the same strategy be crude in one problem and sophisticated in another? Well, the way I see it is that in the first puzzle, equating the two *g*'s was the end of the story — the rest of the letters were simply ignored, which is pretty darn crude — whereas in the second puzzle, equating the two *g*'s was just the start. The next step is to *go to the far ends* of the two strings, where one finds an *e* and an *i*, and to equate *those* letters. Using the *g*'s as identical points of reference or "landmarks" in this way is quite an abstract leap, and brings the full structures of both strings into the picture. Therefore, answers based on this idea seem respectable, even somewhat deep.

Here's one last problem that carries this idea further.

I change *efg* into *dfg*. Can you "do the same thing" to *ghi*?

If one doesn't notice that there is a special relation between the letter *e* and the letter it got changed to — namely, *d* — then this problem is isomorphic to the previous one, with *d* playing the role of *w*, so there are two isomorphic answers, *dhi* and *ghd*. But to ignore the blatant alphabetic proximity of *e* and *d* seems very crude. So if we *do* take this obvious fact into account, the natural step would seem to be to *take the predecessor* either of the leftmost letter (leading to *fhi*), or of the letter at the other end from the *g* (leading to *ghh*). However, the double *h* of the latter answer seems out of keeping with the spirit of the original change, leading us to suspect that something may have gone wrong. And indeed, something vital was overlooked.

Equating *efg*'s *e* with *ghi*'s *i* involves running through *efg* from right to left, and through *ghi* from left to right. In *efg*, we are thus running alphabetically backwards, from a letter to its predecessor, whereas in *ghi* we are running alphabetically forwards, from a letter to its successor. So the key idea we left out is the conceptual slippage from *predecessor* to *successor*. When we take that extra slippage into account, we see that we want to replace *i* not by *h* but by *j*, leading to answer *ghj*. This is arguably a *very* sophisticated answer, taking into account the full nature of both structures (including the alphabetic fabric that binds each of them together, a factor that didn't play any role in the preceding puzzle), yet it is still based on the simplistic, crude-seeming first step of equating the two *g*'s.

Such were the kinds of problems and issues that started proliferating in my mind, and a whole new project started springing to life around them. My first stab at describing the architecture of a program to make these kinds of observations and leaps was in an advanced AI seminar I taught at Indiana University in the spring of 1983. In the fall, I left for MIT on sabbatical, and there I turned these architectural ideas into a grant proposal and a technical report (Hofstadter, 1984a), which became the basis for all further Copycat work.

At the crux of it was a new structure called the *Slipnet*, which was my way of modeling the human mind's subtly context-dependent conceptual halos and conceptual slippages. The trick was to combine this new element in a natural way with the prior notions of Jumbo and related projects — namely, codelets, the "cytoplasm", and temperature. Fortunately, there seemed to be a natural fit, and — at least in my mind — a deep analogy-making architecture was starting to blossom, and I could easily envision a rich harvest of fruits several years down the line. But I knew that this would require the efforts of more than one person.

One fine spring day in 1984, a young woman named Melanie Mitchell turned up at my office at the MIT AI Lab and indicated a strong interest in the lines of research I was pursuing in cognitive science. I had very little to go on, but, exercising my parallel terraced scan in my best possible fashion, I tentatively welcomed her on board. It was the rightest thing I could have done.

Chapter 5

The Copycat Project:
A Model of Mental Fluidity
and Analogy-making

DOUGLAS HOFSTADTER and MELANIE MITCHELL

Copycat and Mental Fluidity

Copycat is a computer program designed to be able to discover insightful analogies, and to do so in a psychologically realistic way. Copycat's architecture is neither symbolic nor connectionist, nor was it intended to be a hybrid of the two (although some might see it that way); rather, the program has a novel type of architecture situated somewhere in between these extremes. It is an *emergent* architecture, in the sense that the program's top-level behavior emerges as a statistical consequence of myriad small computational actions, and the concepts that it uses in creating analogies can be considered to be a realization of "statistically emergent active symbols" (Chapter 26 of Hofstadter, 1985). The use of parallel, stochastic processing mechanisms and the implementation of concepts as distributed and probabilistic entities in a network make Copycat somewhat similar in spirit to certain connectionist systems. However, as will be seen, there are important differences, and we claim that the middle ground in cognitive modeling occupied by Copycat is at present the most useful level at which to attempt to understand the fluidity of concepts and perception that is so clearly apparent in human analogy-making.

Analogy problems in the Copycat domain
The domain in which Copycat discovers analogies is very small but surprisingly subtle. Not to beat around the bush for a moment, here is an example of a typical, rather simple analogy problem in the domain:

1. Suppose the letter-string **abc** were changed to **abd**; how would
you change the letter-string **ijk** in "the same way"?

Note that the challenge is essentially "Be a copycat" — that is, "Do the same
thing as I did", where "same" of course is the slippery term. Almost everyone
answers **ijl**.[1] It is not hard to see why; most people feel that the natural way to
describe what happened to **abc** is to say that *the rightmost letter was replaced by its
alphabetic successor*; that operation can then be painlessly and naturally "ex-
ported" from the **abc** framework to the other framework, namely **ijk**, to yield the
answer **ijl**. Of course this is not the only possible answer. For instance, it is always
possible to be a "smart aleck" and to answer **ijd** (rigidly choosing to replace the
rightmost letter by **d**) or **ijk** (rigidly replacing all **c**'s by **d**'s) or even **abd**
(replacing the whole structure blindly by **abd**), but such "smart-alecky" answers
are suggested rather infrequently, and when they are suggested, they seem less
compelling to virtually everybody, even to the people who suggested them. Thus
ijl is a fairly uncontroversial winner among the range of answers to this problem.

There is much more subtlety to the domain than that problem would
suggest, however. Let us consider the following closely related but considerably
more interesting analogy problem:

2. Suppose the letter-string **aabc** were changed to **aabd**; how would
you change the letter-string **ijkk** in "the same way"?

Here as in Problem 1, most people look upon the change in the first framework
as *the rightmost letter was replaced by its alphabetic successor*. Now comes the tricky part:
should this rule simply be transported rigidly to the other framework, yielding
ijkl? Although rigid exportation of the rule worked in Problem 1, here it seems
rather crude to most people, because it ignores the obvious fact that the **k** is
doubled. The two **k**'s together seem to form a natural unit, and so it is tempting
to change *both* of them, yielding the answer **ijll**. Using the old rule literally will
simply not give this answer; instead, under pressure, one "flexes" the old rule into
a very closely related one, namely *replace the rightmost group by its alphabetic successor*.
Here, the concept *letter* has "slipped", under pressure, into the related concept
group of letters. Coming up with such a rule and corresponding answer is a good
example of human mental "fluidity" (as contrasted with the mental rigidity that
gives rise to **ijkl**). There is more to the story of Problem 2, however.

Many people are perfectly satisfied with this way of exporting the rule (and
the answer it furnishes), but some feel dissatisfied by the fact that the doubled
a in **aabc** has been ignored. Once one focuses in on this consciously, it jumps to

---

1. Though the popularity of this answer can easily be predicted by one's intuition, we have carried
out many surveys, both formal and informal, of people's answers to this and other problems. The
results of the formal surveys are given in Mitchell, 1993.

mind easily that the **aa** and the **kk** play similar roles in their respective frameworks. From there it is but a stone's throw to "equating" them (as opposed to equating the **c** with the **kk**), which leads to the question, "What then is the counterpart of the **c**?" Given the already-established mapping of *leftmost* object (**aa**) onto *rightmost* object (**kk**), it is but a small leap to map *rightmost* object (**c**) onto *leftmost* object (**i**). At this point, we could simply take the successor of the **i**, yielding the answer **jjkk**.

However, few people who arrive at this point actually do this; given that the two crosswise mappings (**aa** ⇔ **kk**; **c** ⇔ **i**) are an invitation to read **ijkk** in reverse, which reverses the alphabetical flow in that string, most people tend to feel that the conceptual role of alphabetical *successorship* in **aabc** is now being played by that of *predecessorship* in **ijkk**. In that case, the proper modification of the **i** would not be to replace it by its successor, but by its alphabetical *predecessor,* yielding the answer **hjkk**. And indeed, this is the answer most often reached by those people who consciously try to take into account *both* of the doubled letters. Such people, under pressure, have flexed the original rule into this variant of itself: *replace the <u>leftmost</u> letter by its alphabetic <u>predecessor</u>.* Another way of saying this is that a very fluid transport of the original rule from its home framework to the new one has taken place; during this transport, two concepts "slipped", under pressure, into neighboring concepts: *rightmost* into *leftmost,* and *successor* into *predecessor.* Thus, being a copycat — that is, "doing the same thing" — has proven to be a very slippery notion, indeed.

Mental fluidity: Slippages induced by pressures

Hopefully, the pathways leading to these two answers to Problem 2 — **ijll** and **hjkk** — convey a good feeling for the term "mental fluidity". There is, however, a related notion used above that still needs some clarification, and that is the phrase "under pressure". What does it mean to say "concept A *slips* into concept B *under pressure*"? It might help to spell out the intended imagery behind these terms. An earthquake takes place when subterranean structures are under sufficient pressure that something suddenly slips. Without the pressure, obviously, there would be no slippage. An analogous statement holds for pressures bringing about conceptual slippage: only under specific pressures will concepts slip into related ones. For instance, in Problem 2, pressure results from the doubling of the **a** and the **k**; one could look upon the doubling as an "emphasis" device, making the left end of the first string and the right end of the second one stand out and in some sense "attract" each other. In Problem 1, on the other hand, there is nothing to suggest mapping the **a** onto the **k** — no pressure. In the absence of such pressure, it would make no sense at all to slip *leftmost* into *rightmost* and then to read **ijk** in reverse, which would in turn suggest a slippage of *successor* into *predecessor,* all of which would finally lead to the downright bizarre answer *hjk*. That would be *unmotivated* fluidity, which is not characteristic of human thought (except in humor, where higher-level considerations often *do* motivate all sorts of normally-unmotivated slippages).

Copycat is a thoroughgoing exploration of the nature of mental pressures, the nature of concepts, and their deep interrelationships, focusing particularly on how pressures can engender slippages of concepts into "neighboring" concepts. When one ponders these issues, many questions arise, such as the following ones: What is meant by "neighboring concepts"? How much pressure is required to make a given conceptual slippage likely? Just how big a slippage can be made — that is, how far apart can two concepts be and still be potentially able to slip into each other? How can one conceptual slippage create a new pressure leading to another conceptual slippage, and then another, and so on, in a cascade? Do some concepts resist slippage more than others? Can particular pressures nonetheless bring about a slippage of such a concept while another concept, usually more "willing" to slip, remains untouched? Such are the questions at the very heart of the Copycat project.

The intended universality of Copycat's microdomain

This project, which sprang out of two predecessors, Seek-Whence (Meredith, 1986) and Jumbo (Hofstadter, 1983a), has been under development since 1983. A casual glance at the project might give the impression that since it was specifically designed to handle analogies in a particular tiny domain, its mechanisms are not general. However, this would be a serious misconception. All the features of the Copycat architecture were in fact designed with an eye to great generality. A major purpose of this article is to demonstrate this generality by describing the features of Copycat in very broad terms, and to show how they transcend not just the specific microdomain, but even the very task of analogy-making itself. That is, the Copycat project is not about simulating analogy-making *per se*, but about simulating the very crux of human cognition: fluid concepts. The reason the project focuses upon analogy-making is that analogy-making is perhaps the quintessential mental activity where fluidity of concepts is called for, and the reason the project restricts its modeling of analogy-making to a specific and very small domain is that doing so allows the general issues to be brought out in a very clear way — far more clearly than in a "real-world" domain, despite what one might think at first.

Copycat's microdomain was designed to bring out very general issues — issues that transcend any specific conceptual domain. In that sense, the microdomain was designed to "stand for" other domains. Thus one is intended to conceive of, say, the *successor* (or *predecessor*) relation as an idealized version of *any* non-identity relationship in a real-world domain, such as "parent of", "neighbor of", "friend of", "employed by", "close to", etc. A *successor group* (*e.g.*, *abc*) then plays the role of any conceptual chunk based on such a relationship, such as "family", "neighborhood", "community", "workplace", "region", etc. Of course, inclusion of the notion of *sameness* needs no defense; sameness is obviously a universal concept, much as is *opposite*. Although any real-world domain clearly contains many more than two basic types of relationship, two types (sameness plus one other one) already suffice to make an inexhaustible variety of structures of arbitrary complexity.

Aside from the idealized repertoire of *concepts* in the domain, there are also the *structures,* such as *ijkk*, out of which problems are made. In particular, allowed structures are linear strings made from any number — usually a small number — of instances of letters of the alphabet. Thus one immediately runs into the *type/token distinction,* a key issue in understanding cognition. The alphabet can be thought of as a very simple "Platonic heaven" in which exactly 26 letter *types* permanently float in a fixed order; in contrast to this, there is a very rudimentary "physical world" in which any number of letter *tokens* can temporarily coexist in an arbitrary one-dimensional juxtaposition. In this extremely simple model of physical space, there are such physical relationships and entities as *left-neighbor, leftmost edge, group of adjacent letters,* and so on (as contrasted with such relationships and entities in the Platonic alphabet as *predecessor, alphabetic starting-point, alphabetic segment,* etc.). Both the Platonic heaven and the physical world of Copycat are very simple on their own; however, the psychological processes of perception and abstraction bring them into intimate interaction, and can cause extremely complex and subtle mental representations of situations to come about.

Copycat's alphabetic microworld is meant to be a tool for exploring general issues of cognition rather than issues specific to the domain of letters and strings, or domains restricted to linear structures with precise distances in them. Thus certain aspects specific to people's knowledge of letters and letter-strings — such as shapes, sounds, or cultural connotations of specific letters, or words that strings of letters might happen to form — have not been included in this microworld. Moreover, problems should not depend on arithmetical facts about letters, such as the fact that *t* comes exactly eleven letters after *i*, or that *m* and *n* flank the midpoint of the alphabet. Arithmetical facts, while they are universal truths, are not common enough in analogy-making to be worthwhile modeling. This may seem to eliminate almost everything about the alphabet, but as Problems 1 and 2 show (and further problems will show even better), there is still plenty left to play with. Reference to the alphabet's *local* structure is fine; for example, it is perfectly legitimate to exploit the fact that *u* comes immediately after *t.* It is also legitimate to exploit the fact that the Platonic alphabet has two distinguished members — namely, *a* and *z*, its starting and ending points. Likewise, inside a string such as *hagizk*, local relationships, such as "the *g* is the right-neighbor of the *a*", can be noticed, but long-distance observations, such as "the *a* is four letters to the left of the *k*", are considered out of bounds.

Although arithmetical operations such as addition and multiplication play no role in the Copycat domain, numbers themselves — small whole numbers, that is — are included in the domain. Thus, Copycat is capable of recognizing not only that the structure *fgh* is a "successor group", but also that it consists of *three* letters. Just as the program knows the immediate neighbors of every letter in the alphabet, it also knows the successors and predecessors of small integers. Under the appropriate pressures, Copycat can even treat small integers as it does letters — it can notice relationships between numbers, can group numbers together, map them onto each other, and so on. However, generally speaking, Copycat tends to resist bringing numbers into the picture, unless there seems to be some compelling reason to do so — and *large* numbers, such as 5, are resisted even more strongly. The idea behind this is to reflect the relative ease humans have of recognizing pairs and perhaps trios of objects, but the relative insensitivity to such things as quintuples, let alone septuples and so on.

Finally, while humans tend to scan strings of roman letters from left to right, are much better at recognizing forwards alphabetical order than backwards alphabetical order, and have somewhat greater familiarity with the beginning of the alphabet than its middle or end, the Copycat program is completely free of these biases. This should not be regarded as a defect of the program, but a strength, because it keeps the project's focus away from domain-specific and nongeneralizable details.

A perception-based, emergent architecture for mental fluidity

When one describes the Copycat architecture in very abstract terms, the focus is not only on how it discovers mappings between situations, but also on how it perceives and makes sense of the miniature and idealized situations it is presented with. The present characterization will therefore read very much like a description of a computer model of *perception*. This is not a coincidence; one of the main ideas of the project is that even the most abstract and sophisticated mental acts deeply resemble perception. In fact, the inspiration for the architecture comes in part from a computer model of low-level and high-level auditory perception: the Hearsay II speech-understanding project (Erman *et al.*, 1980; Reddy *et al.*, 1976).

The essence of perception is the awakening from dormancy of a relatively small number of prior concepts — precisely the relevant ones. The essence of understanding a situation is very similar; it is the awakening from dormancy of a relatively small number of prior concepts — again, precisely the relevant ones — and applying them judiciously so as to identify the key entities, roles, and relationships in the situation. Creative human thinkers manifest an exquisite selectivity of this sort — when they are faced with a novel situation, what bubbles up from their unconscious and pops to mind is typically a small set of concepts that "fit like a glove", without a host of extraneous and irrelevant concepts being consciously activated or considered. To get a computer model of thought to exhibit this kind of behavior is a great challenge.

Following this introductory section, there are six further main sections in this article. The second section is a description of the three main components of the architecture and their interactions. The third section deals with the notion of conceptual fluidity and shows how this architecture implements a model, albeit rudimentary, thereof. The fourth section tackles the seeming paradox of randomness as an essential ingredient of mental fluidity and intelligence. The fifth section views the Copycat program at a distance, summarizing thousands of runs on a few key problems in the letter-string microworld. The sixth section affords a close-up view of Copycat's workings, describing in detail the pathways followed by Copycat as it comes up with subtle answers to two particularly challenging analogy problems. The seventh section concludes the article with a discussion of the generality of Copycat's mechanisms.

The Three Major Components of the Copycat Architecture

There are three major components to the architecture: the Slipnet, the Workspace, and the Coderack. In very quick strokes, they can be described as follows. (1) The Slipnet is the site of all *permanent Platonic concepts*. It can be thought of, roughly, as Copycat's long-term memory. As such, it contains only concept *types,* and no *instances* of them. The distances between concepts in the Slipnet can change over the course of a run, and it is these distances that determine, at any given moment, what slippages are likely and unlikely. (2) The Workspace is the locus of *perceptual activity.* As such, it contains *instances* of various concepts from the Slipnet, combined into *temporary perceptual structures* (*e.g.,* raw letters, descriptions, bonds, groups, and bridges). It can be thought of, roughly, as Copycat's short-term memory or working memory, and resembles the global "blackboard" data-structure of Hearsay II. (3) Finally, the Coderack can be thought of as a "stochastic waiting room", in which small agents that wish to carry out tasks in the Workspace wait to be called. It has no close counterpart in other architectures, but one can liken it somewhat to an *agenda* (a queue containing tasks to be executed in a specific order). The critical difference is that agents are selected *stochastically* from the Coderack, rather than in a determinate order. The reasons for this initially puzzling feature will be spelled out and analyzed in detail below. They turn out to be at the crux of mental fluidity.

We now shall go through each of the three components once again, this time in more detail. (The finest level of detail — complete lists of algebraic formulas, numerical parameters, and their exact values — is not given here, but can be found in Mitchell, 1993.)

The Slipnet — Copycat's network of Platonic concepts

The basic image for the Slipnet is that of a network of interrelated concepts, each concept being represented by a *node* (caveat: what a concept is, in this model, is actually a bit subtler than just a pointlike node, as will be explained shortly), and each conceptual relationship by a *link* having a numerical length, representing the "conceptual distance" between the two nodes involved. The shorter the distance between two concepts is, the more easily pressures can induce a slippage between them.

Some of the main concepts in Copycat's Slipnet are: *a*, *b*, *c*, ..., *z*, *letter, successor, predecessor, alphabetic-first, alphabetic-last, alphabetic position, left, right, direction, leftmost, rightmost, middle, string position, group, sameness group, successor group, predecessor group, group length, 1, 2, 3, sameness,* and *opposite*. In all, there are roughly 60 concepts.

The Slipnet is not static; it dynamically responds to the situation at hand as follows: Nodes *acquire* varying levels of activation (which can be thought of as a measure of relevance to the situation at hand), *spread* varying amounts of activation to neighbors, and over time *lose* activation by decay. Activation is not an on-and-off affair, but varies continuously. However, when a node's activation crosses a certain critical threshold, the node has a probability of jumping discontinuously into a state of *full* activation, from which it proceeds to decay. In sum, the activation — the perceived relevance — of each concept is a sensitive, time-varying function of the way the program currently understands the situation it is facing.

Conceptual links in the Slipnet adjust their lengths dynamically. Thus, conceptual distances gradually change under the influence of the evolving perception (or conception) of the situation at hand, which of course means that the current perception of the situation enhances the chance of certain slippages taking place, while rendering that of others more remote.

Conceptual depth

Each node in the Slipnet has one very important static feature called its *conceptual depth*. This is a number intended to capture the generality and abstractness of the concept. For example, the concept *opposite* is deeper than the concept *successor,* which is in turn deeper than the concept *a*. It could be said roughly that the depth of a concept is how far that concept is from being directly perceivable in situations. For example, in Problem 2, the presence of instances of *a* is trivially perceived; recognizing the presence of *successorship* takes a little bit of work; and recognition of the presence of the notion *opposite* is a subtle act of abstract perception. The further away a given aspect of a situation is from direct perception, the more likely it is to be involved in what people consider to be the *essence* of the situation. Therefore, once aspects of greater depth are perceived, they should have more influence on the ongoing perception of the situation than aspects of lesser depth.

Assignment of conceptual depths amounts to an *a priori* ranking of "best-bet" concepts. The idea is that a deep concept (such as *opposite*) is normally relatively hidden from the surface and cannot easily be brought into the perception of a situation, but that once it *is* perceived, it should be regarded as highly significant. There is of course no guarantee that deep concepts will be relevant in any particular situation, but such concepts were assigned high depth-values precisely because we saw that they tend to crop up over and over again across many different types of situations, and because we noticed that the best insights in many problems come when deep concepts "fit" naturally. We therefore built into the architecture a strong drive, if a deep aspect of a situation is perceived, to use it and to try to let it influence further perception of the situation.

Note that the hierarchy defined by different conceptual-depth values is quite distinct from abstraction hierarchies such as

poodle $\Rightarrow$ *dog* $\Rightarrow$ *mammal* $\Rightarrow$ *animal* $\Rightarrow$ *living thing* $\Rightarrow$ *thing*.

These terms are all potential descriptions of a particular object at different levels of abstraction. By contrast, the terms *a*, *successor*, and *opposite* are not descriptions of one particular *object* in Problem 2, but of various aspects of the situation, at different levels of abstraction.

Likewise, conceptual depth is not the same as Gentner's notion of "abstractness" (Gentner, 1983). In Gentner's theory, attributes (*e.g.*, "the leftmost letter has value *a*") are invariably less abstract than relations (*e.g.*, "the next-to-leftmost letter is the successor of the leftmost letter"), which are in turn invariably less abstract than relations between relations (*e.g.*, "*successor* is the opposite of *predecessor*"). This heuristic, based on syntactic structure, often agrees with our conceptual-depth hierarchy, but in Copycat certain "attributes" are considered to be conceptually deeper than certain "relations" — for example, *alphabetic-first* has a greater depth than *successor* because we consider the former to be less directly perceivable than the latter. (In the following chapter, we go into considerably more detail in contrasting Gentner's work with ours.)

Conceptual depth has a second important aspect — namely, the deeper a concept is, the more resistant it is (all other things being equal) to slipping into another concept. In other words, there is a built-in propensity in the program to prefer slipping shallow concepts rather than deep concepts, when slippages have to be made. The idea of course is that insightful analogies tend to link situations that share a deep *essence,* allowing shallower features to slip if necessary. This basic idea can be summarized in a motto: *Deep stuff doesn't slip in good analogies.* There are, however, interesting situations in which specific constellations of pressures arise that cause this basic tendency to be overridden.

Activation flow and variable link-lengths

Some details about the flow of activation: (1) each node spreads activation to its neighbors according to their distance from it, with near neighbors getting more, distant neighbors less; (2) each node's conceptual-depth value sets its *decay rate,* in such a way that deep concepts always decay slowly and shallow concepts decay quickly. This means that, once a concept has been perceived as relevant, then, the deeper it is, the longer it will remain relevant, and thus the more profound an influence it will exert on the system's developing view of the situation — as indeed befits an abstract and general concept likely to be close to the essence of the situation.

Some details about the Slipnet's dynamical properties: (1) there are a variety of *link types,* and for each given type, all links of that type share the same *label*; (2) each label is itself a concept in the network; (3) every link constantly adjusts its length according to the activation level of its label, with high activation giving rise to short links, low activation to long ones. Stated another way: If concepts A and B have a link of type L between them, then as concept L's relevance goes up (or down), concepts A and B become conceptually closer (or further apart). Since this is happening all the time all throughout the network, the Slipnet is constantly altering its "shape" in attempting to mold itself increasingly accurately to fit the situation at hand. An example of a label is the node *opposite,* which labels the link between nodes *right* and *left,* the link between nodes *successor* and *predecessor,* and several other links. If the node *opposite* gets activated, all these links will shrink in concert, rendering the potential slippages they represent more probable.

The length of a link between two nodes represents the conceptual proximity or degree of association between the nodes: the shorter the link, the greater the degree of association, and thus the easier it is to effect a slippage between them. There is a probabilistic "cloud" surrounding any node, representing the likelihood of slippage to other nodes; the cloud's density is highest for near-neighbor nodes and rapidly tapers off for distant nodes. (This is reminiscent of the quantum-mechanical "electron cloud" in an atom, whose probability density falls off with increasing distance from the nucleus.) Neighboring nodes can be seen as being included in a given concept probabilistically, as a function of their proximity to the central node of the concept.

Concepts as diffuse, overlapping clouds

This brings us back to the caveat mentioned above: Although it is tempting to equate a concept with a pointlike node, a concept is better identified with this probabilistic "cloud" or halo *centered* on a node and extending outwards from it with increasing diffuseness. As links shrink and grow, nodes move into and out of each other's halos (to the extent that one can speak of a node as being "inside" or "outside" a blurry halo). This image suggests conceiving of the Slipnet not so much as a hard-edged network of points and lines, but rather as a space in which many diffuse clouds overlap each other in an intricate, time-varying way.

Conceptual proximity in the Slipnet is thus context-dependent. For example, in Problem 1, no pressures arise that bring the nodes *successor* and *predecessor* into close proximity, so a slippage from one to the other is highly unlikely; by contrast, in Problem 2, there is a good chance that pressures will activate the concept *opposite,* which will then cause the link between *successor* and *predecessor* to shrink, bringing each one more into the other's halo, and enhancing the probability of a slippage between them. Because of this type of context-dependence, concepts in the Slipnet are *emergent,* rather than explicitly defined.

The existence of an explicit core to each concept is a crucial element of the architecture. Specifically, slippability depends critically on the discrete jump from one core to another. Diffuse regions having no cores would not permit such discrete jumps, as there would be no specific starting or ending point. Even an explicit *name* attached to a coreless diffuse region could serve as a substitute for a core — it would permit a discrete jump. In any case, however, slippage requires each concept to be attached to some identifiable "place" or entity. One might liken the core of a concept to the official city limits of a large city, and the halo to the much vaguer metropolitan region surrounding the city proper, stretching out in all directions, and clearly far more subjective and context-dependent than the core.

It may be useful to briefly compare Copycat's Slipnet with connectionist networks. In localist networks, a concept is equated with a node rather than with a diffuse region centered on a node. In other words, concepts in localist networks lack halos. This lack of halos implies that there is no counterpart to slippability in localist networks. In distributed systems, on the other hand, there would seem to be halos, since a concept is equated with a diffuse region, but this is somewhat misleading. The diffuse region representing a concept is not explicitly centered on any node, so there is no explicit *core* to a concept, and in that sense no halo. But since slippability depends on the existence of discrete cores, there is no counterpart to slippability even in distributed connectionist models.

The lack of any explicit center to a concept would probably be found to be quite accurate if one could examine concepts on the neural level. However,

Copycat was not designed to be a neural model; it aims at modeling cognitive-level behavior by simulating processes at a subcognitive but superneural level. We believe that there is a subcognitive, superneural level at which it is realistic to conceive of a concept as having an explicit core surrounded by an implicit, emergent halo.

Another temptation might be to liken Copycat's context-dependent link-lengths to the changing of inter-node weights as a connectionist net adapts to training stimuli. One might even liken the effect of a label node in Copycat to a multiplicative connection (where some node's activation is used as a multiplicative factor in calculating the new weight of a link). To be sure, there is a mathematical analogy here, but conceptually there is a significant difference. As connectionist networks adapt and "learn" by changing their weights, there is no sense of departing from a norm and no tendency to return to an earlier state. By contrast, in Copycat, any changing of link-lengths takes place in response to a temporary context, and when that context is removed, the Slipnet tends to revert to its "normal" state. The Slipnet is thus "rubbery" or "elastic" in this sense; it responds to context but has a built-in tendency to "snap back" to its original state. We know of no corresponding tendency in connectionist networks.

Note that whereas the Slipnet changes over the course of a single run of Copycat, it does not retain changes from run to run, or create new permanent concepts. The program starts out in the same initial state on every run. Thus Copycat does not model *learning* in the usual sense. However, this project does concern learning, if that term is taken to include the notion of adaptation of one's concepts to novel contexts.

Although the Slipnet responds sensitively to events in the Workspace (described in a moment) by constantly changing both its "shape" and the activations of its nodes, its fundamental topology remains invariant. That is, no new structure is ever built, or old structure destroyed, in the Slipnet. The next subsection discusses a component of the architecture that provides a strong contrast to this type of topological invariance.

The Workspace — Copycat's locus of perceptual activity

The basic image for the Workspace is that of a busy construction site in which structures of many sizes and at many locations are being worked on simultaneously by independent crews, some occasionally being torn down to make way for new, hopefully better ones. (This image comes essentially from the biological cell; the Workspace corresponds roughly to the cytoplasm of a cell, in which enzymes carrying out diverse tasks all throughout the cell's cytoplasm are the construction crews, and the structures built up are all sorts of hierarchically-structured biomolecules.)

At the start of a run, the Workspace is a collection of unconnected raw data representing the situation with which the program is faced. Each item in the Workspace initially carries only bare-bones information — that is, for each letter token, just its alphabetic type is provided, as well as — for those letters at the very edges of their strings — the descriptor *leftmost* or *rightmost*. Other than that, all objects are absolutely barren. Over time, through the actions of many small agents "scouting" for features of various sorts (these agents, called "codelets", are described in the next subsection), items in the Workspace gradually acquire various *descriptions*, and are linked together by various *perceptual structures*, all of which are built entirely from concepts in the Slipnet.

The constant fight for probabilistic attention

Objects in the Workspace do not by any means all receive equal amounts of attention from codelets; rather, the probability that an object will attract a prospective codelet's attention is determined by the object's *salience*, which is a function of both the object's *importance* and its *unhappiness*. Though it might seem crass, the architecture honors the old motto "The squeaky wheel gets the oil", even if only probabilistically so. Specifically, the more descriptions an object has and the more highly activated the nodes involved therein, the more important the object is. Modulating this tendency is the object's level of unhappiness, which is a measure of how integrated the object is with other objects. An unhappy object is one that has few or no connections to the rest of the objects in the Workspace, and that thus seems to cry out for more attention. Salience is a dynamic number that takes into account both of these factors, and this number determines how attractive the object in question will appear to codelets. Note that salience depends intimately on both the state of the Workspace and the state of the Slipnet.

A constant feature of the processing is that pairs of *neighboring objects* (inside a single framework — *i.e.,* letter-string) are probabilistically selected (with a bias favoring pairs that include salient objects) and scanned for similarities or relationships, of which the most promising are likely to get "reified" (*i.e.,* realized in the Workspace) as inter-object *bonds*. For instance, the two *k*'s in *ijkk* in Problem 2 are likely to get bonded to each other rather quickly by a *sameness* bond. Similarly, the *i* and the *j* are likely to get bonded to each other, although not as fast, by a *successorship bond* or a *predecessorship bond*.

The existence of differential rates of speed of bond-making is meant to reflect realities of human perception. In particular, people are clearly quicker to recognize two neighboring objects as identical than as being related in some abstract way. Thus the architecture has an intrinsic speed-bias in favor of sameness bonds: it tends to spot them and to construct them more quickly than it spots and constructs bonds representing other kinds of relationships. (How the speeds of rival processes are dynamically controlled will be dealt with in more detail in the next subsection.)

Any bond, once made, has a dynamically varying *strength*, reflecting not only the activation and conceptual depth of the concept representing it in the Slipnet (in the case of *kk*, the concept *sameness,* and in the case of *ij*, either *successor* or *predecessor*) but also the prevalence of similar bonds in its immediate neighborhood. The idea of bonds is of course to start weaving unattached objects together into a coherent mental structure.

The parallel emergence of multi-level perceptual structures

A set of objects in the Workspace bonded together by a uniform "fabric" (*i.e.,* bond type) is a candidate to be "chunked" into a higher-level kind of object called a *group.* A simple example of a *sameness group* is *kk,* as in Problem 2. Another simple group is *abc,* as in Problem 1. This one, however, is a little ambiguous; depending on which direction its bonds are considered to go in, either it is perceived as having a left-to-right *successorship* fabric and is thus seen as a left-to-right *successor group,* or it is perceived as having a right-to-left *predecessorship* fabric and is thus seen as a right-to-left *predecessor group.* (It cannot be seen as both at once, although the program can switch from one vision to the other relatively easily.) The more salient a potential group's component objects and the stronger its fabric, the more likely it is to be reified.

Groups, just like more basic types of objects, acquire their own descriptions, salience values, and strengths, and are themselves candidates for similarity-scanning, bonding to other objects, and possibly becoming parts of yet higher-level groups. As a consequence, hierarchical perceptual structures get built up over time, under the guidance of biases emanating from the Slipnet. A simple example would be the successor (or predecessor) group *ijkk* in Problem 2, made up of three elements: the *i,* the *j,* and the short sameness group *kk.*

Another constant feature of the processing is that pairs of objects in *different* frameworks (*i.e.,* strings) are probabilistically selected (again with a bias favoring salient objects) and scanned for similarities, of which the most promising are likely to get reified as *bridges* (or *correspondences*) in the Workspace. Effectively, a bridge establishes that its two end-objects are considered each other's counterparts — meaning either that they are intrinsically similar objects or that they play similar roles in their respective frameworks (or hopefully both).

Consider, for instance, the *aa* and *kk* in Problem 2. What makes one tempted to equate them? One factor is their intrinsic similarity — both are doubled letters (sameness groups of length 2). Another factor is that they fill similar roles, since one sits at the left end of its string, the other at the right end of its string. If and when a bridge gets built between them, concretely reifying this mental correspondence, it will be explicitly based on both these facts. The fact that *a* and *k* are unrelated letters of the alphabet is simply ignored by most people. Copycat is constructed to behave similarly. Thus, the fact that *aa* and *kk* are both sameness groups will be embodied in an *identity mapping* (here, *sameness ⇔ sameness*); the fact that one is leftmost while the other is rightmost will be embodied in a *conceptual slippage* (here, *leftmost ⇔ rightmost*); the fact that nodes *a* and *k* are far apart in the Slipnet is simply ignored.

Whereas identity mappings are always welcome in a bridge, conceptual slippages always have to overcome a certain degree of resistance, the precise amount of which depends on the proposed slippage itself and on the circumstances. The most favored slippages are those whose component concepts not only are shallow but also have a high degree of overlap (*i.e.*, are very close in the Slipnet). Slippages between highly overlapping *deep* concepts are more difficult to build, but pressures can certainly bring them about.

Once any bridge is built, it has a *strength*, reflecting the ease of the slippages it entailed, the number of identity mappings helping to underpin it, and its resemblance to other bridges already built. The idea of bridges is of course to build up a coherent mapping between the two frameworks.

To form a clear image of all this hubbub, it is crucial to keep in mind that all the aforementioned types of perceptual actions — scanning, bond-making, group-making, bridge-building, and so forth (as well as all the spreading and decaying of activation and so on in the Slipnet) — take place in parallel, so that independent perceptual structures of all sorts, spread about the Workspace, gradually emerge at the same time, and all the biases controlling the likelihood of this concept or that one being brought to bear are constantly fluctuating in light of what has already been observed in the Workspace.

The drive towards global coherence and towards deep concepts

As the Workspace evolves in complexity, there is increasing pressure on new structures to be *consistent*, in a certain sense, with pre-existent structures, especially with ones in the same framework. For two structures to be consistent sometimes means that they are instances of the very same Slipnet concept, sometimes that they are instances of very close Slipnet concepts, and sometimes it is a little more complex. In any case, the Workspace is not just a hodgepodge of diverse structures that happen to have been built up by totally independent codelets; rather, it represents a coherent vision built up piece by piece by many agents all indirectly influencing each other. Such a vision will henceforth be called a *viewpoint*. A useful image is that of highly coherent macroscopic structures (*e.g.*, physical bridges) built by a colony of thousands of myopic ants or termites working semi-independently but nonetheless cooperatively. (The "ants" of Copycat — namely, codelets — will be described in the next subsection.)

There is constant competition, both on a local and a global level, among structures vying to be built. A structure's likelihood of beating out its rivals is determined by its *strength,* which has two facets: a context-independent facet (a contributing factor would be, for instance, the depth of the concept of which it is an instance) and a context-dependent facet (how well it fits in with the rest of the structures in the Workspace, particularly the ones that would be its neighbors). Out of the rough-and-tumble of many, many small decisions about which new structures to build, which to leave intact, and which to destroy comes a particular global viewpoint. Even viewpoints, however, are vulnerable; it takes a very powerful rival to topple an entire viewpoint, but this occasionally happens. Sometimes these "revolutions" are, in fact, the most creative decisions that the system as a whole can carry out.

As was mentioned briefly above, the Slipnet responds to events in the Workspace by selectively activating certain nodes. The way activation comes about is that any discovery made in the Workspace — creation of a bond of some specific type, a group of some specific type, etc. — sends a substantial jolt of activation to the corresponding concept in the Slipnet; the amount of time the effect of such a jolt will last depends on the concept's decay rate, which depends in turn on its depth. Thus, a deep discovery in the Workspace will have long-lasting effects on the activation pattern and "shape" of the Slipnet; a shallow discovery will have but transient effects. In Problem 2, for example, if a bridge is built between the groups *aa* and *kk,* it will very likely involve an *opposite* slippage (*leftmost* $\Leftrightarrow$ *rightmost*). This discovery will reveal the hitherto unsuspected relevance of the very deep concept *opposite,* which is a key insight into the problem. Because *opposite* is a deep concept, once it is activated, it will remain active for a long time and therefore exert powerful effects on subsequent processing.

It is clear from all this that the Workspace affects the Slipnet no less than the Slipnet affects the Workspace; indeed, their influences are so reciprocal and tangled that it is hard to tell the chicken from the egg.

Metaphorically, one could say that *deep concepts* and *structural coherency* act like strong magnets pulling the entire system. The pervasive biases favoring the realization of these abstract qualities in the Workspace imbues Copycat with an overall goal-oriented quality that *a priori* might seem surprising, given that the system is highly decentralized, parallel, and probabilistic, thus far more like a swarm of ants than like a rigid military hierarchy, the latter of which has more standardly served as a model for how to realize goal-orientedness in computer programs. We now turn to the description of Copycat's "ants" and how they are biased.

The Coderack — source of emergent pressures in Copycat

All acts of describing, scanning, bonding, grouping, bridge-building, destruction, and so forth in the Workspace are carried out by small, simple agents called *codelets*. The action of a single codelet is always but a tiny part of a run, and whether any particular codelet runs or not is not of much consequence. What matters is the collective effect of many codelets.

There are two types of codelets: *scout codelets* and *effector codelets*. A scout merely looks at a potential action and tries to estimate its promise; the only kind of effect it can have is to create one or more codelets — either scouts or effectors — to follow up on its findings. By contrast, an effector codelet actually creates (or destroys) some structure in the Workspace.

Typical *effector* codelets do such things as: attaching a description to an object (*e.g.*, attaching the descriptor *middle* to the *b* in *abc*); bonding two objects together (*e.g.*, inserting a *successor* bond between the *b* and *c* in *abc*); making a group out of two or more adjacent objects that are bonded together in a uniform manner; making a bridge that joins similar objects in distinct strings (similarity being measured by proximity of descriptors in the Slipnet); destroying groups or bonds, and so on.

Before any such action can take place, preliminary checking-out of its promise has to be carried out by *scout* codelets. For example, one scout codelet might notice that the adjacent *r*'s in *mrrjjj* are instances of the same letter, and propose a sameness bond between them; another scout codelet might estimate how well that proposed bond fits in with already-existing bonds; then an effector codelet might actually *build* the bond. Once such a bond exists, scout codelets might then check out the idea of subsuming the two bonded *r*'s into a sameness group, after which an effector codelet could go ahead and actually build the group.

Each codelet, when created, is placed in the *Coderack,* which is a pool of codelets waiting to run, and is assigned an *urgency value* — a number that determines its probability of being selected from that pool as the next codelet to run. The urgency is a function of the estimated importance of that codelet's potential action, which in turn reflects the biases embodied in the current state of the Slipnet and the Workspace. Thus, for example, a codelet whose purpose is to seek instances of some lightly activated Slipnet concept will be assigned a low urgency and will therefore probably have to wait a long time, after being created, to get run. By contrast, a codelet likely to further a Workspace viewpoint that is currently strong will be assigned a high urgency and will thus have a good chance of getting run soon after being created.

It is useful to draw a distinction between *bottom-up* and *top-down* codelets. Bottom-up codelets (or "noticers") look around in an unfocused manner, open to what they find, whereas top-down codelets (or "seekers") are on the lookout for a particular kind of phenomenon, such as successor relations or sameness groups. Codelets can be viewed as *proxies* for the pressures in a given problem. Bottom-up codelets represent pressures present in *all* situations (the desire to

make descriptions, to find relationships, to find correspondences, and so on).
Top-down codelets represent specific pressures evoked by the specific situation
at hand (*e.g.*, the desire, in Problems 1 and 2, to look for more successor
relations, once some have already been discovered). Top-down codelets can
infiltrate the Coderack only when triggered from "on high" — that is, from the
Slipnet. In particular, activated nodes are given the chance to "spawn" top-down
scout codelets, with a node's degree of activation determining the codelet's
urgency. The mission of such a codelet is to scan the Workspace in search of
instances of its spawning concept.

Pressures determine the speeds of rival processes

It is very important to note that the calculation of a codelet's urgency
takes into account (directly or indirectly) numerous factors, which may include
the activations of several Slipnet nodes as well as the strength or salience of
one or more objects in the Workspace; it would thus be an oversimplification
to picture a top-down codelet as simply a proxy for the particular concept that
spawned it. More precisely, a top-down codelet is a proxy for one or more
pressures evoked by the situation. These include *workspace pressures,* which
attempt to maintain and extend a coherent viewpoint in the Workspace, and
conceptual pressures, which attempt to realize instances of activated concepts. It
is critical to understand that pressures, while they are very *real,* are not
represented *explicitly* anywhere in the architecture; each pressure is spread out
among urgencies of codelets, activations and link-lengths in the Slipnet, and
strengths and saliences of objects in the Workspace. Pressures, in short, are
implicit, emergent consequences of the deeply intertwined events in the
Slipnet, Workspace, and Coderack.

Any run starts with a standard initial population of bottom-up codelets
(with preset urgencies) on the Coderack. At each time step, one codelet is
chosen to run and is removed from the current population on the Coderack.
As was said before, the choice is probabilistic, biased by relative urgencies in the
current population. Copycat thus differs from an "agenda" system such as
Hearsay II, which, at each step, executes the waiting action with the highest
estimated priority. The urgency of a codelet should not be conceived of as
representing an estimated *priority*; rather, it represents the estimated relative
speed at which the pressures represented by this codelet should be attended to.
If the highest-urgency codelet were always chosen to run, then lower-urgency
codelets would never be allowed to run, even though the pressures they
represent have been judged to deserve *some* amount of attention.

Since any single codelet plays but a small role in helping to further a given
pressure, it never makes a crucial difference that a particular codelet be
selected; what really matters is that each *pressure* move ahead at roughly the

proper speed over time. Stochastic selection of codelets allows this to happen, even when judgments about the intensity of various pressures change over time. Thus allocation of resources is an emergent statistical result rather than a preprogrammed deterministic one. The proper allocation of resources could not be programmed ahead of time, since it depends on what pressures emerge as a given situation is perceived.

The shifting population of the Coderack

The Coderack would obviously dwindle rapidly to zero if codelets, once run and removed from it, were not replaced. However, replenishment of the Coderack takes place constantly, and this happens in three ways. Firstly, *bottom-up* codelets are continually being added to the Coderack. Secondly, codelets that run can, among other things, add one or more *follow-up* codelets to the Coderack before being removed. Thirdly, active nodes in the Slipnet can add *top-down* codelets. Each new codelet's urgency is assigned by its creator as a function of the estimated promise of the task it is to work on. Thus the urgency of a follow-up codelet is a function of the amount of progress made by the codelet that posted it, as gauged by that codelet itself, while the urgency of a top-down codelet is a function of the activation of the node that posted it. The urgency of bottom-up codelets is context-independent.

As a run proceeds, the population of the Coderack adjusts itself dynamically in response to the system's needs, as judged by previously-run codelets and by activation patterns in the Slipnet, which themselves depend on the current structures in the Workspace. This means there is a *feedback loop* between perceptual activity and conceptual activity, with observations in the Workspace serving to activate concepts, and activated concepts in return biasing the directions in which perceptual processing tends to explore. There is no top-level executive directing the system's activity; all acts are carried out by ant-like codelets.

The shifting population of codelets on the Coderack bears a close resemblance to the shifting enzyme population of a cell, which evolves in a sensitive way in response to the ever-changing makeup of the cell's cytoplasm. Just as the cytoplasmic products of certain enzymatic processes trigger the production of new types of enzymes to act further on those products, structures built in the Workspace by a given set of codelets cause new types of codelets to be brought in to work on them. And just as, at any moment, certain genes in the cell's DNA genome are allowed to be expressed (at varying rates) through enzyme proxies, while other genes remain essentially repressed (dormant), certain Slipnet nodes get "expressed" (at varying rates) through top-down codelet proxies, while other nodes remain essentially repressed. In a cell, the total effect is a highly coherent metabolism that emerges without any explicit top-down control; in Copycat, the effect is similar.

Note that though Copycat runs on a serial computer and thus only one codelet runs at a time, the system is roughly equivalent to one in which many independent activities are taking place in parallel and at different speeds, since codelets, like enzymes, work locally and to a large degree independently. The speed at which an avenue is pursued is an *a priori* unpredictable statistical consequence of the urgencies of the many diverse codelets pursuing that avenue.

The Emergence of Fluidity in the Copycat Architecture

Commingling pressures — the crux of fluidity

One of the central goals of the Copycat architecture is to allow many pressures to simultaneously coexist, competing and cooperating with one another to drive the system in certain directions. The way this is done is by converting pressures into flocks of very small agents (*i.e.*, codelets), each having some small probability of getting run. As was stated above, a codelet acts as a proxy for several pressures, all to differing degrees. All these little proxies for pressures are thrown into the Coderack, where they wait to be chosen. Whenever a codelet is given the chance to run, the various pressures for which it is a proxy make themselves slightly felt. Over time, the various pressures thus "push" the overall pattern of exploration different amounts, depending on the urgencies assigned to their codelets. In other words, the "causes" associated with the different pressures get advanced in parallel, but at different speeds.

There is a definite resemblance to classical time-sharing on a serial machine, in which any number of independent processes can be run concurrently by letting each one run a little bit (*i.e.*, giving it a "time slice"), then suspending it and passing control to another process, and so forth, so that bit by bit, each process eventually runs to completion. Classical time-sharing, incidentally, allows one to assign to each process a different speed, either by controlling the *durations* of its time slices or by controlling the *frequency* with which its time slices are allowed to run. The latter way of regulating speed is similar to the method used in Copycat; however, Copycat's method is probabilistic rather than deterministic (comments on why this is so follow in brief order).

This analogy with classical time-sharing is helpful but can also mislead. The principal danger is that one might get the impression that there are pre-laid-out *processes* to which time slices are probabilistically granted — more specifically, that any codelet is essentially a time slice of some preordained process. This is utterly wrong. In the Copycat architecture, the closest analogue to a classical process is a pressure — but the analogy is certainly not close. A pressure is nothing like a determinate sequence of actions; in very broad brushstrokes, a *conceptual* pressure can be portrayed as a concept (or cluster of closely related concepts) trying to impose itself on a situation, and a *workspace* pressure as an established viewpoint trying to entrench itself further while keeping rival viewpoints out of the picture. Whereas classical processes are cleanly distinguishable from one another, this is not at all the case for pressures. A given codelet, by running, can advance (or hinder) any number of pressures.

There is thus no way of conceptually breaking up a run into a set of distinct foreordained processes each of which advances piecemeal by being given time slices. The closest one comes to this is when a series of effector codelets' actions *happen* to dovetail so well that the codelets *appear* to have been parts of some predetermined high-level construction process. However, what is deceptive here is that scattered amongst the actions constituting the visible "process", a lot of other codelets — certainly many scouts, and probably other effectors — have played crucial but less visible roles. In any case, there was some degree of luck because randomness played a critical role in bringing about this particular sequence of events. In short, although some large-scale actions tend to look planned in advance, that appearance is illusory; patterns in the processing are all *emergent*.

A useful image here is that of the course of play in a basketball game. Each player runs down the court, zigzagging back and forth, darting in and out of the enemy team as well as their own team, maneuvering for position. Any such move is simultaneously *responding* to a complex constellation of pressures on the floor as well as slightly *altering* the constellation of pressures on the floor. A move is thus fundamentally deeply ambiguous. Although the crowd is mostly concerned with the sequence of players who have the ball, and thus tends to see a localized, serial process unfolding, the players who seldom or never have the ball nonetheless play pivotal roles, in that they mold the globally-felt pressures that control both teams' actions at all moments. A tiny feint of the head or lunge to one side alters the probabilities of all sorts of events happening on the court, both near and far. After a basket has been scored, even though sports announcers and fans always try to account for the structure of the event in clean, spatially local, temporally serial terms (thus trying to impose a *process* on the event), in fact the event was in an essential way distributed all over space and time, amongst all the players. The event consisted of distributed, swiftly shifting pressures pushing *for* certain types of plays and *against* others, and impositions of locality and seriality, though they contain some truth, are merely ways of simplifying what happened for the sake of human consumption. The critical point to hold onto here is the *ambiguity* of any particular action en route to a basket; each action contributes to many potential continuations and cannot be thought of as a piece of some unique "process" coexisting with various other independent "processes" supposedly taking place on the court.

Much the same could be said for Copycat: an outside observer is free, after a run is over, to "parse" the run in terms of specific, discrete processes, and to attempt to impose such a vocabulary on the system's behavior; however, that parsing and labeling is not intrinsic to the system, and such interpretations are in no way unique or absolute, any more than in a basketball game. In other words, a long sequence of codelet actions can add up to what could be perceived, *a posteriori* and by an outsider, as a single coherent drive towards a particular goal, but that is the outsider's subjective interpretation.

The parallel terraced scan

One of the most important consequences of the commingling of multiple pressures is the *parallel terraced scan*. The basic image is that of many "fingers of exploration" simultaneously feeling out various potential pathways at different speeds, thanks to the coexistence of pressures of different strengths. These "fingers of exploration" are tentative probes made by scout codelets, rather than actual events realized by effector codelets. In the Workspace, there is only one *actual* viewpoint at any given time. However, in the background, a host of nearby variants of the actual viewpoint — *virtual* viewpoints — are constantly flickering probabilistically. If any virtual viewpoint is found sufficiently promising by scouts, then they create effector codelets that, when run, will attempt to realize that alternative viewpoint in the Workspace. This entails a "fight" between the incumbent structure and the upstart; the outcome is decided probabilistically, with the weights being determined by the strength of the current structure as opposed to the promise of the rival.

This is how the system's actual viewpoint develops with time. There is always a probabilistic "halo" of many *potential* directions being explored; the most attractive of these tend to be the *actual* directions chosen. Incidentally, this aspect of Copycat reflects the psychologically important fact that conscious experience is essentially unitary, although it is of course an outcome of many parallel unconscious processes.

A metaphor for the parallel terraced scan is provided by the image of a vast column of ants marching through a forest, with hordes of small scouts at the head of the column making small random forays in all directions (although exploring some directions more eagerly and deeply than others) and then returning to report; the collective effect of these many "feelers" will then determine the direction to be followed by the column as a whole. This is going on at all moments, of course, so that the column is constantly adjusting its pathway in slight ways.

The term "parallel terraced scan" comes from the fact that scouting expeditions are structured in a *terraced* way; that is, they are carried out in stages, each stage contingent upon the success of the preceding one, and probing a little more deeply than the preceding one. The first stage is computationally cheap, so the system can afford to have many first-stage scouts probing in all sorts of directions, including quite unlikely directions. Succeeding stages are less and less cheap; consequently the system can afford fewer and fewer of them, which means it has to be increasingly selective about the directions it devotes resources to looking in. Only after a pathway has been deeply explored and found to be very promising are effector codelets created, which then will try to actually swerve the whole system down that pathway.

The constellation of top-down pressures at any given time controls the biases in the system's exploratory behavior, and also plays a major role in determining the actual direction the system will move in; ultimately, however, top-down pressures, no matter how strong, must bow to the reality of the situation itself, in the sense that prejudices alone cannot force inappropriate concepts to fit to reality. Top-down pressures must adapt when the pathways they have urged turn out to fail. The model is made explicitly to allow this kind of intermingling of top-down and bottom-up processing.

Time-evolving biases

At the very start of a run, the Coderack contains exclusively bottom-up similarity-scanners, which represent no situation-specific pressures. In fact, it is their job to make small discoveries that will then start generating such pressures. As these early codelets run, the Workspace starts to fill up with bonds and small groups and, in response to these discoveries, certain nodes in the Slipnet are activated. In this way, situation-specific pressures are generated and cause top-down codelets to be spawned by concepts in the Slipnet. Thus top-down codelets gradually come to dominate the Coderack.

At the outset of a run, the Slipnet is "neutral" (*i.e.*, in a standard configuration with a fixed set of concepts of low depth activated), meaning that there are no situation-specific pressures. At this early stage, all observations made in the Workspace are very local and superficial. Over the course of a run, the Slipnet moves away from its initial neutrality and becomes more and more biased toward certain organizing concepts — *themes* (highly activated deep concepts, or constellations of several such concepts). Themes then guide processing in many pervasive ways, such as determining the saliences of objects, the strengths of bonds, the likelihood of various types of groups to be made, and in general, the urgencies of all types of codelets.

It should not be imagined, incidentally, that a "neutral" Slipnet embodies no biases whatsoever; it certainly does (think of the permanent inequality of various nodes' conceptual depths, for instance). The fact that at the outset, a sameness group is likely to be spotted and reified faster than a successor group of the same length, for instance, represents an initial bias favoring sameness over successorship. The important thing is that at the outset of a run, the system is more open than at any other time to *any* possible organizing theme (or set of themes); as processing takes place and perceptual discoveries of all sorts are made, the system loses this naïve, open-minded quality, as indeed it ought to, and usually ends up being "closed-minded" — that is, strongly biased towards the pursuit of some initially unsuspected avenue.

In the early stages of a run, almost all discoveries are on a very small, local scale: a primitive object acquires a description, a bond is built, and so on. Gradually, the scale of actions increases: small groups begin to appear, acquire their own descriptions, and so on. In the later stages of a run, actions take place on an even larger scale, often involving complex, hierarchically structured objects. Thus, over time there is a clear progression, in processing, from locality to globality.

Temperature as a regulator of open-mindedness

At the start of a run, the system is open-minded, and for good reason: it knows nothing about the situation it is facing. It doesn't matter all that much *which* codelets run, since one wants many different directions to be explored; hence decision-making can be fairly capricious. However, as swarms of scout codelets and local effector codelets carry out their jobs, that status gradually changes; in particular, as the system acquires more and more information, it starts creating a coherent viewpoint and focusing in on organizing themes. The more informed the system is, the more important it is that top-level decisions not be capriciously made. For this reason, there is a variable that monitors the stage of processing, and helps to convert the system from its initial largely bottom-up, open-minded mode to a largely top-down, closed-minded one. This variable is given the name *temperature*.

What controls the temperature is the *degree of perceived order* in the Workspace. If, as at the beginning of every run, no structures have been built, then the system sees essentially no order, which translates into a need for broad, open-minded exploration; if, on the other hand, there is a highly coherent viewpoint in the Workspace, then the last thing one wants is a lot of voices clamoring for irrelevant actions in the Workspace. Thus, temperature is essentially an inverse measure of the *quality of structure* in the Workspace: the more structures there are, and the more coherent they are with one another (as measured by their strengths), the lower the temperature. Note that although the overall trend is for temperature to wind up low at the end of a run, a *monotonic* drop in temperature is not typical; often, the system's temperature goes up and down many times during a run, reflecting the system's uncertain advances and retreats as it builds and destroys structures in its attempts to home in on the best way to look at a situation.

What the temperature itself controls is the *degree of randomness used in decision-making*. Decisions of every sort are affected by the temperature — which codelet to run next, which object to focus attention on, which of two rival structures should win a fight, and so on. Consider a codelet, for instance, trying to decide where to devote its attention. Suppose that Workspace object A is exactly twice as salient as object B. The codelet will thus tend to be more attracted to A than to B. However, the precise discrepancy in attractive power between A and B will depend on the temperature. At some mid-range temperature, the codelet will indeed be twice as likely to go for A as for B. However, at very *high* temperatures, A will be hardly any more attractive than B to the codelet. By contrast, at very *low* temperatures, the probability of choosing A over B will be much greater than two to one. For another example, consider a codelet trying to build a structure that is incompatible with a currently existing strong structure. Under low-temperature conditions, the strong structure will tend to be very stable (*i.e.,* hard to dislodge), but if the temperature should happen to rise, it will become increasingly susceptible to being swept away. In "desperate times", even the most huge and powerful structures and worldviews can topple.

The upshot of all this is that at the start of a run, the system explores possibilities in a wild, scattershot way; however, as it builds up order in the Workspace and simultaneously homes in on organizing themes in the Slipnet, it becomes an increasingly conservative decision-maker, ever more deterministic and serial in its style. Of course, there is no magic crossover point at which nondeterministic parallel processing turns into deterministic serial processing; there is simply a gradual tendency in that direction, controlled by the system's temperature.

Note that the notion of temperature in Copycat differs from that in simulated annealing, an optimization technique sometimes used in connectionist networks (Kirkpatrick, Gelatt, & Vecchi, 1983; Hinton & Sejnowski, 1983; Smolensky, 1983). In simulated annealing, temperature is used exclusively as a top-down randomness-controlling factor, its value falling monotonically according to a predetermined, rigid "annealing schedule". By contrast, in Copycat, the value of the temperature reflects the current quality of the system's understanding, so that temperature acts as a *feedback mechanism* that determines the degree of randomness used by the system. Thus, the system itself controls the degree to which it is willing to take risks.

Long after the concept of temperature had been conceived and implemented in the program, it occurred to us that temperature could serve an extra, unanticipated role: the *final* temperature in any run could give a rough indication of how good the program considered its answer to be (the lower the temperature, of course, the more desirable the answer). The idea is simply that the quality of an answer is closely correlated with the amount of strong, coherent structure underpinning that answer, and temperature is precisely an attempt to measure that quantity. From the moment we realized this, we kept track of the final temperatures of all runs, and those data provided some of the most important insights into the program's "personality", as will be apparent when we discuss in detail the results of runs.

Overall trends during a run

In most runs, despite local fluctuations here and there, there is a set of overall tendencies characterizing how the system evolves in the course of time. These tendencies, although they are all tightly linked together, can be roughly associated with different parts of the architecture, as follows.

- In the Slipnet, there is a general tendency for the initially activated concepts to be *conceptually shallow*, and for concepts that get activated later to be increasingly *deep*. There is also a tendency to move from *no themes* to *themes* (*i.e.,* clusters of highly activated, closely related, high-conceptual-depth concepts).
- In the Workspace, there is a general tendency to move from a state of *no structure* to a state with *much structure,* and from a state having *many local, unrelated objects* to a state characterized by *few global, coherent structures.*
- As far as the processing is concerned, it generally exhibits, over time, a gradual transition from *parallel* style toward *serial* style, from *bottom-up* mode to *top-down* mode, and from an initially *nondeterministic* style toward a *deterministic* style.

The Intimate Relation between Randomness and Fluidity

It may seem deeply counterintuitive that randomness should play a central role in a computational model of intelligence. However, careful analysis shows that it is inevitable if one believes in any sort of parallel, emergent approach to mind.

Biased randomness gives each pressure its fair share

A good starting point for such analysis is to consider the random choice of codelets (biased according to their urgencies) from the Coderack. The key notion, stressed in earlier sections, is that the urgency attached to any codelet represents the estimated proper *speed* at which to advance the pressures for which it is a proxy. Thus it would make no sense at all to treat higher urgencies as higher *priorities* — that is, always to pick the highest-urgency codelets first. If one were to do that, then lower-urgency codelets would never get run at all, so the effective speeds of the pressures they represent would all be zero, which would totally defeat the notion of commingling pressures, the parallel terraced scan, and temperature.

A more detailed analysis is the following. Suppose we define a "grass-roots" pressure as a pressure represented by a large swarm of low-urgency codelets, and an "elite" pressure as one represented by a small coterie of high-urgency codelets. Then a policy to select high-urgency codelets most of the time would arbitrarily favor elite pressures. In fact, it would allow situations wherein any number of grass-roots pressures could be entirely squelched by just *one* elite pressure — even if the elite pressure constituted but a small fraction of the *total* urgency (the sum of the urgencies of all the codelets in the Coderack at the time), as it most likely would. Such a policy would result in a very distorted image of the overall makeup of the Coderack (*i.e.,* the distribution of urgencies among various pressures). In summary, it is imperative that during a run, low-urgency codelets get mixed in with higher-urgency codelets, and in the right proportion — namely, in the proportions dictated by urgencies, no more and no less. As was said earlier, only by using probabilities to choose codelets can the system achieve (via statistics) a *fair* allocation of resources to each pressure, even when the strengths of various pressures change as processing proceeds.

Randomness and asynchronous parallelism

One might well imagine that the need for such randomness (or biased nondeterminism) is simply an artifact of this architecture's having been designed to run on a sequential machine; were it redesigned to run on parallel hardware, then all randomness could be done away with. This turns out to be not at all the case, however. To see why, we have to think carefully about what it would mean for this architecture to run on parallel hardware. Suppose that there were some large number of parallel processors to which tasks could be assigned, and that each processor's speed could be continuously varied. It is certainly not the case that one could assign *processes* to *processors* in a one-to-one manner, since, as has been stressed, there is no clear notion of "process" in this architecture. Nor could one assign one *pressure* to each processor, since codelets are not univalent as to the pressures that they represent. The only possibility would be to assign a processor to every single codelet, letting it run at a speed defined by that codelet's urgency. (Note that this requires a very large number of co-processors — hundreds, if not thousands. Moreover, since the codelet population varies greatly over time, the number of processors in use at different times will vary enormously. However, on a conceptual level, neither of those poses a problem in principle.)

Now notice a crucial consequence of this style: since all the processors are running at speeds that are completely independent of one another, they are effectively carrying out *asynchronous* computing, which means that relative to one another, the instants at which they carry out actions in the (shared) Workspace are totally decoupled — in short, entirely random relative to one another. This is a general fact: asynchronous parallelism is inseparable from processors' actions being random relative to one another (as pointed out in Hewitt, 1985). Thus parallelism provides no escape from the inherent randomness of this architecture. When it runs on serial hardware, some *explicit* randomizing device is utilized; when it runs on parallel hardware, the randomness is *implicit,* but no less random for that.

The earlier image of the swiftly-changing panorama of a basketball game may help to make this necessary connection between asynchronous parallelism and randomness more intuitive. Each player might well feel that the snap decisions being made constantly inside their own head are anything but random — that, in fact, their decisions are rational responses to the situation. However, from the point of view of *other* players, what any one player does is not predictable — a player's mind is far too complex to be modeled, especially in real time. Thus, because all the players on the court are complex, independent, asynchronously-acting systems, each player's actions *necessarily* have a random (*i.e.,* unpredictable) quality from the point of view of all the other players. And obviously, the more unpredictable a team seems to its opponents, the better.

A seeming paradox: Randomness in the service of intelligence

Even after absorbing all these arguments, one may still feel uneasy with the proposition that greater intelligence can result from making *random* decisions than from making *systematic* ones. Indeed, when the architecture is described this way, it sounds nonsensical. Isn't it always wiser to choose the *better* action than to choose at *random*? However, as in so many discussions about mind and its mechanisms, this appearance of nonsensicality is an illusion caused by a confusion of levels.

Certainly it would seem extremely counterintuitive — in fact, downright nonsensical — if someone suggested that a melody-composition program (say) should choose its next note by throwing dice, even weighted dice. How could any global coherence come from such a process? This objection is of course totally valid — good melodies cannot be produced in that way (except in the absurd sense of millions of monkeys plunking away on piano keyboards for trillions of years and coming up with "Blue Moon" once in a blue moon). But our architecture in no way advocates such a coarse type of decision-making procedure!

The choice of next note in a melody is a *top-level* macro-decision, as opposed to a low-level act of "micro-exploration". The purpose of micro-exploration is to efficiently explore the vast, foggy world of possibilities lying ahead without getting bogged down in a combinatorial explosion; for this purpose, randomness, being equivalent to non-biasedness, is the *most efficient* method. Once the terrain has been scouted out, much information has been gained, and in most cases some macroscopic pathways have been found to be more promising than others. Moreover — and this is critical — the more information that has been uncovered, the more the temperature will have dropped — and the lower the temperature is, the less randomness is used. In other words, the more confidently the system believes, thanks to lots of efficient and fair micro-scouting in the fog, that it has identified a particular promising pathway ahead, the more certain it is to make the macro-decision of picking that pathway. Only when there is tight competition is there much chance that the favorite will not win, and in such a case, it hardly matters since even after careful exploration, the system is not persuaded that there is a clear best route to follow.

In short, in the Copycat architecture, hordes of random forays are employed on a microscopic level when there is a lot of fog ahead, and their purpose is precisely to get an evenly-distributed sense of what lies out there in the fog rather than simply plunging ahead blindly, at random. The foggier things are, the more unbiased should be the scouting mission, hence the more randomness is called for. To the extent that the scouting mission succeeds, the temperature will fall, which in turn means that the well-informed macroscopic decision about to be taken will be made *non*-randomly. Thus, randomness is used *in the service of,* and not in opposition to, intelligent nonrandom choice.

A subtle aspect of this architecture is that there are all shades between complete randomness (much fog, high temperature) and complete determinism (no fog, low temperature). This reflects the fact that one cannot draw a clean, sharp line between micro-exploratory scouting forays and confident, macroscopic decisions. For instance, a smallish, very local building or destruction operation carried out in the Workspace by an effector codelet working in a mid-range temperature can be thought of as lying somewhere in between a micro-exploratory foray and a well-informed macroscopic decision.

As a final point, it is interesting to note that non-metaphorical fluidity — that is, the physical fluidity of liquids like water — is inextricably tied to random microscopic actions. A liquid could not flow in the soft, gentle, *fluid* way that it does, were it not composed of tiny components whose micro-actions are completely random relative to one another. This does not, of course, imply that the top-level action of the fluid *as a whole* takes on any appearance of randomness; quite the contrary! The flow of a liquid is one of the most nonrandom phenomena of nature that we are familiar with. This does not mean that it is by any means *simple*; it is simply familiar and natural-seeming. Fluidity is an emergent quality, and to simulate it accurately requires an underlying randomness.

Copycat's Performance: A Forest-level Overview

The statistically emergent robustness of Copycat

Now that the architecture of the Copycat program has been laid out, we can take a tour through the program's performance on a number of problems in its letter-string microworld. As was discussed earlier, Copycat's microworld was designed to isolate, and thus to bring out very clearly, some of the essential issues in high-level perception and analogy-making in general. The program's behavior on the problems presented here demonstrates how it deals with these issues, how it responds to variations in pressures, and how it is able, starting from exactly the same state on each new problem, to fluidly adapt to a range of different situations.[2] (The program's performance on a much larger set of problems and some comparisons with people's performance on the same problems are given in Mitchell, 1993.)

On any given run on a particular problem, the program settles on a specific answer; however, since the program is permeated with nondeterminism, different answers (to the same problem) are possible on different runs. The nondeterministic decisions the program makes (*e.g.*, which codelet to run next, which objects a codelet should act on, etc.) are all at a microscopic level, compared with the macroscopic-level decision of what answer to produce on a given run. Every run is different at the microscopic level, but statistics lead to far more deterministic behavior at the macroscopic level. For example, there are a huge number of possible routes (at the microscopic level of individual codelets and their actions) the program can take to arrive at the solution *ijl* to Problem 1, and a large number of micro-biases tend to push the program down one of those routes rather than down one of the huge number of possible routes to *ijd*. Thus in this problem, at a macroscopic level, the program is very close to being deterministic: it gets the answer *ijl* almost all the time.

The phenomenon of macroscopic determinism emerging from microscopic nondeterminism is often demonstrated in science museums by means of a contraption in which several thousand small balls are allowed to tumble down, one by one, through a regular grid of horizontal pins that run between two parallel vertical plexiglass sheets. Each ball, as it falls, bounces helter-skelter off various pins, eventually winding up in one of some 20 or 30 adjacent equal-sized bins forming a horizontal row at the bottom. As the number of balls that have

---

2. The current version of the Copycat program can deal only with problems whose initial change involves a replacement of at most one letter (*e.g.*, *abc* ⇒ *abd*, or *aabc* ⇒ *aabd*; of course the *answer* can involve a change of more than one letter, as in *aabc* ⇒ *aabd*; *ijkk* ⇒ *ijll*). This is a limitation of the program as it now stands; in principle, the letter-string domain is much larger. But even given this limitation, a very large number of interesting problems can be formulated, requiring considerable mental fluidity. (For a good number of examples of such problems, see Hofstadter, 1984b or Mitchell, 1993.)

fallen increases, the stacks of balls in the bins grow. However, not all bins are equally likely destinations, so different stacks grow at different rates. In fact, the heights of the stacks in the bins at the bottom gradually come to form an excellent approximation to a perfect gaussian curve, with most of the balls falling into the central bins, and very few into the edge bins. This reliable buildup of the mathematically precise gaussian curve out of many unpredictable, random events is fascinating to watch.

In Copycat, the set of bins corresponds to the set of different possible answers to a problem, and the precise pathway an individual ball follows, probabilistically bouncing left and right many times before "choosing" a bin at the bottom, corresponds to the many stochastic micro-decisions made by the program (at the level of individual codelets) during a single run. Given enough runs, a reliably repeatable pattern of answer frequencies will emerge, just as a near-perfect gaussian curve regularly emerges in the bins of a "gaussian pinball machine".

Copycat's "personality" is revealed through bar graphs

We present these patterns in the form of bar graphs, one for each problem, giving the frequency of occurrence (representing surface appeal) and the average end-of-run temperature (representing quality) for each different answer. For each problem, a bar graph is given, summarizing 1,000 runs of Copycat on that problem. The number 1,000 is somewhat arbitrary; after about 100 runs on each problem, the basic statistics do not change much. The only difference is that as more and more runs are done on a given problem, certain bizarre and improbable "fringe" answers, such as *ijj* in Problem 1 (see Figure V–1), begin to appear very occasionally; if 2,000 runs were done on Problem 1, the program would give perhaps one or two other such answers, each once or twice. This allows the bar graphs to make a very important point about Copycat: even though the program has the potential to get strange and crazy-seeming answers, the mechanisms it has allow it to steer clear of them almost all of the time. It is critical that the program (as well as people) be allowed the *potential* to follow risky (and perhaps crazy) pathways, in order for it to have the flexibility to follow *insightful* pathways, but it also has to avoid following bad pathways, at least most of the time.

In the bar graph of Figure V–1, each bar's height gives the relative frequency of the answer it corresponds to, and printed above each bar is the actual number of times that answer was given. The average final temperature appears below each bar. The frequency of a given answer can be thought of as an indicator of how *obvious* or *immediate* that answer is, given the biases of the program. For example, *ijl*, produced 980 times, is much more immediate to the program than *ijd*, produced 19 times, which is in turn much more obvious than the strange answer *ijj*, produced only once. (To get the latter answer, the program decided to replace the rightmost letter by its predecessor rather than

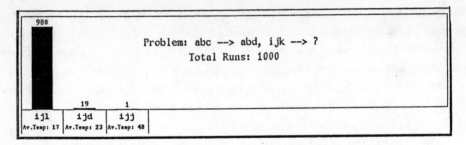

*Figure V–1. Bar graph summarizing 1,000 runs of the Copycat program on the analogy problem "**abc** ⇒ **abd**; **ijk** ⇒ ?".*

its successor. This slippage is always possible in principle, since *successor* and *predecessor* are linked in the Slipnet. However, as can be seen by the rarity of this answer, it is extremely unlikely in this situation: under the pressures evoked by this problem, *successor* and *predecessor* are almost always considered too distant for a slippage to be made between them.)

Although the frequencies shown in Figure V–1 seem quite reasonable, it is not intended that they should precisely reproduce the frequencies one would find if this problem were posed to humans, since, as we said earlier, the program is not meant to model all the domain-specific mechanisms people use in solving these letter-string problems. Rather, what is interesting here is that the program does have the potential to arrive at very strange answers (such as *ijj*, but also many others), yet manages to steer clear of them almost all the time.

As we said earlier, the average final temperature of an answer can be thought of as the program's own assessment of that answer's *quality*, with lower temperatures meaning higher quality. For instance, the program assesses *ijl* (average final temperature 17) to be of somewhat higher quality than *ijd* (temperature 23), and of much higher quality than *ijj* (temperature 48).

One can get a sense for what a given numerical value of temperature represents by seeing how various sets of perceptual structures built by the program affect the temperature. This will be illustrated in the next section, when a detailed set of screen dumps from a run of Copycat is presented. Roughly speaking, an average final temperature below 30 indicates that the program was able to build a fairly strong, coherent set of structures — that it had, in some sense, a reasonable "understanding" of what was going on in the problem. Higher final temperatures usually indicate that some structures were weak, or perhaps that there was no coherent way of mapping the initial string onto the target string.

The program decides probabilistically when to stop running and produce an answer, and although it is much more likely to stop when the temperature is low, it sometimes stops before it has had an opportunity to build all appropriate structures. For example, there are runs on Problem 1 in which the program stops before the target string has been grouped as a whole; the answer is still often *ijl*, but the final temperature is higher than it would have been if the program had continued. This kind of run increases the average final temperature for this answer. The lowest possible temperature for answer *ijl* is about 7, which is about as low as the temperature ever gets.[3]

Systematically studying the effects of variant problems

Systematic studies can be done in which a given problem is slightly altered in various ways. Each such variant tampers with the pressures that the original problem evokes, and one can expect effects of this to show up in the bar graph for that problem. For instance, Problem 2, discussed above, is a variant of Problem 1 in which the doubling of letters shifts the "stresses" in the strings *abc* and *ijk*; one might expect this to make the *aa* and the *kk* far more salient and more similar to each other than the *a* and *k* were in Problem 1, thus pushing towards a crosswise mapping in which the two double letters correspond.

In Figure V–2, one sees that despite the pressure towards a crosswise mapping, the "Replace rightmost group by successor" answer (*ijll*) is still the most common answer and the "Replace rightmost letter by successor" answer (*ijkl*) is second, indicating the lingering appeal of the straightforward *leftmost* ⇒ *leftmost*, *rightmost* ⇒ *rightmost* view, even here. However, the pressure is felt to some extent: *jjkk* makes a good showing and *hjkk* has some representatives too, as well as having by far the lowest average temperature. (This is to be contrasted with the results on Problem 1: note that in 1,000 runs, the program *never* gave an answer involving a replacement of the leftmost letter.) The answers on the fringe here include *jkkk* (which is similar to *jjkk*, but results from grouping the *two* leftmost letters — a far-fetched and, to most people, unappealing way of "parsing" the string); *ijkd* and *ijdd* (both based on the rule "Replace the rightmost letter by *d*", but flexed in different ways because of different bridges built from the *c*); *ijkk* (replacing all *c*'s by *d*'s); and *djkk* (replacing the *i* by a *d* instead of by its successor or predecessor).

Another variant on Problem 1 is the following:

3. Suppose the letter-string *abc* were changed to *abd*; how would you change the letter-string *kji* in "the same way"?

Here a literal application of the original rule ("Replace rightmost letter by successor") would yield *kjj*, which ignores an abstract similarity between *abc*

---

3. There is a problem with the way temperature is calculated in the program as it now stands. As can be seen, the answer *ijd* has an average final temperature almost equal to that of *ijl* (even though it is much less frequent), whereas most people feel it is a far worse answer. This, along with other problems with the current program, is discussed in detail in Mitchell, 1993.

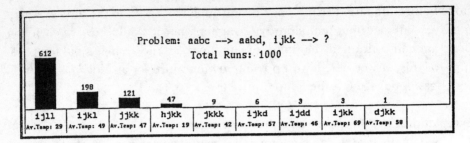

Figure V–2. Bar graph summarizing 1,000 runs of the Copycat program on the analogy problem "aabc ⇒ aabd; ijkk ⇒ ?".

and *kji*. An alternative many people prefer is *lji* ("Replace the leftmost letter by its successor"), which is based on seeing both strings in terms of a *successorship* fabric, in one string running to the right and in the other one to the left; thus there is a slippage from the concept *right* to the concept *left,* which in turn gives rise to the "cousin" slippage *rightmost ⇒ leftmost.* Another answer given by many people is *kjh* ("Replace the rightmost letter by its predecessor"), in which one string is seen as having a *successorship* fabric and the other as having a *predecessorship* fabric (both viewed as moving in the same spatial direction), thus involving a slippage of the concept *successor* into the concept *predecessor.*

As can be seen in Figure V–3, there are three answers that predominate, with *kjh* being the most common (and having the lowest average final temperature), and *kjj* and *lji* almost tying for second place (the latter being a bit less common, but having a much lower average final temperature). The answer *kjd* comes in a very distant fourth, and then there are two "fringe" answers with but one instance apiece: *dji* (an implausible blend of insight and rigidity in which the opposite spatial direction of the two successor groups *abc* and *kji* was seen, but instead of the leftmost letter being replaced by its successor, it was replaced by a *d* — and yet, notice the relatively low temperature on this answer, indicating that a strong set of structures was built!), and *kji* (reflecting the literal-minded rule "Replace *c* by *d*", where there are no *c*'s in *kji*), which has a very high temperature of 89, indicating that on this run, almost no structures were built before the program chanced, against very high odds, to stop.

How hidden concepts emerge from dormancy

Consider now the following problem, which involves a very different set of pressures from those in the previous problems:

4.   Suppose the letter-string *abc* were changed to *abd*; how would you change the letter-string *mrrjjj* in "the same way"?

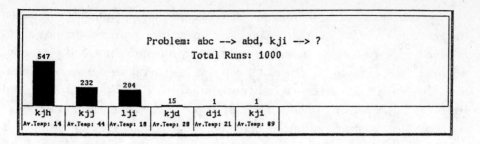

*Figure V–3. Bar graph summarizing 1,000 runs of the Copycat program on the analogy problem "**abc** ⇒ **abd; kji** ⇒ ?".*

There is a seemingly reasonable, straightforward solution: *mrrkkk*. Most people give this answer, reasoning that since *abc*'s rightmost letter was replaced by its successor, and since *mrrjjj*'s rightmost "letter" is actually a *group* of *j*'s, one should replace *all* the *j*'s by *k*'s. Another possibility is to take the phrase "rightmost letter" literally, thus replacing only the rightmost *j* by *k*, giving *mrrjjk*. However, neither answer is very satisfying, since neither takes into account the salient fact that *abc* is an alphabetic sequence (*i.e.*, a successor group). This fabric of *abc* is an appealing and seemingly central aspect of the string, so one would like to use it in making the analogy, but there is no obvious way to do so. No such fabric seems to weave *mrrjjj* together. So either (like most people) one settles for *mrrkkk* (or possibly *mrrjjk*), or one looks more deeply. But where to look, when there are so many possibilities?

The interest of this problem is that there happens to be an aspect of *mrrjjj* lurking beneath the surface that, once recognized, yields what many people feel is a deeply satisfying answer. If one ignores the *letters* in *mrrjjj* and looks instead at *group lengths,* the desired successorship fabric is found: the lengths of groups increase as "1–2–3". Once this hidden connection between *abc* and *mrrjjj* is discovered, the rule describing the change *abc* ⇒ *abd* can be adapted to apply to *mrrjjj* as "Replace the length of the rightmost group by its successor", yielding "1–2–4" at the abstract level, or, more concretely, *mrrjjjj*.

Thus this problem demonstrates how a previously irrelevant, unnoticed aspect of a situation can emerge as relevant in response to pressures. The crucial point is that the process of perception is not just about deciding which *clearly apparent* aspects of a situation should be ignored and which should be taken into account; it is also about the question of how aspects that were initially considered to be irrelevant — or rather, that were initially so far out of sight that they were not even recognized as being irrelevant! — can *become* apparent and relevant in response to pressures that emerge as the understanding process is taking place.

Sometimes, given certain pressures, a concept that one initially had no idea was germane to the situation will emerge seemingly from nowhere and turn out to be exactly what was needed. In such cases, however, one should not feel upset about not having suspected its relevance at the outset. In general, far-out ideas (or even ideas *slightly* past one's defaults) ought not continually occur to people for no good reason; in fact, a person to whom this happens is classified as crazy or crackpot.

Time and cognitive resources being limited, it is vital to resist nonstandard ways of looking at situations without strong pressure to do so. You don't check the street sign at the corner, every time you go outdoors, to reassure yourself that your street's name hasn't been changed. You don't worry, every time you sit down for a meal, that perhaps someone has filled the salt shaker with sugar. You don't worry, every time you start your car, that someone might have stuck a potato in its tailpipe or attached a bomb to its chassis. However, there are pressures — such as receiving a telephone threat on your life — that would make such a normally unreasonable suspicion start to seem reasonable. (These ideas overlap with Kahneman & Miller's 1986 treatment of counterfactuals, and are also closely related to the frame problem in artificial intelligence, as discussed in McCarthy & Hayes, 1969.)

Not only is pressure needed to evoke a dormant concept in trying to make sense of a situation, but the concepts brought in are often clearly related to the source of the pressure. For example, when one looks carefully at Problem 4 (as we will do in the next main section), one can see how certain aspects of it create pressures that, acting in concert, stand a decent chance of evoking the concept of *group length*. Some critical aspects of the story (not in any particular order) are these: (1) once successor relations have been noticed in *abc*, there arises a top-down pressure to look for them in *mrrjjj* as well; (2) once the *rr* and *jjj* sameness groups have been perceived in *mrrjjj*, the normally dormant concept *length* becomes weakly active and lingers in the background; (3) the perception of these sameness groups leads to top-down pressure to perceive other parts of the same string as sameness groups as well, and the only way this can be done is the unlikely possibility of perceiving the *m* as a sameness group consisting of *only one letter*; and (4) after standard concepts have failed to yield progress in making sense of the situation at hand, resistance to bringing in nonstandard concepts decreases.

People occasionally give the answer *mrrkkkk*, replacing *both* the letter category *and* the group length of the rightmost group by their successors (*k* and 4, respectively). Despite its interest, this answer confounds aspects of the two situations. What counts in establishing the similarity of *abc* and *mrrjjj* is their shared successorship fabric. In *mrrjjj*, that fabric has nothing to do with the specific letters *m, r,* and *j*; the letter sequence *m–r–j* is merely acting as a *medium* in which the *numerical* sequence "1–2–3" can be expressed. It is thus misguided

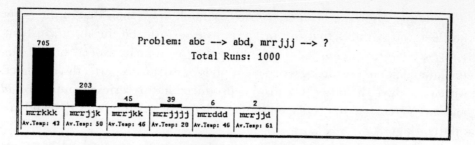

Figure V–4. Bar graph summarizing 1,000 runs of the Copycat program on the analogy problem "**abc** ⇒ **abd**; **mrrjjj** ⇒ ?".[4]

to focus on the *alphabetic* level of **mrrjjj** when one has just established that the essence of that string, in this context, is not its letters but its higher-level *numerical* structure. If the relations between lengths are perceived, then the answer at the level of lengths is 1–2–4. Translated back into the language of the carriers, this yields **mrrjjjj**. Additionally converting the four *j*'s into *k*'s is gilding the lily: it simply blends the alphabetic view with the numerical view in an inappropriate manner.

People have also proposed answers such as **mrryyyy**, where the three *j*'s are replaced by four copies of an arbitrary letter — here, *y*. The reasoning is that since the successorship fabric in **mrrjjj** has nothing to do with the specific letters *m*, *r*, and *j*, it doesn't matter *which* letter-value is used to replace the *j*'s. Such reasoning is too sophisticated for the current version of Copycat, which does not have the concept "arbitrary letter" — but even if Copycat could produce such an answer, we would still argue that **mrrjjjj** is the best answer to this problem. The letters *m*, *r*, and *j* serve as the medium for expressing the message "1–2–3", and we feel the most elegant solution is one that *preserves* that medium in expressing the modified message "1–2–4". Otherwise, why wouldn't an answer involving *total* replacement of the medium, such as **uggyyyy**, be just as good as, if not better than, **mrryyyy**?

As can be seen in Figure V–4, by far the most common answer is the straightforward **mrrkkk**, with **mrrjjk** coming in a fairly distant second. For Copycat, these are the two most immediate answers; however, the average final temperatures associated with them are fairly high, because of the lack of any coherent structure tying together the target string as a whole.

Next come two answers with roughly equal frequencies: **mrrjkk**, a rather silly answer that comes from grouping only the rightmost two *j*'s in **mrrjjj** and viewing this group as the object to be replaced; and **mrrjjjj**. The average final

---

4. The differences between the frequencies given in this figure and those given for the same problem in Mitchell & Hofstadter, 1990b are due to several improvements in the program — in particular, improvements in the way letter-groups and bridges are constructed.

temperature associated with this answer is much lower than that of the other answers, which shows that the program assesses it to be the most satisfying answer, though far from the most immediate. As in many aspects of real life, the immediacy or obviousness of a solution is by no means perfectly correlated with its quality. The other two answers produced in this series of runs, *mrrddd* and *mrrjjd*, come from replacing either a letter or a group with *d*'s, and are on the fringes.

In Problem 4, the successorship fabric is between group lengths rather than between letters, and is thus not immediately apparent. A simple variant on Problem 4 involves a successorship fabric both at the level of letters and at the level of lengths:

> 5. Suppose the letter-string **abc** were changed to **abd**; how would
>   you change the letter-string **rssttt** in "the same way"?

The strong pressures evoked in Problem 4 by the lack of any alphabetical fabric are missing in this variant, and the effect on Copycat can be seen in the bar graph given in Figure V–5: the program gave the *length* answer **rsstttt** only once in 1,000 runs (as contrasted with 39 instances of **mrrjjjj** in Problem 4). In Problem 5, the program is much more satisfied with the *letter-level* answer (**rssuuu**), which dominates and has a relatively low average final temperature. The other answers are similar to the answers given in the previous problem (plus there are a few additional answers based on strange groupings of the target string).[5]

Paradigm shifts in a microworld

A different set of issues comes up in the following problem:

> 6. Suppose the letter-string **abc** were changed to **abd**; how would
>   you change the letter-string **xyz** in "the same way"?

Naturally, the focus is on the letter *z*. One immediately feels challenged by its lack of successor — or more precisely, by the lack of a successor to *Platonic z*, the abstract concept (as opposed to the instance thereof found inside the string *xyz*). Many people, eager to *construct* a successor to Platonic *z*, invoke the commonplace notion of circularity, thus conceiving of *a* as the successor of *z*, much as January can be considered the successor of December, the digit '0' the successor of '9', an ace the successor of a king, or, in music, the note A the successor of G. This would yield *xya*.

Invoking circularity in this way to deal with Problem 6 is a small type of creative leap, and not to be looked down upon. However, the general notion of

---

5. The current version of Copycat is not able to create two simultaneous bonds between two given objects (*e.g.*, both the *alphabetical* and *numerical* successorship bonds between *r* and *ss*), so the program is at present unable to get what many people consider to be the best answer — namely, *rssuuuu.*

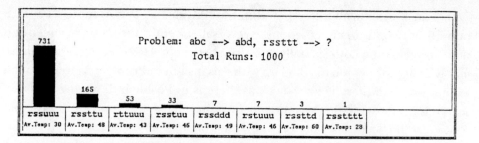

*Figure V–5. Bar graph summarizing 1,000 runs of the Copycat program on the analogy problem "**abc ⇒ abd; rssttt ⇒?**".*

circularity is not available to the program, as it is to people, for borrowing and insertion into the alphabet world. In fact, for important reasons, it is strictly stipulated that Copycat's alphabet is linear and just stops dead at z, immutably. We deliberately set this roadblock, because one of our main goals from the very conception of the project was to model the process whereby people deal with impasses.

In response to the z's lack of successor, quite a variety of thoughts can and do occur to people, such as: replace the z by nothing at all, thus yielding the answer **xy**. Or replace the z by the literal letter **d**, yielding answer **xyd**. (In other circumstances, such a resort to literality would be considered a rigid-minded and rather crude maneuver, but here it suddenly appears quite fluid and certainly reasonable.) Other answers, too, are possible, such as **xyz** itself (since the z cannot move farther along, just leave it alone); **xyy** (since you can't take the *successor* of the z, why not take its *predecessor*, which seems like second best?); **xzz** (since you can't take the successor of the z itself, why not take the successor of the letter sitting next to it?); and many others.

However, there is one particular way of looking at things that, to many people, seems like a genuine insight, whether or not they come up with it themselves. Essentially this is the idea that **abc** and **xyz** are "mirror images" of each other, each one being "wedged" against its own end of the alphabet. This would imply that the z in **xyz** corresponds not to the c but to the a in **abc**, and that it is the x rather than the z that corresponds to the c. (Of course, the **b** and the **y** are each other's counterparts as well.) Underlying these *object* correspondences (*i.e.,* bridges) is a set of three conceptually parallel slippages: *alphabetic-first ⇒ alphabetic-last, rightmost ⇒ leftmost,* and *successor ⇒ predecessor.* Under the profound conceptual reversal represented by these slippages, the raw rule flexes exactly as it did in Problem 2 — namely, into *replace the leftmost letter by its alphabetic predecessor.* This yields the answer **wyz**, which many people (including the authors) consider elegant and superior to all the other answers proposed above. More than any other answer, it seems to result from *doing the same thing* to **xyz** as was done to **abc**.

Note how similar and yet how different Problems 2 and 6 are. The key idea in both of them is to effect a double reversal (*i.e.,* to reverse one's perception of the target string both spatially and alphabetically). However, it seems considerably easier for people to come up with this insight in Problem 2, even though in that problem there is no "snag", as there is in Problem 6, serving to *force* a search for radical ideas. The very same insight is harder to come by in Problem 6 because the cues are subtler; the resemblance between *a* and *z* lurks far beneath the surface, whereas the resemblance between *aa* and *kk* is quite immediate.

In a sense, answer *wyz* to Problem 6 seems like a miniature "conceptual revolution" or "paradigm shift" (Kuhn, 1970), whereas answer *hjkk* to Problem 2 seems elegant but not nearly as radical. Any model of mental fluidity and creativity must faithfully reflect this notion of distinct "levels of subtlety". We will return to these issues of snags, cues, radical perceptual shifts, and levels of subtlety in the next main section, where we discuss the way in which Copycat succeeds, at least occasionally, in carrying out this miniature paradigm shift.

As can be seen in Figure V–6, the most common answer by far is *xyd*, for which the program decides that if it can't replace the rightmost letter by its *successor,* the next best thing is to replace it by a *d*. This is also an answer that people frequently give when told the *xya* avenue is barred.

A distant second in frequency, but the answer with the lowest average final temperature, is *wyz*, which, as was said above, is based on simultaneous spatial and alphabetic reversal in perception of the target string. This discrepancy between rank-order by obviousness and rank-order by quality is characteristic of problems where creative insight is needed. Clearly, brilliance will distinguish itself from mediocrity only in situations where deep ideas are elusive.

To bring about pressures that get the idea of the double reversal to bubble up, radical measures must be taken upon encountering the "z-snag" (the moment when the attempt to take the successor of *z* fails). These include sharply focusing attention upon the trouble spot, and raising the temperature from its rather low value just before the z-snag is hit to its maximum possible value of 100, opening up a far wider range of possible avenues of exploration. Only with the special combination of a sharp focus of attention and an unusually "broad-minded" attitude could *wyz* ever be found. (We will discuss all this in more detail in the next section.)

The next answer, *yyz*, reflects a view that sees the two strings as mapping to each other in a crosswise fashion, but ignores their opposite alphabetic fabrics; thus, while it considers the *leftmost* letter as the proper one in *xyz* to be changed, it clings to the notion of replacing it by its *successor,* since the letter changed in *abc* was replaced by *its* successor. (It is Problem 6's analogue to the answer *jjkk* in Problem 2.) Although this view seems somewhat inconsistent, like a good

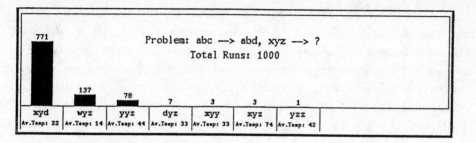

Figure V–6. Bar graph summarizing 1,000 runs of the Copycat program on the analogy problem "abc ⇒ abd; xyz ⇒ ?".

idea carried out only halfway, people very often come up with it. Indeed, such blends and half-completed trains of thought are very characteristic of human cognition (Hofstadter & Moser, 1989 gives examples of many types of cognitive blends and discusses their origin).

The other four answers are much farther out on the low-frequency fringes. The answer **dyz** (much like **dji** in Problem 3) is a highly implausible blend of insight and simple-mindedness, the insight being the subtle perception of the abstract symmetry linking **abc** and **xyz**, and the simple-mindedness being the extremely concrete and unimaginative way of conceiving the **abc** ⇒ **abd** change. Amusingly, this answer is self-descriptive, in that **dyz** can be pronounced "dizzy". Indeed, some people find this answer so dizzy in its style of thought that it evokes laughter. In Hofstadter *et al.* (1989), this and several other Copycat analogies are mapped onto real-world jokes, and are thereby used to suggest a theory of "slippage humor", one of whose tenets is that there is a continuum running from sensible through "sloppy" answers and winding up in "dizzy" answers, where "sloppy" and "dizzy" can be given semi-precise definitions in terms of the degree of consistency with which conceptual slippages are carried out.

The answer **xyy** allows that the two strings are to be perceived in opposite *alphabetic* directions (thus a *successor* ⇒ *predecessor* slippage), yet refuses to give up the idea that the strings have the same *spatial* direction; it thus insists on changing the *rightmost* letter, as was done to **abc**. It is amusing to note that **ijj** — Problem 1's analogue to this answer[6] — was produced one time in 1,000 runs, even without the pressure of a snag.

---

6. It is ironic that a claim of analogousness of answers to different Copycat problems (such as the offhand remark made in the text that **ijj** in Problem 1 is "the analogue" to **xyy** in Problem 6) comes across as objective and unproblematic to most people, despite the fact that many people express doubt about the notion of "rightness" or "wrongness" of letter-string analogies. The fact is, most people *do* have a strong intuitive sense of right and wrong analogies — it's just that when the psychological context is "Solve this analogy puzzle", they put their guard up and become wary of any claims, whereas when the context is "commentary on our program's behavior", they lower their guard and go with their intuitions, without even realizing the change in their attitude.

The answer *xyz*, whose very high temperature of 74 indicates that the program did not "like" it at all, comes from interpreting the *abc* ⇒ *abd* change as "Replace *c* by *d*". (This is not nearly as clever as positing that the *z* might serve as its *own* successor, which is an entirely different way of justifying this same answer, and one that people fairly often suggest. In fact, when *xya* is barred, *xyz* is what people come up with most often.) Note that *ijk*, the analogous answer[6] to Problem 1, was *never* produced. It takes the "desperation" caused by the *z*-snag to allow such strange ideas any chance at all.

Finally, answer *yzz* is a peculiar, almost pathological, variant of the above-discussed answer *yyz*, in which the *x* and *y* in *xyz* are grouped together as one object, which is then replaced *as a whole* by its "successor" (the successor of each letter in the group). Luckily, it was produced only once in 1,000 runs, and was considered a poor answer.

In Problem 6, pressure for a crosswise mapping (leading to the answer *wyz*) comes both from the existence of an impasse and from the possibility of an appealing way out of that impasse — namely, a high-quality bridge linking instances of the two "distinguished" Platonic letters, *a* and *z*. Suppose that the impasse was retained while the appeal of the "escape route" was greatly reduced — what would be the effect on Copycat's behavior? The following variant explores that question.

> 7. Suppose the letter-string *rst* were changed to *rsu*; how would you change the letter-string *xyz* in "the same way"?

As Figure V–7 shows, *wyz* was produced on only 1 percent of the runs, whereas in Problem 6 it was given on almost 14 percent of the runs. Here, there is very little to suggest building a crosswise bridge, because the *r* and *z* have almost nothing in common, aside from the rather irrelevant fact that the *r* is leftmost in its string and the *z* rightmost in *its* string — hardly a powerful reason to make an *r*–*z* bridge.

For some perspective, compare this to Problem 1. How much appeal is there to the idea of mapping the leftmost letter of *abc* onto the rightmost letter of *ijk*? Such a crosswise *a*–*k* bridge would result either in the answer *hjk* (characterized at the beginning of this article as "unmotivated fluidity"), or possibly in *jjk* or *djk*. However, in 1,000 runs on Problem 1, Copycat never produced any of those answers, nor have we ever run into a human who has suggested any of them as an answer to Problem 1. (Actually, one person once *did* propose *hjk* in response to Problem 1, but this was under the influence of having just seen the *wyz* answer to Problem 6.) In colloquial terms, answers to Problem 1 based on a crosswise mapping seem completely "off the wall".

In Problem 7, of course, things are different, because there is, after all, a snag, and hence a kind of "desperation". The various emergency measures —

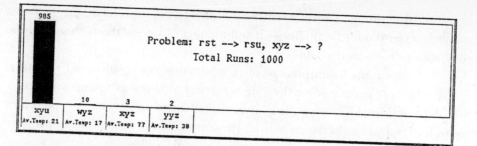

Figure V–7. Bar graph summarizing 1,000 runs of the Copycat program on the analogy problem "rst ⇒ rsu; xyz ⇒ ?".

especially the persistent high temperature — make the normally unappealing *r–z* bridge a bit more tempting, and so, once in a while, it gets built. From that point on, the whole paradigm shift goes through exactly as in Problem 6, and *wyz* is the outcome.

In some sense, Problem 7 lies halfway between Problems 1 and 6, so answer *wyz* in Problem 7 represents an intermediate stage between unmotivated and motivated fluidity. It is most gratifying to us that Copycat responds to the different constellations of pressures in these problems in much the way that our intuition feels it ought to.

Families of problems as a "miniature Turing Test"

It cannot be overestimated how critical we feel this method of probing Copycat through various families of subtly related problems is. When we began running Copycat on a large number of problems, we had no clear idea of what its performance would be, and we were, frankly, somewhat nervous. To us, the experience of watching Copycat reacting to each new problem had the feel of a kind of "miniature Turing Test", in the sense that each new problem posed to Copycat was like a question and answer in the Turing Test that would inevitably bring out some new and unanticipated aspect of the program's personality (this analogy to the Turing Test is further discussed in French, 1995; see also the Epilogue of this book). Copycat's mechanisms were truly put to the test by the families of problems we challenged it with, and by and large it came through with flying colors. A thorough discussion of those tests is found in Chapters 4 and 5 of Mitchell (1993).

The bar graphs just presented suggest the range of Copycat's abilities, and show how diverse constellations of pressures affect its behavior. They show the degree to which the program exhibits rudimentary fluid concepts able to adapt to different situations in a microworld that, though idealized, captures much of the essence of real-world analogy-making. They also reveal, by displaying *bad* analogies Copycat makes, some of the program's flaws and weaknesses. But they also show that though Copycat has the *potential* to get far-fetched answers — a potential essential for flexibility — it still manages to *avoid* them almost all the time, which shows its robustness.

It is important to emphasize once again that our goal is not to model specifically how people solve these letter-string analogy problems (it is clear that the microworld involves only a very small fraction of what people know about letters and might use in solving these problems), but rather to propose and model mechanisms for fluid concepts and analogy-making *in general*. These mechanisms, described earlier on, will be illustrated in detail in the next section, which follows the temporal progression of Copycat as it solves two problems. First we take a particular run of the program on Problem 4 and present it through a series of screen dumps; then we move to Problem 6 and discuss the abstract pathway followed by almost all runs that lead to answer *wyz*.

Copycat's Performance: A Tree-level Close-up

A problem where perception plays a crucial role is chosen as a focus

We now illustrate the mechanisms described in this article by presenting a detailed set of screen dumps from a single run of Copycat on Problem 4. As was discussed above, this problem has a seemingly reasonable, straightforward solution, *mrrkkk*, but neither this answer nor the more literal *mrrjjk* is very satisfying, since neither reflects an underlying successorship structure in *mrrjjj* analogous to that in *abc*. Such a successorship fabric can be found only if relationships between group lengths are perceived in *mrrjjj*. But how can the notion of *group length*, which in most problems remains essentially dormant, come to be seen as relevant by Copycat?

Length is certainly in the halo of the concept *group*, as are other concepts, such as *letter category* (*e.g., j* for the group *jjj*), *string position* (*e.g., rightmost*), and *group fabric* (*e.g., sameness*). Some of these concepts are more closely associated with *group* than others; in the absence of pressure, the notion of *length* tends to be fairly far away from *group* in conceptual space. Thus in perceiving a group such as *rr*, one is virtually certain to notice its letter category (namely, *r*), but not very likely to notice, or at least attach importance to, its length (namely, 2). However, since the concept *length* is in *group*'s halo, there is some chance that lengths will be noticed and used in trying to make sense of the problem. One might, for instance, consciously notice a group's length at some point, but if this doesn't turn out useful, *length*'s relevance will diminish after a short while. (This might happen in the variant problem "*abc ⇒ abd; mrrrrjj ⇒ ?*".) This dynamic aspect of relevance is very important: even if a new concept is at some point brought in as relevant, it is counterproductive to continue spending much of one's time exploring avenues involving that concept if none seems promising.

The story in quick strokes

One way Copycat arrives at **mrrjjjj** is now sketched (of course, since the program is nondeterministic, there are many possible routes to any given answer). The input consists of three "raw" strings (here, **abc**, **abd**, and **mrrjjj**) with no preattached bonds or preformed groups; it is thus left to the program to build up perceptual structures constituting its understanding of the problem in terms of concepts it deems relevant.

On most runs, the groups **rr** and **jjj** get built (the program tends to see sameness groups quite fast). Each group's letter category (**r** and **j**, respectively) is noted, as the concept *letter category* is relevant by default. Although there is some chance for length to be noticed when a group is made, it is low, as *length* is only weakly associated with *group*. Once **rr** and **jjj** are made, *sameness group* becomes very relevant, which creates top-down pressure to describe other objects, especially in the same string, as sameness groups, if possible. The only way to do this here is to describe the **m** as a sameness group having just one member. But this is resisted by a strong opposing pressure: a single-member group is an intrinsically weak and far-fetched construct. It would be disastrous if Copycat were willing to bring in unlikely notions such as single-member groups without strong pressure: it would then waste huge amounts of time exploring ridiculous avenues in every problem. However, the prior existence of two other strong sameness groups in the same string, coupled with the system's unhappiness at its failure to incorporate the lone **m** into any large, coherent structure (revealed by a persisting high temperature), pushes against this intrinsic resistance.

These opposing pressures fight; the outcome is decided only as a statistical result of probabilistic decisions made by a large number of codelets. If the **m** chances to be perceived as a single-letter sameness group, that group's length will very likely be noticed (single-letter groups are noteworthy precisely because of their abnormal length), making *length* more relevant in general, and thus increasing the probability of noticing the other two groups' lengths. Moreover, *length*, once brought into the picture, has a good chance of staying relevant, since descriptions based on it turn out to be useful. (Without reinforcement, a node's activation decays over time. Thus, for instance, had the target string been **mrrrrjj**, *length* might get brought in at some point, but it would not turn out useful, so it would likely fade back into obscurity.)

In **mrrjjjj**, once lengths are seen, the (numerical) successor relations among them might be spotted by bottom-up codelets, ever-present in the Coderack, continually seeking new relations in the Workspace. (Note that such spontaneous bottom-up noticing could happen only in a parallel architecture where many types of properties can be continually being looked for at once without a need for explicit prompting.) Another way, perhaps more likely, that the noticing of successor relations in *mrrjjj* could occur is through top-down pressure caused by the already-seen successor relations in *abc*. In any case, as soon as the numerical successorship relations are seen and a much more satisfying view of *mrrjjj* begins to emerge, interest in the groups' letter categories fades and *length* becomes their most salient aspect. Thus the crux of finding this solution lies in the triggering of the concept *length*.

Screen dumps tell the story in detail

Figure V–8 is a series of screen dumps from a run of Copycat, showing one way it arrives at the answer *mrrjjjj* (note that this answer is not very typical: according to Figure V–4, this answer is given only about 4 percent of the time).

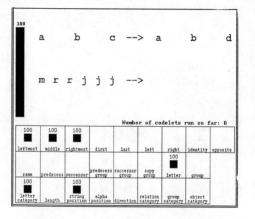

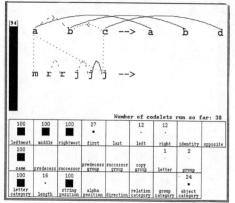

1. The problem is presented. Temperature, shown on a "thermometer" (at the left), is at its maximum of 100, since no structures have yet been built. At the bottom, some Slipnet nodes are displayed. (Note: links are not shown. Also, due to limited space, many nodes are not shown, *e.g.,* those for *a, b,* etc.) A black square represents a node's current activation level (the numerical value, between 0 and 100, is shown above the square).

Nodes here displayed include *leftmost, middle,* and *rightmost* (the possible *string positions* of objects in the Workspace); *first* and *last* (the distinguished *alphabetic positions* of Platonic letters *a* and *z*); *left* and *right* (the possible *directions* for bonds and groups); *identity* and *opposite* (two of the possible relations between concepts); *same, predecessor,* and *successor* (the possible *bond categories* for bonds between Workspace objects); *predecessor group, successor group,* and *copy-group*[1] (the various *group categories*); *letter* and *group* (the possible *object categories* for Workspace objects); and in row 3, nodes representing these various categories of descriptions, including *length.*

Every letter comes with some preattached descriptions: its *letter category* (*e.g., m*), its *string position* (*leftmost, middle, rightmost,* or none — *e.g.,* the fourth letter in *mrrjjj* has no string-position description), and its *object category* (*letter,* as opposed to *group*). These nodes start out highly activated.

2. The 30 codelets so far run have begun exploring many possible structures. *Dotted* lines and arcs represent structures in early stages of consideration; *dashed* lines and arcs represent structures in more serious stages of consideration; finally, *solid* lines and arcs represent structures actually built, which can thus influence temperature as well as the building of other structures. Various bonds and bridges between letters are being considered (*e.g.,* the dotted *a–j* bridge, which is based on the relatively long *leftmost–rightmost* Slipnet link; being implausible, it won't be pursued much further).

Bridges connecting letters in *abc* with their counterparts in *abd* have been built by bottom-up codelets, as has a *j–j* sameness bond at the right end of *mrrjjj*; this latter discovery activated the node *same,* resulting in top-down pressure (*i.e.,* new codelets) to seek instances of sameness elsewhere.

Some nodes have become lightly activated via spreading activation (*e.g.,* the node *first,* via activation from the node *a* [not shown]). The slight activation of *length* comes from its weak association with *letter category* (letters and numbers form linear sequences and are thus similar; numbers are associated with *length*). The temperature has fallen in response to the structures so far built. It should be pointed out that many fleeting explorations are constantly occurring (*e.g.,* "Are there any relations of interest between the *m* and its neighbor *r*?"), but they are not visible here.

---

1. Note that the term *copy-group,* in this set of screen dumps and captions, means *sameness group,* in the terminology of the text.

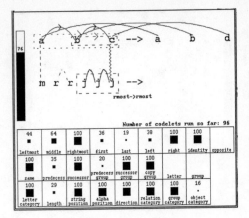 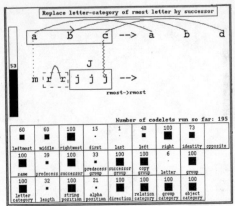

3. The successorship fabric of **abc** has been observed, and two rival groups based on it are being considered: **bc** and **abc**. Although the former got off to an early lead (it is dashed while the latter is only dotted), the latter, being intrinsically a stronger structure, has a higher chance of actually getting built.

Exploration of the crosswise **a–j** bridge was aborted, since it was (probabilistically) judged to be too weak to merit further consideration. A more plausible **c–j** bridge has been built (jagged vertical line); its reason for existence (namely, both letters are *rightmost* in their respective strings) is given beneath it, in the form of an identity mapping.

Since *successor* and *sameness* bonds have been built, these nodes are highly active; they in turn have spread activation to *successor group* and *copy-group* (*i.e., sameness group*), which creates top-down pressure to look for such groups. Indeed, a **jjj** copy-group is being strongly considered (dashed box). Also, since *first* was active, *alphabetic position* became highly active (a probabilistic event), making alphabetic-position descriptions likely to be considered.

4. Groups **abc** and **jjj** have been built (the bonds between their letters still exist, but for the purposes of graphical simplicity are no longer being displayed). An **rr** copy-group is being considered. The already-built copy-group **jjj** strongly supports this potential move, which accelerates it, in the sense that codelets investigating the potential structure will be assigned higher urgency values.

Meanwhile, a *rule* (shown at the top of the screen) has been constructed to describe how **abc** changed. The current version of Copycat assumes that the example change involved the replacement of exactly one letter, so rule-building codelets fill in the template "Replace ____ by ____", choosing probabilistically from descriptions that the program has attached to the changed letter and its replacement, with a probabilistic bias toward choosing more abstract descriptions (*e.g.,* usually preferring *rightmost letter* to *c*).

Since the nodes *first* and *alphabetic position* didn't turn out useful, they have faded. Also, although *length* received additional activation from *group*, it is still not very activated, and so lengths are still unlikely to be noticed.

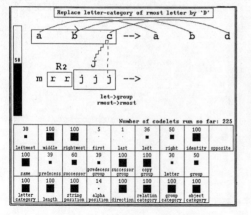

Replace letter-category of rmost letter by 'D'

a b c --> a b d

58

R2
J

m r r j j j -->

let->group
rmost->rmost

Number of codelets run so far: 225

leftmost	middle	rightmost	first	last	left	right	identity	opposite
38	100	100	5	1	36	50	100	

same	predecess	successor	predecess group	successor group	copy group	letter	group
100	39	60	39	100	100	30	50

letter category	length	string position	alpha position	direction	relation category	group category	object category
100	100	100	14	100	100	100	100

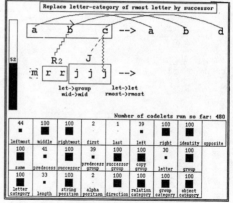

Replace letter-category of rmost letter by successor

a b c --> a b d

52

R2 J

m r r j j j -->

let->group   let->let
mid->mid   rmost->rmost

Number of codelets run so far: 480

leftmost	middle	rightmost	first	last	left	right	identity	opposite
44	100	100	2	1	39	100	100	

same	predecess	successor	predecess group	successor group	copy group	letter	group
100	41	100	39	100	100	30	100

letter category	length	string position	alpha position	direction	relation category	group category	object category
100	33	100	2	100	100	100	100

5. Now, some 225 codelets into the run, the letter-to-*letter* *c–j* bridge has been defeated by the stronger letter-to-*group* *c–J* bridge, although the former possibility still lurks in the background. Meanwhile, an *rr* copy-group has been built whose length (namely, 2) happened to be noticed (a probabilistic event); therefore, a "2", along with the group's letter category (namely, *r*), is displayed just above the group. *Length* is now fully active, and for this reason the "2" is a salient Workspace object (indicated by boldface).

A new rule, "Replace the letter category of the rightmost letter by *d*", has replaced the old one at the top of the screen. Although this rule is weaker than the previous one, fights between rival structures (including rules) are decided probabilistically, and this one simply happened to win. However, its weakness has caused the temperature to go up.

If the program were to stop now (which is quite unlikely, since a key factor in the program's probabilistic decision when to stop is the temperature, which is now quite high), the rule would be adapted for application to the string *mrrjjj* as "Replace the letter category of the rightmost *group* by *d*" (the *c–J* bridge establishes that the role of *letter* in *abc* is played by *group* in *mrrjjj*), yielding the answer *mrrddd* (an answer that Copycat does indeed produce, on rare occasions).

6. The previous, stronger rule has been restored (again the result of a fight having a probabilistic outcome), but at the same time, the strong *c–J* bridge also happened to get defeated by its weaker rival, the *c–j* bridge. As a consequence, if the program were to stop at this point, its answer would be *mrrjjk*. This, incidentally, would also have been the answer in screen dump #4.

In the Slipnet, the activation of *length* has decayed a good deal, since the length description given to *rr* wasn't found to be useful. In the Workspace, the diminished salience of the *rr*'s length description "2" is represented by the fact that the "2" is no longer in boldface.

The temperature is still fairly high, since the program is having a hard time making a single, coherent structure out of *mrrjjj*, something that it did easily with *abc*. That continuing difficulty, combined with strong top-down pressure from the two copy-groups that have been built inside *mrrjjj*, makes it now somewhat tempting for the system to flirt with the *a priori* extremely unlikely idea of making a single-letter copy-group (this flirtation is represented by the dashed rectangle around the letter *m*).

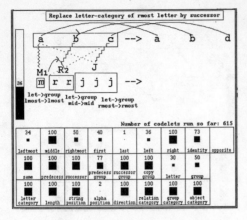

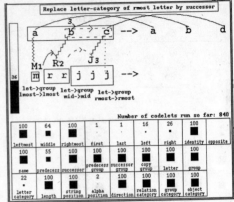

7. As a result of these combined pressures, the *a priori* extremely unlikely single-letter copy-group **m** happened to get built, and its length of 1, being very noteworthy, has been attached to the group as a description. A successorship bond between that "1" and its right-neighbor "2" has already been built; all of this is helping *length* to stay active. A consistent trio of *letter* ⇒ *group* bridges has now been made, and as a result of these promising new structures, the temperature has fallen to the relatively low value of 36, which in return helps to lock in this emerging view.

If the program were to halt in this screen dump or in the following one, it would produce the answer **mrrkkk**, which is its most frequent answer.

8. As a result of *length*'s continued activity, length descriptions have been attached to the remaining two groups in the problem (*jjj* and **abc**), and a successorship bond between the "2" and the "3" (for which there is much top-down pressure coming from both **abc** and the emerging view of **mrrjjj**) is being considered (dashed arc). *Letter category* has decayed, indicating that it hasn't lately been of use in building structures.

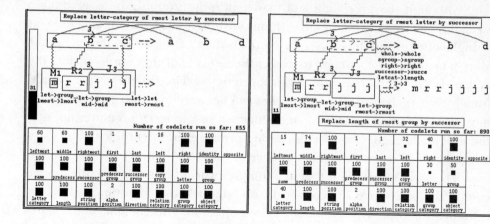

9. The "2"–"3" bond was built, whereupon a very abstract, high-level numerical successor group involving the group lengths in *mrrjjj* was perceived (large solid rectangle surrounding the three copy-groups). Also, a bridge (dotted vertical line to the right of the two strings) is being considered between strings *abc* and *mrrjjj* in their entireties.

Ironically, just as these sophisticated ideas seem to be converging toward a highly insightful answer, a small renegade codelet, totally unaware of the global momentum, has had some good luck: its bid to knock down the *c–J* bridge and replace it with a *c–j* bridge was accepted. Of course, this is a setback on the global level. If the program were forced to stop at this point, it would answer *mrrjjk* — the same rather dull answer as it would have given in screen dumps #6 and #4. However, at either of those stages, that answer would have been far more excusable than it is now, since the program hadn't yet made the subtle discoveries it has now made about the structure of *mrrjjj*. It would seem a shame for the program to have gotten this far and then to "drop the ball" and answer in a relatively primitive way. However, 31 is a high enough temperature that there is a good chance that the program will get back on track and be allowed to explore the more abstract avenue to its logical conclusion.

10. Indeed, the aberrant bridge from the *c* was quickly destroyed and replaced by a rebuilt *c–J* bridge, in keeping with the emerging sophisticated view. Also, the high-level bridge between *abc* and *mrrjjj* as wholes, which in the previous screen dump was merely a dotted candidate, has now been promoted through the dashed state and actually built. Its six component concept-mappings, including identity mappings (such as *right* ⇒ *right*, meaning that both strings are seen as flowing rightwards) as well as conceptual slippages (such as *letter category* ⇒ *length*, meaning that letters are mapped onto numbers — specifically, onto group lengths), are listed in the middle of the screen, somewhat obscuring the bridge itself.

The original rule has been translated, according to the slippages, for application to the target string *mrrjjj*. The translated rule, "Replace the length of the rightmost group by its successor", appears just above the Slipnet, and the answer *mrrjjjj* at the right. The very low final temperature of 11 reflects the program's unusually high degree of satisfaction with this answer.

Although the preceding run may look quite smooth, there were many struggles involved in coming up with this answer: it was hard not only to make a single-letter group, but also to bring the notion of *length* into the picture, to allow it to persist long enough to trigger the noticing of all three group lengths, and to build bonds between the group lengths. The program, like people, usually gives up before all these hurdles can be overcome, and gives one of the more obvious answers. Arriving at the deeper answer *mrrjjjj* requires not only the insights brought about by the strong pressures in the problem, but also a large degree of patience and persistence in the face of uncertainty.

The moral of all this is that in a complex world (even one with the limited complexity of Copycat's microworld), one never knows in advance what concepts may turn out to be relevant in a given situation. This dilemma underscores the point made earlier: it is imperative not only to avoid dogmatically open-minded search strategies, which entertain all possibilities equally seriously, but also to avoid dogmatically closed-minded search strategies, which in an ironclad way rule out certain possibilities *a priori*. Copycat opts for a middle way, in which it quite literally takes calculated risks all the time — but the *degree* of risk-taking is carefully controlled. Of course, taking risks by definition opens up the potential for disaster — and indeed, disaster occurs once in a while (as was evidenced by some of the far-fetched answers displayed earlier). But this is the price that must be paid for flexibility and the potential for creativity.

People, too, occasionally explore and even favor peculiar routes. Copycat, like people, has to have the potential to concoct strange and unlikely solutions in order to be able to discover subtle and elegant ones like *mrrjjjj*. To rigidly close off any routes *a priori* would necessarily remove critical aspects of Copycat's flexibility. On the other hand, the fact that Copycat so rarely produces strange answers demonstrates that its mechanisms manage to strike an effective balance between open-mindedness and closed-mindedness, imbuing it with both flexibility and robustness.

Hopefully, these screen dumps have made clearer the fundamental roles of nondeterminism, parallelism, non-centralized and simple perceptual agents (*i.e.*, codelets), the interaction of bottom-up and top-down pressures, and the reliance on statistically emergent (rather than explicitly programmed) high-level behavior. Large, global, deterministic decisions are never made (except perhaps towards the end of a run). The system relies instead on the accumulation of small, local, nondeterministic decisions, none of which alone is particularly important for the final outcome of the run. As could be seen in the screen dumps, large-scale effects occur only through the statistics of the lower levels: the ubiquitous notion of a "pressure" in the system is really a shorthand for the statistical effects over time of a large number of actions carried out by codelets, on the one hand, and of activation patterns of nodes in the Slipnet, on the other.

As was seen in the screen dumps, as structures are formed and a global interpretation coalesces, the system gradually makes a transition (via gradually falling temperature) from being quite parallel, random, and dominated by bottom-up forces to being more serial, deterministic, and dominated by top-down forces. We believe that such a transition is characteristic of high-level perception in general.

The importance of such a transition in degree of risk-taking was confirmed by two experiments we performed on the model (described in Mitchell, 1993). Temperature was clamped throughout a run, in one experiment at a very high value and in the other at a very low value. In each experiment, 1,000 runs were made. In neither test did the hobbled Copycat ever come up with the answer *mrrjjjj* — not even once.

Micro-anatomy of a paradigm shift

We now turn our attention from Problem 4 to Problem 6, and give a sketch of how Copycat can, on occasion, come up with the answer *wyz*. It turns out to be a surprisingly intricate little tale. The reason for this is that it is an attempt to show in slow motion how a human mind, under severe pressure, can totally transform its perception of a situation in a blinding flash (colloquially termed the "Aha!" phenomenon). Since such paradigm shifts are often found at the core of deeply creative acts, one should expect their microstructure to be very complex (otherwise, the mystery of creativity would long ago have been revealed and put on a mass-produced microchip). Indeed, the challenge of getting Copycat to produce *wyz properly* — faithfully to what we believe really goes on in a human mind at the most informative subcognitive level of description — has, from the very outset, been the central inspiration in guiding the development of the Copycat architecture.

The very cursory justification for *wyz* given in the previous section lies at far too high and coarse-grained a level to be informative about the mental mechanisms responsible for paradigm shifts. The following detailed story, by contrast, evolved hand in hand with the architecture itself, and is intended not only as a description of Copycat, but hopefully as an accurate description of the underpinnings of a typical paradigm shift in a human mind. (An annotated series of screen dumps of a particular run on Problem 6 is given in Mitchell & Hofstadter, 1990a.)

Emergency measures convert a serious snag into a set of exploratory pressures

Things start out essentially analogously to a typical run on Problem 1, in terms of bonding, grouping, bridge-building, and such — that is, both source and target strings come quite quickly to be perceived as successor groups, and the raw rule "Replace rightmost letter by its successor" is effortlessly produced.

Everything thus proceeds pretty smoothly up to the point of trying to take the successor of *z*, which is impossible. This serious snag causes several coordinated "emergency measures" to be taken:

- the *physical* trouble spot — here, the instance of *z* in the Workspace — is highlighted, in the sense that its salience is suddenly pumped up so high that, to codelets, it becomes the most attractive object in the entire Workspace;
- the *conceptual* trouble spot — here, the node *z* in the Slipnet — is highlighted, in the sense that a huge jolt of activation is pumped into it, and as a consequence, its halo broadens and intensifies, meaning that related concepts are more likely to be considered, at least fleetingly;
- the temperature is pumped up to its maximum value of 100 and temporarily clamped there, thus encouraging a broader and more open-minded search;
- the high temperature enables previously dormant "breaker" codelets to run, whose purpose is to arbitrarily break structures that they find in the Workspace, thus reducing the system's attachment to a viewpoint already established as being problematic.

Note the generality of these "impasse-handling" mechanisms: they have nothing to do with this snag itself, with the particular problem, with the alphabetic domain, or even with analogy-making! The reason for this is of course that running into an impasse is a critical and common event that any cognitive system must be capable of dealing with. To be sure, no set of mechanisms can be guaranteed to resolve *all* snags (otherwise we would be dealing with omniscience, not intelligence). The best that can be hoped for is that the impasse itself can be "read" as a source of *cues* — possibly very subtle ones — that may launch tentative forays down promising new avenues. A "cue", in the Copycat architecture, is essentially the creation of a *pressure* that pushes for exploration along a certain direction. Thus the idea of *interpreting the snag as a source of pressures* is the philosophy behind the four mechanisms above, especially the first two.

Although these emergency measures are not powerful enough to guide Copycat to coming up with *wyz* all that often, when it does get there, it does so essentially according to the following scenario.

The spotlight focused on Platonic *z* has the effect of making all concepts in *z*'s halo — including the closely-related concept *alphabetic-last* — somewhat more likely to be looked at by description-building codelets. The probability is thus significantly increased that the instance of *z* in the Workspace will get explicitly described as *alphabetic-last*.

Note that in most problems — even ones that involve one or more instances of the letter *z* — there is little or no reason to pay attention to the notion *alphabetic-last,* and therefore, this conceptually deep neighbor of Platonic *z* usually remains — and *should* remain — dormant (as it did in Problems 1–5). After all, as was brought out in the discussion of Problem 4, it is extremely crucial to avoid cluttering up the processing with all sorts of extraneous interfering notions, no matter how conceptually deep they may be. But now, under the emergency measures, unusual avenues are more likely to at least be "sniffed out" a short ways.

If the description *alphabetic-last* does indeed get attached to the *z*, which is a dicey matter, then a further boost is given to the node *alphabetic-last,* as the system has deemed it potentially relevant. So now that *alphabetic-last* (part of the halo of Platonic *z*) has been lifted considerably out of dormancy, concepts in *its* halo will in turn receive more activation, which means codelets will tend to pay more attention to them (probabilistically speaking). One such neighbor-concept is *alphabetic-first,* which is now briefly given the chance to show its relevance. Obviously, if there were no instance of *a* in the problem, *alphabetic-first* would be found to be completely irrelevant and would soon decay back to dormancy, but since there *is* an *a* inside *abc*, it has a fair chance of getting explicitly described as *alphabetic-first,* in much the same way as the *z* in *xyz* got described as *alphabetic-last.*

If both these descriptions get attached — and that is a big "if" — then both letters become even more salient than before; in fact, they almost cry out to be mapped onto each other — not because the system can anticipate the great insight that such a mapping will bring, but simply because both letters are so salient! Once the system tries it out, however, the great appeal of the tentative mapping instantly becomes apparent. Specifically, a pair of conceptual slippages are entailed in the act of "equating" the *a* with the *z* (*i.e.,* building an *a–z* bridge): *alphabetic-first* $\Rightarrow$ *alphabetic-last,* and *leftmost* $\Rightarrow$ *rightmost.*

How resistance to a deep slippage is overcome — a tricky matter

Although normally the deep slippage of *alphabetic-first* into *alphabetic-last* would be quite valiantly resisted (recall the motto given earlier, "Deep stuff doesn't slip in good analogies"), here a special circumstance renders it a bit more probable: the companion would-be slippage *rightmost* $\Rightarrow$ *leftmost* is of the same type — in particular, each of these slippages involves slipping a concept representing an extremity into its opposite concept. These two would-be slippages are thus conceptually parallel, so that each one on its own reinforces the other's plausibility. This fact helps to overcome the usual resistance to a deep slippage. (Incidentally, this is the kind of subtlety that was not apparent to us before the computer implementation was largely in place; only at that point were Copycat's flailings and failures able to give us pointers as to what kinds of additional mechanisms were needed.)

Another fact that helps overcome the usual resistance to the deep slippage in this bridge is that *any* two slippages, whether parallel or not, provide more justification for building a bridge than either one alone would. Altogether, then, there is a fairly good chance that this bridge, once tentatively suggested, will actually get built. Once this critical step has taken place, essentially it's all downhill from there. This is why we have paid particularly close attention to the pathway via which such a bridge can emerge.

Locking-in of a new view

The first thing that is likely to happen as a result of an *a–z* bridge getting built is that the temperature will get unclamped from its value of 100. In general, what unclamps the temperature is the construction of any strong structure different from those that led up to the snag — in other words, a sign that an alternative way of looking at things may be emerging — and this bridge is a perfect example of such a structure. Like most actions in Copycat, the unclamping of temperature is probabilistic. In this case, the stronger the novel structure is, the more likely it is to trigger the unclamping. Since the *a–z* bridge is both novel and very strong, unclamping is virtually assured, which means that the temperature falls drastically right after the bridge is built. And when the temperature falls, decisions tend to get more deterministic, which means that the emerging new view will tend to get supported. In short, there is a powerful kind of *locking-in* effect that is triggered by the discovery of an *a–z* bridge. This is a critical effect.

Another aspect of locking-in is the following idea. The building of this first bridge involving the simultaneous slippage of two concepts into their opposites sends a burst of activation into the very deep concept *opposite*; as a result, all pairs of concepts connected via links labeled *opposite* are drawn much closer together, facilitating the slippage of one into the other. Such slippages will still not happen without reason, of course, but now they will be much easier to make than in ordinary circumstances. Thus in a sense, making *one* bridge based on conceptual opposites sets a tone making it easier to make *more* of them. The emerging theme of the concept *opposite* can fairly be characterized as a kind of "bandwagon".

Given all this, one of the most likely immediate consequences of the crosswise *a–z* bridge is the building of the "mirror" crosswise bridge connecting the *c* with the *x*. It, too, depends on the slippage between *leftmost* and *rightmost*, and is thus facilitated; in addition, once built, it strongly reinforces the emerging relevance of the concept *opposite*. Moreover, the temperature will fall significantly because this bridge, too, will be very strong. Thanks to all of this, the locking-in effect may by now be so strong that it will be hard to stop the momentum towards building a completely new view of the situation.

The reversals taking place become a near-stampede at this point, with significant pressure emerging to flip the direction of the fabric of the group *xyz* from *rightwards* to *leftwards,* which means also that the perceived fabric itself would switch from *successor* to *predecessor.* Thus at this point, Copycat has carried out both a spatial and an alphabetical reversal of its vision of *xyz*. The paradigm shift has been completed. At this point, Copycat is ready to translate the raw rule, and, as was said above, the result is the new rule *replace the leftmost letter by its alphabetic predecessor*, which yields the answer *wyz*.

It must be stressed that all the multifarious activity just described — shifting degrees of activation of various key concepts; deep slippages; interrelated spatial and conceptual reversals — all this takes place in a flash in a human mind. There is no hope of making out all the details of this paradigm shift (or any other) in one's own mind through mere introspection. In fact, it has taken the authors several years to settle on the above account, which represents our current best stab at the true story's intimate details.

How hard is it to make this paradigm shift?

As was pointed out a moment ago, the motto "Deep stuff doesn't slip in good analogies" is violated by the answer *wyz*, in that *alphabetic-first* is a deep concept and yet is allowed to slip into *alphabetic-last* here. This is one reason that makes it so hard for many people to discover it on their own. Yet many people, when they are shown this answer, appreciate its elegance and find it very satisfying. Problem 6 is thus a circumstance where a constellation of pressures can occasionally overcome the powerful natural resistance expressed by the motto; in fact, making such a daring move results in what many people consider to be a deep and insightful analogy.

There is an important irony here. In particular, even though slippages tend to be (and should be) *resisted* in proportion to their depth, once a very deep slippage has been made, then it tends to be (and should be) *respected* in proportion to its depth. We consider this to be characteristic of creative break-throughs in general. More specifically, we consider the process of arriving at answer *wyz* to be very similar, on an abstract level, to the process whereby a full-scale conceptual revolution takes place in science (Kuhn, 1970).

Now we come back to the point raised in our earlier discussion of Problem 6 about "levels of subtlety" of answers. Specifically, we claimed above that, because finding the answer *wyz* to Problem 6 is far subtler *for people* than finding the similar answer *hjkk* to Problem 2, any model of mental fluidity should respect this difference in levels of subtlety. Yet when one compares the bar graphs for these problems, one discovers that *wyz* was found far more often than *hjkk* was found (a ratio of 137 to 47, when both problems were run 1,000 times). This seems to completely contradict the claim that the former answer is subtler than the latter. How can one account for this unexpected ratio?

There are two basic factors that explain it. The first has to do with the fact that there was a snag in one problem and no snag in the other. In attacking Problem 6, Copycat was *forced* to look for solutions other than taking the successor of the rightmost letter, because that route turned out to be impossible. By contrast, all sorts of superficially attractive routes led directly to solutions in Problem 2. There was no snag that prevented any attractive route from being taken all the way to its natural conclusion. Had all or most of the easy routes been barred, then of course *hjkk* would have constituted a much larger percentage of the answers found.

The second factor is that the average *length of time* taken to find various solutions (measured in terms of number of codelets run) is a key notion. This fact is not apparent, because average run-lengths are unfortunately not represented in the bar graphs. When Copycat came up with *hjkk* in Problem 2, it was essentially always a relatively direct process involving no backtracking or getting stuck for a while in a loop. To be specific, the average number of codelets taken to get *hjkk* was 899. By contrast, the average number of codelets taken to get *wyz* was 3,982 — over four times as long. The reason for this is that in most runs, the program came back time and time again to the standard way of looking at *xyz*, and thus hit the snag over and over again: it was stuck in a kind of rut. This means that on runs where Copycat was lucky enough to come across the double reversal, by the time it did so it had usually tried out all sorts of other pathways in vain beforehand. In this sense of *time needed to make the discovery*, *wyz* was an extremely elusive answer for the program, whereas *hjkk* was not at all elusive. In sum, *wyz* was indeed far subtler for Copycat than *hjkk* was, as ought to have been the case.

Conclusion: The Generality of Copycat's Mechanisms

The crucial question of scaling-up

As was stated at the outset, the Copycat project was never conceived of as being dependent in any essential way on specific aspects of its small domain, nor even on specific aspects of analogy-making *per se*. Rather, the central aim was to model the emergence of insightful cognition from fluid concepts, focusing on how slippages can be engendered by pressures.

One of the key questions about the architecture, therefore, is whether it truly is independent of the small domain and the small problems on which it now works. It would certainly be invalidated if it could be shown to depend on the relative smallness of its repertoire of Platonic concepts and the relatively few instances of those concepts that appear in a typical problem. However, from the very conception of the project, every attempt has been made to ensure that Copycat would not succumb to a combinatorial explosion if the domain were enlarged or the problems became bigger. In some sense, Copycat is a caricature of genuine analogy-making. The question is, what makes a caricature faithful? What is the proper way to construct a cognitive model that will scale up?

Shades of gray and the mind's eye

Real cognition of course occurs in the essentially boundless real world, not in a tiny artificial world. This fact seems to offer the following choice to would-be "cognition architects": either have humans scale down all situations by hand in advance into a small set of sharp-edged formal data-structures, so that a brute-force architecture can work, or else let the computer effectively do it instead — that is, use a heuristic-based architecture that at the outset of every run makes a sharp and irreversible cut between concepts, pathways, and methods of attack that might eventually be brought to bear during that run, and ones that might not. There seems to be no middle ground between these two types of strategy, because either you must be willing to give *every* approach a chance (the brute-force approach), or you must choose some approaches while *a priori* filtering others out (the "heuristic-chop" approach).

The only way out would seem to involve a notion of "shadedness", in which concepts, facts, methods of attack, objects, and so on, rather than being ruled "out" or "in" in a black-and-white way, would be present in shades of gray — in fact, shades of gray that change over time. At first glance, this seems impossible. How can a concept be invoked only *partially*? How can a fact be neither fully ignored nor fully paid attention to? How can a method of attack be merely "sort of" used? How can an object fall somewhere in between being considered "in the situation" and being considered "not in the situation"?

Since we believe that these "shades of gray" questions lie at the crux of the modeling of mind, they merit further discussion. A special fluid quality of human cognition is that often, solutions to a problem — especially the most ingenious ones, but even many ordinary ones — seem to come from far outside the problem as conceived of originally. This is because problems — or more generally, *situations* — in the real world do not have sharp definitions; when one is in, or hears about, a complex situation, one typically pays no conscious attention to the question of what counts as "in" the situation and what counts as "out" of it. Such matters are almost always vague, implicit, and intuitive.

Using the metaphor of the "mind's eye", we can liken the process of considering an abstract situation to the process of visually perceiving a physical scene. Like a real eye, the mind's eye has a limited field of vision, and cannot focus on several things simultaneously, let alone large numbers of them. Thus one has to choose where to have the mind's eye "look". When one directs one's gaze at what one feels is the situation's core, only a few centrally located things will come into clear focus, with more tangential things being less and less clear, and then at the peripheries there will be lots of things of which one is only dimly aware. Finally, whatever lies beyond the field of vision seems by definition to be outside of the situation altogether. Thus "things" in the mind's eye are definitely shaded, both in terms of how clear they are, and in terms of how aware one is of them.

The very vague term "thing" was used deliberately above, with the intent of including both *abstract Platonic concepts* and *concrete specific individuals* — in fact, to blur the two notions, since there is no hard-and-fast distinction between them. To make this clearer, think for a moment of the very complex situation that the Watergate affair was. As you do this, you will notice (if you followed Watergate at all) that all sorts of different events, people, and themes float into your mind with different degrees of clarity and intensity. To make this even more concrete, turn your mind's eye's gaze to the Senate Select Committee, and try to imagine each different senator on that committee. Certainly, if you watched the hearings on television, some will emerge vividly while others will remain murky. Not just *Platonic abstractions* like "senator" are involved, but many *individual* senators have different degrees of mental presence as you attempt to "replay" those hearings in your mind. Needless to say, the memory of anyone who watched the Watergate hearings on television is filled to the brim both with Platonic concepts of various degrees of abstractness (ranging from "impeachment" to "coverup" to "counsel" to "testimony" to "paper shredder") and with specific events, people, and objects at many levels of complexity (ranging from the "Saturday night massacre" to the Supreme Court, from Maureen Dean to the infamous 18½-minute gap, and all the way down to the phrase "expletive deleted" and even Sam Ervin's gavel, with which every session of the committee was rapped to order). When one conjures up one's memories of Watergate, all of these "things" have differential degrees of mental presence, which change as one's mind's eye scans the "scene".

Note that in the preceding paragraph, all the "things" mentioned were carefully chosen so that readers — at least readers who remember Watergate reasonably well — would give them unthinking acceptance as genuine "parts" of Watergate. However, now consider the following "things": England, France, communism, socialism, the Viet Nam War, the Six-Day War, the Washington Monument, the *New York Times,* Spiro Agnew, Edward Kennedy, Howard Cosell, Jimmy Hoffa, Frank Sinatra, Ronald Reagan, the AFL–CIO, General Electric, the electoral college, college degrees, Harvard University, helicopters, keys, guns, flypaper, Scotch tape, television, tape recorders, pianos, secrecy, accountancy, loyalty, and so on. Which of these things are properly thought of as being "in" Watergate, and which ones as "out" of it? It would obviously be ludicrous to try to draw a sharp line. One is forced to accept the fact that for a model of a mind to be at all realistic, it must be capable of imbuing all concrete objects and individuals, as well as all abstract Platonic concepts, with shaded degrees of mental presence — and of course, those degrees of presence must be capable of changing over time.

Like a real eye, the mind's eye can be attracted by something glinting in the peripheries, and shift its gaze. When it does so, things that were formerly out of sight altogether now enter the visual field. By a series of such shifts, "things" that were totally outside of the situation's initial representation can eventually wind up at the very center of attention. This brings us back, finally, to that special fluid quality of human thought whereby initially unsuspected notions occasionally wind up being central to one's resolution of a problem, and reveals how intimately such fluidity is linked with the various "shades of gray" questions given above.

Copycat's shaded exploration strategy

Let us thus return to the list of "shades of gray" questions: How can a concept be invoked only *partially*? How can a fact be neither fully ignored nor fully paid attention to? How can a method of attack be merely "sort of" used? How can an object fall somewhere in between being considered "in the situation" and being considered "not in the situation"? These questions were not asked merely rhetorically; in fact, it was precisely to respond to the challenges that they raise that the probabilistic architecture of Copycat was designed.

Copycat's architecture has in common with brute-force architectures the fact that every possible concept, fact, method, object, and so on is *in principle* available at all times;[7] on the other hand, it has in common with heuristic-chop architectures the fact that out of all available concepts, facts, methods, objects, and so on, only a few will get very intensely drawn in at any given moment, with most being essentially dormant and an intermediate number having a status somewhere in between. In other words, virtually all aspects of the Copycat architecture are riddled by shades of gray instead of by hard-edged, black-and-white cutoffs. In particular, *activation* (with continuous values rather than a binary on/off distinction) is a mechanism that gives rise to shadedness in the Slipnet, while *salience* and *urgency* serve similar purposes in the Workspace and Coderack, respectively. These are just three of a whole family of related "shades-of-gray mechanisms" whose entire *raison d'être* is to defeat the scaling-up problem.

---

7. Note that the claim is not that every single concept imaginable to humans is available, but simply that all concepts *within the system's dormant repertoire* are in principle accessible at any point during a run. The fact that Copycat cannot reach beyond its own conceptual repertoire, thus effectively "transcending itself", is not a defect, but simply a fact of existence that it shares with every finite cognitive system, such as human minds. Put another way, if this property is a defect of Copycat, then it is a defect that Copycat shares with human minds.

An architecture thus pervaded by shades of gray has the very attractive property that although no concept or object or pathway of exploration is ever *strictly* or *fully* ruled out, only a handful of them are at any time *seriously* involved. At any given moment, therefore, the system is focusing its attention on just a small set of concepts, objects, and pathways of exploration. However, this "searchlight of attention" can easily shift under the influence of new information and pressures, allowing *a priori* very unlikely concepts, objects, or pathways of exploration to enter the picture as serious contenders.

The chart below summarizes the various mechanisms in the Copycat architecture that incorporate shades of gray in different ways. In it, the term "shaded" should be understood as representing the opposite of a binary, black/white distinction; it often means that one or more *real numbers* are attached to each entity of the sort mentioned, as opposed to there being an on/off distinction. The term "dynamic" means that the degree of presence — the "shade", so to speak — can change with time.

Shades of gray in the Slipnet
- shaded, dynamic presence of Platonic concepts (via dynamic activation levels)
- shaded, dynamic conceptual proximities (via dynamic link-lengths)
- shaded, dynamic spreading of activation to neighbor concepts (giving rise to "conceptual halos")
- shaded conceptual depths of nodes
- shaded decay rates of concepts (determined by conceptual depths)
- shaded, dynamic emergence of abstract themes (stable activation patterns of interrelated conceptually deep nodes)

Shades of gray in the Workspace
- shaded, dynamic number of descriptions for any object
- shaded, dynamic importance of each object (via activation levels of descriptors in Slipnet)
- shaded, dynamic unhappiness of each object (determined by degree of integration into larger structures)
- shaded, dynamic presence of objects (via dynamic salience levels)
- shaded, dynamic tentativity of structures (via dynamic strengths)

Shades of gray associated with the Coderack
- shaded, dynamic degrees of "promise" of pathways

- shaded, dynamic emergence of pressures (via urgencies of codelets and shifting population of Coderack)
- shaded, dynamic degree of willingness to take risks (via temperature)
- shaded, dynamic mixture of deterministic and nondeterministic modes of exploration
- shaded, dynamic mixture of parallel and serial modes of exploration
- shaded, dynamic mixture of bottom-up and top-down processing

There is one further aspect of shadedness in Copycat that is not localized in a single component of the architecture, and is somewhat subtler. This has to do with the fact that, over time, higher-level structures emerge, each of which brings in new and unanticipated concepts, and also opens up new and unanticipated avenues of approach. In other words, as a run proceeds, the field of vision broadens out to incorporate new possibilities, and this phenomenon feeds on itself: each new object or structure is subject to the same perceptual processes and chunking mechanisms that gave rise to it. Thus there is a spiral of rising complexity, which brings new items of ever-greater abstraction into the picture "from nowhere", in a sense. This process imbues the Copycat architecture with a type of fundamental unpredictability or "openness" (Hewitt, 1985) that is not possible in an architecture with frozen representations. The ingredients of this dynamic unpredictability form an important addendum to the list of shades of gray given above.

Dynamic emergence of unpredictable objects and pathways
- creation of unanticipated higher-level perceptual objects and structures
- emergence of *a priori* unpredictable potential pathways of exploration (via creation of novel structures at increasing levels of abstraction)
- creation of large-scale viewpoints
- competition between rival high-level structures

By design, none of the mechanisms in the lists presented above has anything in the least to do with the size of the situations that Copycat is currently able to deal with, or with the current size of Copycat's Platonic conceptual repertoire. Note, moreover, that none of them has anything whatsoever to do with the subject matter of the Copycat domain, or even with the task of analogy-making *per se*. Yet these mechanisms and their emergent consequences — especially commingling pressures and the parallel terraced scan — are what Copycat is *truly* about. This is the underpinning of our belief in the cognitive generality of the Copycat architecture.

Preface 6:

Two Early AI Approaches to Analogy

In the chapter to follow, Copycat is compared with recent work on the modeling of analogy-making — specifically, work dating from the past decade. However, I felt it would be interesting and informative to discuss here two efforts in that direction that go much further back in time.

Thomas Evans' Program ANALOGY

Indisputably, the most famous early attempt to model analogy-making on a computer was Thomas Evans' program ANALOGY, described in Evans (1968). As was customary back in those days, this program dealt with a microworld — the rather elegant one defined by the geometric-analogy problems found in traditional IQ tests. Such problems are of the form "A is to B as C is to 1, 2, 3, 4, or 5?", where each letter or number stands for a picture containing a small number of related geometric figures (*e.g.*, a small triangle above a large square, with a dot inside the square). The "B" figure is always made from the "A" figure by some kind of simple transformation, such as moving one object, deleting one object, replacing one object by another, and so on. The challenge is to "do the same thing" to the "C" figure, which would presumably result in one of the five numbered figures. Unlike Copycat, ANALOGY did not *produce* its answer; rather, it *selected* its answer from a list of five candidates. Moreover, it did so by inspecting all five candidates and rank-ordering them. Although this hardly sounds like an exhaustive search when there are just five pictures to look at, it nonetheless typifies the philosophy of the architecture, which was based on the idea of deterministically exploiting the raw power of computers.

To my mind, there are two striking differences between Evans' early work and almost all recent work on modeling analogy-making. First is the fact that ANALOGY built up its own representations of the eight pictures in each problem. Admittedly, it did so not from a pixel-by-pixel scan of each picture, but from a set of Lisp data-structures describing lines, curves, and dots.

Nonetheless, Evans made a clear attempt to merge a primitive kind of visual or perceptual processing with the more conceptual act of mapping. Unfortunately, in his program, these two aspects of analogy-making were temporally cleanly separated; perception was a stage that completely preceded mapping, and there was no way to go back and undo perceptual decisions on the basis of observations made during the mapping phase. Still, both types of process were explicitly modeled, and this was a foresighted strategy that somehow got left by the wayside in further research.

The second noteworthy aspect of Evans' work is that his model in no way conceived of analogy-making as a kind of "big gun" to be brought out to help solve some problem. Most recent work takes it as an unquestioned axiom that that is *by definition* what analogy-making is about and for. In fact, the very term "analogical reasoning", which is the standard one used in most articles and books, reveals the bias towards *reasoning* as opposed to simply *understanding*. The idea that people constantly see situations in terms of other situations simply because that is human nature, and not because they wish to solve some problem, is almost totally ignored these days.

I certainly do not wish to argue that Evans' program was a full and deep model of human analogy-making. In many ways, it was very primitive. It had almost no representation of concepts (which, by contrast, form the core of our model) and only the slightest sense of slippage. Its architecture was rigid, brute-force, deterministic, and serial. It was not touted as a cognitive model, although perhaps it was conceived of as being one. (In those days, the distinction between a successful AI program and a good cognitive model was hardly ever considered.) However, despite its weaknesses, Evans' work still stands out as an interesting approach to analogy-making, in part because no one really followed up on the many interesting ideas that he pioneered.

Walter Reitman's Argus Program

The other early project — earlier, in fact, than ANALOGY — that I feel deserves mention is Walter Reitman's Argus, a computer program developed in the early 1960's in large part as a reaction against the rigidity of Newell, Shaw, and Simon's famous General Problem Solver. The entire project, including its background and philosophy, is described in Reitman's book *Cognition and Thought* (1965).

As implemented, Argus could solve only the simplest and most uninteresting of "analogies problems", as Reitman termed them. Curiously, these problems also smack of IQ tests, but they are verbal rather than geometric. One of two sample problems Reitman cites as having been solved by Argus is this: "bear : pig :: chair : {foot, table, coffee, strawberry}" — that is, " 'Bear' is to 'pig' as 'chair' is to what?", where the brackets give a fourfold choice. The desired answer, rather amazing in its simple-mindedness, is "table", the justification being simply that just as "bear" and "pig" share a *superordinate* concept (namely, "animal"), so do "chair" and "table" (namely, "furniture") — end of story. Note that in a single-word-based "analogies problem" of this sort, there is no need for dynamic build-up of perceptual structures in a Workspace; all one needs is a static repertoire of stored concepts with some kinds of static interrelationships built in. In this case, for example, the program didn't need to know anything — and *didn't* know anything — about bears and pigs themselves; all it knew was that the words "bear" and "pig" were both linked, via a superordinate link, to the word "animal". Thus using English words for familiar things constituted the usual kind of deceptive *façade* of real-world knowledge, but Reitman made no attempt to claim that his program *understood* the concepts it was, or seemed to be, dealing with.

The only other example given as a solved problem is this one: "hot : cold :: tall : {wall, short, wet, hold}". Clearly, the virtue of Argus was not its impressive level of performance. Rather, it was its extremely forward-looking architecture, which was inspired more than anything by the notion of Hebbian cell-assemblies and their intrinsic parallelism (Hebb, 1948). Reitman was disturbed by the single-mindedness of GPS — the fact that, unlike people, it was uninterruptible and undistractable. It appealed very much to Reitman to model these "imperfect" aspects of human cognition, and this is why he turned to a parallel-processing vision of the mind — almost unheard-of in computer models those days. This was in itself a remarkably daring and innovative step to take. While emphasizing numerous times that his model should not be thought of as a simulation of *neural-level* activity, Reitman referred over and over to the notion of cell-assemblies that "fire" and send signals to other cell-assemblies, thus implementing a kind of spreading activation at a *superneural* level, perhaps even at a conceptual level.

The heart of Argus involved the mutual interaction of a serial process trying to carry out a given task (*e.g.*, an "analogies problem") and a semantic network in which activation propagated around in parallel fashion. Very loosely speaking, this image maps onto Copycat's interaction between codelets working on a problem in the Workspace, and the Slipnet with its spreading activation. In Argus, nodes with high activation introduced dynamic biases into the serial process that was working away on the given task.

One other way in which Argus' inventor saw far ahead was in his utter rejection, at least in principle, of the notion of human-constructed and fixed representations of situations fed into a "mapping engine". Citing the sample problem "Sampson : hair :: Achilles : {strength, shield, heel, tent}", Reitman points out that one way of answering it correctly is simply to recognize that both Sampson and Achilles are people, and that both hair and heels are body parts of people. This is enough to single out "heel" as the probable answer. However, a more sophisticated solution, and one more typically produced by people, would take into account specific facts about Sampson and Achilles as individuals — namely, that Sampson lost his strength when his hair was cut, and that Achilles would lose his power if his heel were wounded. Thus Sampson's hair and Achilles' heel share the extremely abstract and *a priori* unlikely description "point of susceptibility to major negative influence from the environment". Such a notion, if it existed explicitly in the semantic network before the problem was seen, would of course render solution of this problem no harder than solution of the "hot/cold" problem. Reitman points this out and says bluntly, "That is the trouble.... Argus finds the one problem no more difficult than the other because, as it gets them, they are formally identical." Reitman clearly felt that a program, in order to be scientifically interesting, would need to be able to *create* such notions on the fly. A bit later, the passage continues as follows: "If Argus solves the 'Sampson : hair' problem now, it is not because of the capability for intelligent behavior *in the program*. It is only because *human* intelligence is applied beforehand to precode the information into a particular special-purpose form.... In so doing, we ignore the really interesting psychological problem." (italics in original). When he states it, it seems so obvious — yet it is a point that has gone largely unheeded since the time it was written.

Although Argus was unfortunately carried no further than this early stage, much of its inspiration came from thinking about how people go about working on ill-defined "problems" such as composing a piece of music, and clearly Reitman was pulled on by the distant goal of trying to model that kind of activity. One of his book's most interesting chapters is in fact a careful description of what went on (at least at a conscious level) in the mind of a particular (but unnamed) professional composer who spoke aloud while composing a fugue. The constant interaction between high-level abstract principles and low-level concrete structures of specific notes represented a fusion of bottom-up and top-down processing, and clearly this kind of thing was behind Reitman's vision of a parallel, interruptible, distractable system.

Unlike much of today's research, which focuses on modeling pragmatic activities such as problem-solving (Chapter 6 will present a couple of examples, but the tendency is far broader than that), Reitman's research was deeply motivated by artistic and esthetic concerns. In many places in his book one finds discussions of what many people would consider "pointless" activities, such as riddles, games, puzzles, joke invention, musical composition, and so forth. Reitman's attitude is in sharp contrast to the no-nonsense attitude of many of today's researchers, who seem to feel that if an answer to an analogy problem can be justified only by appeal to personal artistic taste or a personal sense of esthetics, then it is hopelessly subjective, incapable of being discussed scientifically, and hence not worthy of being modeled. The fact that the difference between run-of-the-mill and top-notch mathematical and scientific research is often determined by just such intangible, esthetic decisions is something that such objectors pay no attention to, since they would undoubtedly find it uncomfortable to admit that science itself rests on ill-defined, murky foundations in which judgments about beauty, elegance, symmetry, and so forth play a crucial role. A strong interest in just such elusive yet central aspects of thought was very influential in shaping the philosophy behind Argus, which is reason to regret the fact that Argus died so young.

In a curious way, Copycat seems to be a kind of merger of these two early projects in analogy-making. Take the architectural directions that Argus was heading in, and combine that with ANALOGY's idea of integrating perceptual processing with mapping, and doing so in an elegant microworld, and you have something loosely resembling Copycat.

Chapter 6

___

Perspectives on Copycat: Comparisons with Recent Work

MELANIE MITCHELL and DOUGLAS HOFSTADTER

How to Judge Copycat?

How is the performance — both tree-level and forest-level — that has been described in the previous chapter to be judged? Of course, the ultimate criterion of its validity has to be whether our results jibe with general and universal characteristics of human analogy-making (we do not mean a precise numerical comparison with human data on the same problems, since as we have repeatedly said, precise simulation of human performance in this specific microdomain is *not* our goal). However, there is another way our work can and ought to be set in perspective: by comparison with other projects. Therefore, we turn now to a description of various types of related research.

Much effort has been devoted in artificial intelligence and cognitive science to the construction of computer models of analogy-making, with most of it concentrating on the use of analogical reasoning in problem-solving. The majority of models concentrate on how a mapping is made from a source problem (whose solution is known) to a target problem (whose solution is desired), with some kind of representation of the various objects, descriptions, and relations in the source and target problems given to the program in ready-made and fixed form. Very few (if any) other computer models focus, as Copycat does, on the *construction* of representations for the source and target situations, and on how this construction interacts with the mapping process. Few (if any) other models attempt to deal with the question of how new, previously unincluded concepts can be brought in and can come to be seen as relevant, in response to pressures that emerge as processing proceeds. In short, no other current computer model of analogy-making of which we are aware attempts, as Copycat does, to integrate high-level perception, concepts, and conceptual slippage.

In what follows, rather than giving a broad survey of computer models of analogy-making, we focus on just two projects chosen for their relevance to our work. (Descriptions of many other models can be found in Hall, 1989 and Kedar-Cabelli, 1988b.) Then comes a discussion of Copycat's place in the spectrum of computer models of mind, ranging from high-level symbolic models to low-level subsymbolic models. (Further comparisons with related work are given in Mitchell, 1993.)

SME, the Structure Mapping Engine

Psychologist Dedre Gentner's research is perhaps the best-known work in cognitive science on analogy. She has formulated a theory of analogical mapping, called the "structure-mapping" theory (Gentner, 1983), and she and her colleagues have constructed a computer model of this theory: the Structure Mapping Engine, or SME (Falkenhainer, Forbus, & Gentner, 1990). The structure-mapping theory describes how mapping is carried out from a source situation to a (sometimes less familiar) target situation. The theory gives two defining criteria for what counts as true analogical mapping: (1) *relations* between objects rather than *attributes* of objects are mapped; and (2) relations that are part of a coherent interconnected system are preferentially mapped over relatively isolated relations (the "systematicity" principle). Gentner's very definition of analogy in effect presupposes these properties. According to her, there is a continuum of kinds of

Figure VI–1. Pictorial representations of two situations between which the SME program constructs a mapping. On the left, water flows between containers filled to different levels; on the right, heat flows between bodies at different temperatures.

comparison: an *analogy* is a comparison in which only systematic relations are mapped, whereas a comparison in which both attributes and relations are mapped is a *literal similarity,* not an analogy.

One of Gentner's examples of an analogy is illustrated in Figure VI–1 (from Falkenhainer, Forbus, & Gentner, 1990). The idea "heat flow is like water flow" is illustrated by constructing correspondences between a situation in which water flows from a beaker to a vial through a pipe and a situation in which heat flows from coffee in a cup to an ice cube through a metal bar.

The predicate-logic representations given by Falkenhainer, Forbus, and Gentner for these two situations are displayed in Figure VI–2.

The idea is that the causal-relation tree on the left (representing the fact that greater pressure in the beaker causes water to flow from the beaker to the vial through the pipe) is a *systematic* structure and should thus be mapped to the heat-flow situation, whereas the other facts ("the diameter of the beaker is greater than the diameter of the vial", "water is a liquid", "water has a flat top", etc.) are irrelevant and should be ignored. Ideally, mappings should be made between *pressure* and *temperature*; between *coffee* and *beaker*; between *vial* and *ice cube*; between *water* and *heat*; between *pipe* and *bar*; and more obviously, between *flow* and *flow*. Once these mappings are made, a conjecture about the cause of heat flow in the situation on the right can be made by analogy with the causal structure in the situation on the left. Gentner claims that if people recognize that this causal structure is the deepest and most interconnected system for this analogy, then they will favor it for mapping.

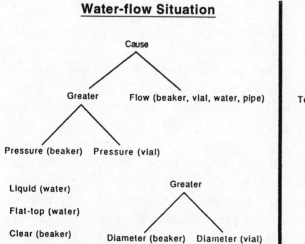
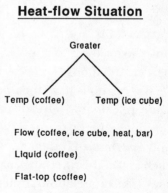

Figure VI–2. Predicate-calculus representations for the two situations shown in the preceding figure.

Gentner gives the following (possibly conflicting) criteria for judging the quality of an analogy:

- *clarity* — a measure of how clear it is which things map onto which other things;
- *richness* — a measure of how many things in the source are mapped to the target;
- *abstractness* — a measure of how abstract the things mapped are, where the degree of "abstractness" of an attribute or relation is its "order": attributes (*e.g.*, "flat-top" in the example above) are of the lowest order, relations whose arguments are objects or attributes (*e.g.*, "flow") are of higher order, and relations whose arguments are relations (*e.g.*, "cause") are of even higher order;
- *systematicity* — the degree to which the things mapped belong to a coherent interconnected system.

The computer model of this theory (SME) takes a predicate-logic representation of two situations (such as the representation given in Figure VI–2), makes a mapping between objects, attributes, and relations in the two situations, and then makes inferences from this mapping (such as "the greater temperature of the coffee causes heat to flow from the coffee to the ice cube"). The only knowledge the program has of the two situations consists of their syntactic structures (*e.g.*, the tree structures given for the water-flow and heat-flow situations displayed above); it has no knowledge of any of the *concepts* involved in the two situations. In other words, there is no prior structure containing any facts about water, liquids, heat, flow, or even physical objects. All processing is based solely on syntactic structural features of the two given representations.

SME first uses a set of "match rules" (provided to the program ahead of time) to make all "plausible" pairings between objects (*e.g.*, *water* and *heat*) and between relations (*e.g.*, *flow* of water and *flow* of heat). Typical match rules are: "If two relations have the same name, then pair them"; "If two objects play the same role in two already-paired relations (*i.e.*, are arguments in the same position), then pair them"; "Pair any two functional predicates" (*e.g.*, *pressure* and *temperature*). It then gives a score to each of these pairings, based on factors such as: Do the two things paired have the same name? What kind of things are they (objects, relations, functional predicates, etc.)? Are they part of systematic structures? The kinds of pairings allowed and the scores given to them depend on the set of match rules given to the program; different sets can be supplied.

Once all plausible pairings have been made, the program makes all possible sets of consistent combinations of these pairings, making each set (or "global match") as large as possible. "Consistency" here means that each element can match only one other element, and a pair (*e.g.*, *pressure* and *temperature*) is allowed to be in the global match only if all the arguments of its two paired elements are also paired up in the global match. Such consistency ensures "clarity" (in Gentner's sense), and the fact that the sets are maximal induces a preference for "richness". After all possible global matches have been formed, each is given a score based on the individual pairings it is made up of, the inferences it suggests, and its degree of systematicity. Gentner and her colleagues have compared the relative scores assigned by the program with the scores people give to the various analogies (Skorstad, Falkenhainer, & Gentner, 1987).

Points of agreement between Copycat and SME

Analogy-making as modeled in the Copycat program is in agreement with several aspects of Gentner's theory. We agree with the main idea of systematicity: that in general, the essence of a situation — the part that should be mapped — is a high-level coherent whole, not merely a collection of isolated low-level facts. In Copycat, there is a definite pressure toward systematicity, which is itself an emergent result of several pressures:

- the pressure, coming from codelets, to perceive relations and groupings within strings;
- the pressure to see things abstractly (which itself emerges from the preference for using descriptions of greater conceptual depth, and from the tendency of deeper concepts to stay active longer);
- the pressure to describe the change from the initial to the modified string in terms of relationships and roles, since these tend to be deeper than attributes (*e.g.*, in formulating a rule for the change *abc* $\Rightarrow$ *abd*, it is in general better to describe the *d* as "the successor of the rightmost letter" rather than as "an instance of *d*");
- the greater salience of larger relational structures (*e.g.*, a group consisting of an entire string), which makes them more likely to be paid attention to by codelets, and hence mapped;
- the high strength of correspondences between large relational structures (such as whole-string groups): such correspondences are strong not only because they involve large structures, but also because they are based on many concept-mappings;
- the pressure toward forming a set of compatible correspondences that, taken together, form a coherent worldview;
- the system's overall drive to achieve as low a temperature as possible, which pushes for the building-up of deep hierarchical perceptual structures, and thus for an abstract, conceptually deep view to be produced of the situations.

Gentner captures several important notions in her four-point characterization of a "good" analogy, and corresponding pressures exist in Copycat:
her pressure toward "clarity" is enforced by Copycat's prohibition of many-to-
one or one-to-many mappings without first chunking the "many" into a group;
her pressure toward "richness" corresponds to Copycat's preference for having
many correspondences and many concept-mappings underlying a correspondence; and Copycat's drives toward abstraction and systematicity were just
described above. But note that Gentner's syntax-based definition of "abstraction" (namely, the order of a relation) is very different from Copycat's notion
of "conceptual depth". Conceptual depths are not rooted in syntax (in fact, the
original name for them was "semanticity values"); rather, these values are
assigned by hand, with quite high values sometimes going to concepts that
Gentner might call "attributes" (such as *alphabetic-first,* which could be seen as
an attribute of the letter *a*).

Points of disagreement between Copycat and SME

Despite some points of agreement, there are several fundamental issues
on which our approach and Gentner's disagree; moreover, some of the most
important aspects of analogy-making addressed in the Copycat project are not
dealt with in any way in Gentner's theory and computer model.

Genter's abstractness and systematicity principles capture something important about analogy-making, but there are often other pressures in an analogy: both superficial and abstract similarities that may not be parts of systematic
wholes, but are still strong contenders in a competition. An example of this in
the Copycat domain is Problem 2 from Chapter 5:

$$aabc \Rightarrow aabd;\ ijkk \Rightarrow ?$$

The abstractness and systematicity principles would, it seems, argue for the
answer *ijll*, since the label *sameness group* describing the group of *a*'s and the group
of *k*'s is firstly merely an *attribute,* and secondly is not related to the systematic set
of successor relations in each string; according to the systematicity principle, it
should thus not be mapped, but should be ignored. However, as was discussed
earlier, many people feel that the two groups should map onto each other
nonetheless, and that the best answer is *hjkk*, in spite of what would be, it seems,
an *a priori* dismissal by the structure-mapping theory. Making any analogy involves
a competition between rival views, and one cannot be certain ahead of time that
the mapping with the highest degree of systematicity (in Gentner's sense) will
have the highest quality (*i.e.,* lowest average final temperature).

Another problem with Gentner's theory is that for any complex situation,
there are many possible sets of relations that exhibit systematicity, and it is not
explained how certain ones are considered for mapping and not others, on

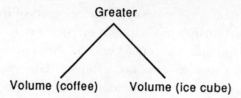

Figure VI–3. A complicating factor that might be added to the predicate-calculus representation of the heat-flow situation, shown in Figure VI–2.

syntactic grounds alone. For example, suppose the heat-flow domain had contained the relation displayed in Figure VI–3.

There would be no reason, based on syntax alone, to prefer the structure concerning temperature over this structure for mapping; if this structure were chosen, the analogy-maker would mistakenly conclude that, just as the pressure differential causes the water flow, the volume differential causes the heat flow. The program has no semantic connection between *temperature* and *heat* that might guide it to suspect the possible role of temperature in the phenomenon of heat flow; semantic connections play no role in the structure-mapping theory. However, as this example shows, syntactic structure alone is insufficient to determine which facts are part of a relevant systematic whole, and which are isolated and irrelevant.

By contrast, in Copycat, the mechanisms for deciding which things to concentrate on and which mappings to make involve *semantics*: they involve activation of concepts in the Slipnet in response to perception of instances of those concepts in the letter-strings, and through spreading activation to associated concepts; they also involve competition among objects clamoring to be noticed and among various descriptions of objects and relationships between objects; and they involve certain *a priori* notions of conceptual depth. For SME, not only are the attributes and relations in each situation laid out in advance, but there is no notion of differential relevance among them: which ones get used in an analogy is entirely a function of the syntactic structure connecting them. In Copycat, the notions of differential relevance and "shaded" (as opposed to black-or-white) inclusion of concepts in a situation (via probabilities as a function of differential activation in the Slipnet) — and of the program *itself* bringing in the concepts to be used to describe the situation — are fundamental, since Copycat is a model of how situations are interpreted as well as of how mappings are made between them, and of how the two processes interact.

The philosophy of Gentner and her colleagues is that the interpretation stage and the mapping stage can be modeled independently; that there are, in effect, separate "modules" for each. In contrast, a philosophy underlying the Copycat project is that the two are inextricably intertwined; the way in which the two situations are understood is affected by how they are mapped onto each other, as well as vice versa. (This issue of the necessity of integrating these two processes is discussed further in Chapter 4 of this book.)

Another fundamental difference between our approach and Gentner's is that her theory does not include any notion of conceptual similarity or of slippage, notions absolutely central to the Copycat project. In the water-flow/heat-flow example given above, the representations of the two situations are sufficiently abstract to make the analogy a virtual isomorphism. For example, the distinct concepts of *water flow* and *heat flow* have both been abstracted in advance into a general notion of *flow*. Likewise, in another analogy that Gentner describes, in which the hydrogen atom is mapped onto the solar system (see Figure IV–1, on page 183), all the important predicates in both situations have been given identical labels (*e.g., attracts, revolves around, mass*). This is necessary because of the theory's reliance on syntax alone. If this "identicality" constraint were to be relaxed, semantics and context-dependence (*i.e.,* some knowledge of conceptual proximity and how it is affected by context) would have to be brought in. But at present, since the concepts contained in the preconstructed representations always come in a sufficiently abstract form, there is no need for a Slipnet-like structure in which various concepts flexibly *become* more or less similar to one another in response to context. The analogy is already effectively given in the representations.

Yet another problem with Gentner's theory (and with SME) is that it relies on a precise and unambiguous representation of situations in the language of predicate logic. The structure-mapping theory's reliance on syntax alone requires that situations be broken up very clearly into objects, attributes, functions, first-order relations, second-order relations, and so on. For example, the water-flow/heat-flow analogy includes the following correspondences:

$$water \Leftrightarrow heat$$
(both are objects);
$$coffee \Leftrightarrow beaker$$
(both are objects);
flow (beaker, vial, water, pipe) $\Leftrightarrow$ *flow (coffee, ice cube, heat, bar)*
(both are 4-place relations).

But suppose that, in the heat-flow situation, *heat* had been described not as an object, but as an attribute of *coffee,* as in *holds-heat (coffee)*; or suppose that *heat flow* had been given as a 3-place rather than a 4-place relation: *flow (coffee, ice cube, heat)*, where the medium through which the heat flows is considered to be irrelevant; or suppose that, in the water-flow situation, *water flow* had been given as a 5-place relation: *flow (beaker, vial, water, pipe, 10 cc/second)*, where the rate of flow is included. Any of these quite plausible changes would totally block a successful application of the structure-mapping theory.

The problem is that in the real world, the labels "object", "attribute", and "relation" are very blurry, and people (if they make such distinctions at all) have to use them very flexibly, allowing initial classifications to slide if necessary at the drop of a hat. And to do this, semantics must be taken into account (this point is also made by Johnson-Laird, 1989).

For example, the judgment that *heat* belongs to the category "object" is necessary for the water-flow/heat-flow analogy to work, but it is arbitrary; it is rather dubious that all analogy-makers would independently come up with this judgment. Rather, it seems likely that any two people (or even one person, at different times) would produce very different predicate-logic representations of, say, the water-flow situation, no doubt differing on which things were considered to be objects, which were attributes, which were relations, how many arguments a given relation has, and so on. Thus, a serious weakness of the structure-mapping theory is its inability to deal with any flexibility in the representation of situations.

Admittedly, there are ways in which Copycat also breaks up a situation's representation too cleanly into object-attributes (descriptions) and relations between objects (bonds), where many people would not do so. However, we believe that this defect is not a deep aspect of Copycat's architecture, but that it can be remedied without large conceptual revisions. Mitchell (1993) describes problems in which there is a need for structures that fluidly change from being descriptions to being bonds in response to real-time perceptual pressures, and we believe that Copycat can be made to handle such problems. By contrast, the possibility for real-time representational flexibility is *fundamentally* lacking in a program like SME, which relies solely on the syntax of predicate-logic representations that are supplied to it before the fact. For such a program to work, the representations have to be tailored carefully. This discrepancy between architectures reflects a significant philosophical difference about analogy-making.

Although SME is meant to simulate human analogy-making, in that it models which types of structures people tend to map from one situation to another and which of the various possible mappings people tend to prefer, it doesn't attempt to model the underlying concepts or the dynamic perceptual processes in the psychologically realistic manner that Copycat does; certainly, the exhaustive search SME performs through all consistent mappings is psychologically implausible. This brings up one last, and very important, difference between SME and Copycat.

In Copycat, there is a clear distinction between answers that are immediately appealing (high-frequency answers) and answers that are insightful but hard to uncover (low-frequency but also low-temperature answers, such as *wyz* in Chapter 5's Problem 6 and *mrrjjjj* in Problem 4). This distinction falls out naturally from the nature of the parallel terraced scan, in which, roughly speaking, obvious routes are explored before nonobvious ones. Thus in Problem 6, there is a definite temporal order that is generally followed, in which an attempt to take the successor of *z* is first made, the snag is hit, and then, thanks to this pressure, other, less-obvious possibilities wind up getting explored. Some of the time, the deeply hidden but also deeply insightful answer *wyz* is discovered. We believe that the story sketched above of this process mirrors the natural stages in human exploration. Likewise, in Problem 4, answers such as *mrrjjk* and *mrrkkk* "pop to mind" for Copycat much more readily than does *mrrjjjj*; the relative ease and the relative rarity of these different answers are comparable to what happens when people consider this problem.

In a brute-force system like SME, by contrast, the distinction between answers of superficial and obvious appeal and answers of deeper but more hidden appeal does not exist. Indeed, in such systems, this type of temporality makes no sense, because *all* answers — that is, all answers that lie within the system's capability — are generated together and evaluated together, and then simply ranked according to their algorithmically-assigned scores. If, for instance, such a system were given Problem 6, then not only would it fail to perceive any snag, but if *wyz* were within its conceptual range, it would get it effortlessly, and its high ranking (which it presumably — and hopefully — would receive) would give no indication of its counterintuitive twist. This is psychologically inaccurate; *wyz*, if reached at all, is reached only after many other routes have been explored, and this critical fact needs representation if the model is intended to reflect human-style cognition.

Problem 4 presents a similar dilemma for a brute-force exploration strategy such as that of SME, since in order for the answer *mrrjjjj* to be created, a novel concept (*group length*) has to be brought into the picture. The difficulty for a brute-force system is that the only way for it to be able to discover the aptness of one *particular* initially-absent concept is for it to bring in *all* concepts in its repertoire, regardless of their initial plausibilities, giving equal chances to all of them to prove their worth. Again, this is utterly unrealistic from a psychological point of view, but then so is the opposite strategy — namely, not ever bringing in novel concepts at all. (Note that this latter is the strategy of SME.) A subtler middle way is needed, poised delicately between the absurdity of exploring *every* possible idea with equal enthusiasm, and the sterility of not daring to explore *anything* new. Such a middle way is provided by a probabilistic architecture such as Copycat's, which was inspired specifically by the ubiquitous need for such shades of gray (as discussed in the final section of Chapter 5).

In summary, both the architecture and purpose of the Structure Mapping Engine are extremely different in spirit from those of Copycat. SME is an algorithmic but psychologically implausible way of finding what the structure-mapping theory would consider to be the best mapping between two given representations, and of rating various mappings according to the structure-mapping theory, allowing such ratings then to be compared with those given by people. And as far as the structure-mapping theory itself is concerned, it makes a number of useful points about what features appealing analogies tend to have, but in dealing only with the end result of a mapping while leaving aside the problem of how situations become understood and how this process of interpretation interacts with the mapping process, it leaves out some of the most important aspects of how analogies are made.

ACME, the Analogical Constraint Mapping Engine

Psychologist Keith Holyoak and philosopher Paul Thagard have built a computer model of analogical mapping (Holyoak & Thagard, 1989), based in part on theoretical and experimental work by Holyoak and his colleagues (Gick & Holyoak, 1983; Holland *et al.*, 1986), and inspired in part by research by Marr and Poggio on constraint-satisfaction networks used to model stereoscopic vision. The computer model, ACME (Analogical Constraint Mapping Engine), is similar to SME in that it uses representations of a source situation and a target situation given in formulas of predicate logic, and makes an analogical mapping consisting of pairs of constants and predicates from the representations. In fact, ACME has been tested on several of the same predicate-logic representations of situations that SME was given, including the water-flow and heat-flow representations.

The model takes as input a set of predicate-logic sentences containing information about the source and target domains (*e.g.*, water flow and heat flow), and it constructs a network of nodes (taking into account a number of constraints) where each node represents a syntactically allowable pairing between one source element and one target element (a constant or a predicate).[1]   A node is made for every such allowable pairing. For example, one node might represent the *water* ⇔ *heat* mapping, whereas another node might represent the *water* ⇔ *coffee* mapping. Links between nodes in the network represent "constraints": a link is weighted positively if it represents mutual support of two pairings (*e.g.*, there would be such a link between the *flow* ⇔ *flow* node and the *water* ⇔ *heat* node, since *water* and *heat* are counterparts in the argument lists of the two *flow* relations), and negatively if it represents mutual disconfirmation (*e.g.*, there would be such a link between the *flow* ⇔ *flow* node and the *water* ⇔ *coffee* node).

---

1. Here, "syntactically allowable" means adhering to the "logical-compatibility constraint", which specifies that a mapped pair has to consist of two elements of the same logical type. That is, constants are mapped onto constants and *n*-place predicates are mapped onto *n*-place predicates.

The network can optionally be supplemented with a "semantic unit": a node that has links to all nodes representing pairs of predicates. These links are weighted positively in proportion to the "prior assessment of semantic similarity" (*i.e.*, assessed by the person constructing the representations) between the two predicates.

In addition, ACME has an optional "pragmatic unit": a node that has positively weighted links to all nodes involving objects or concepts (*e.g.*, *flow*) deemed ahead of time (again by the person constructing the representations) to be "important".

Once all the nodes in the network have been constructed, a spreading-activation relaxation algorithm is run on it, which eventually settles into a final state with a particular set of activated nodes representing the winning matches.

Agreements and disagreements between Copycat and ACME

There are several points of agreement between the philosophy of this model and that of the Copycat program. We share the idea that analogy-making is closely related to perception and should be modeled with techniques inspired by models of perception. We also share the belief that analogies emerge out of a parallel competition among pressures (or "soft constraints"), involving a large number of local decisions that give rise to a larger coherent structuring. And we agree with the philosophy of Holyoak and Thagard that an analogy-making system's pressure toward systematicity (as described by Gentner) should emerge from other pressures.

There are, however, deep differences between Copycat and ACME, related to Copycat's differences with SME discussed in the previous subsection. First, like SME, ACME tries all syntactically plausible pairings, a method that is both computationally infeasible and psychologically implausible in any realistic situation. For example, in making an analogy between the Watergate and Iran–Contra scandals, would a human consider a mapping between Nixon and every person involved in the Iran–Contra situation, including Fawn Hall, Daniel Inouye, Ed Meese, and Dan Rather? Or even less plausibly, would a human consider mapping Gerald Ford onto the Contras' base camp in Honduras, or onto the cake taken as a gift to Iran by United States emissaries? Yet these are *all* plausible, according to the logical-compatibility constraint, in which semantics plays no role at all.

The existence of this exhaustive (though parallel) search through all possible mappings shows that ACME is not attempting to model how people very selectively search through such possibilities, whereas this is one of the Copycat project's main focuses. In Copycat, although any object in one string can in principle be compared with any object in the other string, an exhaustive search is avoided thanks to the parallel terraced scan, in which comparisons, if they are made at all, are made at different speeds and to different levels of depth, depending on constantly-evolving estimates of their promise.

ACME shares with SME the following major problem, discussed above: the representations of knowledge used are rigid, and are also tailored specially for each new analogy. ACME uses the same representation as did SME for the water-flow/heat-flow analogy, so the same issues discussed with respect to SME apply here. Again, the program has no ability to restructure its descriptions or to add new descriptions in the course of making an analogy; the descriptions are constructed by a human ahead of time and are frozen.

ACME differs from SME in that it has an optional "semantic unit" giving semantic similarities, which correspond in some sense to those embodied in Copycat's Slipnet, but the similarities are also decided in advance by the programmer for the purposes of the given analogy, and are frozen. Unlike the developers of SME, Holyoak and Thagard recognize the necessity of considering semantics as well as syntax, but the problem is that it is impossible in general to have a "prior assessment of similarities" (as encoded in ACME's semantic unit); rather, as has been shown in several ways above, analogy-making is all about how conceptual similarities shift dynamically in response to pressures that weren't apparent ahead of time.

ACME also sidesteps the question of how certain objects and concepts come to be seen as important in response to pressures; this is instead taken care of by the pragmatic unit, which encodes the *programmer's* prior assessment of what is important in the given situations. It might be claimed that the pragmatic unit encodes information corresponding to the activation of Slipnet nodes and to the importance values of objects in the Workspace. But again, unlike in Copycat, where these values *emerge* dynamically in response to what the program perceives, in ACME, the pragmatic unit is set up by a human and then *frozen* for each new problem. Thus, like SME, ACME does not deal with another of Copycat's main focuses: how concepts adapt to different situations.

In comparing Copycat with ACME, Holyoak and Thagard point out that Copycat lacks a pragmatic unit, as if this were a glaring defect. But our philosophy was not to *tell* our program what mattered in a situation; it was, rather, to let the program *itself* figure that out. What Copycat deems important changes over time, and depends on many interacting factors, rather than being a fixed set of numbers given to the program before the run starts. If we nonetheless did wish to simulate ACME's pragmatic unit, it would be a trivial matter of clamping the activations of some fixed set of Slipnet nodes and/or clamping the salience values of some fixed set of Workspace objects. Needless to say, this would affect the course of the processing, and experiments could be done to see what would happen. But we felt that replacing dynamic activity with static values was of little interest, and consequently did not advertise the possibility of node-clamping as a "feature" of our model.

ACME, like SME, models only the "mapping stage" of analogy-making, but, as was said before, a philosophy underlying Copycat is that the mapping process cannot be separated from the processes of perceiving and of reformulating perceptions and assessments of similarities in response to pressures. Holyoak and Thagard (1989) themselves point out that their model does not address this issue — they call it the issue of "re-representation" — and acknowledge that it will often be necessary to interleave mapping with manipulation of the representations, taking into account top-down pressures — which is essentially just what Copycat does.

Since ACME's knowledge is set up ahead of time, the program's success, like that of SME, is totally dependent on the representations it is given. In the examples given by Holyoak and Thagard (1989), the representations of the source and target matched each other almost perfectly; the essence had been distilled in exactly the right form for making an analogy (see Preface 4 in this book for further discussion of this). As with SME, it is very doubtful that the representations given to ACME could have been made by someone who didn't already have the mapping in mind, and it is very likely that the program would not succeed if the representations of the source and target situations were made independently by two different people.

Finally, the same criticisms as were leveled at SME regarding the absence of temporality hold equally for ACME. That is to say, ACME, no less than SME, is based on brute-force full-width exploration strategies, and in it there simply is no counterpart to the natural distinction in Copycat between immediately obvious answers and insightful answers whose discovery requires deeper, longer search. More specifically, ACME, like SME, has no counterpart to the snags and paradigm shifts that can occur in Copycat runs on Problem 6, and no counterpart to the highly selective awakening of novel concepts from dormancy that can occur in Copycat runs on Problem 4. This "temporal flatness" of both ACME and SME is psychologically inaccurate. Those programs' creators might reply that they are not trying to model this aspect of analogy-making; our response would be that we see such temporality as central to the nature of analogy, and therefore critical to include in one's model.

How Real Are These "Real-world" Analogies?

A criticism often leveled against Copycat is that it makes analogies in an idealized microworld, whereas other analogy-making programs work in more complex, real-world domains. On the surface it would seem that SME and ACME make real-world analogies that are much more complex than the "toy" problems

Copycat deals with. But if one looks below the surface (as we did here for the water-flow/heat-flow example, and as was also done in Preface 4 for the Socrates/midwife example), it is clear that the knowledge possessed by these programs (that is, the knowledge given to them for each new problem, in the form of sentences of predicate logic), in spite of the real-world aura of words like "pressure" and "heat flow", is even more impoverished than Copycat's knowledge of its letter-string microworld.

The programs know virtually nothing about concepts such as *heat* and *water* — much less than Copycat knows about, say, the concept *successor group*, which is embedded in a network and can be recognized and used in an appropriate way in a large variety of diverse situations. For example, each one of the strings **abc, aabbcc, cba, abbbc, mrrjjj, mmrrrjjjj, jjjrrm, abbccc, xpqefg**, and **k** can be recognized as an instance of *successor group*, given the appropriate pressures.[2] Nothing like this is the case for, say, SME's and ACME's notion of *heat* as given for the purpose of making a water-flow/heat-flow analogy. There, the notion of *heat* has essentially no semantic content and certainly cannot be adapted to any other situation. Nor can these programs recognize heat or a heat-like phenomenon.

These programs are purported to make very high-level analogies involving the concepts *heat* and *water*, and yet, ironically, the programs have absolutely no sense of the meaning of *heat* or *water*; that is, the much more fundamental ability to recognize *instances* of these categories is completely absent from the programs.

This can be put in a slightly different way. Consider a computer program that manipulates complex data-structures in which the Lisp atoms "NUCLEUS", "ELECTRON", "SUN", and "PLANET" (among others) occur, and suppose the data-structures are intended to encode a few basic propositions about the hydrogen atom and the solar system. Now suppose the program comes up with a mapping in which the Lisp atom "NUCLEUS" corresponds to the atom "SUN", and similarly, "ELECTRON" to "PLANET". The computer would of course have done its job as easily if the names of Greek letters, say, had been used in the place of English words. Then the program would have made the following "scientific analogy": "NU maps to SIGMA; EPSILON maps to PI". Neither reading this result nor any amount of watching the program's behavior, no matter how long, would give a human any reason to suspect that the Lisp atom "NU" stood, in the programmer's mind, for the idea of an atomic nucleus, and that "SIGMA"

---

2. Myriad other examples of successor groups, with different degrees of abstruseness, can be formed in the letter-string domain. Many are beyond Copycat's current recognition capabilities, although the same perceptual mechanisms the program has now could, we believe, be extended to recognize more complex instances, such as **ace** (a "double-successor" group), **aababc** (which can be seen as a "coded" version of **abc** when parsed as **a–ab–abc**), **kmxxrreeejjj** (which could be described as "11–22–33"), **axbxcx** (where the *x*'s form a "ground" for the "figure" **abc**), **abcbcdcde** (which could be parsed **abc–bcd–cde**), and so on.

stood for the concept of the sun. There is simply nothing in the program itself, or in its behavior, that makes these symbols stand for anything in a recognizable way. Only the person who used them to encode "pieces of knowledge" sees them as standing for anything.

Why can't this exact argument be turned word for word against Copycat? Could one not argue that when *successor* slips to *predecessor* in Problem 2, for instance, it might as well simply be the Lisp atom "SIGMA" slipping to the Lisp atom "PI"? What would the machine care? Of course the machine wouldn't care at all, but that misses the point entirely. An astute human watching the performance of Copycat and seeing the term "SIGMA" evoked over and over again by the presence of successor relationships and successorship groups in many diverse problems would be likely to make the connection after a while, and then might say, "Oh, I get it — it appears that the Lisp atom 'SIGMA' stands for the idea of successorship."

In other words, whether or not an English *façade* is used, the symbols in Copycat acquire at least some degree of genuine meaning, thanks to their correlation with actual phenomena, even if those phenomena take place in a tiny and artificial world. By contrast, in other analogy-making programs, there is no tiny and artificial world (let alone a big and real one) that serves to imbue the symbols they manipulate with any meanings.[3]

The claim that Copycat's microworld is a "toy domain" while these other programs are solving real-world problems is truly unfounded, and is based on

3. (Footnote by DRH) This argument to the effect that Copycat's symbols owe their non-zero semantic content to their dynamic ties to entities in an external world might remind some readers of Stevan Harnad's notion (Harnad, 1989 and 1990) of "grounding" a cognitive system's symbols by links to an external world in order to ensure that they have genuine meanings. Unfortunately, however, Harnad's distinction between "grounded" and "ungrounded" symbols seems to me to be essentially a reaction to the fear, instilled by such biochauvinistic philosophers as John Searle, that there could be two fundamentally different ways in which the unrestricted Turing Test could be passed: not only by a system with genuine meanings and understanding (*i.e.*, a system possessing intentionality or "aboutness"), but also by a system whose symbols were utterly empty and content-free — a "zombie" system, in other words. A zombie system's external behavior would, most troublingly to Harnad, be indistinguishable from that of a system possessing the *élan mental* of "grounded symbols". That is the root fear that arises when one falls for Searle's smoothly-worded yet self-contradictory "Chinese room" scenario (see Searle, 1980, also reprinted with comments by myself in Hofstadter & Dennett, 1981). In order not to be taken in by a zombie system, Harnad seems to want *two* levels of certification of a program: firstly, high-quality performance, but *above and beyond performance*, he also wants a "pedigree" for its symbols (as if performance, no matter at what level, were simply a *façade*). Here I part company with Harnad. If I were given the first level of certification, I would feel no need for the second level: to me, passing the Turing Test is no more and no less than a proof that the symbols behind the scenes are meaningful. Undoubtedly, those symbols *would* be linked to dynamic entities in the real world, but that fact would already be perceptible in the high level of the system's performance, and not in need of further, independent verification. Or, to turn this around, if those links to the real world were absent, that lack would be detectable through suitable probing in the Turing Test. (For more on this idea of using the Turing Test to probe behind-the-scenes mechanisms, see the Epilogue.)

a tendency of people to attribute much more intelligence to a program than it deserves based simply on real-world-sounding words it uses (such as "heat") — words that are extremely rich in imagery and connotations for people, but are almost completely empty as far as the program is concerned. Any program that uses words that exude real-world connotations but that are in reality completely devoid of semantic content as far as the program is concerned has the potential to be highly misleading — misleading not just to lay people, but to professionals as well. (This point about AI research has been trenchantly made by McDermott, 1976.) An "all the cards are on the table" quality is one of the advantages of using explicit microworlds for research in artificial intelligence.

Copycat's Position along the Symbolic/Subsymbolic Spectrum

The philosophy behind the Copycat project is similar in many ways to that of various connectionist or "parallel distributed processing" (PDP) models (Rumelhart & McClelland, 1986) and to that of classifier systems (Holland, 1986; Holland *et al.*, 1986). Connectionist networks and classifier systems are examples of *subsymbolic* (also called *subcognitive*) architectures. Smolensky (1988) characterizes the difference between the symbolic and subsymbolic paradigms as follows. In the symbolic paradigm, descriptions used in representations of situations are built of entities that are *symbols* both in the semantic sense (they refer to categories or external objects) and in the syntactic sense (they are operated on by "symbol manipulation"). In the subsymbolic paradigm, such descriptions are built of *subsymbols*: fine-grained entities (such as nodes and weights in connectionist networks, or classifiers in a classifier system) that give rise to symbols.

In a symbolic system, the symbols used as descriptions are explicitly defined (*e.g.*, a single node in a semantic network represents the concept *dog*). In a subsymbolic system, symbols are statistically emergent entities, represented by complex patterns of activation over large numbers of subsymbols. (Similar characterizations have been made in Chapter 26 of Hofstadter, 1985, and by McClelland, Rumelhart, & Hinton, 1986.) Smolensky makes the point that subsymbolic systems are not merely "implementations, for a certain kind of parallel hardware, of symbolic programs that provide exact and complete accounts of behavior at the conceptual level" (page 7). Symbolic descriptions are too rigid or "hard", and a system will be fluid enough to model human cognition only if it is based on the more flexible and "soft" descriptions that emerge from a subsymbolic system.

The faith of the subsymbolic paradigm is that human cognitive phenomena are emergent statistical effects of a large number of small, local, and distributed subcognitive events with no global executive. This is the philosophy underlying connectionist networks, classifier systems, and Copycat as well. Fine-grained parallelism, local actions, competition, spreading activation, and distributed and emergent concepts are essential to the flexibility of all three types of architecture (although in classifier systems, spreading activation is not explicit, but rather emerges from the joint activity of many classifiers).

Some connectionist networks (*e.g.*, Boltzmann machines, Hinton & Sejnowski, 1986, and Harmony-Theory networks, Smolensky, 1986) have an explicit notion of computational temperature with some similarity to Copycat's (although, as was explained in Chapter 5, there is a significant difference between the use of temperature in Copycat and in simulated annealing, which is essentially the temperature notion used by Hinton & Sejnowski and by Smolensky).

In classifier systems, something akin to a parallel terraced scan emerges from probabilistically-decided competitions among classifiers and from the genetic algorithm's implicit search through schemata (*i.e.*, templates for classifiers) at a rate determined by each schema's estimated promise (Holland, 1975). In addition, the interaction of top-down and bottom-up forces is central both in connectionist systems (see, for example, McClelland and Rumelhart's interactive-activation model of letter perception, 1981) and in classifier systems (for example, as discussed in Chapter 2 of Holland *et al.,* 1986).

The philosophy underlying the Copycat project is more akin to that of the subsymbolic paradigm than that of the symbolic paradigm, but the actual *program* fits somewhere in between (although as we said at the outset of Chapter 5, Copycat is not a hybrid of the two paradigms). As an oversimplified but nonetheless useful characterization of the two ends of the spectrum, one could say that concepts in subsymbolic systems are highly distributed, being made up of individual nodes that have no semantic value in and of themselves, whereas in symbolic systems, concepts are represented as simple unitary objects (*e.g.*, as Lisp atoms). Concepts in Copycat could be thought of as "semi-distributed", since a concept in the Slipnet is probabilistically distributed over only a small number of nodes — a central node (*e.g.*, *successor*) and the nodes in its probabilistic halo (*e.g.*, *predecessor*), to which it can probabilistically slip.

The basic units of subsymbolic systems such as connectionist networks are meant to model mental phenomena further removed from the cognitive, conscious level than those modeled by Copycat's codelets and Slipnet nodes (the basic units out of which pressures and concepts — some of Copycat's main higher-level phenomena — emerge). It may be that connectionist networks are neurologically more realistic than Copycat, but their remoteness from the cognitive level makes the problem of controlling their high-level behavior quite difficult. At this time it is dubious whether it would be possible to use connectionist systems similar to those currently in existence to model the types of high-level cognitive behavior exhibited by Copycat. Ideally, a cognitive model should have the property that a high-level semantic structure such as Copycat's Slipnet demonstrably arises as an implicit consequence of a low-level, distributed model, but this is beyond the achievements of current research in connectionism.

Likewise, in classifier systems, several extremely basic notions implanted directly in Copycat (such as nodes, links, and spreading activation) would have to emerge automatically, which would make a high-level task such as Copycat's quite difficult for classifier systems as they are currently conceived. Thus Copycat models concepts and perception at an intermediate level, in terms of the degree to which concepts are distributed and the extent to which high-level "semantic" behavior emerges from lower-level "syntactic" processing.

A major difference between Copycat's architecture and that of connectionist networks is the presence in Copycat of both the Slipnet, containing Platonic concept *types,* and the Workspace, in which structures representing concept *tokens* (*i.e.,* instances of concepts) are dynamically constructed and destroyed. Connectionist networks have no such separate working area; both types and tokens are represented in a single network. This has led to a great deal of research in connectionism on the so-called "variable-binding problem", which is related to the larger question of the relationship between concept types and concept tokens.[4] One reason researchers in connectionism hesitate to make such a separation is that neural plausibility is a central aspect of their research program, and a structure like Copycat's Workspace — a mental region in which representational structures are constantly being created and destroyed — does not have a clear neural underpinning.

In contrast, for the purposes of Copycat and related projects, we have been influenced more by *psychological* than *neurological* findings. Based on many pieces of evidence for a separation between a Platonic-concept memory and a workspace where instances reside (see, *e.g.,* Treisman, 1988, which reports on many experiments on visual perception that strongly point to the existence of a site of temporary episodic structures), we simply *assume* the existence of something like Copycat's Workspace even though we do not know its neural basis, and we investigate how a spreading-activation network with distributed and permanent concept *types* interacts with a working area in which ephemeral concept *tokens* can be arranged in complex structures. The lack of such a working area in connectionist networks is another reason why it may turn out to be very difficult to use such systems to model concepts and high-level perception in the way Copycat does.

---

4. Philosophers have a long-standing distinction between types and tokens, but it is a highly formal one, characterizable with precision and no ambiguity. The kind of type/token distinction we are talking about here, however, is much blurrier, because it has to do, ultimately, with how such things are realized in biological "wetware". See the discussion of letter types and letter tokens in Preface 2 for a sense of how different our distinction is from the philosophers' distinction.

Connectionist networks and classifier systems learn over time, while Copycat does not. Copycat is not meant to be a model of learning in this strict sense, though it does model some fundamental aspects of learning: how concepts adapt to new situations that are encountered, and how the shared essence of two situations is recognized.

The belief underlying our methodology is that building a model at the level of Copycat's architecture is essential not only for the purpose of providing an account of the mental phenomena under study at its intermediate level of description, but also as a step towards understanding how these phenomena can emerge from even lower levels.

The "connectionist dream" — that of modeling all of cognition using a subsymbolic, neurologically plausible architecture — may be too ambitious at this point in the development of cognitive science. If there is any hope for understanding how intelligence emerges from billions of neurons, or even how it might emerge from connectionist networks, we need to understand the nature of *concepts,* fundamental entities whose operating principles seem to lie somewhere between those of highly parallel neural networks and those of highly symbolic serial cognition.

The question "What is a concept?" could be said to lie at the crux of cognitive science (See Chapters 12, 23, and 26 of Hofstadter, 1985), and yet concepts still lack a firm scientific basis. The long-term goal of the Copycat project and related research is to use computer models to help provide such a scientific basis. The hope is that the understanding that results from this approach not only will contribute in its own right to answering long-standing questions about the mechanisms of intelligence, but also will provide a guide to connectionists studying how such intermediate-level structures can emerge from neurons or cell-assemblies in the brain.

In summary, the architecture of Copycat is very different from the more traditional symbolic artificial-intelligence systems, both in its parallel and stochastic processing mechanisms, and in its representation of concepts as distributed and probabilistic entities in a network. These features make it more similar in spirit to connectionist systems, though again there are important differences. The high-level behavior of connectionist systems emerges statistically from a lower-level substrate, as in Copycat. However, the fundamental processing units in connectionist systems are more primitive; also, concepts in such networks are distributed to a much higher degree than in Copycat; and finally, concept types and tokens are required to reside in the same network. Consequently, there has not been much success so far in using connectionist systems as models of high-level cognitive abilities such as analogy-making.

Copycat explores a middle ground in cognitive modeling, between the high-level symbolic systems and the low-level connectionist systems; the claim made for this approach is that this intermediate level is at present the most useful one at which to grapple with the fluid nature of concepts and perception, central aspects of mind that emerge with maximal clarity in analogy-making.

*   *   *

Postscript (by DRH).

Some time after the preceding material was written, I came across two quite new and refreshing approaches to metaphor and analogy and the computer-modeling thereof. I couldn't see leaving them out of the discussion, so I have added this brief section about them.

Indurkhya's View of Creativity, and the PAN Model

Computer scientist Bipin Indurkhya, in his penetrating book *Metaphor and Cognition: An Interactionist Approach* (1992), writes at length on what he terms "similarity-creating metaphors" — metaphors that supposedly bring into being similarities that literally did not exist before they were made. An example he gives is Stephen Spender's poem "Seascape", in which sunlight reflecting on waves is likened to someone strumming a harp. Indurkhya argues that metaphors of this special type can and ought to be distinguished from "similarity-based metaphors", the far more frequent and more mundane metaphors that merely recognize similarities that were always there to be noticed. I find this distinction dubious; in my view, all analogies (and metaphors) are cases of *discovered* rather than *invented* similarity, with some similarities simply being harder to recognize than others, sometimes so much so that the act of finding the resemblance seems to reveal a point of view that never existed before. (The discovery-versus-creation controversy as it applies to analogies, as well as to other products of the mind running the gamut from theorems to novels, is explored at length in Hofstadter, 1987a.)

Despite my disagreement with Indurkhya on this philosophical point, I sympathize with his desire to put his finger on just what it is that distinguishes mundane and straightforward analogies (or metaphors) from ones that bring powerful and unexpected insights to a situation. For me, this is essentially the distinction between "creative" analogies — that small minority of analogies that are deep yet very hard to find (of the *wyz* and *mrrjjjj* type) — and all the rest, which includes strong-but-obvious analogies, mediocre-and-obvious analogies, mediocre-yet-hidden analogies, and even "sloppy" and "dizzy" analogies (as described in Chapter 5). In my way of looking at things, the two dimensions of *average final temperature* and *frequency* can be used to roughly define a zone of creative or insightful analogies in the Copycat domain — say, answers that are characterized by a final temperature of 20 or below, as well as a relatively low frequency of occurrence (somewhat arbitrarily, one could perhaps define the cutoff as 20 percent of the time).

What Indurkhya sees as critical for coming up with "similarity-creating" analogies (and what he sees as lacking in most current models of analogy-making) is *the ability to perceive a given structure in vastly different ways, under the influence of context*. With this thesis I fully agree. In his book, Indurkhya gives numerous examples in which a very familiar object (*e.g.*, a paintbrush), through mental juxtaposition with some unanticipated idea, suddenly reveals unexpected and novel aspects involving one or more concepts (*e.g.*, "pump") that one would *a priori* have had no reason to think were at all relevant. This effect and its converse are summarized in the phrase "making the familiar strange and the strange familiar", which could in fact serve as the motto of Indurkhya's book.

In order to model the creative invention of analogies, Indurkhya and his student Scott O'Hara selected two microdomains, thus running against the conventional dogma that only "real-world" domains will lead to genuine progress in cognitive science. One of their microdomains involves geometric figures, and is reminiscent of the domain used by Evans in ANALOGY. However, Indurkhya and O'Hara are far more concerned than Evans was with the extreme ambiguity possessed by geometric figures. The other microdomain they selected is our own letter-string domain, in which they focus on many of the same analogy problems as we have, as well as on a few problems that are close cousins to those in our core set of challenges. In strong contrast to us, however, O'Hara and Indurkhya argue for the viability of a fully symbolic and deterministic approach, in which an exhaustive breadth-first search of ways of perceiving a given structure is carried out in a terraced manner — first very shallowly, then a bit more deeply, then slightly more deeply, and so on, regulated by a variable called CL (for "change level"), which is likened to a leash that starts out tight and is gradually loosened (is this a similarity-creating analogy?).

Their proposed PAN model (O'Hara and Indurkhya, 1993; O'Hara, 1992) would start out with an initial structure — a geometric shape or a letter-string — already having a description provided by a human; it then would embark on a highly systematic series of modifications thereto in an effort to come up with a new description that applies to the target structure. The terraced strategy mentioned above thus amounts to tampering with the original description, first in a very shallow way, then in slightly more radical ways, and so forth, so that obvious or "close" variants are explored first, and then the search space is gradually opened up more and more by the loosening of the "leash" until, presumably, a modified description is eventually reached that exactly fits the target structure. In Copycat terms, the space of descriptions gets exhaustively searched in a "least-slippage-first" order. In this manner, the program would eventually come to see the target structure "in terms of" the initial structure.

The brute-force aspect of this search strategy strikes me as highly implausible, not only psychologically but also computationally. There is no counterpart to our notions of salience, strength, conceptual depth, and urgency, all of which serve as powerful probabilistic biases towards certain directions of exploration and away from others, and which thereby drastically restrict the searching that takes place.

Recently, in fact, I learned from O'Hara (O'Hara, 1994a) that some time ago, he came to the reluctant conclusion that this architecture would inevitably succumb to combinatorial explosions, and therefore abandoned it after making only a very rudimentary implementation. In its place he is now developing a new architecture called INA (O'Hara, 1994b), which has several points of commonality with Copycat, including much more emphasis on perceptual activity carried out by a mixture of bottom-up and top-down processing, although it is still serial and deterministic. I am gratified by this significant shift in direction, caused by the failure of the earlier ideas to pan out, one might say, in the form of a successful program.

In summary, I am in agreement with many of Indurkhya's keen observations about what really matters about analogy and metaphor, as well as his choice to work in microdomains; I feel that O'Hara and Indurkhya's old PAN architecture is highly implausible, a judgment on which they now apparently concur; and I look forward to seeing what results the new INA model brings.

Kokinov's AMBR System

Cognitive scientist Boicho Kokinov has developed an ambitious theoretical cognitive architecture called DUAL (Kokinov, 1994a), which he has concretely implemented in a computer program called AMBR (Associative Memory-Based Reasoning), whose purpose is to make analogies (Kokinov, 1994b).

Processing in AMBR is realized collectively by many agents that carry out various small symbolic tasks. These agents can be likened to codelets, in that they are small, operate on representational structures, and do so in parallel; also like codelets, their joint activity constitutes the emergent performance of the system. On the other hand, unlike codelets, AMBR agents are linked together to form a network in which every node — that is, every agent — has a time-varying degree of activation. This network features the spreading of activation between neighboring nodes, as well as decay of activation over time. This is of course reminiscent of our Slipnet.

The level of activation of any agent in AMBR indicates its degree of *perceived relevance* to the matter at hand, and as such, it controls the *speed* with which that particular agent carries out its symbolic task. Agents with activation below a certain threshold do not run at all, and hence contribute nothing to the system's overall behavior. At any time, only a relatively small number of AMBR agents have above-threshold activation, and this subset of the whole network gradually changes over time, much as in Copycat, the Coderack's population of codelets evolves over time. The fact that many agents operate in parallel with their speeds varying dynamically according to their perceived relevance is strikingly reminiscent of the way in which processing works in Copycat.

Despite all these notable similarities, there are many fundamental differences between AMBR and Copycat. Perhaps the most salient is the lack of any kind of perceptual workspace in the AMBR architecture. The idea of building up hierarchical perceptual representations of situations is in fact completely absent from the architecture, since it, very much like ACME and SME, deals exclusively with prepackaged and fixed representations of situations. Thus in AMBR there is no counterpart to the building of bonds or groups, the attaching of descriptions, the differential degree of interest in different entities, and so on — in other words, no perceptual activity.

Moreover, AMBR shares with ACME and SME the philosophy of working in hollow "real-world" domains. The quintessential example that Kokinov gives of his program's analogy-making abilities involves a human being in a forest figuring out, on the basis of an analogy to the urban concept of an immersion heater, that it is possible to boil water in a carved-out piece of wood by heating up a big rock over a fire and then plopping the hot rock into the water. However, when one looks at the trivial representations of these ideas that are given to the program, one sees that the immense amount of imagery summoned up instantly and automatically by these words in any human mind is completely lacking. For instance, the sum total of the idea of an immersion heater is, when paraphrased into English, this:

> A *hot* Immersion Heater that is *inside* some Water, which is *inside* a *glass* Container, causes the Water *inside* the Container to get *hot*.

But to any human who knows English, even this single sentence cannot help but suggest infinitely more imagery than the program has. AMBR has no knowledge base providing the italicized and capitalized words with any kind of semantics; for this reason, it would be more accurate to replace them with symbols that connote nothing to people, and simply say that the following is what AMBR "knows" about IH's (immersion heaters):

> An IH having property h and relationship i to W, which has relationship i to C, which has property g, causes W, in relationship i to C, to have property h.

In a moment of weakness, I retained the verb "causes", just to make the stripped-down version easier to read; I should have replaced it by something more vacuous, like "is such that". All the comments about the semantic hollowness raised above in connection with ACME and SME apply equally to AMBR, unfortunately.

On the other hand, one noteworthy virtue of AMBR, lacking in Copycat, is that it carries out retrieval of relevant episodes or sets of related facts from long-term episodic memory, whereas Copycat has no episodic memory and therefore does no such thing.

In summary, AMBR contains many rich and stimulating ideas, but like so many other approaches to analogy-making, it completely leaves out high-level perception of situations, which we consider to be the core of analogical thinking.

Preface 7:

Retrieval of Old and Invention of New Analogies

The Lifelike Appearance of Copycat

The following chapter was written as the afterword to Melanie Mitchell's book *Analogy-Making as Perception* (Mitchell, 1993). That book is "required reading" for anyone who really wants to understand Copycat — definitely a worthwhile goal, in my book.

I might also mention that Melanie is very happy to provide a running version of Copycat (source code in Common Lisp, running on a Sun Sparc-station) for anyone interested in it, either via electronic mail or on tape. She can be reached at the Santa Fe Institute in Santa Fe, New Mexico, and electronically as mm@santafe.edu. I myself am also willing to provide a limited number of videotapes of Copycat runs, provided the requestor covers the expenses of copying the tape and mailing it. (Interested readers are requested to contact Helga Keller at CRCC for details.) Undoubtedly, however, the majority of readers will never see an actual run of Copycat, which is a shame. For their benefit, I would like to try to give a feel for what it is like to watch Copycat running on the screen.

The screen looks just like the screen dumps shown in the previous chapter, except for the fact that it is essentially in continual flux. Dotted and dashed lines are constantly flickering all over the place, showing the program tentatively trying out scads of potential pathways, reminiscent of how a bolt of lightning ultra-rapidly shoots out a myriad serpent-tongue-like little "feeler boltlets" in all directions in space, testing the atmosphere to try to find as good a pathway to the ground as it possibly can. Simultaneously, the little squares showing the amount of activation of all the different concepts in the Slipnet are constantly jiggling as they get a little bigger or smaller through spreading activation and related mechanisms. Here and there in the Workspace, solid lines and arcs (bonds and bridges) and rectangular boxes (groups) are established, stay a while, maybe suddenly poof out of existence, only to be replaced by other ones in the same place or elsewhere. Meanwhile, various letters and numbers and words in the Workspace change in shade, being sometimes light, sometimes bold, as the program's estimation of their importance fluctuates up and down. Obviously, there are many time scales involved here, some far too fast for the eye to follow, others rather slow and easy to see, others somewhere in between.

As a run progresses, one eventually starts to see a coherent macro-view of the situation emerging out of all the micro-caprice. Often it is the view that one intuitively favors and hopes to see Copycat find, but sometimes it is an alternative way of looking at things. Occasionally, one observes the program (or rather, tiny agents inside it) making tentative little explorations that would lead it away from its current, entrenched view and toward a subtler, harder-to-find view that one prefers. In such cases, though the chances may be slim, one can't help rooting for Copycat to veer toward the dark-horse underdog. Sometimes one even witnesses a "power struggle" between two such views, in which the upstart knocks the older one out, then the older one stages a comeback and reclaims its former status. Back and forth it can go for a while, as in a boxing match.

The overall effect is, if I can be trusted to fairly summarize the reactions of many observers, one of great dynamism and lifelikeness. If anyone had doubts as to whether Copycat is best thought of as a parallel system, those doubts certainly vanish after seeing one run.

An Axed Pet Analogy

This mention of Copycat's "lifelikeness" brings me to an analogy that I devised and dearly wished to incorporate in the previous article, but since Melanie was uncomfortable with it, it was axed. However, Melanie herself suggested that I might include it elsewhere in the book, so here it is.

There's more than one way to copy a cat

Although the following analogy is doubtless somewhat biased, it gives the flavor of the difference between Copycat and other computer models of analogy-making with which we are familiar. Suppose one wanted to create an exhibit explaining the nature of feline life to an intelligent alien creature made of, say, interstellar plasma or some substrate radically different from that of terrestrial life. The Copycat approach might be likened to the strategy of sending a live ant along with some commentary aimed at relating this rather simple creature to its far larger, far more complex feline cousins. The rival approach might be likened to the strategy of sending along a battery-operated stuffed animal — a cute and furry life-sized toy kitty that could meow and purr and walk. This strategy preserves the surface-level size and appearance of cats, as well as some rudimentary actions, while sacrificing faithfulness to the deep processes of life itself, whereas the previous strategy, sacrificing nearly all surface appearances, concentrates instead on conveying the abstract processes of life in a tiny example and attempts to remedy that example's defects by explicitly describing some of what changes when you scale up the model.

Now the world knows precisely whom to blame and whom not to blame for this shameless and inflammatory comparison!

As a matter of fact, I showed this "pet analogy" of mine to Dedre Gentner, whose work certainly represents one of the approaches being likened to the "toy kitty" strategy, and to my surprise and relief, she was not offended by the analogy, or at least did not seem to be. In fact, she said it was curious and thought-provoking, or something to that effect.

To be sure, I myself am not *totally* convinced by my own analogy, fond of it though I am. I can easily poke holes in it. In the first place, it overstates its case by contrasting an actual living animal and an inanimate toy. The difference between Copycat and other computer models may be large, but certainly it is not *that* vast or dramatic. Moreover, I realize that the language I used to describe the toy animal itself is filled with innuendoes of triviality: loaded words like "cute", "furry", "toy kitty" — even such objectively true characterizations as "stuffed animal" and "battery-operated" — are all subliminal slurs of a sort.

When the analogy is stripped of all its surface trickery, the skeleton that remains is still an interesting, if bold, claim: that somehow, the true processes involved in analogy-making are present in some essential form in Copycat but are lacking in its rivals.

Any careful analysis of my ant-and-toy-kitty analogy will bear much resemblance to the judgmental processes that Copycat and its rivals are all about: how much weight to give to attributes as opposed to relational features, how important the different aspects of the situations being mapped are, whether certain slippages are too distant or too conceptually deep to ring true, and so on. The whole process of coming to a judgment about this analogy, or about any other real-world analogy, is amazingly subtle and intangible.

Meta-Analogies, Caricature Analogies, and Memory Retrieval

The above analogy typifies analogies made at a very high level of abstraction. It also has the curious property of being *about* analogy-making — a meta-analogy. Making meta-analogies might seem a rarefied and highly intellectual kind of activity, but in fact they crop up surprisingly often in very mundane contexts. An unremarkable exchange in an everyday conversation might go like this:

> *A:* All those American processed breads — Wonder Bread,
> Kilpatrick's, all the rest — they're all interchangeable.
>
> *B:* Are you kidding? That's like saying Chevrolet is identical to Ford!

Speaker A has effectively made an analogy among various types of bread, lumping them all into a single category (a slight extension of analogy-making, some might say), and Speaker B responds to this cognitive act by coming up with an analogy likening Speaker A's analogy to an analogy that Speaker B obviously thinks is deserving of nothing but scorn.

I call an analogy like this — an analogy concocted spontaneously for purposes of ridicule — a *caricature analogy*. Here is another example of the phenomenon:

> *Doug:* The Germans call a tortoise a *Schildkröte* — literally, a "shield-toad".
>
> *Carol:* "Shield-toad"?! Come on! That's like calling an eagle a "feather-cow"!

Of course, my ant-and-toy-kitty analogy is another example of a caricature analogy — one that has been worked out much further.

We all come up with caricature analogies all the time, but how in the world do we do it? Obviously, one must have a vast storehouse of experiences on which to draw. Equally obviously, one must have a strong sense — probably unconscious, nonverbal, in the back of one's mind, but nonetheless very confident and strong — for what *really* matters about the original situation. This "conceptual skeleton" of the situation functions as a kind of memory trigger.

Take the Chevrolet/Ford example above. Speaker B hears two things being equated, which to B seem quite different. So B searches for a stereotypical domain — ideally a domain somewhat related to the original domain but somehow more canonical or central — in which to mock this equation. In America, bread is a fairly random element of "consumer-goods space", whereas cars indisputably lie at that space's dead center. For the purposes of caricature, the domain should shift towards the center, so "cars" becomes the new domain. Then, carrying the process of "centralization" further, one looks for two items at the dead center of "car space" — absolute stereotypes of the make of a car — and there one effortlessly finds Ford and Chevrolet (at least in America). Now the shift toward the center has been carried out at all necessary levels (two, in this case), and the manufacture of the caricature analogy is done! The hope is, of course, that when the whole situation is "centralized" in this manner, the silliness of the claimed equation will be much clearer and easier to see. That is the strategy of a caricature analogy.

I am not trying to trivialize the process in any sense. It is extremely complicated, and requires a very deep knowledge, tantalizingly close to being consciously accessible but not completely so, of how items are categorized and how concepts are structured and interrelated in the mind. How this comes about is as deep a problem of creative cognition as any. I am merely giving a sketch of a charicature of a theory.

Much good work in cognitive science has gone into these types of question about the organization of memory, and especially about how retrieval of analogically related past episodes is triggered by the current context (*e.g.*, a situation one is currently facing, or a story one has just heard). Some of the most elaborate work in these directions was initiated by Roger Schank (see Schank, 1982) and has been continued by many other researchers coming out of his school. This whole approach is now generally known as "case-based reasoning" (see Riesbeck & Schank, 1989 and Kolodner, 1993 for book-length expositions of the field). In many ways, I am skeptical of this direction of work, but I do not dispute that it deals with issues at the true crux of cognition. Certainly its claims are provocative and challenging.

Analogy-puzzle Invention as a Cognitive Challenge

Making up a brand-new analogy that meets one or more desirable criteria (*e.g.*, a caricature analogy) is not the same as retrieving a past memory, although there are borderline cases that belong to both types of process. Making up new analogies is not only more inventive than retrieval is, but is also more playful. I will never forget the delight that David Rogers and I took, when we were collaborating at the MIT AI Laboratory and the Copycat project was just in its infancy, in making up new analogies in the Copycat domain — analogies that we hoped that Copycat might someday be able to tackle and even solve.

Together, David and I uncovered many beautiful surprises in the austere-seeming Copycat domain. From one good analogy puzzle there would inevitably sprout a set of variants, out of which we would choose one or two favorites and make new variants of them. Around and around went the inventive wheel, often leaving our heads spinning with the diversity and complexity of what we had found. (Note the similarity to the "variations on a theme" game discussed in Chapter 1.)

One of David's most original finds in this vast abstract space of analogy puzzles was this one:

If *eqe* changes to *qeq*, how does ***abbbc*** change?

The initial change reminded us both of turning something inside-out, and so the challenge he had set was to somehow turn ***abbbc***, the target string, inside-out as well. Readers may well enjoy pondering this.

I remember struggling hard with this challenge, trying to make the ***b***'s go to the outside in a nice symmetric manner, and to have the ***a*** and ***c*** come to the center. Two of my more successful attempts were **bbbacbbb** and **babcb** — but nothing I thought of deeply satisfied my sense of esthetics. When I saw David's solution, I was full of admiration.

His idea was simply to strip off the surface level of *letters,* revealing a hidden *numerical* level of structure that went as follows: 1–3–1. This mapped perfectly onto **eqe**, with the resulting inside-out number-string being 3–1–3, of course. When the surface-level "clothes" are put back on, the elegant answer is revealed: **aaabccc.** The letters stay completely in place, while a hidden number-string turns inside-out. There are, of course, other defensible solutions, but this, to me, was by far the most pleasing.

A very interesting perspective is cast on this problem by this minor variant, differing just by a single letter:

> If *eqe* changes to *qeq*, how does **abbba** change?

Here the numerical strategy is of course still possible, yielding **aaabaaa**, but in this context, using that strategy seems like overkill. This is because there is a much simpler idea that now works — the interchange of letter-types **a** and **b**, which gives **baaab**. The pressures in this variant are completely different from those in the original, because the new target **abbba** contains only two distinct letter-types (just as does **eqe**), whereas the original target **abbbc** contained *three* distinct letter-types. This little fact changes everything! David and I took much delight in finding just these kinds of subtle contrasts between apparently very similar problems. Each such discovery gave us new insights into what analogy-making was really about, and deepened our respect for the depth of the Copycat domain, in bringing so many issues to the fore with such clarity.

Over a period of months, David and I discussed all sorts of problems and all sorts of variants of them, carefully comparing their merits and defects and trying to find the best ones — the most interesting, the most surprising, the trickiest, the deepest, the most confusing, the most humorous, the most frustrating, those with the most different answers, and so on. It was very challenging and very enjoyable work/play.

At the time we were making up these many problems for our own delectation as well as for potential challenges to a future Copycat program, it certainly didn't occur to us in the slightest that this very process of puzzle-invention, puzzle-evaluation, and puzzle-comparison would one day become a focus of our research efforts. But that is indeed what has come to pass. A whole new level of self-awareness during and about the process of analogy-making, as well as the ability to manipulate concepts fluently at this meta-level, has now emerged as a set of critical aspects of mind to simulate. Clearly, *making up high-quality analogy problems* requires a much deeper understanding of what analogies are all about than does merely *making analogies.* This is the basic theme dealt with in the following chapter.

Chapter 7

Prolegomena to Any Future Metacat[1]

DOUGLAS HOFSTADTER

An Incipient Model of Fluidity, Perception, Creativity

In her book *Analogy-Making as Perception*, Melanie Mitchell has described with great precision and clarity the realization of a long-standing dream of mine — a working computer program that captures what, to me, are many of the central features of human analogy-making, and indeed, of the remarkable fluidity of human cognition.

First and foremost, the Copycat computer program provides a working model of *fluid concepts* — concepts with flexible boundaries, concepts whose behavior adapts to unanticipated circumstances, concepts that will bend and stretch — but not without limit. Fluid concepts are necessarily, I believe, *emergent* aspects of a complex system; I suspect that conceptual fluidity can only come out of a seething mass of subcognitive activities, just as the less abstract fluidity of real physical liquids is necessarily an emergent phenomenon, a statistical outcome of vast swarms of molecules jouncing incoherently one against another. In previous writings I have argued that nothing is more central to the study of cognition than the nature of concepts themselves, and yet surprisingly little work in computer modeling of mental processes addresses itself explicitly to this issue. Computer models often study the *static* properties of concepts — context-independent judgments of membership in categories, for instance — but the question of how concepts stretch and bend and adapt themselves to unanticipated situations is virtually never addressed.

Perhaps this is because few computer models of higher-level cognitive phenomena take perception seriously; rather, they almost always take situations

---

1. This chapter was originally written as the Afterword to *Analogy-Making as Perception* by Melanie Mitchell, and was published therein.

as static givens — fixed representations to work from. Copycat, by contrast, draws no sharp dividing line between perception and cognition; in fact, the entirety of its processing can be called *high-level perception*. This integration strikes me as a critical element of human creativity. It is only by taking fresh looks at situations thought already to be understood that we come up with truly insightful and creative visions. The ability to *reperceive,* in short, is at the crux of creativity.

This brings me to another way of describing Copycat. Copycat is nothing if not a model, albeit incipient, of human creativity. When it is in trouble, for instance, it is capable of bringing in unanticipated concepts from out of the blue and applying them in ways that would seem extremely far-fetched in ordinary situations. I am thinking specifically of how, in the problem "*abc* ⇒ *abd*; *mrrjjj* ⇒ ?", the program will often wake up the concept "sameness group" and then, under that unanticipated top-down pressure, will occasionally perceive the single letter *m* as a *group* — which *a priori* seems like the kind of thing that only a crackpot would do. But as the saying goes, "You see what you want to see." It is delightful that a computer program can "see what it wants to see", even if only in this very limited sense — and doing so leads it to an esthetically very pleasing solution to the problem, one that many people would consider both insightful and creative.

All these facets of Copycat — fluid concepts, perception blurring into cognition, creativity — are intertwined, and come close, in my mind, to being the crux of that which makes human thought what it is. Connectionist (neural-net) models are doing very interesting things these days, but they are not addressing questions at nearly as high a level of cognition as Copycat is, and it is my belief that ultimately, people will recognize that the neural level of description is a bit too low to capture the mechanisms of creative, fluid thinking. Trying to use connectionist language to describe creative thought strikes me as a bit like trying to describe the skill of a great tennis player in terms of molecular biology, which would be absurd. Even a description in terms of the activities of muscle cells would lie at far too microscopic a level. What makes the difference between bad, good, and superb tennis players requires description at a high functional level — a level that does not belong to microbiology at all.

If thinking is a many-tiered edifice, connectionist models are its basement and the levels that Copycat is modeling are much closer to the top. The trick, of course, is to fill in the middle levels so that the mechanisms posited out of the blue in Copycat can be justified (or "cashed out", as philosophers tend to say these days) in lower-level terms. I believe this will happen, eventually, but I think it will take a considerable length of time.

Copycat: Self-aware, But Very Little

Cognition is an enormously complex phenomenon, and people look at it in incredibly different ways. One of the hardest things for any cognitive scientist to do is to pick a problem to work on, because in so doing one is effectively choosing to ignore dozens of other facets of cognition. For someone who wishes to be working on fundamental problems, this is a gamble — one is essentially putting one's money on the chosen facet to be the *essence* of cognition. My research group is gambling on the idea that the study of concepts and analogy-making is that essence, and the Copycat program represents our first major step toward modeling these facets of cognition. I think Copycat is an outstanding achievement, and I am very proud of this joint work by Melanie and myself.

But of course, this work, however good it is, falls short of a full explanation of the phenomena it is after. After all, no piece of scientific work is ever the last word on its topic — especially not in cognitive science, which is just beginning to take significant strides toward unraveling the mind's complexity. In this afterword, I would therefore like to sketch out some of my hopes for how the Copycat project will be continued over the next few years.

One of the prime goals of the Copycat project is, of course, to get at the crux of creativity, since creativity might be thought of as the ultimate level of fluidity in thinking. I used to think that the miniature paradigm shift in the problem "$abc \Rightarrow abd$; $xyz \Rightarrow$?", wherein a is mapped onto z and as a consequence, a sudden dramatic perceptual reversal takes place, was really getting at the core of creativity. I still believe that this mental event as carried out in a *human* mind contains something very important about creativity, but it now seems to me that there is a significant quality lacking in the way this mental event is carried out in the "mind" of Copycat.

I would say that Copycat's way of carrying out the paradigm shift that leads to $xyz \Rightarrow wyz$ is too *unconscious*. It is not that there is no awareness in the program of the *problem* it is working on; it is more that Copycat has little awareness of the *processes* that it is carrying out and the *ideas* that it is working with. For instance, Copycat will try to take the successor of z, see that it cannot do so, go into a "state of emergency", try to follow a new route, and wind up hitting exactly the same impasse again. This usually occurs several times before Copycat discovers a way out of the impasse — not necessarily a clever way out, but just *some* way out. By contrast, people working on this problem do not get stuck in such a mindless mental loop. After they have hit the z-snag once or twice, they seem to know how to avoid it in the future. Copycat's brand of awareness thus seems to fall quite short of *people's* brand of awareness, which includes a strong sense of what they themselves are doing. One wants a much higher degree of *self*-awareness on the program's part.

Shades of Gray along the Consciousness Continuum

There is a clear danger, whenever one thinks about the "awareness" or "consciousness" of a computer model of any form of mentality, of getting carried along by the intuitions that come from thinking about computers at the level of their arithmetical hardware, or even at the level of ordinary deterministic symbol-manipulating programs, such as word-processing programs, graphics programs, and so on. Virtually no one believes that a word processor is conscious, or that it has any genuine understanding of notions such as "word", "comma", "paragraph", "page", "margin", etc. Although such a program *deals* with such things all the time, it no more *understands* what they are than a telephone understands what voices are. One's intuition says that a word processor is just a user-friendly but deceptive *façade* erected in front of a complex dynamic process — a process that, for all its complexity and dynamism, is no more alive or aware than a raging fire in a fireplace is alive.

This intuition would suggest that *all* computer systems — no matter what they might do, no matter how complex they might be — must remain stuck at the level of zero awareness. However, this uncharitable view involves an unintended double standard: one standard for machines, another for brains. After all, the physical substrate of brains, whether it is like that of computers or not, is still composed of nothing but inert, lifeless molecules carrying out their myriad minuscule reactions in an utterly mindless manner. Consciousness certainly seems to vanish when one mentally reduces a brain to a gigantic pile of individually meaningless chemical reactions. It is this *reductio ad absurdum* applying to *any* physical system, biological or synthetic, that forces (or ought to force) any thoughtful person to reconsider their initial judgment about both brains and computers, and to rethink what it is that seems to lead inexorably to the conclusion of an in-principle lack of consciousness "in there", whether the referent of "there" is a machine or a brain.

Perhaps the problem is the seeming need that people have of making black-and-white cutoffs when it comes to certain mysterious phenomena, such as life and consciousness. People seem to want there to be an absolute threshold between the living and the nonliving, and between the thinking and the "merely mechanical", and they seem to feel uncomfortable with the thought that there could be "shadow entities", such as biological viruses or complex computer programs, that bridge either of these psychologically precious gulfs. But the onward march of science seems to force us ever more clearly into accepting intermediate levels of such properties.

Perhaps we jump just a bit too quickly when we insistently label even the most sophisticated of today's "artificial life" products as "absolutely unalive" and the most sophisticated of today's computational models of thought as "absolutely unconscious". I must say, the astonishing subtlety of Terry Winograd's SHRDLU program of some 20 years ago (Winograd, 1972) always gives me pause when I think about whether computers can "understand" what is said or typed to them. SHRDLU always strikes me as falling in a very gray area. Similarly, Thomas Ray's computational model of evolution, "Tierra" (Ray, 1992), can give me eerie feelings of looking in on the very beginnings of genuine life, as it evolved on earth billions of years ago.

Perhaps we should more charitably say about such models of thought as SHRDLU and Copycat that they might have an unknown degree of consciousness — tiny, to be sure, but not at an absolute-zero level. Black-and-white dogmatism on this question seems as unrealistic, to me, as black-and-white dogmatism over whether to apply the label "smart" or "insightful" to a given human being.

If one accepts this somewhat disturbing view that perhaps machines — even today's machines — should be assigned various shades of gray (even if extremely faint shades) along the "consciousness continuum", then one is forced into trying to pinpoint just what it is that makes for *different* shades of gray.

The Key Role of Self-monitoring in Creativity

In the end, what seems to make brains conscious is *the special way they are organized* — in particular, the higher-level structures and mechanisms that come into being. I see two dimensions as being critical: (1) the fact that brains possess *concepts,* allowing complex representational structures to be built that automat- ically come with associative links to all sorts of prior experiences, and (2) the fact that brains can *self-monitor,* allowing a complex internal self-model to arise, allowing the system an enormous degree of self-control and open-endedness. (These two key dimensions of mind — especially their role in creativity — are discussed in Chapters 12 and 23 of Hofstadter, 1985.) Now Copycat is fairly strong along the first of these dimensions — not, of course, in the sense of having many concepts or complex concepts, but in the sense of rudimentarily modeling what concepts are really about. On the other hand, Copycat is very weak along the second of these dimensions, and that is a serious shortcoming.

One might readily admit that self-monitoring would seem to be critical for *consciousness* and yet still wonder why self-monitoring should play such a central role in *creativity.* The answer is: to allow the system to *avoid falling into mindless ruts.*

The animal world is full of extremely complex behaviors that, when analyzed, turn out to be completely preprogrammed and automatized. (A particular routine by the *Sphex* wasp provides a famous example, and indeed, forms the theme song in the second of the two chapters cited above.) Despite their apparent sophistication, such behaviors possess almost no flexibility. The difference between a human doing a repetitive action and a more primitive animal doing a repetitive action is that humans *notice* the repetition and get bored; most animals do not. Humans do not get caught in obvious "loops"; they quickly perceive the pointlessness of loopy behavior and they *jump out of the system*. This ability of humans (humorously dubbed "antisphexishness" in the aforementioned chapter) requires more than an *object-level* awareness of the *task* they are performing, but also a *meta-level* awareness — an awareness of their own actions. Clearly, humans are not in the slightest aware of their actions at the *neural* level; the self-monitoring carried out in human brains is at a highly chunked *cognitive* level, and it is this coarse-grained kind of self-monitoring that seems so critical if one is to imbue a computer system with the same kind of ability to choose whether to remain *in* a given framework or to *jump out* of that framework.

In my above-mentioned chapter on self-watching, I wound up surprising myself by citing, in an approving manner, somebody with whom I had earlier thought I had absolutely no common ground at all — the British philosopher J. R. Lucas, famous for his strident article "Minds, Machines, and Gödel" (Lucas, 1961), in which he claims that Gödel's incompleteness theorem proves that computers, no matter how they are programmed, are intrinsically incapable of simulating minds. Let me briefly give Lucas the floor:

> At one's first and simplest attempts to philosophize, one becomes entangled in questions of whether when one knows something one knows that one knows it, and what, when one is thinking of oneself, is being thought about, and what is doing the thinking....
>
> The paradoxes of consciousness arise because a conscious being can be aware of itself, as well as of other things, and yet cannot really be construed as being divisible into parts.... A machine can be made in a manner of speaking to 'consider' its performance, but it cannot take this 'into account' without thereby becoming a different machine, namely the old machine with a 'new part' added. But it is inherent in our idea of a conscious mind that it can reflect upon itself and criticize its own performances, and no extra part is required to do this: it is already complete, and has no Achilles' heel.

This passage suggests the vital need for what might be called "reflexivity" (*i.e.*, the quality of a system that is "turned back" on itself, and can watch itself) if a mechanical system is to attain what we humans have. I am not at all sympathetic to Lucas' claims that machines can never do this — indeed, I shall give below a kind of rough sketch of an architecture that could do something of this sort; rather, I am sympathetic to the flavor of his argument, which is one that many lay people would resonate with, yet one that very few people in cognitive science have taken terribly seriously.

Another idea that resonates with the flavor of Lucas' article is captured by the title of a posthumous book of papers by the uniquely creative Polish–American mathematician Stanislaw Ulam: *Analogies Between Analogies*. The obvious implication of the title is that Ulam delighted in meta-level thinking: thinking about his own thoughts, and thinking about his thoughts about his thoughts, etc. etc., *ad nauseam*, as Lucas might say. Spelling out the next level implied by this title would be superfluous — everybody sees where it is heading — and the feeling is of course that the more intelligent someone is, the more levels of "meta" they are comfortable with.

A Stab at Defining Creativity

This sets the stage for me to describe my long-term ambitions for Copycat. The goals to be described below have emerged in my mind over the past several years, as I have watched Copycat grow from a metaphorical embryo into a baby and then a toddler. I was led to summarize these goals succinctly at a lecture I was giving on Copycat as a model of creativity, when somebody asked me point-blank if I thought that Copycat really captured the essence of creativity. Was there anything left to do? Of course I felt there was much more to do, and so, prompted by this question, I tried to articulate, in one short phrase, what I think the creative mind does, as opposed to more run-of-the-mill minds. Here is the phrase I came up with:

> Full-scale creativity consists in having a keen sense for what is interesting, following it recursively, applying it at the meta-level, and modifying it accordingly.

This was too terse and cryptic, so I then "unpacked" it a little. Here is roughly how that went. Creativity consists in:

- *Having a keen sense for what is interesting:* that is, having a relatively strong set of *a priori* "prejudices" — in other words, a somewhat narrower, sharper set of resonances than most people's, to various possibilities in a given domain. It is critical that the peak of this particular *individual's* resonance curve fall close to the peak of the curve representing the average over *all people,* ensuring that the would-be creator's output will please many people. An example of this is a composer of popular tunes, whose taste in melodies is likely to be much more sharply peaked than an average person's. This aspect of creativity could be summarized in the phrase *central but highly discriminating taste.*

- *Following it recursively:* that is, following one's nose not only in choosing an *initially* interesting-seeming pathway, but also *continuing* to rely on one's "nose" over and over again as the chosen pathway leads one to more and more new choice-points, themselves totally unanticipated at the outset. One can imagine trying to find one's way out of a dense forest, making one intuitive choice after another, each choice leading, of course, to a unique set of further choice-points. This aspect of creativity could be summarized in the term *self-confidence.*

- *Applying it at the meta-level:* that is, being aware of, and carefully watching, one's pathway in "idea space" (as opposed to the space defined by the domain itself). This means being sensitive to unintended patterns in what one is producing as well as to patterns in one's own mental processes in the creative act. One could perhaps say that this amounts to sensitivity to *form* as much as to *content.* This aspect of creativity could be summarized in the term *self-awareness.*

- *Modifying it accordingly:* that is, not being inflexible in the face of various successes and failures, but modifying one's sense of what is interesting and good according to experience. This aspect of creativity could be summarized in the term *adaptability.*

Note that this characterization implies that the system must go through its own experiences, just as a person does, and store them away for future use. This kind of storage is called "episodic memory", and Copycat entirely lacks such a thing. Of course, during any *given* run, Copycat retains a memory of what it has done — the Workspace serves that role. But once a given problem has been solved, Copycat does not store memories of that session away for potential help in attacking future problems, nor does it modify itself permanently in any way. Amusingly, babies and very young children seem similarly unable to lay down permanent memory traces, which is one reason that adults virtually never have memories of their infancy. If Copycat is to grow into an "adult", it must acquire that ability that adults have: the ability to commit to permanent memory episodes one has experienced.

Five Challenges Defining What Any Future Metacat Must Do

The third point listed above stresses the importance of self-watching — making explicit representations not just of objects and relationships in a situation before one, but also representations of one's own actions and reactions. To a very limited extent, Copycat already has a self-watching ability. It is described at the end of Chapter 7 of Mitchell's book, in a section entitled

"Self-Watching". However, the degree of reflexivity that I envision goes far beyond this. Indeed, it would alter Copycat so radically that the resulting program ought probably to be given some other name, and for want of a better one, I tentatively use the name "Metacat". Here, then, are several ways in which the hypothetical Metacat program would go beyond Copycat:

(1) We humans freely refer to the "issues involved in" or "pressures evoked by" a given puzzle. For example, the problem "*abc* $\Rightarrow$ *abd*; *xyz* $\Rightarrow$?" is about such issues as: recovery from a snag; bringing in a new concept ("last") under pressure; perceptual reversal and abstract symmetry; simultaneous two-tiered reversal; and a few other things. However, Copycat has no explicit representation of issues or pressures. Although it makes conceptual slippages such as "*successor* $\Rightarrow$ *predecessor*", it does not anywhere *explicitly* register that it is trying out the idea of "reversal". It does not "know what it is doing" — it merely *does* it. This is because, even though the concept of "reversal" (*i.e.*, the node "opposite") gets activated in long-term memory (*i.e.*, the Slipnet) and plays a crucial role in guiding the processing, no explicit reflection of that activation and that guiding role ever gets made in the Workspace.

A Metacat, by contrast, should be sensitive to any sufficiently salient event that occurs inside its Slipnet (*e.g.*, the activation of a very deep concept) and should be able to explicitly turn this recognition into a clear characterization, in its Workspace, of the issues that the given problem is really "about". Additionally, the program should be able to take note of the most critical actions that occur in its Workspace, and to create a *record* of those actions. This way, the program would leave an explicit coarse-grained temporal trail behind it.

The way in which such self-monitoring would take place would roughly be this. Copycat, as it currently stands, is pervaded by numerical measures of "importance", roughly speaking. There are important *objects,* important *concepts,* important *correspondences,* and so on. We would simply add one further numerical measure — a rough-and-ready measure of the import of an *event* in the Workspace or of a *change in activation* in the Slipnet. For actions in the Workspace, import would reflect such features as the *size* of the object acted upon (the bigger the better) and the *conceptual depth* of its descriptions, among other things; for actions in the Slipnet, import would reflect the *conceptual depth* of the node activated, among other things. The details don't matter here; the main thing is that events' import values would be spread out over a spectrum, allowing one to filter out all those events whose import was below some threshold, leaving one with a highly selective view of what has happened.

This high-level view of events taking place, once it is explicitly represented in some part of the Workspace (the "Lucas part", it might be called!), would then itself be subject to perceptual processing by codelets looking for patterns. This would thus allow the system to become aware of regularities in its own actions, and perhaps to get a hold of the *pressures* in a given problem, which would lead to a characterization of what a given problem is "about". Of course, what a problem is considered to be about depends on what answer one comes up with, so in a sense this would be a description not of what the *problem* as a whole is about, but of the issues that a given *answer* is about. This is but the crudest approximation to what people do, but it is at least a first stab.

(2) We humans readily see how a given answer to a given analogy puzzle could make sense to someone else, even if we ourselves did not think of it, and might well never have thought of it on our own. The current version of Copycat, however, has no such capability. It needs to be given the capacity to work backwards from a given answer suggested by an outside agent. If the program has the capacity to see through to the issues that a given answer is about, this working-backwards capacity would allow it to size up an answer quickly and to put it in mental perspective. From that point on, it would be able to engage in "banter" of sorts with a human about the merits and demerits of a given solution.

(3) We humans do not forget what we have done right after doing it. Rather, we store our actions in episodic memory. So too, Metacat should store a trace of its solution of a problem in an episodic memory. There are two important types of consequence of this ability.

The first type of consequence is that, during a single problem-solving session, the program should be able to avoid falling into mindless loops, and to "jump out of the system" (meaning, for instance, that it should be able to make an explicit decision, based on its failures, to focus on previously-ignored objects or concepts). The second type of consequence is that, over a longer time span, it should be able to be "reminded" of previously encountered problems by a new problem. At present, of course,

Copycat does not in any way try to model retrieval of episodes in its past, because it simply does not *have* any past, no matter how many problems it has solved. (Of course, *during* a run, the present version of Copycat has a short-term past, but all that is lost once the run is over.) With an episodic memory, that would all be changed.

Metacat's search for analogous episodes would be governed by many of the same principles that pervade Copycat's architecture: activation would spread from concepts involved in the current problem (*e.g.,* "symmetry", "reversal") to problems in episodic memory that were indexed under those concepts. Needless to say, all of this would be heavily biased by conceptual depth, so that surface-level remindings would be kept to a minimum.

(4) We humans have a clear "meta-analogical" sense — that is, an ability to see *analogies between analogies,* as in the title of Ulam's book. An episode-retrieval ability as just described would endow Metacat with the capacity to map one Copycat analogy problem (and its answers) onto another, thus making an analogy not between two letter-strings, but a meta-level analogy between two puzzles, based on issues and pressures that they evoke.

Going even further, it would be hoped that the ability to make such meta-level analogies would automatically entail the ability to make meta-meta-analogies, and so forth. Thus a Metacat program would hopefully be able to relate the way in which two specific analogy puzzles were early on noticed to resemble each other to the way in which two *other* analogy puzzles have *just now* been noticed to resemble each other. (Thus we are moving beyond the title of Ulam's book!) Achieving this multi-leveled type of self-reflectiveness would, I firmly believe, constitute a major milestone en route to a theory of how consciousness emerges from the interaction of many small subcognitive agents in a system.

(5) Finally, we humans not only enjoy *solving* these kinds of puzzles, but can enjoy *making up* new puzzles. It takes a keen explicit sense of the nature of pressures involved in problems to make up a really new and high-quality Copycat analogy puzzle. What makes a problem good is often the fact that it has two appealing answers, of which one is deep but elusive while the other is shallow but easy to see. An elegant example is the puzzle "*apc* $\Rightarrow$ *abc*; *opc* $\Rightarrow$?", which admits of the easy-to-find answer *opc* $\Rightarrow$ *obc* (based on a letter-by-letter imitation of what

happened to *apc*) and the elusive answer $opc \Rightarrow opq$ (based on an abstract vision according to which the defect in a flawed successor group is removed). While both answers make perfect sense, the latter is clearly far more elegant than the former.

Inventing a problem of this sort, delicately poised at the balance point between two rival answers, requires an exquisite internal model of how people will perceive things, and often requires exploration of all sorts of variants of an initial problem that are close to it in "problem space", searching for one that is optimal in the sense of packing the most issues into the smallest and "cleanest" problem. This is certainly a type of esthetic sense. (Incidentally, I feel no need to apologize for the inclusion of esthetic qualities, with all the subjectivity that they imply, in the modeling of analogy-making. Indeed, I feel that responsiveness to *beauty* and its close cousin, *simplicity,* plays a central role in high-level cognition, and I expect that this will gradually come to be more clearly recognized as cognitive science progresses.) Needless to say, Copycat as it currently stands has nothing remotely close to such capabilities.

Work towards this type of Metacat program is just beginning at the Center for Research on Concepts and Cognition at Indiana University. If the effort to impart these sorts of abilities and intuitions to a Metacat program is a success, then, I would say, Metacat will be *truly* insightful and creative. I make no pretense that the above description is a clear recipe for an architecture, although to be sure, what is in my mind is considerably more fleshed-out than this vague sketch. These wildly ambitious ideas are unlikely ever to be realized in full, but they can certainly play the role of a pot of gold at the end of the rainbow, pulling me and my co-workers on toward a perhaps chimerical goal.

I have been privileged to travel a long way in search of the pot of gold in company of Melanie Mitchell. Her beautiful work has deeply inspired me, and I hope that it will similarly inspire a new generation of questers after the mysteries of mind. As we go forward, the more clearly we see how long a road remains ahead of us.

Preface 8:

Analogy-making in a Coffeehouse

Singin' the Copycat Blues

There came a day when, having explained the Copycat domain and a few canonical analogy problems in it for the umpteenth time, I could no longer bring myself to say, "Well, suppose **abc** goes to **abd**…" once again. Umpteen plus one is just one too many. I was fed up, for the time being at least, with explaining Copycat.

Now it happened that quite often when I was called upon to explain my research, I found myself sitting across a table from my interlocutor — a dinner table or a coffeehouse table or something of the sort. Without planning it out, I gradually developed a routine of touching various table objects and asking people to "Do this!" from their vantage point. This little stratagem worked like a gem at getting across the idea that doing analogies involves dealing with any number of conflicting pressures of different strengths. It was also fun because as soon as one problem had been done, I could move one of the objects on the table a little bit, and the constellation of pressures would change significantly, giving a very different problem. This way, a dozen closely related yet highly contrasting analogy problems could be explored in five minutes or less. The fascination of the domain emerged quickly and effortlessly.

The more I used this metaphor for explaining Copycat, the more it appealed to me, and the larger a mental collection I built up of interesting stock problems I could use on any tabletop. But I found it more fun to improvise, and just about every time, some fascinating new example would crop up in the discussion, resulting from some peculiarity of the setup on the tabletop before us. I eventually realized that aside from having found a way of explaining Copycat indirectly, I had unwittingly uncovered a very nice second micro-domain for studying analogy-making.

Mimicking "Real" Analogies in Tabletop's Mini-universe

As did Copycat's, Tabletop's microdomain aspired to "universality", in the sense that I tried to mimic in it, albeit crudely, something of the spirit of analogies

Henry

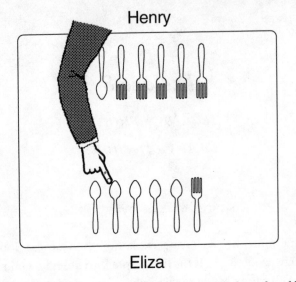

Eliza

Figure VIII–0. A "translation" into the Tabletop domain of a real-world analogy.

in the most far-flung of real-world domains (*e.g.*, the "First Lady of England" problem). For instance, one day I was talking with a friend from Spain — let's call her "Eliza", to protect the innocent — about how, when I was a graduate student, I had lived for a while in Germany and had studiously avoided the company of the Americans I met there partly because I tended to find them boring, but more importantly because I wanted to learn German well and to get to know Germany. Eliza said, "That's exactly how I feel here." I at first assumed that she was making a fluid mapping of her situation onto mine and that she therefore meant, "I, too, try to stay away from my Spanish compatriots and to immerse myself in American culture, given that I'm studying in America." This seemed like the natural interpretation of her "me-too". However, as she went on, it soon became apparent that she meant quite the contrary — namely, "Here in America, I, too, find Americans boring and generally prefer the company of people from Spain." Eliza's me-too was more of a *rigid* analogy than a fluid one — although admittedly, "German people" did slip to "Spanish people". Perhaps it was "semi-fluid".

I was highly amused by this little exchange, and thought it would be interesting to try to "inject" it in some abstract way into the Tabletop domain. So I imagined a situation in which there are, say, five forks and one spoon on my side of the table, and on the other side — Eliza's side, let's say — five spoons and one fork (see Figure VIII–0). The idea of this is that forks represent the *natives* of my country, with the spoon being a lonely foreigner, and of course the reverse on Eliza's side. Thus, to mimic my traveling to Europe, I reach all the way across the table — and to mimic my preferring to associate with Europeans, I touch one of the spoons — the local natives — on Eliza's side. The question now is, what should Eliza do, in order to "do what I did"?

One option for her would be to reach across to my side and touch one of my forks. This seems to me like the most analogous, or the "most same", act, and it would map onto my initial supposition that Eliza, when in America, preferred associating with Americans and avoiding Europeans.

Another way of "doing what I did" would obviously be for her to touch some spoon on her own side — probably the one I touched, but possibly a different one. That, however, would be a most unimaginative, stick-in-the-mud kind of thing to do. It would map onto Eliza staying at home in Spain and never coming to America at all — and moreover, avoiding all contact with foreigners (especially Americans) in her own land.

Eliza's true situation, expressed in her me-too remark, was that she had indeed come to America (which maps onto her reaching across the table to my side), but having arrived, was nonetheless sticking with her own kind (which maps onto her touching the single spoon on my side).

It was an interesting challenge and stimulation to see how well the essence of a "real" analogy such as this one could be abstracted and then reconstructed in this tiny domain. Not all real-world analogies were as easy or as natural to translate as this example, to be sure, but it was gratifying to see how often some sort of rough equivalent could be found. Actually, there is one flaw to this "translation" of Eliza's analogy: the tabletop configuration by which I simulated it looks highly unrealistic. Part of the idea behind the domain was to keep it close to the flavor of real configurations that one might encounter on a coffeehouse or restaurant tabletop.

As time went on, I waxed quite enthusiastic about studying analogies in the tabletop domain on a serious level. Fortunately, it wasn't too hard to entice Bob French to work on this for his Ph.D. project. Bob was very excited by the Copycat architecture, but the project of implementing Copycat had already been given over to Melanie Mitchell. Since Bob wanted to do something Copycat-like on his own, Tabletop came along at just the right time for him. All I needed to do was point at Melanie's work and say, "Do this!" Bob knew exactly what I meant.

Henry

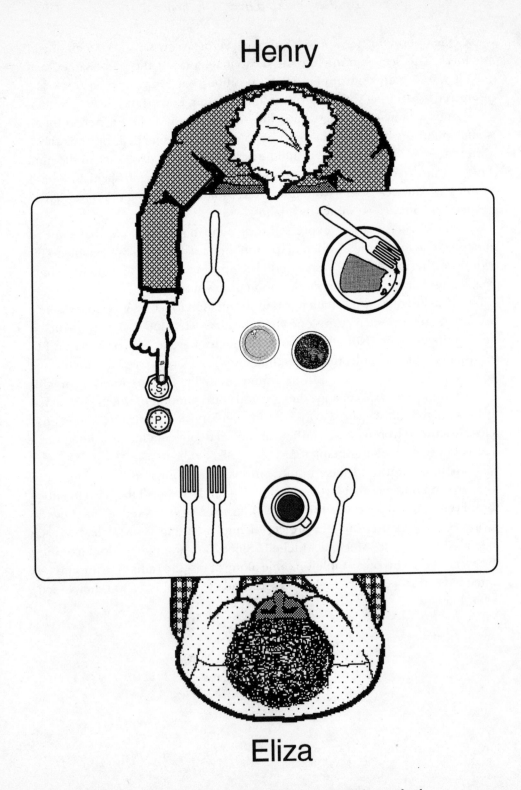

Eliza

Figure VIII–1. Henry and Eliza sitting across a table from each other.

Chapter 8

Tabletop, BattleOp, Ob–Platte, Potelbat, Belpatto, Platobet

DOUGLAS HOFSTADTER and ROBERT FRENCH

Analogy Problems in Microdomains

> *Doug touched his three-year-old son's nose and said to him, "Do this! Do this, Danny!" Danny looked at Doug's face for a moment, then at Doug's hand. Then he reached out with his fingertip, and touched Doug's fingertip — the finger that had just touched Danny's nose.*

"Do this!" puzzles on a tabletop

Across from each other, over a somewhat rickety little wooden table in the Runcible Spoon, a venerable old coffeehouse located just a literal stone's throw away from a street that is called, most uninspiringly, "Seventh Street", but that would have been called "Hoagy Carmichael Street", in honor of the town's most famous citizen, had the renaming proposal not been turned down several years ago now, in a typical display of lack of local pride, by the city council of Bloomington, Indiana, the small and charming university town where the authors of this article, notable for, among other things, its moderately droll title as well as an opening sentence that, seen from the standpoint of parsing, possesses an almost Proustian thorniness, reside (or rather, did reside, at the time of its writing), are seated, as can be seen in Figure VIII–1, two friends, Henry and Eliza. During a momentary lull in their conversation, Henry reaches out, touches a salt shaker, and says to Eliza, "Do this!" Eliza reaches over toward the middle of the table, where her hand hovers for a moment above the pepper

Henry

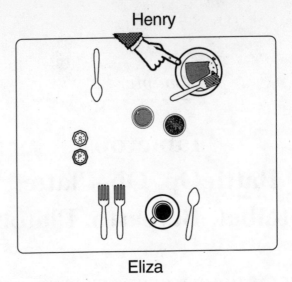

Eliza

Figure VIII–2. A simple Tabletop analogy problem.

shaker. But then it moves on and lands on the salt shaker. A look of disappointment crosses Henry's face, but he quickly banishes it and tries again.

"Can you do *this*?", he asks, touching his plate (Figure VIII–2). Eliza surveys the table; then, after some hesitation, she touches her own saucer. Henry appears more satisfied.

Finally (Figure VIII–3), Henry reaches over to Eliza's side of the table and touches her coffee cup, saying, "And now do *this*!"

Henry

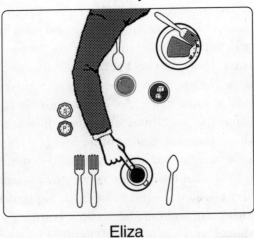

Eliza

Figure VIII–3. A more complex Tabletop analogy problem.

Eliza's finger at first inches gingerly toward her own cup but then suddenly swerves and darts across the table to Henry's side, and touches, of all things, his spoon! Henry says, "How did you come to pick *that* object?" Eliza replies, "Well, your two glasses formed a little unit to me, which mapped quite naturally onto my pair of forks. So this suggested a temporary association between the concepts 'glass' and 'fork'. Then," she continues, "since you reached over to *my* side to touch my coffee cup, and since a cup is *almost* a glass, I reasoned that I should reach over to *your* side and touch something that is almost a *fork* — and that could only mean your spoon!" This is a bit stretched, perhaps far-fetched, but certainly an interesting viewpoint.

These three puzzles posed to Eliza by Henry are representative puzzles from the Tabletop domain. The basic idea is simply to take an ordinary coffeehouse tabletop and to pose "Do this!" puzzles involving the touching of various objects on the table. Although it is always possible to interpret the command "Do this!" absolutely literally and to touch exactly the object touched, most people — most adults, that is — tend automatically to interpret "this" with unconscious conceptual slippages (*e.g.*, not taking a category like "cup" rigidly, but being willing to touch a glass instead), and to touch what they perceive as being a "corresponding" object, as seen from their vantage point. For very small children, making such slippages is a little advanced, and they tend to touch the literal object touched.

Occasionally, one also runs into an adult who, when first told "Do this!", takes the request very literally. For instance, our colleague Dave was explaining the Tabletop project to a visiting cognitive scientist named Peter in, of all places, the Runcible Spoon. Dave, picking up his coffee cup, said, "The idea is, can you do this?" Peter obediently reached over and picked up Dave's coffee cup. Dave said, "No, that's not what I meant. Here — do *this*!", and he touched his own nose. Peter's finger started out in a trajectory aimed straight at Dave's nose, but before it reached its target, enough social pressures built up in Peter's mind that his finger effected a 180-degree turnaround somewhere above the table, and landed on his own nose. Peter had seen the light! From then on, he interpreted all of Dave's "Do this!" commands in terms of counterparts from his side of the table. In a way, it seems amazing that a sophisticated adult could be so literal-minded as Peter was at the outset, but there is often more to trivial-seeming mental activity than one at first suspects.

Incidentally, we are not in any sense claiming that Peter was *wrong* in touching Dave's cup, or even in formulating a plan to touch Dave's nose. It is just that some responses to these simple-seeming "Do this!" challenges are far more common than others, and seem to represent more typical human — or rather, *adult* human — thought processes. Those answers that involve automatic unconscious slippages are the result of what we call *mental fluidity*. Some answers, of course, are quite strange and perhaps overly fluid, such as little Danny's curious response to his father's "Do this" challenge, described in the epigraph above.

Realistically modeling analogy-making on a highly idealized tabletop

The Tabletop project is concerned with developing a computer program to do these sorts of trivial-seeming analogy problems involving objects on a coffeehouse tabletop (not noses or other parts of faces, although that presumably is not a fundamental limitation of the project). As if the project weren't already trivial-seeming, it is rendered even more so by the fact that in the computer model, the cups and forks and so on are all idealized; that is, no object has any idiosyncratic distinguishing properties above and beyond its location on the table and its category membership. Thus, whereas real coffee cups come in various shapes and sizes and often have fancy designs and slogans printed on them that can refer to any imaginable aspect of the real world, a cup in the Tabletop domain is simply *a cup* and nothing more — it is shapeless, sizeless, and decoration-free. Similarly austere conventions hold for all other objects in the Tabletop domain. Thus we are dealing with such a bare-bones domain that on one level, it cannot help but give an impression of triviality.

On the other hand, the "Platobet" — that is, the Tabletop domain's list or "alphabet" of abstract Platonic categories — is fairly rich in structure. Thus, for instance, Platonic "plate" and Platonic "saucer" are conceptually quite close, since plates and saucers are round and are used for putting objects on; on the other hand, in the absence of extreme contextual pressures, both concepts are very far from "salt shaker". This is explicitly put into the program. Another concept that is fairly close to "saucer" is "cup", but they are close in a different way — their closeness comes from the fact that cups are used with saucers. There is a fairly intricate web of such interrelationships crisscrossing the Platobet. Although the amount of information contained in this web is by no means huge (perhaps two pages' worth of text), it adds a degree of complexity and subtlety to the domain that might not be appreciated at first.

Also contributing to the project's nontriviality is the fact that each different table situation evokes a unique constellation of interacting mental pressures. By "tweaking" a given table situation one way and then another, one can subtly manipulate the pressures. Whereas in one configuration, touching object X may seem the best, in a very slight variant of that configuration, the pressures may have shifted such that object Y now appears better than X, and in another close variant, object Z might appear better than both X and Y. Vast numbers of close variants of a given problem may be produced by moving an object slightly on the table, by "tweaking" an object's category membership (a glass becomes a cup, say), by adding or taking away an object on the table, and of course by making combinations of several such "tweaks". The space of *problems* is thus closely associated with a space of *mental pressures,* and ultimately, it is these pressures and their interactions that the Tabletop project — indeed, that analogy-making in general — is all about.

The tangible goal of the project is to build a computer program to do "pointing analogies" in this small domain in a psychologically realistic manner. This means that to any "Do this!" problem, one or more different answers must be producible by the program, but certain answers have to be favored above others, in a manner clearly reflecting adult human preferences. Just how hard is it to build such a program, and what should be its underlying principles?

A brute-force approach to Tabletop analogy problems

A skeptic could argue that the number of mental pressures in this microworld is so small that a rather simple and mechanical computer program could solve all problems easily. The idea would be to exploit the fact that each specific attribute of objects on the table, such as location, category membership, and so on, gives rise to a specific type of mental pressure. Since there are only a few such attributes, there are only a few pressures. Thus, just make an algebraic formula that takes into account however many pressures there are, giving them different weights. Using this formula, one could quickly select the "winning" object, and that's all there is to it.

Let us spell out this argument a bit more carefully. Since each object on a coffeehouse table has only a few significant attributes — the *category* to which it belongs (*e.g.,* "cup", "fork", "salt shaker", etc.), its *location* on the table, its *orientation,* its *owner,* and its *spatial neighbors* — it would seem next to trivial to characterize any object fully. Choose the object you want to characterize, sequentially list the values of each of these attributes, and presto! — you have a complete dossier for the object. Given this standard way of representing all objects, it would seem to be a piece of cake to select the winner (or the top contenders) in a problem wherein Henry touches one object and Eliza must "do the same thing". To each object's dossier one could simply apply a standard algebraic formula, which would numerically rate the object for similarity to the touched object. The formula would presumably be a weighted sum of terms, one for each attribute. Then the top-scoring object would be the "winner". Obviously, adding or deleting a few attributes or a few objects wouldn't make much of a difference to the feasibility of this strategy.

Such a formulaic approach seems to provide a fittingly trivial solution to what seems, after all, to be a rather trivial kind of problem. So is this the end of the story? Of course, the authors believe it is not. In fact, the purpose of this article is firstly to argue that, given the psychological goals of the Tabletop project, such an approach is fundamentally misguided, and secondly to hint at a contrasting architecture — an emergent, stochastic one — that is much more satisfying, in that it works not only in the Tabletop domain, but also far beyond it.

Ironically, however, the first program written to solve Tabletop problems *was* a pure brute-force formula-based program of just the sort described above. As such programs represent the diametric opposite of our philosophy, we shall henceforth refer to the actual one we created (or to any similar formula-based program) as a "Potelbat" ("tabletop" backwards). All the arguments that we shall present below against the relevance of a Potelbat-type architecture were known to us at the time we made our Potelbat, but it just seemed to us that to keep ourselves honest, *some* kind of Potelbat had to be actually realized, so that we could see how well it worked. We needed, in short, some direct knowledge of "the enemy". To no one's surprise, our quickly-built Potelbat performed reasonably well on simple Tabletop problems, although on some trickier ones (such as are shown in Figures VIII–7 and VIII–8 on pages 353 and 354, for example), it did poorly. To be sure, we could have added on more mechanisms and worked at tuning the many internal parameters of our Potelbat so that it would do well on those problems too, but from our point of view, doing so would have been a waste of valuable time.

Why developing a brute-force program would miss the point

Development of a "perfect" Potelbat would totally miss the fundamental purpose of the Tabletop project — why it was thought of in the first place. It would be a little like spending years and years on developing a chess-playing program (as well as super-fast special-purpose hardware to run it on), and finally having one's program defeat all human (and computer) rivals, thus gaining the title of World Chess Champion. This would be an admirable achievement — indeed, it seems a good bet that the designers of Deep Thought or perhaps some rival team will carry it off in the next few years — but it would reveal nothing about *how people play chess*. What it *would* reveal is an unanticipated property of the formal domain of chess itself — namely, that it is small enough to be conquered by brute-force techniques. This is something that few people would have suspected 50 or even 20 years ago, and so there is considerable intellectual interest to the act of "debunking" or "trivializing" chess in this way. On the other hand, a similar "trivialization" of the Tabletop domain would surprise no one — it was *deliberately* created as a microdomain. The purpose of our computer-modeling project is not, *à la* Deep Thought, to "conquer" the domain; rather, the idea is to use the domain as a laboratory in which to study general mechanisms of fluid analogy-making, whatever the domain. To us, any architecture designed to solve problems in the Tabletop domain should be evaluated with the question of generality, or perhaps we should say *generalizability*, in mind.

Generality: the fundamental goal of the Tabletop project

Our goal in Tabletop and related projects has been to develop a program that behaves "intelligently" or "reasonably" in a small domain, and whose basic architecture is highly general, in the sense that it bears few traces of the idiosyncratic domain in which it actually functions. Such a program, if realized, could be adapted without serious trouble from one domain to another. It would be of interest not so much because of its domain-specific power, but because of its general mechanisms.

It is interesting and challenging for us to confront the skeptic's question, "But how is your admittedly admirable goal of generality, adaptability, and so on any different from the proclaimed goal of software companies that develop expert-system shells? Don't they use exactly the same kind of phrases to describe what they are doing?" We do not have a precise answer to this question, and yet we feel a million miles away from such groups in philosophy. Whereas expert-system-shell developers seem to find even the largest of real-world domains to be rather tame beasts that readily fit into predefined frameworks, we seem, without even trying, to constantly run into what we consider to be truly wild problems in even the smallest of microdomains. It is an amazing contrast of attitudes.

Our attitude has led us to attempting to model *concepts* and their inter-relationships in a psychologically realistic way, and we see no way to do this other than in very stripped-down domains. As outsiders, we guess that high-level expert-system performance in non-stripped-down domains must be dependent on particularities of those domains, much as the spectacular success of chess-playing programs turns out to depend on the unexpected computer-tractability of the game of chess itself. By contrast, the game of Go has not yielded to computerization to any significant degree. Some real-world domains are more tractable, like chess, and some are less so, like Go; finding out which is the case is not a fundamental discovery about intelligence, but simply a discovery about the particular domains in question, and about the kinds of questions one is asking in them. In conclusion, we admit to not really knowing the full answer to this skeptical question, but we suspect it has to do with what kinds of questions one considers worth answering, and what one considers to be an adequate level of performance.

In any event, in our attempt to design a very general analogy-making architecture, we felt that it was necessary to constantly jump back and forth in our minds between several rather different potential domains, even while developing a specific program in a specific domain. Such mental juggling of domains is a "keep 'em honest" exercise — that is, its aim is to ensure that the central mechanisms of one's developing program do not wind up depending on any special properties of its specific domain. If the mental juggling is done properly, although the final program itself will ultimately work in just one domain, its *principles* will be abstract and general: valid for many domains.

In the case of Tabletop, one of the most obvious features of its specific domain is its tiny size, and so we felt it was crucial to keep in mind not just some other microdomains, but also some larger, real-world domains, as theoretical alternate domains. Our prime concern was that the Tabletop architecture should *scale up* to much larger domains — that is, that it should be designed in such a way that it intrinsically circumvented all the various causes of combinatorial explosion. Much of what we describe below, therefore, focuses on the sources of combinatorial explosion and the radical difficulties they pose to the designers of an analogy-making program.

Analogy Problems in Scaled-up Domains

> *Kenya is rapidly earning a reputation*
> *as the Hollywood of Africa.*
> — article in an airline magazine

Tabletop gets mentally scaled up to global proportions

We feel that the kind of mental domain-juggling described above is a very important aspect of the Tabletop project. One cannot really understand the project without considering its relationship to the scaled-up alternate domains that we kept in mind (and there were quite a few). In what follows, we shall therefore present and analyze in detail one particular alternate domain — a much larger one than the Tableworld — and we shall even describe the route by which we came to this alternate domain.

It is critical, however, to begin our discussion by emphasizing that by "domain", we do not mean a single isolated real-world analogy, such as the analogy between the solar system and the hydrogen atom, or the analogy between heat flow and liquid flow (two of the analogies most studied in recent years). Focusing on a single isolated example (such as either of these) does not allow one to make a comparison with a host of slight variants in which diverse mental pressures are brought to bear in different proportions. We strongly believe that it is only when one looks at a problem together with its "halo" of variant problems that one can come to understand what makes one mapping appealing and another one not. For this reason, then, we feel it is essential to focus on a domain that is chock-full of analogy problems that form a kind of tight web, so that every problem is surrounded by a halo of closely-related problems.

We now explain our route to a particularly challenging (and amusing) alternate domain. We were struck by the "tit for tat" feel of what goes on in

Tabletop's "Do this!" analogies. The whole thing felt somehow like a type of *retaliation*. This abstract idea brought to mind a number of concrete real-world retaliation scenarios, all of them in the following vein. A war breaks out between California and Indiana over the former's attempt to divert rain clouds from soggy Indiana to the parched San Joaquin Valley. Unfortunately the conflict goes nuclear, and California obliterates Bloomington. The war council in Indianapolis, wishing to be appropriately punitive but not to risk further escalation, must then decide what Californian entity to annihilate in retaliation. Thus — what is the Bloomington of California?

Given the size of the act of aggression committed by California, it would be nonsense to blast Los Angeles, a city with a population over 100 times that of Bloomington. Attacking San Diego would be precluded because of its world-famous zoo. And detonating an H-bomb in the Pacific so as to cause a tidal wave to destroy Carmel would be ruled out because an attack mounted on that jewel of a city would be likely to enrage Californians to a too-risky degree. After some consideration, then, the war council might reason that the Hoosier Armed Forces would best achieve "the same result" not by destroying a city, but by offering all the migrant workers in California one dollar an hour more to come and work in Indiana.

Although the above example is obviously facetious, many real battlefield situations are, in fact, decided on the basis of precisely this kind of "What is the Bloomington of California?" reasoning. It often happens, for example, that a country wishes to retaliate for some belligerent act on the part of one of its neighbors but has no desire to open the door to full-scale war. This tit-for-tat strategy of retaliation "in proportion to" the original aggression requires the retaliating party to find and attack "the same" target as the one attacked.

One does not have to look far for examples of this: a bomb, presumably of Libyan origin, explodes in a discothèque in Germany, killing an American soldier. The U.S. retaliates in a presumably proportional manner by carrying out a pinpoint bombing raid on Tripoli. For years, incidents along the vast border between China and the Soviet Union were settled in this tit-for-tat manner. And of course we all have nostalgic memories of the Cold War, in which on one day, an American journalist is accused of spying and expelled from the Soviet Union, and the next day, allegedly by pure coincidence, three low-level Soviet delegates to the United States are told to leave.

International analogies are of great import and influence, far beyond the cases where a country is looking to retaliate for some act of aggression against itself. Consider, for instance, what happens when a conflict suddenly flares up in some unexpected portion of the world. As soon as this happens, every noninvolved country, no matter how far away, is forced to scrutinize the conflict for possible resemblances to its *own* situation, and then must take a stand according to any analogies perceived — and the stronger the analogy, the more compelling the argument. If a country's stance concerning another country's behavior is inconsistent with its own behavior in an obviously analogous situation, it runs the risk of appearing hypocritical in the court of world opinion, and thus undermining the legitimacy of its own behavior. A sufficiently blatant analogy will override any other pressures, including prior ideological commitments.

To make all this more concrete, take the case of Greece during the Falkland Islands conflict. One would think that this rather poor and backward third-world-leaning, socialist-oriented nation would have instantly sided with poor, backward, underdeveloped Argentina when the latter tried to reclaim the insignificant *Islas malvinas,* right off its coast, from the clutches of the rich, industrialized, right-wing, anachronistically colonial, and — most important of all — enormously *distant* nation of Great Britain. And yet no. The Greeks were in fact staunchly in solidarity with the British on this issue. Why? Because Greece's position *vis-à-vis* the Falkland Islands conflict could hardly be determined without taking into account the blatantly obvious existence, to the world community, of "the Falkland Islands of Greece" — namely Cyprus, an island to which Greece lays claim but that is in fact much closer to *another* country (Turkey) that also claims sovereignty over it. Given this obvious analogy in the eyes of the world, how could Greece conceivably have sided with Argentina, no matter how much it would have liked to do so for other reasons? It *had* to side with Britain, because siding with Argentina over the Falklands would have been so analogous to siding with Turkey over Cyprus that doing so would have totally undermined whatever legitimacy Greece's claims to Cyprus may have.

In summary, analogy is a pervasive and hugely potent, even though often covert and denied, force throughout international affairs. Indeed, for analogous reasons, analogy plays an analogous role in interpersonal affairs, but that is another story. Let us return to analogical thought as a strategy for determining retaliation.

"BattleOp" seemed a fitting name for a retaliation algorithm that, given any act of military aggression, scanned not only the surface of the earth but indeed the entire repertoire of political options, searching not only for the optimal *place* to respond, but also for the optimal *action* to take. Obviously the number of true historical situations of this sort is very large — and when one expands the domain to include hypothetical scenarios of all sorts (or even just mildly "tweaked" versions of true historical incidents), the domain becomes astronomical in size. (What should President Clinton have done if ex-president George Bush, on a visit to Kuwait, had been injured by Iraqi agents? What if Dan Quayle had gone along and been hit in the face by a gooey cream pie intended for Bush?) Indeed, the main problem with the BattleOp domain is that it is *too* huge and open-ended to contemplate — BattleOp is Tabletop scaled up with a vengeance.

Thus in the end, we were led to trying to scale *down* the BattleOp domain in order to come up with an intermediate-sized domain that on the one hand was fairly simple to think about, but on the other hand was clearly a full-fledged real-world domain, filled to the brim with challenging analogy puzzles. At some point we realized that the class of puzzles we were looking for had been tacitly suggested when we asked the very first BattleOp question, "What is the Bloomington of California?"

BattleOp, in turn, gets cut down to reasonable proportions

What, indeed, *is* the Bloomington of California? Readers are encouraged to take a few moments to consider their own reactions to this particular question — not just what their final answer is, but what ideas (even if silly) come to mind along the way, and in what order. This puzzle is of course a specific question of the general form "What is the A of Y?" with A being a geographical entity such as a city or mountain range, and Y being a geographical region such as a state or country. Thus a shivering Siberian contemplating emigration to the Great Plains of the United States might worriedly inquire, "But what is the Ob of Nebraska?" Knowing, of course, that the Ob is the mighty river traversing Siberia, any red-blooded Nebraskan would proudly reply, "The Platte, of course!" Because of this classic example (Belpatto, 1890), such geographical analogy questions have traditionally been called "Ob–Platte puzzles", and we shall not disrespect that worthy tradition.

It is perhaps too ambitious to deal with the vast variety of types of geographical features, such as lakes, forests, glaciers, islands, cities, airports, national parks, and so on. Thus one further domain simplification seemed reasonable: the restriction of A-values to nothing but *cities*. However, even under this restriction, the variety of Ob–Platte puzzles remains enormous, as the following sampler shows:

- What is the Athens of Georgia?
- What is the West Point of Maryland?
- What is the Hobart of India?
- What is the Colombo of Australia?
- What is the Colombo of Greenland?
- What is the Tijuana of Texas?
- What is the Calexico of California?
- What is the Calexico of Mexico?
- What is the Mexicali of Mexico?
- What is the Mexicali of Michigan?
- What is the New York City of Connecticut?
- What is the New York City of New York City?

- What is the Newark of Delaware?
- What is the Toronto of Uruguay?
- What is the Honolulu of Oklahoma?
- What is the Hollywood of Africa?
- What is the Carmel of Indiana?
- What is the Gettysburg of Hawaii?
- What is the Kansas City of Missouri?
- What is the Pittsburgh of the Midwest?
- What is the Pittsburgh of the East?
- What is the Vatican City of Indiana?

Figure VIII–4 shows one possible answer to the first puzzle in this list. Our answer to the last of these puzzles is perhaps worth discussing briefly. The tiny country of Vatican City, seat of world Catholicism, is completely surrounded by Rome, the capital of Italy. Its main edifice and only genuine attraction is Saint Peter's cathedral, a huge construction. People refer to "Roman Catholicism" even though the Holy See is, strictly speaking, "outside" of Rome. Analogously, the small town of Speedway, Indiana, site of the world-famous Indianapolis 500, is completely surrounded by Indianapolis, the capital of Indiana. Its main edifice and only genuine attraction is the Indianapolis Racetrack, a huge construction. People refer to "the Indianapolis 500" even though the great race is, strictly speaking, "outside" of Indianapolis. One could even look upon America's love affair with cars as the counterpart of Italy's relationship with Catholicism, though this might seem a bit forced. In any case, Speedway is a very strong, if somewhat obscure, answer to the puzzle.

Some Ob–Platte problems seem to have strong, almost perfect, answers like this, while most of course do not, but in any case, suggested answers are all instances of a general class of analogies that might be called "supertranslations". A supertranslation is an analogy of the form "X is the A of Y", which implicitly refers to a role that A plays relative to some unmentioned but presumably obvious entity B. Using the traditional signs denoting proportionality, one can write, "A : B :: X : Y". By leaving out two elements of a full supertranslation and asking "What is the A of Y?", one can turn it into a riddle. In this article, we will mostly concentrate on a very special class of geographical supertranslation riddles — namely, Ob–Platte problems in which A is a city or town and Y is a state of the United States.

Are Ob–Platte puzzles about analogical mapping or analogical retrieval?

At first glance, Ob–Platte puzzles seem very much like Tabletop analogy problems, just in a much larger domain. Specifically, both types of problem involve focusing on a designated region — California (say), or Eliza's side of

Figure VIII–4. A cartoon based on the idea of Ob–Platte analogies.

the table — and "touching" an object in it that plays the role of an already-touched object in a different region. Despite this, one could object that the resemblance, while fairly strong, masks a more fundamental difference: in Tabletop, one is concerned with two situations *present in their entireties* before one's eyes, and the task is merely to map these situations onto each other, whereas in an Ob–Platte puzzle, the two "situations" — the large geographical regions containing the source and target cities — are not perceptually available in their entireties, but instead lie *largely dormant in memory,* and the task is to retrieve from memory just one item, or at most a few items, in the target situation, rather than to make a full mapping of the two situations onto each other. Once this has been pointed out, it seems like a significant distinction between the two types of puzzle. However, is it really that important a distinction?

The fact that a situation is perceptually available before one's eyes does not by any means imply that in the first instant of exposure it will get fully represented mentally or fully understood. For instance, to relate to a painting or a photograph or an electric-circuit diagram may take anywhere from a few seconds to many minutes — and often, a page of text is not fully fathomed until one has read it and thought about it, sometimes for hours. Furthermore, in order to arrive at a similarly rich representation of a page of text in a language one does not know, one would first have to learn the language, which could take years! Generally speaking, then, mental absorption of a visual pattern before one's eyes takes time — even as simple a visual pattern as those in Tabletop problems. A full representation of a visual situation is built up piecemeal, with various areas proceeding at different rates.

For a human, the process of perceptually scanning a tabletop situation — whether on a real tabletop or in the idealized Tabletop microworld — involves focusing briefly on one area and then on another, and having one's attention gradually drawn in more and more to specific areas. Early on, for instance, one might casually notice a glass in a remote corner of the table but pay it little attention. Then, however, after a cup has been noticed somewhere else, the overlap of the concepts "cup" and "glass" may make one wish to revisit the neighborhood of the glass to see how it maps onto the neighborhood of the cup. If indeed it turns out that these local neighborhoods map fairly well, the concept "glass" may at that point become much more salient, which fact will have further repercussions concerning which further areas of the table are chosen to focus on, and how intensely. Around and around this kind of process goes, causing the focus of attention of the mind's eye to wander in an erratic but far from arbitrary path, with different areas receiving vastly different levels of interest.

The upshot of this is that the "all there before your eyes" appearance of Tabletop problems is something of an illusion. In reality, different areas are "perceptually there" to different degrees, ranging from very strong to essentially zero, and moreover their amount of "thereness" changes over time.

This becomes particularly important when one remembers that the Tabletop project's purpose is not merely to study how analogies are done on coffeehouse tabletops, but to highlight the central mechanisms of analogy-making in general. In the case of very complex situations, whether they are largely stored in memory or largely perceptually available in front of one's eyes, there will be a strong effect of "shaded presence" of different regions or facets of the situations involved, and there will be a kind of scanning, whether literal or figurative, wherein certain portions are highlighted especially strongly for a brief time.

This effect can certainly be observed in the Ob–Platte domain. In particular, if one works on an Ob–Platte problem for a long time, one's mental attention — one's mind's eye — glides over a kind of mental map in a complex pattern that is a function both of where one has just "been" on the map, and of the desirable criteria for a candidate town that are currently most active in one's mind. Some towns drift gradually into focus while others drift gradually out. The bigger and more complex the situations concerned, the bigger a role will be played by the phenomenon of shaded presence — and this principle holds whatever the domain is. (For more on this, see the subsection of Chapter 5 entitled "Shades of gray and the mind's eye".)

In sum, because of the highly selective, dynamic, and focused nature of perception, Tabletop analogy problems have much more in common with problems involving selective retrieval of analogues from memory — such as Ob–Platte puzzles — than might be suspected at first glance.

Fundamental Obstacles Facing a Formula-based Architecture

A brute-force approach to Ob–Platte puzzles

Now that we have described a scaled-up domain that can serve as a contrasting point of reference for the Tabletop domain, let us carefully consider what types of challenges a brute-force, formula-based approach would encounter in this domain. Then we will carry the lessons learned back into the Tabletop domain.

By definition, a brute-force, formula-based approach to solving Ob–Platte puzzles would require the prior establishment of (1) a context-independent and fixed data base of *cities,* and (2) a context-independent and fixed list of *criteria* that could be used to characterize cities and towns, and that could be applied in a mechanical way to any city in the program's data base. Given such a pre-existent criterion list and data base, then as soon as the specific values of A and Y had been set — in the specific puzzle, to "Bloomington" and "California" — the strategy for finding a solution would presumably consist of the following steps:

Step 1:   Run down the *a priori* list of city-characterization criteria and characterize the "source town" A according to each of them.

Step 2:   Retrieve an *a priori* list of "target towns" inside target region Y from the data base.

Step 3:   For each retrieved target town X, run down the *a priori* list of city-characterization criteria again, calculating X's numerical degree of match with A for every criterion in the list.

Step 4:   For each target town X, sum up the points generated in Step 3, possibly using *a priori* weights, thus allowing some criteria to be counted more heavily than others.

Step 5:   Locate the target town with the highest overall rating as calculated in Step 4, and propose it as "the A of Y".

Any plausible *a priori* list of city-characterization criteria would be long. How would it look? Presumably, among the more heavily-weighted criteria would be items of the following sort (proposed in descending order of weight, very roughly):

- the population of the candidate town;
- the physical geography (*i.e.,* proximity to hills, forests, rivers, lakes, etc.);
- site of an important historical event;
- the average income of the residents;
- characteristic architectural style;
- the climate;

- the racial/ethnic mixture of the population;
- the political environment (*i.e.,* conservative, liberal, etc.);
- having a striking name (*e.g.,* "Truth or Consequences");
- the distance from the nearest large metropolis;
- the crime rate;
- site of a famous natural disaster;
- presence of sports teams;
- home of a famous company;
- site of a military installation;
- presence of heavy industry;
- presence of light industry;
- presence of major museums or orchestras;
- presence of a university;
- birthplace of a celebrity.

Obviously, this is not the full list — indeed, there is no precise "full list" — but it gives a sense for what types of *a priori* criteria one might feel are reasonable. Clearly it is a far longer list than the corresponding list for the Tabletop domain, which consisted of just a handful of aspects for each table object. Each of these many geographical criteria, however, plays a pressure-generating role similar to those played by location and category membership for table objects.

Exhaustive search is deeply wrong

Imagine you are developing a computer program to solve Ob–Platte puzzles in a formula-based mold. Imagine further that your model's skill, as well as its psychological realism, will be put to the test by throwing it a series of "curve balls" it has never before encountered. Your task is of course to try to draw up as complete a geographical data base and list of city-characterization criteria as you can, in order to handle all the weird kinds of curves that might get thrown at you. What kinds of difficulties would you have to anticipate? Here is the first of a series of difficulties to be considered.

> *Difficulty 1:* It is psychologically unrealistic to explicitly consider *all* the towns one knows in a given region in order to come up with a reasonable answer.

It is clear that people, in order to come up with a Californian counterpart to Bloomington, Indiana, don't apply a battery of tests to each and every one of the several thousand towns in California. One obvious reason is that practically nobody's mind contains that long a list of towns, or anything approaching it in size. Even a long-time California resident is unlikely to know more than a few hundred California towns. No one, no matter what the size of their mental gazetteer, be it of length 15 or length 300, runs down the whole thing in order to come up with a Californian counterpart to Bloomington. In fact, only a handful of towns are likely to be considered before one of them is proposed. Then, if one continues to ponder the matter after proposing an initial answer, still probably fewer than a dozen towns will actually be considered altogether, even if one knows hundreds of towns in California. It is certainly not necessary to do extensive psychological experimentation to make this claim with absolute assurance.

Could pruning help matters?

The use of a *pruning* algorithm might be proposed, to quickly eliminate most cities in the target area. In other words, given a source town A and a target town X, the machine, rather than applying the *entire* list of criteria to X, could run down the list of criteria one by one, and as soon as a single criterion on the list was not met sufficiently well, town X would be rejected without any consideration of criteria further down on the list. We can call this the *snap-judgment-rejection* method of pruning. In the case of our puzzle, this would mean that target towns Los Angeles, San Diego, and San Francisco would instantly be rejected for being "too big"; Peanut, Igo, and Ono for being "too small"; Death Valley Junction for being "too dry" and Truckee for being "too high"; Pacific Palisades and Carmel for being "too affluent"; Vacaville, Calexico, and Yreka for "having no university"; and so on.

The snap-judgment-rejection method of pruning will certainly be somewhat more efficient, but does that make it any more psychologically realistic? Not really; after all, it still requires at least *brief* explicit consideration of every single town on the list, which from a cognitive point of view is still nonsense.

Aside from the fact that it explicitly considers far too many candidate towns, the snap-judgment-rejection method of pruning has another problem — namely, it seriously risks throwing the proverbial baby out with the bathwater. That is, a superb candidate town may be rejected for trivial reasons, simply because of literal-mindedness. For example, suppose that, unlike California, target state Y consists mostly of small farming communities with a handful of larger towns distributed here and there, but there is simply no medium-sized town with a university. In the town of Smithville, however, there is a large and nationally famous dental school with a beautiful wooded campus, limestone buildings, and roughly as many dental students as there are students at Indiana University in Bloomington. It sounds like a very good candidate — but is a dental school a university?

Of course not! Therefore, according to the criterion "presence of a university", Smithville would have to be mercilessly pruned from the list. This would happen even if Smithville had a population of 50,000 and an annual bicycle race and was located in hilly, wooded surroundings; even if it had a popular bar named Nick's and the "townies" spoke with a slight drawl. It wouldn't matter how many ways Smithville and Bloomington resembled each other; if "presence of a university" is interpreted rigidly, Smithville would be *out,* period. And conversely, University City, the home of a tiny and fundamentalist Christian university awarding degrees only in "creation science", would remain in the running.

Another complexity involves the hypothetical town of Oral City, a kind of splice of Smithville and University City. Let us suppose that Oral City has both a famous and large dental school and a tiny and insignificant fundamentalist Christian university. Clearly, the only way that Oral City maps well onto Bloomington involves mapping Indiana University not onto the Christian university but onto the dental school. However, a literal-minded program that worked by sorting towns according to strict criteria such as listed above would have no chance of spotting this resemblance. This raises the critically important issue that what really matters in analogical mapping is not *literal* similarity, but something far more intangible — similarity in *spirit*.

Shades of gray are needed — but they quickly lead far outside the domain

The discussion above reveals the criticality, in analogy-making, of not being tricked by rigid categories and of respecting the many shades of gray that make up real-world situations. For instance, rigid categories could lead to Oakland getting rejected as "the Newark of northern California" because the psychological proximity of the concepts *bay* and *river* was not recognized, or to Annapolis getting rejected as "the West Point of Maryland" because the similarity between the Navy and the Army was not recognized.

The only way around this problem is to give the program the ability to judge *approximate* matches. This, however, opens a Pandora's Box. Specifically, it requires an ability to judge *conceptual similarities* between aspects of the source and target towns, which is no more and no less than an ability to do analogies. We can articulate this as follows:

> *Difficulty 2:*   Comparison of a target town and a source town according to a specific city-characterization criterion is not a hard-edged mechanical task, but rather, can itself constitute an analogy problem as complex as the original top-level puzzle.

Note that an appeal to the technique of recursion will not work here, because the "subproblems" engendered will be of roughly the same complexity as the original problem, and are likely to be many in number, to boot. In other words, the original analogy problem could engender a host of equally difficult analogy problems, which in turn could engender yet further ones, and so on. Furthermore, these secondary and tertiary (etc.) analogy problems are not themselves *geographical* in nature, and can involve concepts in arbitrarily far-removed areas of knowledge. This means that, contrary to one's initial impression, the Ob–Platte domain is not even close to being self-contained. We shall come back to this matter in a moment.

Trying to capture the "essence" of a town

We now turn to another difficulty, which will turn out to be closely related to the problem of approximate matches and the proliferation of spun-off analogy problems.

> *Difficulty 3:*   There will always be source towns A whose "essence" — that is, set of most salient characteristics — is not captured by a given fixed list of city-characterization criteria.

Take Bloomington, Indiana. What is its essence? Many people would say that Bloomington's biggest yearly event is the Little 500, a bicycle race that has been held there since the fifties. The fame of this event has spread well beyond the confines of Indiana. A nationally-acclaimed movie, *Breaking Away,* was even made about the Little 500. In many people's minds, then, that bicycle race is the defining feature — the quintessence — of Bloomington, Indiana. However, in drawing up the *a priori* list of criteria, unless one had anticipated this particular idiosyncrasy of this particular town, it is most unlikely that one would have thought of including "has a major annual amateur bicycle race". After all, the list of criteria is not supposed to be about solving the particular puzzle "What is the Bloomington of California?" or even the more general question "What is the Bloomington of state Y?"; the list's purpose is to enable a machine to find the counterpart of *any* town in *any* state.

If features like the Little 500 are considered idiosyncratic and left out of the list of criteria, then serious problems are liable to ensue. For instance, the hypothetical totally anti-intellectual little logging town of Bikeville, California might have a well-known annual bike race and the people of the area might all come out to watch it and celebrate afterwards. To some people, Bikeville and Bloomington would thus *share essence,* and differ merely on the surface. Thus if one wishes one's program to be able to propose Bikeville as a candidate for "the Bloomington of California", then it will be necessary to include "has a major annual amateur bicycle race" on the *a priori* list.

Needless to say, Bloomington has more than one claim to fame. Consider the fact that, thanks to the Indiana University men's basketball team and its notorious coach Bobby Knight, Bloomington is basketball-crazy. Certainly in the minds of many sports fans, *this* fact constitutes Bloomington's essence. This strongly suggests that "enthusiasm for college basketball" ought to be included on the *a priori* list of city-characterization criteria. Otherwise, one again runs the risk of the list's not being rich enough to capture the essence of the source town.

Some other salient facets of Bloomington are the following: Indiana University has the biggest and one of the best music schools in America. Bloomington was once selected as an All-America City, and is the national center for the Children's Organ Transplant Association, as well as the site of the main plant for the famous Otis Elevator Factory. In addition, Bloomington is the town where Crest toothpaste was invented, and the town where Hoagy Carmichael wrote the song "Stardust". One could go on and on. Each of these unexpected and noteworthy idiosyncrasies seems to require a new line to be added to the list of city-characterization criteria. And unfortunately, virtually every town in America has its own idiosyncrasies that play an important role in defining its essence ("site of a phonograph museum"; "named after a famous outlaw"; "site of a giant roadside coffeepot"; "site where ten old Cadillacs are buried in the ground with their fins sticking up"; "site of a famous Civil War battle"; "highest point in the state"; "birthplace of sitcom star X"; and so on), and each of these idiosyncrasies cries out to be added to an already-burgeoning list. When one considers the vast diversity of cities, it seems as if the *a priori* list of city-characterization criteria will have to contain an incredibly large number of items, most of which will be totally irrelevant to all but a few towns.

Analogy problems at all levels — the nightmare of a potentially infinite regress

Despite its daunting size, we shall assume that a high-quality list of criteria for characterizing cities has been drawn up, and we shall press on with the analysis of further difficulties — this time difficulties involving the comparison of a candidate town X with the source town A.

Consider the hypothetical California towns of Trikeville and Scooterville, which sponsor, respectively, an annual tricycle race and an annual scooter race. A literal-minded program could not recognize any similarity between them and Bloomington on the basis of the races that they sponsor, despite the fact that to a human, the similarity is glaring. The problem is, in a nutshell, that we are dealing with *similarity* rather than *identity* along a specific conceptual dimension. Once again, we are back to Difficulty 2 — the fact that an analogy-making ability seems to be required even to make a comparison between two towns along one single conceptual dimension.

Someone might suggest that this obstacle could be overcome by using broad criteria rather than narrow ones — thus, for instance, instead of a narrow criterion that says "has a major annual amateur bicycle race", one might use the much broader "has a major athletic event on a regular basis". Of course, for this strategy to be usable, the program would need the ability to classify *instances* of general categories — such as the fact that "bicycle race", "tricycle race", and "scooter race" are all members of the category "athletic event". This in itself is a nontrivial task, but we will let that slide, because our main point is that using broad rather than narrow criteria leads to its own truly severe problems.

What if Balloonville, California has an annual balloon race? To a human, this might map rather well onto Bloomington's annual bike race, but is a balloon race an "athletic event"? Few people would *a priori* classify it as such. This means that the already-broad category might have to be made yet broader, in order for it to catch Balloonville in its net. But unfortunately, the broader the category becomes, the vaguer it becomes as well. Even the category as it stands — "athletic event" — doesn't give any advantage to Bikeville, which, like Bloomington, has a *bicycle* race, over Trikeville and Scooterville, which do not. If the category were broadened out even further so as to allow a match between Bloomington and Balloonville, then Bikeville would have no advantage over Balloonville, which seems absurd. In short, the broader the criterion, the looser the resemblances that qualify under it, meaning that no advantage is given to *tight* resemblances.

The complexities are starting to be staggering, but someone might still propose that all these difficulties could be handled by having, for each conceptual dimension, a *range* of criteria having differing degrees of precision, something like a series of concentric circles in conceptual space. Points would be awarded for matching with respect to any of these criteria — that is, for being inside each circle — so that closer matches would get many more points. For instance, Bloomington's Little 500 might be categorized simultaneously under the labels "bike race", "race", "athletic event", and a number of other labels. Under this system, Bikeville would get more points than Scooterville, and Scooterville would get more points than Balloonville, but Balloonville would still get some points.

The problem is still that one would have to make an enormous number of *a priori* decisions about categories, motivated by trying to guess what kinds of unanticipated "curve balls" might get thrown at one's program. For instance, how should one code the fact that Crest toothpaste was invented in Bloomington? Just categorizing the concept of Crest alone is complicated enough. Should it be listed under "oral-hygiene product"? What about the town where Ivory Soap was invented? Ivory Soap doesn't satisfy this description, yet the town probably ought to be awarded some points. So we would need to list Crest as a "bodily-hygiene product" as well. And of course we would want Crest described as a "well-known consumer product". However, wouldn't Chevrolet also match this? How do we get rid of that? And should Crest be listed as a "product sold in tubes"? How *big* do the tubes have to be to count as a good match? Shouldn't *squeezable* tubes get extra points? But let's move on.

How do you encode the fact that Hoagy Carmichael wrote "Stardust" in Bloomington? There are many ways of perceiving this. One would have to try to think of them all in advance. How do you code the fact that the movie *Breaking Away* was made in (and about) Bloomington? Could it map onto the fact of making a *book* in (and about) some town? What about a poem? What about a song? What if the song "New York, New York" was written in New York City? Does that map well onto the writing of "Stardust" in Bloomington? How do you code the fact that Bloomington is the national center for the Children's Organ Transplant Association? What kinds of organizations are good counterparts to this one? Any children-oriented organization? What about Boys Town, Nebraska? Any health-oriented organization? What about the Center for Disease Control in Atlanta? To do justice to these features of Bloomington would require giving each feature dozens, possibly hundreds, of different characterizations of various degrees of sharpness, and with many types of overlap and redundancy, so as to allow matches with all sorts of unanticipated towns with all sorts of unanticipated features.

Town names alone constitute a bottomless barrel of worms

One potential factor we haven't considered yet is the simple property of resemblance of *names*. It so happens that between Los Angeles and San Bernardino there is an obscure little town called "Bloomington". Even if this Bloomington has little else in common with Indiana's Bloomington, its name alone would certainly make it a far better candidate for "the Bloomington of California" than a similar California town named, say, "Uxhaha". All of a sudden, then, this suggests that *name identity* would be a sensible criterion to include on the *a priori* list.

But once we have opened the door to name *identity*, we find that it is yet another Pandora's Box. For example, suppose there were a town whose name had a slightly different spelling — say, "Bloomingtown", "Blumington", or even "Blossomton". Certainly this would be almost as good as being named "Bloomington", which shows that *name proximity* rather than *name identity* is the proper criterion to be on our *a priori* list. But then one must consider in detail how points should be awarded for proximity of name. Certainly it is a more complex matter than just noting a resemblance in spelling. For instance, "Bloodington" would not be nearly as good as "Flowerville", but to see this requires a nontrivial level of semantic understanding.

Suppose there were a town in California called "Pueblo Florido" — Spanish for "blooming town". To someone who knew Spanish, would this fact not elevate that town's chances of being selected as "the Bloomington of California", at least to some degree? To concede this point implies, unfortunately, that *proximity of meaning of translated name* should be included somewhere on the *a priori* list of criteria — perhaps not near the top, but somewhere.

And suppose it turned out that in Navajo, the word "uxhaha" meant "town of many blossoms". For a person who knew it, this fact would be likely to elevate Uxhaha's chances of being selected as "the Bloomington of California". Moreover, it seems that not just Spanish but even Navajo ought to be included in the *a priori* list of languages in which to look for translations. But if Navajo ought to be included, then what language could be excluded? Before this hypothetical example was raised, it might have seemed far-fetched to include such factors *a priori*, but this makes clear that one couldn't afford to drop them from such a list.

Or what if there were a town in which there was a university called "Indiana University"? (This actually is the case for the town of Indiana, Pennsylvania — the site of the university with the unlikely name "Indiana University of Pennsylvania".) Should the *a priori* list of city-characterization criteria include "presence in both towns of universities having the same name"? And what if the name of the second university was *almost* the same — say, "The University of Indiana"? Should the list include "presence in both towns of universities having *similar* names"? This would seem absurd, and yet the existence of a "University of Indiana" in some California town couldn't help but be a strong factor in making that town a serious contender for selection as the Bloomington of California.

What all this shows is that Difficulty 2, concerning *capturing the essence* of a town, constantly pushes for the inclusion of ever more new criteria on the *a priori* list, and Difficulty 3, concerning the *ubiquity of similarity judgments* rather than exact matches, makes us realize that each new criterion added to the list poses a host of fresh new difficulties in making mechanical proximity-judgments. Taken together, the problems suggested by these two objections are formidable.

What constitutes a town?

We now move on to yet another serious problem that is not at all apparent on the surface.

Difficulty 4: What constitutes a "town in region Y" is not *a priori* evident.

It turns out that the assumption one would tend to blithely make at first — namely, that source town A would have to be mapped to a *single* town in region Y — is in fact highly dubious. Some cities, such as Minneapolis/St. Paul and Dallas/Fort Worth, are really compound entities, yet only their components would be found in a gazetteer, since the compound chunk has only an informal status. In fact, there are informal chunks with more components, such as North Carolina's "Research Triangle", consisting of Chapel Hill, Durham, and Raleigh, along with the triangular region they define. It would not be hard to come up with an Ob–Platte puzzle in which this informal chunk would be the best possible answer. Or, for yet more trouble, consider the composite of Kansas City, Missouri and Kansas City, Kansas, or the unofficial entity known as the "Quad Cities", consisting of two towns in Iowa (Davenport and Bettendorf) and two in Illinois (Moline and Rock Island). Each of these composite cities spans state borders, which really blurs matters up. At the other end of the size spectrum, there are relatively well-defined clumps of houses, stores, and churches that have never been incorporated as towns or cities. Although such conglomerations, large or small, would not appear on any official list of cities in state Y, they should certainly not be excluded *a priori* as candidates for "the A of Y". After all, who knows what A is going to be?

For a simple but telling example of the way that unofficial informal chunks make matters blurry, consider the trivial-seeming puzzle "What is the New York City of California?" One is inclined to answer "Los Angeles" without hesitation, but what would really be meant by this? Would one intend precisely that area within the official municipal boundaries of Los Angeles? Or the greater Los Angeles area, all the way out to San Bernardino (roughly)? Or some intermediate conglomeration? What would its boundaries be? Perhaps the five largest cities in the metropolitan area of Los Angeles should be chunked together into a unit, as this might correspond to the five boroughs of New York? Which one would be "the Queens of California", which one "the Brooklyn", which one "the Staten Island"?

The set of potential entities in California to which Bloomington (or New York) could be mapped is far from trivially definable. And who would have suspected, *a priori,* that one day some travel writer would suggest that an entire *country* roughly the size of Texas is the proper African counterpart of a certain small but glamorous American *town* (as in the epigraph above, alluding to the intensity of movie-making activities in both places)? Undeniably, putting together a list of all possible entities to consider as counterparts of a given entity is much more difficult than might have been suspected.

Violations of geographical and conceptual boundaries

All of this might seem bad enough, but there is worse to come. Even the seemingly undeniable assertion that "the Bloomington of California" would have to be located *inside* the state of California starts to get shaky when one looks carefully enough. At first, this claim might seem absurd. The challenge, after all, is to find "the Bloomington *of California*". However, does "of" necessarily mean "in"? For people, state (or country) boundaries are not inviolable, impenetrable membranes. In a sufficiently fluid mind, facts can easily "leak over" state boundaries. For instance, the University of Oregon, because of the very high percentage of California residents in its student body, is occasionally referred to as "the University of California at Eugene" by students who attend it. Of course, this is a jocular reference, but that fact, combined with several other commonalities between Eugene and Bloomington (such as similar population, degree of isolation, local countryside, style of dress, number of coffeehouses, and so on), would make Eugene at least a mildly plausible candidate for "the Bloomington of California", even if not the eventual winner.

Perhaps this will sound incredible to you, but to our surprise, one Indiana University professor, when asked what he would nominate for "the Bloomington of California", actually suggested Portland in all seriousness. He argued that its several colleges, such as Reed College and others, mapped well *as a group* onto Indiana University. When it was pointed out to him that Portland is in Oregon, he simply shrugged and said, "I know, I grew up in California, but that's the best I could do." A town in Oregon as "the Bloomington of California" may not work for you, but for this particular intelligent, well-informed person, it was the top choice. This lends support to our contention that out-of-state answers cannot be ruled out out of hand.

In fact, there are occasions when the choice of a town not in the designated region is totally convincing to just about *anyone*. Consider the question: "If you had to choose, which would be the better Atlantic City of California — Las Vegas or Lompoc?" If borders were always inviolate in people's minds, they would invariably answer "Lompoc" *without hesitation* because Las Vegas (gambling capital of the West in much the same way as Atlantic City is the gambling capital of the East) is not within the borders of California, even though the *only* thing that Lompoc has going for it is that it is in California.

The point can be made even more strongly. Consider the following hypothetical scenario. There is a little independent country called Nuevo Mónaco, five miles on a side, located in an enclave on the Pacific Coast twenty miles south of Los Angeles; the capital of this hypothetical little monarchy is the town of Ciudad Pacífica, world-famous for its gambling casinos and lavish entertainment palaces. Would lumpish little Lompoc, located inland and known only for its prison, *still* be the better choice for "Atlantic City of California"?

The hypothetical example of Nuevo Mónaco has an almost perfect parallel in the real world. If one were to ask, "What is the Atlantic City of France?", a large number of people undoubtedly would answer "Monaco", even knowing full well that Monaco is not a city inside France, but rather an independent, albeit small, *country* that is not in France, but borders it (the gambling city within Monaco is Monte Carlo). However, Monaco is world-famous for its casinos, is small enough to be thought of as a city, and, like Atlantic City, is located on the sea. What possible set of *a priori* criteria would allow a computer to reply, perfectly self-confidently, that the country of Monaco is "the Atlantic City of France"?

Although the proposal of Eugene as "the Bloomington of California" might strain many people's sense of reason, the proposal of Monaco as "the Atlantic City of France" probably would seem quite reasonable to most people. Thus these two examples indicate that there is a continuum of plausibility values for out-of-region answers, ranging from "very strong" through "weak" to "outrageous". As we said earlier, one cannot simply dismiss the idea out of hand.

Several of the entries in our list of Ob–Platte puzzles were intended to suggest this kind of issue. Take just the question "What is the Hobart of India?" We remind readers that Hobart is the capital of the Australian state Tasmania, which is of course an island just south of the mainland of Australia. The way that Tasmania dangles off the bottom of Australia would almost inevitably recall the way that Sri Lanka dangles off the bottom of India, and so the city of Colombo would flash to mind almost instantly — despite the fact that Sri Lanka is a completely separate country from India, whereas Tasmania is an integral part of Australia.

Answers of this sort remind us a bit of the famous old nine-dots puzzle. For readers not acquainted with that warhorse of a puzzle, we restate it here. "Without lifting your pen from the paper, draw four straight line segments that pass through all nine dots in a square three-by-three array." (See Figure VIII–5.) One tries and tries to do it and fails; when one is finally shown the solution, it turns out to involve lines that go *outside the bounds of the three-by-three square.* Nothing in the statement of the problem precluded jumping outside the square, but nothing suggested doing so, either, and one tends to make an extremely hard-to-dislodge default assumption that there is a boundary one cannot cross. Moreover, not being *aware* of having made any such assumption, one has no way to overturn it, since it is not even psychologically available for consideration as a bad assumption. A few people, of course, do catch themselves making this assumption, and once they have recognized it explicitly, it's easy to deliberately violate it, and doing so leads to a good solution.

In Ob–Platte puzzles, too, one has an instinctive default reflex reaction to look only inside the named state, and that reaction is similarly hard to dislodge. In fact, the default of staying in-state is probably harder to dislodge than the default of staying in-square in the nine-dots puzzle. Moreover, if you dare to jump out of the given state's boundaries, then the farther away you go, geographically, from the named state, the more you will experience a twinge — a kind of self-imposed psychological penalty. Thus Portland, Oregon as an answer to the puzzle "What is the Bloomington of California?" might be fairly convincing to a few people, but it's hard to imagine that Portland, Maine would ever

Figure VIII–5. A famous puzzle.

work for anybody. But who knows? There might be a way of justifying even that much of an out-of-state answer.

Local versus locational aspects of a town

Most of the criteria shown on the suggested list concerned *local* properties of a town: its population, ethnic make-up, industries, and so on. Local properties involve only the town itself, and not its location. Only one criterion was included that might be called *locational* — namely, "the distance from the nearest large metropolis". But sometimes, locational properties can be the major defining factors in the perceived identity of a city or town. For instance, consider Oakland, California. To many people, Oakland's primary identifying feature is locational — namely, it is perceived as a "sidekick" to San Francisco, located across the famous bay. A secondary feature might be that it is much poorer than the city to which it is a "sidekick", and a third feature that it has a large minority population. Note that only the last of these three is a purely local fact about Oakland.

Now, suppose a challenge to name "the Oakland of Illinois" were issued. One answer that would probably come to mind for many people is "Gary", despite the fact that Gary is in Indiana, not Illinois. This oddball but completely defensible answer would arise by an indirect process. First, given that Oakland is conceived of as a satellite of San Francisco, one would search for "the San Francisco of Illinois", which would be almost certain to suggest Chicago, and probably only Chicago. Once these two "landmark" cities had been mapped onto one another, one would look for towns that play a "sidekick" role to Chicago, especially ones located across a body of water and having large minority populations. Gary fills this bill very well, since it is located across the Chicago River (even in the "correct" direction — namely, to the east — which is a free bonus) and its population is largely black. The fact that Gary is not strictly speaking inside Illinois constitutes a strike against it, but not a strikeout. As someone once remarked about Whiting, Indiana (a small town very near Gary), "If it weren't for the state line, it would be in Illinois" — a droll counterfactual that holds for Gary as well. Indeed, Gary has so much else going *for* it that it probably would be the eventual winner. Even in the minds of people who wound up ultimately rejecting it, Gary would have to be seen as a far more plausible candidate than, say, Peoria (which in this very special context might be called "the Lompoc of Illinois").

Gary is not the only plausible "Oakland of Illinois". A very different but quite reasonable answer would be East St. Louis, a poor and largely black suburb of St. Louis located across the Mississippi from it, and thus in Illinois. Some people might even feel East St. Louis is a superior answer to Gary, as it is located inside the perimeters of the target state. However, this time the landmark city by reference to which it is defined — St. Louis — is *outside* the state borders. This delightful irony brings out the subtle fact that the search for reasonable answers *inside* the designated state will still have to be allowed to stray across the state lines in order to put in-state cities, especially ones near any border, in their proper perspective.

All of this suggests the following Ob–Platte puzzle, a definite favorite of ours:

What is the East St. Louis of Illinois?

Initially, this might sound either trivial or nonsensical, given that East St. Louis is in Illinois to begin with. However, the mere posing of the puzzle suggests that one could make it make sense — but that to do so will require looking at East St. Louis in a new light. Fortunately, this is not too hard: thanks not only to the physical proximity of St. Louis and East St. Louis but also to the sonoric proximity of their names, East St. Louis is on a certain *conceptual* level more attached to Missouri than to Illinois. This deep association between East St. Louis and Missouri hints that the puzzle makes some sense after all, and that we should begin our search for an answer by solving this subproblem:

What is the St. Louis of Illinois?

This is a piece of cake: Chicago, of course. The next step is to look for Chicago's analogous sidekick — in other words, to attack this problem:

What is the East St. Louis of Chicago?

One's first instinct might be to look within Illinois, but given the relationship between East St. Louis, St. Louis, and the Mississippi River, it would perhaps seem better to look eastwards across the Chicago River into Indiana. This leads us back again to Gary, which is not an awful answer. However, as a potential counterpart to East St. Louis, Gary is a bit hefty (their populations are roughly 50,000 and 150,000), and furthermore, the name "Gary" bears no relationship to that of the landmark city, Chicago. Consideration of these flaws suggests looking further, and fairly quickly, at least if one knows Chicago's eastern suburbs reasonably well, one comes across another potential answer: East Chicago (population: roughly 40,000). This rather small and mostly black suburb of Chicago is, like East St. Louis, close to its landmark city in both location and name. It also lies east, across both the river and the state line from the landmark city. All this leaves little doubt that "the East St. Louis of Illinois" is East Chicago, Indiana.

Of course, from a narrow and rigid point of view, East St. Louis itself ought to be the unhesitating answer, and no cities in Indiana should ever be considered at all. However, narrowness and rigidity are the antithesis of what analogy-making is all about, and in our opinion, East Chicago is a convincing and charming answer.

Bringing These Ideas Back to Tabletop

Downward transfer of lessons from the big domain to the tiny domain

We have spent many pages exploring a variety of geographical analogy problems. Although the discussion has hopefully been amusing, our main purpose in this exploration was to show the pointlessness and the incredibility of a formula-based approach to such problems. But assuming that we have done a reasonable job, the question still remains: What do lessons about the domain of Ob–Platte problems have to do with Tabletop, the actual computer model? Do they transfer downward to its microdomain, or is the gulf between the domains too wide?

Certainly we would not claim that all the properties of the Ob–Platte domain are mirrored in the far simpler Tabletop domain. For instance, the *noncontainability* of the Ob–Platte domain with respect to analogy-making — the fact that spin-off analogy problems are generated that may have to do with essentially *any* aspect of the world, no matter how unlikely *a priori* — is not in any sense mirrored in the Tabletop domain. The Tabletop world is a very small, self-contained world with a clearly delimited amount of information. In that respect, it resembles chess. In chess, all the pieces' moves can be described in a page or two, and that's really all one needs to know, at least in principle. Likewise in Tabletop: as was mentioned earlier, there is precious little to know about any particular table object, and not much more to know about any Platonic category. Because the objects and categories are so stripped-down, one is never forced by a Tabletop analogy problem into solving analogy problems in other, seemingly disconnected, domains.

But the fact that the infinite-regress nightmare utterly vanishes in the tiny Tabletop domain is *irrelevant* to us, as designers of an analogy-making architecture. For us, what matters is that in *real* domains, the infinite-regress nightmare *is* there, and cannot be sidestepped. For just this sort of reason, we forced ourselves to keep alternate domains in mind. Similarly for the other nightmares that became crystal-clear only when we looked carefully at the scaled-up domain of Ob–Platte puzzles: although they may not arise in the Tabletop domain itself, they are facts about the challenge of analogy-making in general, and *as such are one hundred percent relevant* to the Tabletop project. As we stated before, the critical idea is that if one is working in a restricted domain but striving for psychological realism, one must not succumb to the temptation to employ mechanisms or strategies that, in a scaled-up domain, would give rise to intractable difficulties. Taking advantage of the Tabletop domain's deliberate smallness and deliberate simplicity would simply be cheating.

Thus in sum, we would argue that the lessons from the larger domain *perforce* transfer to the smaller domain; indeed, the main purpose behind considering a hugely scaled-up domain was to unmask techniques that, though they would work in the tiny domain, would fail in the scaled-up domain. Such techniques are merely trickery and should be totally banned, if one is trying to build a model that is psychologically realistic.

Another reason we are interested in looking at all sorts of problems in scaled-up domains, such as Ob–Platte puzzles, is that other domains contain fascinating analogy-making challenges that inspire us, and whose flavor we would like to echo in our microdomain. Is it possible to mirror analogy problems belonging to a huge domain in the very small Tabletop microdomain? We have found that it often is. The "translations" are often crude but nonetheless quite interesting, and it is to a few of them that we turn our attention in the next subsection.

Crude mirrorings, in Tabletop, of some tricky Ob–Platte puzzles

In order to "copy" a given analogy problem in a totally disjoint domain, one obviously has to look at the problem at a considerable level of abstraction. The level we feel is the most telling, as far as analogy-making is concerned, is that of *interacting pressures*. As we have seen in earlier sections, there are many types of pressures in common between the Ob–Platte world and the Tabletop world. In particular, there are pressures, either local or locational, having to do with category membership, physical location, grouping, relative salience, and a number of other notions. All of these types of pressures are *universal*, or nearly so, in the sense of existing in virtually any domain in which analogies can be made. The idea is therefore to take advantage of this universal level of analogy-making.

One begins by selecting an interesting Ob–Platte puzzle. Next, one strips it of its surface features, exposing the more hidden (but more essential) brew of interacting pressures. Finally, one creates a Tabletop problem that mimics, as closely as possible, the flavor of that brew. To the extent that the Tabletop domain is interesting, its interest derives from its capacity to reproduce many diverse constellations of interacting pressures, not from the objects themselves.

Let us look at a few concrete examples of "translations" of Ob–Platte analogies into the Tabletop world, involving rich mixtures of pressures brought

Henry

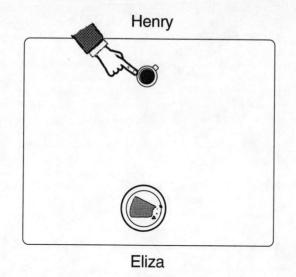

Eliza

Figure VIII–6. Working towards the "East St. Louis of Illinois" problem in the Tabletop domain.

about by rivalries between various types of local and locational properties of objects on the table. For instance, in Figure VIII–6, if Henry touches his cup, Eliza could simply reach across the table and touch the same cup, arguing "I have no choice, since there is no cup over here on my side." This would be a little bit like answering "East St. Louis" to the "East St. Louis of Illinois" puzzle.

One can adjust the pressures slightly away from this literal-minded answer by adding silverware surrounding the cup and plate, as in Figure VIII–7. This

Henry

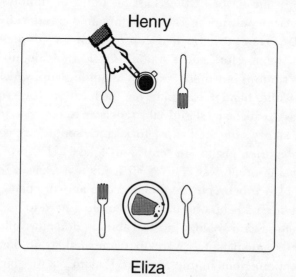

Eliza

Figure VIII–7. Getting closer to the "East St. Louis of Illinois" problem in the Tabletop domain.

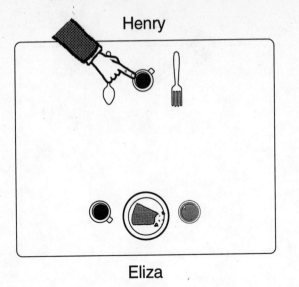

Figure VIII–8. Even closer to the "East St. Louis of Illinois" problem in the Tabletop domain.

encourages perception of Henry's cup in terms of its *locational* properties (*i.e.,* relating it to the "St. Louis" of the situation) above and beyond seeing it simply as a member of the category "cup". Since the plate on Eliza's side has an identical locational description ("between a fork and a spoon"), there is considerably more reason now for Eliza not to reach across the table to Henry's cup, but to touch her own plate. This might correspond roughly to naming a Chicago sidekick *inside* Illinois.

Even the ironic flavor of the "East St. Louis of Illinois" puzzle can be mimicked in a crude way in the Tabletop domain. Consider the configuration in Figure VIII–8. Here, the touched cup on Henry's side has the locational description "between a fork and a spoon". On Eliza's side, no object has this locational description, but there is a very tempting cup, which could be the answer. On the other hand, Eliza's plate has the locational description "between a cup and a glass", which at a slightly abstract level is *analogous* to the locational description of Henry's touched cup ("fork" and "spoon" are considered to be very similar categories, just as are "cup" and "glass"). This would suggest that Eliza's plate is a good answer. This is an ironic twist, given the total lack of *intrinsic* resemblance between the touched cup and the plate, and especially given the existence of a cup on Eliza's side, pulling hard to be touched.

Needless to say, no analogy in the Tabletop domain could come close to reproducing every nuance of a given Ob–Platte analogy. Rather, translation of an analogy from one domain into another domain is a very high-level analogy problem in itself, and at times can even take on aspects of an art form. Similar points have been made in earlier writings (Chapter 24 of Hofstadter, 1985;

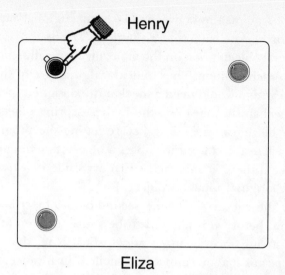

Figure VIII–9. A "blockage" scenario in the Tabletop domain.

Mitchell, 1993), where numerous examples were given of "translations" of real-world analogies into the very small Copycat domain, consisting of linear letter-strings. We feel there is a certain beauty to the idea that a tiny domain can in some sense mirror the more abstract complexities of situations in the real world even if their surface-level features are not in the least mirrored.

Other tricky Tabletop problems and mechanisms tailor-made to handle them

Earlier, we mentioned that the first program that operated in the Tabletop domain was a "Potelbat" — a brute-force, formula-based program — and that it did quite well on simple problems but had trouble with some of the trickier ones. We pointed out that such a program could have been enhanced to handle the trickier problems by the addition of specialized mechanisms. We wish here to give readers a feel for how this could have been done, but why, from our point of view, it would have been an exercise in pointlessness to do so.

Figure VIII–9 illustrates how the discovery of a particular structure — in this case, a diagonal "bridge" (that is, a perceived object-to-object correspondence) between two glasses — can have an unexpected effect. Specifically, the existence of this bridge tends to *block* an otherwise perfectly reasonable answer — namely, touching the glass on Eliza's side of the table — and to push strongly for touching Henry's cup itself. The Tabletop architecture handles this type of situation naturally, with no need for a special "blockage-checking mechanism". The drastically reduced attractiveness of the glass on Eliza's side falls out automatically from the role that bridges play in the program.

Why could a Potelbat-type program not recognize that Eliza's glass was "already taken", and thus opt for touching Henry's cup? The answer is that a

Potelbat sees no relationships between objects, makes no groupings, builds no bridges — in short, a Potelbat has *no ability to build up new perceptual structures*; it just takes the literal objects given on the table, mechanically scans them all, and assigns points to each of them. Thus to deal with this type of "blocked-answer" situation, such a system would need a mechanism designed specifically for this purpose — a mechanism that takes the highest-scoring object (the one that would normally be the winner) and, before letting the program actually go ahead and touch it, rescans the entire table for objects that resemble it, checking specifically to see if there is one back on Henry's side that resembles it more closely than the original touched object does. If so, the would-be winner is declared out of the running, and the second-place object becomes the new front-runner (but before *it* is allowed to be touched, the blockage-checking mechanism must run once again — and so on). Note that since problems featuring this type of situation cannot be "smelled" in advance, this mechanism would have to be *run on every problem,* even though it will have an effect in only a tiny minority of cases.

Adding such a specialized mechanism would allow *this* problem to be solved, but for each such explicit mechanism that one might have remembered to include, there would be many others that one might well have overlooked. To be concrete, any problem that depends in any way on the build-up of groups (or of bridges) will require a specific mechanism to be added on — unless, of course, group-building (or bridge-building) itself is added into the basic architecture. This would be a radical extension of the original notion of a formula-based program, but let us not question it for that reason. The real trouble is that such a program, if it retains its mechanical brute-force style of assigning points to *everything,* would have to build *all conceivable* groups, rather than building only ones that are more "attractive" or "plausible". In a situation with n objects on the table, the number of conceivable groups will be 2^n, a very large number — and needless to say, most of them will be laughably implausible from a human point of view. Note also that building groups out of "raw" table objects is not the end of the line; one can then go on to build bridges between groups, and even to build higher-order groups whose elements are smaller groups. So in fact, we are dealing with a number considerably larger than 2^n. Lest all this sound too theoretical, we should point out that there are many natural table situations that humans instantly and intuitively *do* perceive in terms of such higher-order groups, such as the one shown in Figure VIII–10 (note its similarity to Figure IX–1*e* on page 395).

The situation in this figure is interesting, in that many people are drawn to the salt shaker — *a priori* certainly a very unlikely object to be considered "the same" as a cup, especially when there is a glass on Eliza's side. Obviously, the salt shaker's "similarity" to the cup has nothing to do with category membership;

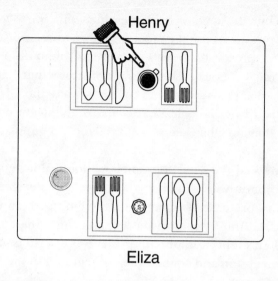

Henry

Eliza

Figure VIII–10. A Tabletop scenario in which people naturally see triply-nested groups.

it comes entirely from paying attention to where these items are located inside their respective six-object groups, each of which breaks down most naturally (from the point of view of human perception) into a fork–fork group and a knife–spoon–spoon group, as well as one further item between them. And as for the two knife–spoon–spoon subgroups, it is actually more plausible to see each of them as consisting not of *three* objects, but rather of one knife and a *group* of two spoons. Thus in this problem, people intuitively see a group (spoon–spoon) inside a larger group (knife–spoon–spoon) inside a yet larger group (containing six table objects). This gives a feeling for how unlikely it would be that a Potelbat would solve this problem. What, indeed, would tempt such a program *at all* to point at the salt shaker?

Concluding Words

Contrary to what one might at first suppose, analogy-making in the Table-top microworld is not trivial; in fact, it poses a variety of deep and interesting challenges. Some of these challenges might appear at first sight to be tractable by special-purpose mechanisms. However, to build into an analogy-making system a large number of special-purpose mechanisms (all of which would have to be brought to bear on every problem) would amount, in scaled-up situations, to computational suicide; moreover, it would be psychological nonsense.

The actual Tabletop program, rather than relying on a host of diverse brute-force special-purpose mechanisms, attempts to use a small set of very general perceptual and cognitive mechanisms. As it scans its little world, it builds up multi-level perceptual structures, from which emerge context-dependent pressures. These pressures push the program to focus on certain concepts, certain areas of the table, and certain objects more heavily than on others. Altogether there is a deep mutual interaction between the processes that build new structures and the processes that focus spatial and conceptual attention. It is the fact that it possesses this type of psychologically reasonable architecture, rather than the smallness of its domain, that keeps the Tabletop program from "exploding" combinatorially.

The smallness of the Tabletop domain should not be misread. It is an attempt to deal honestly with the huge cognitive complexity of small-seeming tasks, rather than a pretense of dealing with the limitless complexity of the real world. On its surface, the Tabletop project is not glamorous, but when one takes the time to look carefully at it, one finds that it is a serious attempt at understanding how perception and cognition are integrated in the human mind.

Preface 9:

The Knotty Problem
of Evaluating Research
in AI and Cognitive Science

Striving to Simulate the Style of an Individual Human Mind

What would make a computer model of analogy-making in a given domain a *good* model? Most cognitive psychologists have been so well trained that even in their sleep they would come up with the following answer: *Do experiments on a large number of human subjects, collect statistics, and make your program imitate those statistics as closely as possible.* In other words, a good model should act very much like Average Ann and Typical Tom (or even better, like an average of the two of them). Cognitive psychologists tend to be so convinced of this principle as essentially the *only* way to validate a computer model that it is almost impossible to talk them out of it. But that is the job to be attempted here.

One of our several counterarguments to this viewpoint comes simply from caricaturing the earlier statement of the goal: a good model should act very much like Boring Bob, Mediocre Melanie, Conventional Carol, and Dull Doug. Anyone — probably even Average Ann and Typical Tom — can see what is being driven at. Who would want to spend their time perfecting a model of the performance of *lackluster* intellects when they could be trying to simulate *sparkling* minds? Why not strive to emulate, say, the witty columnist Ellen Goodman or the sharp-as-a-tack theoretical physicist Richard Feynman?

Indeed, one of our unabashed motivations is a desire to understand what goes into deep esthetic insight. This goal is particularly clear in the Letter Spirit project (see Chapter 10), where just about the only criteria for judgment are esthetic, but it is also the case in Copycat and Tabletop. In domains where there is a vast gulf between the taste of sophisticates and that of novices, it makes no sense to take a bunch of novices, average their various tastes together, and then use the result as a basis for judging the behavior of a computer program meant to simulate a sophisticate.

This might seem to suggest an alternative research methodology for validating an analogy-making program, one that would please cognitive psychologists: Send out an analogy-puzzle questionnaire to your favorite set of elite minds (Nobel Prize winners, MacArthur Award recipients, witty columnists, pithy farmers, and so on), collect and tabulate the results, and make sure your computer model imitates the average of *these* minds.

There is still something deeply wrong with this idea. Averaging a bunch of top-notch minds together is as scramble-brained as mixing a bunch of great recipes together in hopes that the result will itself turn out to be a great recipe. A ridiculous notion! Great recipes and great minds are one of a kind. The act of averaging destroys them. Averaging Richard Feynman's answers together with Ellen Goodman's equally thoughtful answers (whatever they might be) would be senseless.

The point is that creative minds will collectively come up with a spectrum of answers and rankings thereof that, as a whole, may be quite atypical of any one creative mind. Any individual cognitive style would get completely lost in the blur. This suggests that a more reasonable goal for a model of analogy-making would be for it to act like some *particular* creative mind, or even, perhaps, for it to be able to act like various different creative minds when certain critical "cognitive-style parameters" are varied. Of course, when critical parameters are allowed to vary sufficiently, the performance will rapidly move out of the "elite" class and into the "eccentric" area, and from there on, who knows where!

A Throng of Interacting Subcognitive Mechanisms

What kinds of cognitive-style parameters might be involved in the determination of an individual cognitive style in analogy-making? Let us list a number of typical ones, to give a concrete feeling for what is being talked about:

- *role of salience in determining levels of attention paid to various objects*
- *degree to which extra attention is devoted to objects having extremal attributes*
- *differential rates of perceptual processes — e.g., spotting sameness vs. relatedness*
- *desire to build chunks based on samenesses or other types of connections perceived*
- *willingness to adjust or destroy already-solidified perceptual chunks*
- *propensity for a dormant concept to "wake up" when lightly activated*
- *differential rates of spreading of activation from a concept to various semantic neighbors*
- *differential rates of decay of activation of various concepts*
- *degree to which activated concepts bias ongoing perceptual search*
- *ease with which default perceptions can be overridden by drive for perceptual uniformity*
- *size of conceptual halos of concepts*
- *relative levels of abstractness assigned to concepts*
- *degree of preference for abstract descriptions over concrete ones*
- *manner in which perceptions affect inter-concept distances*

- *probability of making a slippage between two concepts at a specific distance*
- *willingness to let a slippage carry related slippages along on its coattails*
- *willingness to accept fragmentary mappings*
- *willingness to consider alternatives even when there is one clear leading viewpoint*
- *degree of resistance to blending rival views, rival rules, rival answers, etc.*
- *degree of attraction to symmetry*
- *intensity of dislike of trivial or boring answers*

By no means do all insightful people have the same settings of such parameters. Each specific combination of numerical values of these (and countless other) hidden and usually totally unconscious variables determines a unique and sharply-defined intellectual style, or taste. Mushing together the responses of a large number of intelligent people would wash out the precision of each personal taste. Most likely the result would be a goulash as tasteless as a mixture of many good recipes that had nothing to do with each other.

But then why do so many cognitive psychologists, upon first hearing about the Copycat and Tabletop projects, tend to insist on experiments in which many people's answers to analogy puzzles are pooled? Because they are used to an extremely different experimental paradigm — one in which *one single cognitive mechanism* (or perhaps the interaction of a couple) is probed. The expectation, and a quite reasonable one in standard experiments, is that the basic mechanism under investigation is reproduced almost deterministically from mind to mind.

For instance, a psycholinguist might seek to understand whether lexical items are stored, associated, and retrieved more by their semantic roots or by their phonetic properties. Such a researcher might therefore design an experiment aimed at determining the degree to which some specific word, when heard, primes (*i.e.,* lightly activates) various other words that are related to the first word in different ways. For instance, how much does the word "decide" prime the word "decision" (with which it shares a root but in a slightly disguised way) as opposed to priming "deciduous" (which looks more like it, though they are unrelated)? (Such an experiment is described in Marslen-Wilson *et al.*, 1992.) There are known techniques for determining the strength of priming effects, such as reaction-time tests (*e.g.,* measuring the length of time a subject takes to decide whether a word displayed on a computer screen is a *real* word or a *nonexistent* word, with shorter-than-usual reaction times being indicative of priming). For each individual subject, one can use such a technique to measure the reaction times in the two different conditions (namely, tight phonetic resemblance but no root in common, as opposed to less phonetic resemblance but shared root). Then a ratio can be taken, showing which type of effect is stronger in the given individual. When such experimental ratios are then compared across many subjects, it usually turns out that not only is there a definite trend, but also the individual differences are not all that great, thus confirming the expected approximate universality.

Experiments of this sort are beautiful and deep probes into the fundamental mechanisms of the human mind, but they probe only one tiny aspect at a time. The Copycat and Tabletop projects, on the other hand, aim to model the way in which a very large number of independent tiny mechanisms, all working in concert, produce high-level emergent behavior. When each of, say, a hundred different mechanisms has some variability on its own, even if just a little bit, their overall interaction has an enormous number of degrees of freedom. When many, many mechanisms are simultaneously involved in this way, it is simply a confusion, brought on by habit and by training, to insist that the validation criteria for models of such complex behavior are identical to those for models of a single mechanism.

How Are Natural-language Programs Validated?

Perhaps the inappropriateness of using statistics to validate a computer model of mental phenomena becomes clearest when the idea is applied to natural-language programs. Consider a lengthy natural-language dialogue between a person and a computer program, such as the famous dialogue involving Terry Winograd's program SHRDLU (Winograd, 1972, reprinted in many places, including Hofstadter, 1979). What would make a reader feel that such a dialogue was psychologically realistic? The answer is, simply, the *feel* of the conversation. Do the remarks make sense? Is this the kind of thing that someone might actually say in such a situation? To be sure, one could theoretically give the first N interchanges of the dialogue, along with the $N+1$st question, to 100 college sophomores and then ask them all to write down how they would reply. For each value of N, all the diverse responses could be collected and tabulated and compared with how the computer program actually replied at the same juncture — but what would such an exercise prove? At each juncture there are millions of plausible answers, and the whole idea of averaging falls apart. (How do you average together 100 different replies that include such things as "I don't know", "The yellow one", "The last one I touched", "I'm not sure what you mean", and "I think it's the one between the two tall ones"?)

In everyday life, we don't judge how well a fellow human being handles language by the degree of stereotypicality and predictability of their replies to our remarks. Rather, our judgments are based on *flexibility* (how fluidly the person reacts to unexpected turns of many different sorts in the conversation), and on *insight* (the degree to which their replies get to the nub of the matter). This is how a natural-language program should be judged, as well.

There is a sense in which we agree with the desire for *some* kind of statistics. Obviously, one single natural-language exchange between a human and a program, no matter how impressive, will not prove anything. It could have all been canned. One needs a great deal of bulk, revealing how the program performs in a wide and open-ended variety of contexts, in order to assess the program's flexibility and thus its validity as a psychological model. The same would go for an analogy-making program.

Some people might ask, "What makes you think that mere *flexibility* over many different contexts is any sort of evidence of the correctness of the underlying psychological mechanisms?" The answer, it seems to us, is that flexibility can come only from the underlying substrate, and therefore if a model is flexible in the same ways as people are, then its substrate must share mechanisms with people. (This thesis of the necessary interrelationship between overt behavior and hidden mechanisms is discussed in greater detail in the Epilogue.)

We would judge a computer model of natural language by watching its performance over a very large number of highly diversified interchanges, with freedom being given to a sophisticated human prober to pose any number of unanticipated situations to the program. As Alan Turing suggested over four deades ago in his famous article on the criteria for recognizing genuine intelligence in an artifact (Turing, 1950), consistent flexibility and insight in response to open-ended probing would be the validation criteria, not some kind of match with human statistics. Why should it be different for analogy-making, which is in many ways on the same level of complexity as use of natural language?

Concrete Criteria for the Validation of Our Computational Models

In the Copycat and Tabletop projects, we have been driven by the following set of qualitative but certainly not quantitative criteria regarding the behavior of the program:

(1) the answers that the program comes up with should almost always seem *plausible* to people (especially answers that crop up with a reasonable frequency — say, more than one time in 100);

(2) the spectrum of answers that people come up with should in large part be *accessible* to the program (in the sense that for a given answer there is some theoretical pathway that could lead to it);

(3) if a given answer seems *obvious* to most people, then it should also be obvious to the program (*i.e.,* it should crop up with high frequency);

(4) if a given answer seems *far-fetched* to most people, then in the model it should crop up with low frequency;

(5) if a given answer seems *elegant but subtle* to most people, then in the model it should crop up with low frequency;

(6) if an answer is considered *ugly* or *silly* by most people, then its quality should be rated low by the program;

(7) if an answer is considered *elegant* by most people, then its quality should be rated high by the program;

(8) the program should occasionally come up with *insightful* or *creative* answers.

Note that these criteria (*e.g.*, is $xyz \Rightarrow xyy$ [or touching Eliza's spoon] a plausible answer? obvious? elegant?) can all be assessed informally in discussions with a few people, without any need for extensive psychological experimentation. None of them involves calculating averages or figuring out rank-orderings from questionnaires filled out by large numbers of people.

The eight validation criteria listed above are already stringent; however, we now list three further criteria that go considerably beyond looking simply at various answers to a fixed problem:

(9) as a situation is gradually "tweaked" in a series of stages that, to people, clearly strengthen the pressures for certain answers and weaken those for others, the program's frequencies and quality judgments for its answers should shift to reflect this tendency;

(10) the most frequent pathways taken to a given answer should seem plausible from a human point of view;

(11) when various key features of the architecture are *disabled* (or "lesioned") by taking them one by one out of the program, the behavior should degrade in ways that reflect the theoretical role that each feature was intended to play.

Although the first two of these refer to judgments by people, such judgments do not need to be discovered by conducting large studies; once again, they can easily be gotten from casual discussions with a handful of friends. For example, concerning criterion (9), it is an obvious fact hardly needing elaborate experimental confirmation that when one changes the category of a given table object so that it becomes further and further removed from the category of the object Henry touched (*e.g.*, Henry touches a glass, and in a series of four problems, a given object starts out as a glass, then becomes a cup, then a saucer, then a knife), a human Eliza will be less and less inclined to touch that object, all other things being equal. To illustrate criterion (10), it is patently clear on an intuitive level that for a human working in the Copycat domain to arrive at the answer $xyz \Rightarrow wyz$ in the problem "$abc \Rightarrow abd$; $xyz \Rightarrow$?" without ever running into the z-snag is highly implausible.

Criterion (11) provides a further and very deep check of the relationship of the architecture to the observed behavior. Here are just a few of dozens of interesting ways of "lesioning" Copycat or Tabletop:

(1) telescope the series of scout and effector codelets for each given type of action into one single codelet, thus making processes far less fine-grained and thereby considerably suppressing the terraced scan;

(2) make all concepts have the same conceptual-depth value (thus effectively abolishing conceptual depth as a motivator);

(3) clamp the temperature at a very high value or a very low value;

(4) prevent link-lengths in the Slipnet from changing at all during a run.

All these and several other modes of lesioning Copycat were tried out by Mitchell, and the results are described in her book (Mitchell, 1993). These were very elegant experiments and certainly deserve careful study, but on the other hand, there was nothing too astonishing in her findings, in the sense that each lesion caused the expected type of serious degradation that it should have. In other words, Mitchell's lesioning experiments helped to confirm the need for the central features of the Copycat/Tabletop architecture.

The Substrate of Cognition is Culture-independent

It has occasionally been suggested that people from other cultures might see Copycat or Tabletop analogy problems very differently from the way people in our culture do, and that this might cast doubt on the universality of the mechanisms in our models. However, this type of objection strikes us as seriously misguided. True, an aborigine from deep in the Outback who had never eaten off a table, let alone seen forks and napkins and such, would hardly respond to Tabletop problems as we do. Similarly, someone who knew only Tibetan would have a radically different way of seeing Copycat analogy problems. But such outlandish examples are simply distractors — they raise a cloud of confusing smoke and thereby divert our attention from where the real issues lie.

The question of interest is, what mechanisms do people use in making analogies concerning domains and situations *familiar* to them? The long but far-from-complete list, given earlier in this preface, of subcognitive mechanisms that play a role in our models shows the level of abstraction that we are exploring. We find it highly implausible, to say the least, that at that high level of abstraction, entirely different mechanisms from our own are engaged when aboriginal hunters or Tibetan monks make analogies. To make such an assumption would be almost as silly as thinking that every physics experiment done in

America must be repeated in Australia and Tibet to check whether the same laws of physics hold there as well.

A Crazy Bazaar

As I expect is clear from the last few pages (as well as from many other sections of this book), one of the deepest problems in AI/cognitive science is that of figuring out universal criteria for judging pieces of research. This is a very murky area. Vastly different types of claims for validity and importance and novelty vie with each other, and often speak completely different languages. AI/cognitive science is trying to understand a phenomenon so complex that it is simply not clear how to judge ideas (it is not even clear in physics or mathematics, and cognitive science is infinitely vaguer than those long-established disciplines).

In a nutshell, AI/cog-sci is a crazy bazaar. Or at the very least, it's a crazy, bizarre discipline. Its participants span an enormous spectrum of backgrounds and are working on the most far-ranging of projects. Just off the top of my head, here are some of the extraordinarily diverse things going on in this field.

There are theoretical physicists who study the properties of neural networks using mathematical techniques of statistical mechanics. There are neurologists who look for the storage-places of concepts in the brain using magnetic-resonance imagery and PET scans. There are cognitive ethologists who observe apes and monkeys in the wild and analyze the nature of their mental representations of their peers. There are music theorists who try to get computers to improvise jazz or compose in the style of Bach. There are cognitive psychologists who make huge collections of slips of the tongue, and then sift through them for cues about the underlying processing in the brain. There are linguists who devise complex grammars to account for abstruse syntactic phenomena in unheard-of languages. There are anthropologists who study how groups of people, by collaborating on a large task, bond together into higher-level cognitive units. There are roboticists who build artificial insects that clamber over rocky moonlike surfaces and cars that drive themselves down city streets. There are artificial-lifers who design fake fish or birds and study how they school or flock. There are philosophers of mind who strive to figure out what, if anything, could imbue symbols, whether in computers or in brains, with "aboutness". There are chess-program designers who build superfast hardware and streamlined algorithms in an attempt to make a machine that will become world champion. There are connectionists who model the perception of faces and the learning of past-tense forms of verbs. There are developmental psychologists who study how infants learn to crawl and walk, and how children learn the concept of number. There are expert-system developers who write programs that not only can diagnose complex eye diseases but also can explain the whole chain of reasoning behind their diagnoses. There are mathematical logicians who prove theorems about metaknowledge and nonmonotonic reasoning. There are genetic-algorithm researchers who make programs that can mate with each other and whose progeny compete for survival by demonstrating their skill in running mazes. There are perceptual psychologists who study the optical illusions to which the human visual system is susceptible. There are philosophers of science who make computer models of the course of scientific revolutions. There are machine-translation researchers who have programs that they claim can translate from Japanese to English with 95 percent accuracy. There are brain researchers who study patients with horrible brain damage or diseases and try to deduce from their tragic deficits the pathways and interconnections on which normal cognition depends. There are computer scientists who write programs that attempt to come up with new fables or to devise sitcom plots. There are AI researchers who write programs that supposedly daydream about dating Hollywood stars or reproduce momentous discoveries by Galileo, Ohm, and Kepler.

This list represents but the tip of the iceberg. I came up with it in a half an hour without consulting any books or journals. With the help of such reference materials, one could easily name hundreds of projects that aim at unraveling the secrets of mind, consciousness, memory, intelligence, creativity, flexibility, versatility, and so on. What in the world to make of all this? How can one judge what's good and what's bad in this hundred-ring circus?

Let me hasten to say that I do not have the answer. In fact, I find it all quite mind-boggling, even disturbing. I occasionally get quite overwhelmed by the enormous diversity and the plethora of impressive claims. Every year, for instance, there is a "Turing Test" in Boston where human judges taken from off the Boston streets frequently mistake programs running on PC's for intelligent people, and intelligent people for PC's. One also hears of programs that compose music that "really sounds like Bach", of books in which consciousness is fully explained, of computers that can supposedly read sloppy handwriting, understand rapid-fire slurred speech, or even write entire novels (of course just potboilers — we haven't quite reached the level of Tolstoy yet).

CYC

A couple of years ago, I heard a lecture by Doug Lenat, well-known for his pioneering projects AM (briefly discussed in the Epilogue; see also Lenat, 1979, 1982) and Eurisko (Lenat, 1983b), which, respectively, involved modeling mathematical discovery by using many heuristics, and the evolution of new heuristics from old ones, carried out by — of course — meta-heuristics. However interesting they once were to him, these projects are now ancient history for Lenat. The topic of his lecture was quite different — namely, his vast CYC project (Guha & Lenat, 1994), upon which he had at that point been working for about nine years, and which he claimed would be done in a mere year or two (right about now, in other words), thus bringing it in for a landing "essentially on schedule, due to a large number of errors canceling each other out", as he humorously put it.

CYC (pronounced as, and related to, the "cyc" in "encyclopedia") involves the encoding of millions upon millions of everyday facts into a uniform representation language based on predicate calculus, in an attempt to imbue a computer with common sense of the kind that any ten-year-old has. The basic message of Lenat's talk was that this type of knowledge-intensity is the only way to go, if you really want to make models of intelligence. Anything less is baby stuff. He said that in his opinion, within twenty years or so, thanks to advances in hardware (both memory size and raw speed), the complexity of programs, and the amount of knowledge available, computers will reach and surpass humans in intelligence level. He seemed perfectly confident and quite happy about this. I was nonplussed that he could say or believe such things, but in private conversation later, he reaffirmed his statements and incidentally remarked that work in microdomains is hopelessly outmoded. It was hard to even know how to reply to this, except to say, "I don't think so."

CBR

Not many months later, I attended another talk, this one given by my colleague and friend David Leake. It was about models of creativity involving case-based reasoning ("CBR"), a field founded by famed AI researcher Roger Schank, and in which Leake himself has worked for many years. Essentially the message conveyed was that CBR models try to understand and explain situations and to devise new ideas by using analogical thinking (although the word "analogy" is conspicuously bypassed, for some reason).

These models (see Riesbeck & Schank, 1989 and Kolodner, 1993) operate in real-world domains of (to me) unbelievable conceptual richness (*e.g.,* speculating about the causes of such events as the Challenger crash and the mysterious death of the racehorse Swale, inventing new ways to combat international terrorism, and many others). They all, to one degree or another, accept descriptions of some very complex and multi-faceted event in one of these domains, search vast memories for an experience that this event "reminds" them of, analogically map the retrieved situation onto the base event, make any number of appropriate conceptual slippages so that the mapping works right (this is called "tweaking"), sometimes splice together pieces of two or more retrieved memories in a kind of frame blend, suggest an explanation of the original event through this tweaked and spliced analogical mapping, and then, to top if off, perform a self-evaluation of the whole result! Now if *this* isn't creative thinking, by God, what is?

I was amazed by the versatility of these programs and the kinds of insights that Leake described them as coming up with. But mixed with this feeling was a considerable degree of bafflement. How could it be that CBR researchers have succeeded in making such stunningly complex cognitive models when I find each little *piece* of what they claim to be doing to be enormously daunting in itself?

During Leake's talk, I was absent-mindedly fiddling with the metal clips I wear while riding my bike to keep my pants legs from getting caught in the chain, and in the question period, I commented, "I just can't relate to the idea of programs knowing all about counterterrorism or shuttle launchings or race-horse deaths — I mean, it seems to me that there's almost an infinite amount of stuff to know just about these little bicycle clips." I then held them up for view while reciting a dozen or so trivial facts about them, attempting to make clear, through such a list, that there were thousands of further trivial facts that could be formulated without coming close to exhausting what people intuitively know about bicycle clips or any everyday object. (These objections are similar to, though not identical to, long-standing objections to the entire discipline of AI raised by Hubert Dreyfus in his 1972 book *What Computers Can't Do*.)

In his reply, Leake admitted that it was true that CBR models don't know too much about physical objects and that in some sense even the most sophisticated of today's CBR models is still really working in a microworld, but he suggested that perhaps if one could combine the reasoning abilities of such models with the vast kind of database that Doug Lenat's group is working on, then maybe one would come close to having a system with genuine intelligence and full understanding. Leake expressed personal skepticism about CYC itself (probably because it is based on predicate calculus and logical inference rather than on Schank's model of how knowledge is organized), but he thought that some kind of hybrid along those lines would turn the trick. This is a very provocative claim. Could it be true? I myself have grave doubts, but how can I know for sure?

Dustbuster

The large school of CBR researchers all take from Schank a fascination with the mysteries of human memory — particularly how *reminding* takes place. I couldn't be more in sympathy with these intellectual goals. It seems to me that the phenomena they are studying do really involve mental fluidity. Let me give a personal example — a reminding experience I had a while ago. It is the kind of thing that lies at the root of my fascination with the mind and that drives my interest in cognitive science.

My daughter Monica, then a bit over a year old, was sitting on our playroom floor, pushing the on–off button of a Dustbuster (a hand-held battery-operated vacuum cleaner), which she loved to do because of the buzzing noise it made. At one point, she noticed a differently-shaped button on a different part of the Dustbuster, so of course she tried pushing that one. Nothing happened. She tried several times, and then gave up. The reason it did nothing was that this was the release button for the lid that holds the trashbag inside the machine, and pushing it does nothing. You have to slide it, and even then, all that happens is that the lid flips open. That was way beyond her, but the feeling of disappointment was not.

When I saw Monica trying that second button and getting nowhere, I went over and showed her what it did. All of a sudden, completely out of the blue, there flashed to my mind an experience from my own childhood. As a child, I always loved mathematics. Something that excited me no end was the operation of exponentiation. I made table after table of squares, cubes, and higher powers of many integers, and I compared the powers and studied their patterns and so on. I was just enchanted by them. One day, when I was about eight, I happened to see one of my father's physics papers lying around on a table in our house, and I looked at the equations. Of course, they were way beyond my grasp, but I did notice a salient feature of the notation: the ubiquitous use of *subscripts*. Of course I knew that *superscripts* represented the beautiful, endlessly deep operation of exponentiation, so I jumped to the conclusion that subscripts, looking so much like superscripts, must likewise represent some kind of marvelously deep mathematical concept, so I asked my father. To my surprise, he said that subscripts were simply used to distinguish one variable from another, and that no arithmetical calculation whatsoever was symbolized by putting a subscript of "3", say, on the letter "x". Thus were dashed my childish hopes of finding some new mathematical treasure.

This was the memory that flashed into my mind when little Monica failed to make a new noise by pushing the second button on the Dustbuster. Monica was me, I was my Dad, the first button was superscripts, the second button was subscripts, the buzzing noise was the thrill of exponentiation, the lack of noise was the meaninglessness of subscripts… When you hear about it, it makes perfect sense — the two events map onto each other very elegantly — fathers, children, disappointment, and all. But how was it that this retrieval occurred? How did the eight-year-old boy store the original memory? How did the adult fish it out, some forty years later, triggered by the event involving his baby daughter?

I am confident that CBR researchers could model this reminding experience fairly easily. They could encode my childhood experience in Schank's conceptual-dependency notation, then likewise encode my perception of Monica's experience, and they could get a program to fish out of a huge memory of diverse experiences just the one about superscripts and subscripts. The key to it all would be the shared memory index code "MS; DH" — "Misleading Similarity; Dashed Hopes" — a little "representational nugget", one might say, symbolizing the common abstraction at the core of both events. (Such representational nuggets, officially called "thematic organization points" — "TOP's" for short — are defined and discussed in Schank, 1982.) And thus would be explained what happened in my mind.

I think that such a model, though it would be impressive and perhaps even correct on some levels, would nonetheless miss the core of the phenomenon. It is hard for me to believe that when I was eight years old, my mind understood the abstraction "Misleading Similarity; Dashed Hopes" and unconsciously stored the memory of my father's explanation of subscripts under that code so that, some forty years later, the event could be, dredged out again. There is something tantalizingly close to this going on, I would guess, but I don't think that explicit representational nuggets of this sort are quite it. I personally feel that Pentti Kanerva's beautiful theory of "sparse distributed memory" (Kanerva, 1988) also comes close, although from the completely different starting point of neural hardware, but still doesn't quite hit the nail on the head. Something in between these two very distant levels of description is needed.

My intuition about what is missing in CBR and CYC, and would even be missing in a hypothetical splice of the two systems' best aspects, is a deep model of *concepts*. I think this lack exists also in Kanerva's theory. For me, a model of concepts implies something like the overlapping, emergent halos that arise in a Copycat- or Tabletop-style Slipnet, with the parallel terraced scan, with fluidly reconformable representations, with commingling bottom-up and top-down pressures, with dynamic probabilistic biases emerging from the interaction of conceptual depths, saliences, and strengths, with perceptual structures having arbitrary shades of tentativity, with computational temperature reflecting perceived order and controlling randomness, with self-watching, and so on. That's my bias.

Soar

Of course not everybody agrees that the nature of concepts is the key mystery to be confronted by cognitive science; indeed, some major researchers even seem to be at the diametrically opposite pole. Take the late Allen Newell, who made many signal contributions to AI and cognitive science, including, in the mid-1950's, the General Problem Solver, and a decade or so later, the notion of production systems, which could be considered a distant ancestor of the

Hearsay II architecture, which so deeply influenced my ideas. He even was a co-developer of a curious analogy-making model called MERLIN (Moore & Newell, 1974). Newell is indisputably one of AI's most famous pioneers, and he spent the last decade or more of his life developing, along with his colleagues Paul Rosenbloom and John Laird, an ambitious cognitive architecture called "Soar". Shortly before his death, the book *Unified Theories of Cognition,* which I suspect Newell considered his *magnum opus,* appeared (Newell, 1990). It was a valiant attempt to describe how Soar integrates virtually every important feature of cognition known. The program's many successes in all sorts of hugely disparate domains were described, and compared in great detail with the results of psychological experiments on human subjects. If you take Newell's word for it, he and his colleagues really have pulled it off — we now know what cognition is all about. The rest is just a matter of filling in details.

And yet I don't see how Soar has anything at all to do with the questions about mind that I find most fascinating and mysterious. I came to this conclusion after a number of encounters with Soar, of which the following one was perhaps the capstone. When Newell's book came out, I was most curious about it, especially given its provocative title. However, since it was very long and since I had already grappled with Soar and not found it richly rewarding, I wasn't about to read this tome from start to finish without having first tried to size it up. (This is of course the principle of the terraced scan.) One way to get a feel for a book is to scan its index, so I tried that. Among the first words I looked up was "concept". To my surprise, I found only one reference to concepts at all, and that entry was "Concept learning, 435". I flipped to that page, which was in a chapter about things Soar hasn't yet done, and the prospects of getting it to do them. There I found one paragraph about concepts — one single paragraph in this 500-page book. The first half of the paragraph cursorily describes the idea of "prototypes" — essentially the idea that membership in categories is not black-and-white but shaded — and then Newell comments:

> Soar must give an account of prototypes and show how they arise when the situations call for them, just as predicates arise when the situations call for them. Here is a place where it would occasion surprise (at least on my part) if additional architectural assumptions were needed to get these different types of concepts, and others as well. That a processing device that is designed, so to speak, to provide the ultimate flexibility in response functions is limited to computing only predicates would seem passing odd.

In other words, this minor aspect of cognition — obviously minor, since it played no role whatsoever in motivating a "unified theory of cognition" — must *surely* be able to be handled by Soar, if somebody ever felt like working on it, because after all, the system already has "the ultimate flexibility".

I found Newell's neglect of the concept of "concept" so stunning — so passing odd, one might say — that I looked elsewhere in his index. Could it be that "concept" was simply the wrong word? I tried "category", but there was nothing there either. There was, admittedly, an entry for "Categorical syllogisms", but when I checked it out, it really was about syllogisms — formal arguments belonging to symbolic logic. This was clearly irrelevant to fluid concepts, at least as I conceive of them. Lastly, I tried "analogy", and again found nothing except "Analogical view of representations", which, it turned out, had to do with the question of whether we have mental pictures or not.

I wound up looking through the book for several hours, reading large sections here and there and, frankly, finding myself quite uninterested by most of the issues Newell discusses. Thus, for instance, in a section devoted to how Soar might read sentences and compare them with pictures of situations, I encountered the following passage:

> If Soar wants to comprehend a sentence, the natural thing for it to do is to invoke a *comprehension operator* [Newell's italics] on the sentence. Since Soar acquires the sentence word by word, the natural thing is to execute a comprehension operator on each word as it arrives.

There follows a brief discussion of what "comprehension operators" do, yet it sounded like very little more than ordinary common sense — basically that a comprehension operator tries to figure out the word's *meaning*. Surprise, surprise! It's not that there's something wrong with this; it's just that it seems extremely prosaic as an explanation of what thought is. It doesn't seem to be grappling with deep issues, but merely with superficia. Moreover, I was struck by the exceedingly simplistic image of reading, according to which it should be modeled by a process of fetching one word at a time in a strictly serial and linear fashion. I couldn't imagine a more wooden, more mechanical image of what reading is.

In Soar, every last aspect of cognition is cast in terms of problem-solving in problem spaces, and it all comes out sounding so neat and rational. But it doesn't seem to address, let alone answer, any of the questions that have intrigued me for the last two decades. Or if I am wrong and Soar somehow does address those issues, as my good friend Daniel Dennett assures me it does, then all I can say is, it certainly is hard to recognize this fact in the book.

Private Hunches versus Public Stances

These are the kinds of things that it is very risky to talk about, especially in print. Informal, anecdotal intuitions and highly personal, even emotional, reactions are things that science is not supposed to be about. But it seems to me that the truth of the matter is that in AI/cognitive science, that is what we all run on. In our technical articles, we dress up our ideas in fancy, starched clothes and pretend that there is rigor everywhere, but in truth, behind the scenes, 90 percent of what is going on is intuition.

To be more precise, if you take any article or book in the field on its own grounds — in other words, if you willingly enter into its unspoken framework (and one is most often unaware that this is happening) — then very likely, you will find it tightly reasoned and internally consistent, and therefore quite convincing. After all, people in our field *do* do experiments carefully and write articles carefully, so there aren't likely to be overt gaping holes in their presentations. To be sure, you may find a few issues here and there to criticize, but these will probably be technical, nitpicky quibbles.

It's only if you somehow manage to step back and see a larger, tacit picture — the project's unspoken framework — that the *important* issues, the areas of *serious* potential disagreement, will emerge. But people who write articles seldom perceive, let alone wish to point out, the many tacit assumptions underlying their work. And surprisingly seldom do others point them out, precisely because these things are very hard to see and very hard to articulate. Doing so would lead to a messy debate about all sorts of ill-defined intangibles. And yet these invisible, unmentioned issues are what, from deep down, guide any research project.

When, at a cognitive-science conference, proponents of competing viewpoints debate in public, they often seem to be talking right past each other, like ships passing in the night. Neither seems to understand what the other is saying, or worse yet, even to grasp the fact that there exists a wide mental chasm between them. This is a peculiar state of affairs.

One would think — I at first did, anyway — that cognitive scientists, of all people, should be experts at recognizing these kinds of communication problems, since, after all, they specialize in thinking about thinking itself. Surely they would know how other people hear what they are saying! Surely they would know how to hear below the surface of what other people are saying! Surely they would know that what matters most of all is to transmit deep imagery, and that words, our best tools in that endeavor, are often slippery and vague and highly personal in meaning — especially abstract ones like "structure", "process", "concept", "representation", "schema", "frame", "syntax", "semantics", "analogy", "prototype", "context", "symbol", "rule", "feature", "model", "pattern", "imagery", "meaning", and dozens more that are constantly bandied about in cognitive science.

But alas, this turns out to be a very naïve assumption: the field of cognitive science, no less than any other intellectual field, is full of people profoundly misinterpreting each other's phrases and images, unconsciously sliding and slipping between different meanings of words, making sloppy analogies and fundamental mistakes in reasoning, drawing meaningless or incomprehensible diagrams, and so on. Yet almost everyone puts on a no-nonsense face of scientific rigor, often displaying impressive tables and graphs and so on, trying to prove how realistic and indisputable their results are. This is fine, even necessary — we do it to some extent in this book — but the problem is that this *façade* is never lowered, so that one never is allowed to see the ill-founded swamps of intuition on which all this "rigor" is based.

This whole situation reminds me of the mighty controversy that swirled about President Reagan's "Star Wars" (or Strategic Defense Initiative) proposal nearly a decade ago. People of all stripes were advancing arguments, *pro* and *con*, of all sorts, most of which went deeply into one or another highly technical issue: throw-weights, interceptor velocities, particle beams, X-ray lasers, kinetic-energy weapons, fault-tolerant computing, electromagnetic pulses, missile-silo hardening, and a myriad others. All these virtuosic, knowledge-intensive arguments were very intimidating to a non-expert like me — how could an outsider presume to evaluate *any* of them, let alone *all* of them?

And yet somehow, non-expert though I was, I got into the fray by writing a "My Turn" piece for *Newsweek* magazine (Hofstadter, 1986), in which I sidestepped all these technical issues and merely relied instead on rather simple, even childlike intuitions — intuitions such as the analogy between the dream of an infallible missile defense and the obviously absurd dream of an absolutely crashproof airplane. No matter what technical arguments might be advanced for a supposedly crashproof airplane, one simply knows, *a priori*, without even hearing the arguments, that they must be flawed, for no one can infallibly anticipate all possible eventualities in advance — especially if there's somebody just as smart as you, working just as hard as you, doing their damnedest to *make* your plane crash!

To me, it was as plain as day that the Russians would not just stand idly by as we developed our "peace shield" against their potential invasion, but rather, would constantly be changing all sorts of elements in their strategy, thus constituting a "moving target" — indeed, a target moving with a vengeance. Not only would they be frantically at work trying to come up with ways of tricking our detection systems by putting up dozens or even hundreds of decoy missiles for every real one, but also they would be infiltrating our laboratories with spies who could report home on just what we were developing, and who could even insert subtle bugs into the computer code being developed. Can a frozen immune system, no matter how well it may work in today's microbial environment, be guaranteed to fend off all future attackers forevermore? Evolution sees to it that every chink in the armor of an immune system is explored and exploited by constantly novel types of invaders, and it would be just the same with any "peace shield".

My strongest anti-SDI arguments thus had no technical expertise or "iron-clad logic" behind them at all. Rather, my personal convictions came from analogies. Indeed, had I been forced to debate throw-weights or interceptor velocities with some star-spangled general in full military regalia flipping through one fancy graph after another, I would certainly have gone down in smoke and flames; I knew nothing of those things. But such a media show would have been entirely beside the point. The point was, one had to avoid getting trapped in a debate on technical issues, because the *true* arguments about SDI were not at all rigorous, virtuosic, or knowledge-intensive — they lay at a much more down-to-earth, even nonverbal, level. At this level, a thorough search for good analogies, not a long chain of logical arguments, provided the only way to come to grips with and find the essence of the issues.

Of course, SDI is only one of thousands of such political debates in which virtuosity and posturing on easily-discussed issues take the place of serious confrontation of the deeper, harder issues. This whole idea was gotten across excellently by Jack Valenti, a former assistant to Lyndon Johnson, writing on the Op-Ed page of the *New York Times* in November of 1992, about the many difficulties that newly-elected President Clinton would soon be facing. Valenti wrote:

> Bill Clinton will find that no decision during his tenure will ever be made with enough information. Ever. As he ponders a decision, he will first walk down a corridor clearly mapped with facts, fiscal arithmetic, precedents, and data ground out by computers. Then the corridor will grow darker as he nears the outer rim of decision time. Soon the corridor is unlighted, uninviting, no guideposts, no arithmetic, no further facts, no alternatives weighted with data.
>
> But, as L.B.J. used to say, at 9 A.M. the next morning the President must decide. What does he do? He must call on intuition — the stuff of political judgment and the final arbiter of a President's place in history.

And so it is in cognitive science, I am afraid. All research must be based on intuitions that are very hard to articulate, let alone to defend in a debate (and don't even think of trying to defend them rigorously!). Though few seem to recognize or admit it, ours is a field still searching for its foundations.

I have felt compelled to spell out some of these messy issues, because I am convinced that such things badly need airing in this complex, amazingly speculative field. There are so many competing claims and points of view that no one could ever hope to understand them all, and yet each of us has to struggle to sort good from bad, deep from shallow, right from wrong, promising from unpromising, and so on. Perhaps "confessions" like those just expressed can serve in a small way to help prompt further candid self-examinations by the field.

Chapter 9

The Emergent Personality of Tabletop, a Perception-based Model of Analogy-making

DOUGLAS HOFSTADTER and ROBERT FRENCH

Striving for Cognitive Generality in a Mundane Microdomain

Tabletop, the computer program described in this paper, is a model of analogy-making. Unlike many analogy programs, it does not attempt to discover analogies between political situations, between scientific phenomena, or between literary plots. Rather, it operates in a restricted version of an everyday domain: that of simple place-settings on a small table.

The Tabletop project is based on the fundamental premise that analogy-making is not a fancy add-on to basic cognition, but comes as a standard feature — in fact, is an automatic by-product of high-level perception. Thus the program scans situations on the table and builds representations of them, and, as a natural part of this process, analogies simply fall out. The cognitive activity of scrutinizing tabletop situations falls somewhere between the purely perceptual and the purely abstract levels of thought. This intermediate quality of the domain is one of its virtues, since it helps clarify the inseparability of perception and analogy-making.

Imagine two people, Henry and Eliza, facing each other across a table. Henry touches some object and says to Eliza, "Do this!" She must respond by touching some object. The program plays the role of Eliza, with *selection* playing the role of touching. One obvious possibility for Eliza would be to "touch" literally the same object as Henry touched. Such an option is always open, no matter what the configuration and no matter what object Henry touches. More often, though, there are perceived aspects of the situation — *pressures* — that tend to make the literal-sameness option less appealing than touching some other object. The possibility of any number of pressures coexisting, and their often subtle interactions, lend the domain considerable complexity and depth.

For instance, suppose both individuals have coffee cups before them. Most people would perceive the two cups, even if only unconsciously, as *counterparts*. Thus, if Henry touches *his* cup, it would probably seem more natural for Eliza to touch *her* cup than to reach across the table to touch his. But now suppose the "counterparthood" is weakened by changing Eliza's cup to a glass (as in Figure IX–2*a*, page 396). Here, although the two objects remain counterparts in terms of their *positions*, their *categories* no longer match exactly. However, since "cup" and "glass" are closely related categories, there remains a strong pressure for Eliza to see her glass and Henry's cup as counterparts. Chances are good that Eliza will rank touching her glass higher than touching his cup. Of course, if the category mismatch is further increased — give Eliza a fork, not a glass — the sense of counterparthood will be so diminished that Eliza may well revert to the literal-sameness option (touching Henry's cup).

The types of pressures that might influence Eliza's choice include:

- the nature and location of the specific object that Henry touched;
- the category memberships of the other objects;
- the physical locations of objects;
- the "ownership" of objects;
- the orientations of objects;
- the sizes of objects;
- common functional associations between objects;
- perceptual groupings on one or more hierarchical levels.

By "common functional association", we mean such things as the fact that cups and saucers often go together, or that knives and forks are often used together. Such inter-category relationships are crucial aspects of individual categories, reminding us that categories are not isolatable structures but rather are deeply interdependent, overlapping structures whose identities are mutually determined.

Perceptual groupings play a major role in determining what items Eliza is prone to perceive as counterparts. The plausibility of seeing a set of objects as constituting a single higher-level entity is a subtle matter, and depends on both the physical and the conceptual proximity of the items involved. Obviously, for reasons of efficiency, not all theoretically possible groupings can be considered by a person or by a program — after all, with just a handful of objects on the table, there are hundreds of potential ways of grouping them. Moreover, if small perceptual chunks are allowed to be members of larger chunks (a key feature of human perception, which routinely builds up such hierarchical representations), then the number of potential chunks is far higher.

Of course, not only efficiency but cognitive plausibility militates strongly against a computer model in which brute-force strategies of any sort play any role. Thus a critical design philosophy of the Tabletop program is that it does not routinely invoke all possible pressures in each situation; rather, it lets a limited number of context-dependent pressures *emerge* — and to different degrees — as each situation is perceptually processed. The central challenge of the Tabletop project is to model the way in which a complex situation selectively evokes, in a human mind, many unrelated pressures of different intensities, whose interaction over time gives rise to an emergent collection of perceptual and conceptual structures, ranging from simple judgments about the relative importance of simple objects, to the building of a plausible group out of several related objects or a plausible correspondence between two objects, all the way up to the most complex structural analogies, involving a set of mutually-reinforcing correspondences between a number of different hierarchically structured groups.

We contend that the Tabletop program should be judged not only on the accuracy with which it mimics human performance in its narrow domain (we have carried out some psychological experiments along these lines), but also — in fact, even more so — on its general principles, which were intended to apply to *any* domain, irrespective of size. For example, the Copycat program, Tabletop's forerunner, uses a similar architecture to carry out perception and analogy-making in a completely unrelated microdomain. (See Chapter 5 of this book and Mitchell, 1993 for detailed presentations and discussions of Copycat.)

Perceptual Processes in Tabletop

As was stated above, our philosophy is that analogies emerge automatically as a result of the high-level perception of situations (a theme elaborated further in Chapter 4 of this book). This philosophy is the essence of Tabletop. However, we would not want to convey the impression that the project involves modeling visual perception from the ground up. In fact, analogy problems on the tabletop are communicated to the program by a graphical interface whereby a user drags highly stylized icons around on a computer screen until satisfied with the configuration. At that point, the coordinates and orientation of each placed object, along with its category membership, are conveyed to the program, and that, along with an indication of which object was touched by Henry, is what constitutes the raw input. Thus low-level object perception is completely bypassed. Nonetheless, a great deal of high-level perception of the tabletop situation remains to be done.

In particular, "high-level perception" means the carrying-out of all the following activities (many of which must be carried out partially or wholly concurrently, because they are highly interdependent):

- labeling of table objects in terms of surface-level local attributes (*e.g.*, size, sharpness, open versus closed, liquid-holder, food-holder, receptacle, etc.);
- further labeling of table objects in terms of perceived relational properties and groupings (*e.g.*, neighbor relations, functional associations, relative sizes and orientations, position inside group, position relative to salient objects, etc.);
- the hierarchical building-up of *groups* (tentative perceptual chunks) on the basis of:
  * physical proximity of component items (*i.e.*, table objects or already-built groups);
  * conceptual proximity of descriptions of component items;
  * structural similarity of component subgroups;
  * prior existence of similar groups;
- the building-up of *correspondences* (links between two items establishing them tentatively as each other's counterparts) on the basis of:
  * corresponding physical positions of the two items concerned;
  * conceptual proximity of descriptions of the two items concerned;
  * structural similarity of the two items concerned (if they are groups);
  * prior existence of similar correspondences;
- the assignment of a time-varying *salience* to each perceived item (object, group, or correspondence);
- the assignment of a time-varying *strength* to each perceived group or correspondence;
- competition among rival perceptual structures, giving rise to a pruning-out of weaker structures and of structures that do not cohere with one another.

These activities all involve the creation, destruction, or modification of temporary representational structures. All such structures reside in the *Workspace,* which is a kind of short-term perceptual memory.

Complementary to the Workspace is the long-term memory called the *Slipnet,* a network of permanent nodes joined by permanent associative links.

Each node serves as the nucleus of a particular concept, and the surrounding "halo" of other nearby nodes constitutes the rest of the concept. The Slipnet is the repository of all information about concepts, including their *proximities,* their levels of *activation* (activation being essentially an indicator of perceived relevance), their preassigned *depths* (depth being essentially a measure of abstraction and generality), and so on.

The 47 concepts in Tabletop's Slipnet are as follows: *fork, knife, spoon, cup, saucer, plate, big glass, small glass, receptacle, salt shaker, pepper shaker, soup bowl, silverware, crockery, liquid holder, food holder, sharp object, open object, closed object, object-that-goes-in-mouth, object-that-is-eaten-from, used-together, neighbor, bigger than, smaller than, similar-shape, group, number, one, two, three, many, above, below, horizontal, vertical, left, right, direction, position, middle, end, symmetry, diagonal symmetry, mirror symmetry, same, opposite.* These concepts are potential descriptors, both concrete and abstract, of individual objects, of relations between objects, of pairs of objects, of groups of objects (or of groups of groups), and of correspondences between objects (or between groups). There is of course nothing sacrosanct about this exact set of concepts. Many others could easily be added, but such additions would require no changes to the architecture of the program.

During a run, nothing is built or destroyed in the Slipnet, but there is activity of another sort: the waxing and waning of conceptual activations, and the concomitant stretching and shrinking of conceptual distances (conceptual depths do not change). Thus the following list of Slipnet activities completes the characterization of high-level perception in Tabletop:

- the sending of a jolt of activation to a concept, at the moment an instance of it is perceived in the Workspace;
- the gradual decay of activation of any concept, at a rate determined by its depth;
- the flow of activation from any concept to its immediate neighbors, gated by conceptual proximity;
- the adjustment of interconceptual distances in accordance with the activation levels of certain key concepts.

As was said earlier, these diverse types of perceptual activity in the Workspace and Slipnet are not carried out serially; rather, each type of activity occurs at the same time as many other activities. In other words, the architecture is deeply parallel. It is also *probabilistic,* with biased decisions of all sorts being made quickly with the aid of a random-number generator. The biases pushing for one choice or another are constantly changing, and come from many different sources (to be discussed in a later section).

Structures, Strengths, and Fights for Survival

Since correspondences (often referred to as "bridges" in articles about Copycat) are the building blocks out of which full analogies are made, we give here a few words about how they come into being and what they represent. A correspondence between two simple table objects is likely to be built if their categories are identical, or are conceptually close (as measured in the Slipnet). In the case of conceptual proximity but not identity, the forgiven mismatch is called a *conceptual slippage.* Conceptual slippages come in various types, since conceptual proximity comes about for many different reasons (*e.g.,* objects looking similar, objects having similar functions, objects being used together, and so on).

Another factor that enhances the likelihood of building a correspondence between two items is if the items are situated in mirror-image or diagonally opposite locations on the table, or nearly so. Also, if the two items are not table objects but groups, then either having similar items for members or having similar abstract structures counts in favor of building the correspondence. Thus in sum, a correspondence built between two items symbolizes the program's commitment, at least tentatively, to a view of those items as each other's *counterparts,* and along with the correspondence itself are stored the specific reasons underpinning that vision of "counterparthood".

Every correspondence, once built, is assigned a *strength* that reflects, metaphorically speaking, the "comfort" that the program feels in "equating" two items that, of course, are not equal to each other. A correspondence's strength is not fixed but is constantly being re-evaluated in light of the latest perceptual events. For example, the building of a new correspondence that resembles an old one, especially if the two are near each other on the table, enhances the strength of the old one. On the other hand, the building of a very different kind of correspondence right next to a prior one tends to weaken the older one. Thus a once-shaky correspondence can gradually become solidly entrenched, and conversely, a once-powerful correspondence can grow feeble and endangered. When a rival correspondence is proposed, the existing one and the upstart must "fight", and the greater the discrepancy between their strengths, the more likely the stronger one is to win the fight.

Perceptual groups, like correspondences, are assigned strengths when they are made, and those strengths likewise undergo constant re-evaluation in light of recent perceptions. The kinds of factors that affect a group's strength include the following, some of which are fairly obvious, others of which are rather subtle:

- its physical dimensions;
- number of constituent items;

- uniformity of composition;
- distance between constituent items;
- the prior existence of similar groups.

Of course, a proposed perceptual group will often be incompatible with an already-existing group (*i.e.*, it would lay claim to items already "taken" by the pre-existing group), and thus the two will be in implicit competition. As with correspondences, a fight will sooner or later take place, in which the probability of victory of either competitor is determined by its current strength relative to its rival's.

In sum, Tabletop's perceptual processing is a rough-and-tumble contest among conflicting interpretations (often just fragmentary and local), the outcome of which, in the end, is hopefully a strong set of mutually-reinforcing global-level perceptual structures.

Incidentally, conceptual slippages are the core of analogy-making, and as such, they play a curiously ambiguous role. On the one hand, when the making of an analogy between two given situations requires a slippage to take place, the analogy is obviously somewhat weaker than if the two aspects had been identical. On the other hand, it is precisely the various discrepancies between the two situations that make an analogy interesting, subtle, and intellectually powerful. Thus in every crisp and strong analogy, there is certainly *some* conceptual slippage, but there cannot be too much because then the analogy would start to become sloppy and weak. Yet sometimes, two "parallel" slippages — slippages of the same type — can be better than just one, because they reinforce each other. (This principle is brought out especially clearly by the analogical dilemmas featured in Figures IX–3*d* and IX–3*e*, page 399.) Thus it is not by counting slippages, or by summing the conceptual distances in all the slippages, or by any technique nearly that simple that one can measure the quality or depth or interest of an analogy.

For reasons like this, finding the precise balance point where the amount of conceptual slippage is optimal is a perceptual act deeply rooted in esthetics, and, in our opinion, only a cognitive architecture such as Tabletop's or Copycat's, specifically designed to handle arbitrary sets of interacting pressures, many of which come to light only as the understanding of the situation develops (and therefore are unanticipatable), can accomplish this task. This brings us to the question of how Tabletop can handle many pressures at once, including ones that arise dynamically.

Codelets and the Interleaving of Perceptual Processes

A key idea of the architecture is that each type of perceptual activity is implemented as a sequence of small, independent *codelets*. Building a group, for instance, involves an escalating series of "micro-tests" that check the physical and conceptual distances between prospective group members, as well as the compatibility of the prospective group with other already-established groups. If any such test fails, the potential group is aborted; if it succeeds, the way is clear for further tests to be carried out. If all these prerequisite hurdles are cleared, the group gets officially created by a specific codelet.

To carry out all the above high-level perceptual activities all over the table requires a very large number of codelets, which are interleaved stochastically so as to bring about parallelism of higher-level processes. For instance, one micro-test checking out the attractiveness of some potential group on Eliza's side might run, followed at random by another codelet that proposes attaching an abstract label to some object on Henry's side, followed at random by another micro-test that checks out some other aspect of Eliza's potential group, followed at random by a codelet that tests some aspect of a proposed correspondence elsewhere on the table, and so on. In sum, many different sorts of small things happen, one after another, at different places all over the table. Through this kind of dense interleaving of many scattered micro-actions, *large-scale perceptual structures gradually emerge in parallel in different areas all around the table.* And because all these processes exert influences on one another, there is a strong tendency for the perceptual structures that they build to form conceptually coherent sets.

We emphasize that the Tabletop program runs on a serial computer, so that its smallest actions — its codelets — do not run in parallel with one another. We nonetheless feel perfectly justified in calling Tabletop a "parallel architecture" because to our minds, it is most easily pictured as consisting of simultaneously occurring, mutually influencing *perceptual activities,* which take place, of course, over time-scales much longer than the duration of a single codelet. Tabletop's brand of parallelism is thus *effective* parallelism, somewhat like classical time-sharing on a mainframe computer, rather than *hardware* parallelism.

The Pervasiveness of Dynamic Perceptual Biases

A completely unbiased random selection of codelets would result in all large-scale activities getting carried out, on average, at the same speed — a completely "fair" division of perceptual attention among objects all over the table, and among processes of all types. However, such perceptual egalitarianism is far from the strategy employed in Tabletop. On the contrary, Tabletop accelerates avenues of exploration that offer promise, while retarding ones that appear uninteresting. This strategy, first developed in the Jumbo project and continued in related projects, is known as the *parallel terraced scan.*

For instance, Tabletop is not equally likely to inspect all objects on the table; at any given moment, probabilities strongly bias its choice of what to look at. Metaphorically speaking, certain objects and areas of the table are perceptually "hot" while others are "cool", and these biases are dynamic: they change as new perceptions are made. What makes an item — an object or group, that is — salient? Many factors are involved, some of the main ones being:

- the item's location relative to the touched object;
- the item's conceptual proximity to the touched object;
- the current level of activation of the object's category;
- the item's being a group (as opposed to a primitive object);
- if the item is a group, its size and its type;
- if the item is inside a group, its physical position inside the group;
- prior perception of other items having similar attributes;
- the item's already having a counterpart.

Obviously, some of these factors vary over time, so that saliences also change, which implies that various areas of the table become stronger or weaker probabilistic foci of the program's attention.

Not only the saliences of objects and groups but also the strengths of correspondences (described above) play a role in dynamically determining an area's (probabilistic) perceptual attractiveness.

Tabletop's expressly nonegalitarian brand of parallel processing is realized by assigning each codelet an *urgency* — effectively its probability of being chosen next. The urgency assigned to a given codelet is a function of its action's perceived promise. Obviously, the more promising the action, the higher the codelet's urgency should be. For instance, a proposed codelet involving a salient object would be given a higher urgency than one involving an object of low salience (all other things being equal).

In the assignment of urgency to a codelet, numerous factors are taken into account, such as:

- the type of action that the codelet is supposed to take;
- the location on the table where the proposed action will occur;
- the types of items that the proposed action will involve;
- the strengths and/or saliences of the objects or correspondences involved;

and so on. Since codelets having high urgencies tend to be chosen swiftly, sequences of logically related high-urgency codelets will tend to be accelerated, while sequences of low-urgency codelets will tend to be slowed down. In this way, different perceptual structures emerge naturally at different speeds, depending on the program's best *a priori* estimate of their significance.

The Gradual Emergence of Coherence

At the start of a run, the program's sole focus of interest is the object touched by Henry — its nature and its location. All the other table objects are, of course, in its visual field and can potentially be scanned and paid attention to, but in a very real sense, the program knows nothing about them. In other words, the input data inside the computer's memory representing the categories, positions, and orientations of all the table objects should *not* be thought of as what the program is aware of, but rather as the physical situation itself, or perhaps as extremely low-level visual information (something like retinal information) about the physical situation. Tabletop, like a creature in a complex real-life visual environment, cannot possibly attend to all the details before its "eyes" at once, and in fact ignores everything except the most salient features.

The first codelets that run are thus inspired by the touching-action, and they scan the table in a biased manner, giving probabilistic preference to certain areas of the table. For example, there are codelets that look directly across the table from the object Henry touched to see what, if anything, is there. Other codelets look diagonally across the table, and then other codelets will look in the immediate neighborhood of the touched object. All these codelets can be thought of as *scouts* probabilistically scanning the table for information such as what objects are in interesting areas, what objects are neighbors of interesting objects, what objects might potentially be grouped together into chunks, and what objects resemble each other.

As objects are found and identified as being of interest (*e.g.*, an object of the same category as the touched object, a neighbor of the touched object, etc.), the diffuse searchlight of attention wanders around the table in a semi-purposive manner and many items are at least given the once-over. But there is nothing resembling a systematic or patterned scouring of the table — for example, from the upper left to the lower right corner of the table. It is quite plausible, therefore, that some areas of the table will be left nearly totally unexplored, or explored on a superficial level and then essentially abandoned, out of lack of interest.

The neighborhoods of objects considered interesting fill up rapidly with perceptual information, so that recognition of relationships between objects, followed by the building of groups of objects, is likely to take place there earliest. Once any group is built, the program devotes resources to looking for identical or similar groups in other areas of the table.

In this way, directed or *top-down* perceptual activities are engendered by the results of observations made in a less directed, more *bottom-up* manner. However, there is no sharp or absolute distinction between bottom-up and top-down processing in Tabletop, since every small action is subject to many probabilistic biases, and each type of bias — the salience of an object, the strength of a structure, the urgency of a codelet, the attractiveness level of a table area — is computed by an algebraic formula that takes into account pressures of all sorts, whether they are bottom-up or top-down. (The exact formulas are available in French, 1995.)

Tabletop Gauges its Own Progress

Gradually, a coherent set of perceptual structures emerges in the Workspace; this is called the *Worldview* (this notion is discussed further in the section comparing the architectures of Tabletop and Copycat). Of course, there are many possible ways of perceiving a tabletop configuration, with some being full and rich, others sparse and simple, some being tightly coherent, others jumbled and confused. It is crucial that at all times during a run, Tabletop itself have a sense for how its Worldview is developing along these dimensions — a sense for how well it is currently doing in making sense of the situation before it. To this end, there is a dynamically varying number called the *structure-value,* which is intended to summarize the total strength of the current Worldview.

This all-important number is essentially a sum of the individual strengths of all the structures in the Worldview, with penalties for incoherence. By "incoherence", we mean a situation in which there are two or more correspondences emanating from within a single group (that is, from different items inside the group) but terminating in unrelated spots on the table. To put this positively instead of negatively, when correspondences reinforce each other by linking two groups together on an element-by-element basis, there is a "reward" to the Worldview in the sense that its structure-value is increased. Moreover, when two or more correspondences are supported by the same type of conceptual slippage, there is also a reward. These two ways of rewarding Worldviews for coherence effectively encourage tightly linked *systems* of correspondences to arise and survive. Altogether, then, structure-value is a constantly changing number reflecting *how much structure* has been found, and *how well knitted-together* that structure is. It thus gives Tabletop a sense of how it is doing, perceptually.

At the end of a run, the value of this number can serve a useful secondary function — namely, that of being a "quality measure" of the answer produced by the program. Metaphorically, we think of the structure-value at the end of a run as the program's way of telling us how well it "likes" the answer it has come up with. An important pressure on Tabletop is therefore to *maximize structure-value*; counterbalancing this, however, is a competing pressure — time pressure — that pushes for the program to finish within a reasonable amount of time.

As perceptual structures build up in parallel around the table, *mappings* emerge as well. After all, a mapping is simply a special, quite global, type of perceptual structure: a family of one or more mutually compatible (often mutually reinforcing) correspondences. Such a mapping, needless to say, is an analogy. This brings us back to the basic premise of Tabletop — namely, that analogy-making is simply a high-level by-product of perception. In other words, analogies represent the highest (*i.e.*, most abstract, most global) level of perception.

We feel, therefore, that it is not an exaggeration to describe Tabletop as *a model of high-level vision*. Of course, the raw input to Tabletop must be thought of as being the output of some prior module that carries out visual processing at a lower (and more modality-specific) level. Tabletop is certainly not a model of all of vision, but of vision's high end — the end that interfaces with *concepts* at various levels of abstraction.

Some Differences between Tabletop's and Copycat's Domains

The alphabetic microdomain of the Copycat project, because of its inherent simplicity and elegance, lends itself to the design of problems featuring symmetry, uniformity, and other sorts of precise, almost mathematical, patterns. Of course Copycat problems need not exude this flavor, but of the ones that have been concentrated on, a large fraction do. In the Tabletop domain, by contrast, such notions play only a small role, and consequently the problems tend to be considerably "mushier" and vaguer. The clear sense of elegance and "rightness" that so frequently accompanies certain answers in Copycat is much less frequent in the Tabletop domain.

Of course, one could artificially design tabletop configurations featuring geometric symmetries of various sorts, and thereby imitate, after a fashion, the precision of Copycat problems, but that would be unnatural and would even run against the spirit of the project, which is to remind people of a realistic tabletop in a coffeehouse or restaurant. After all, the Tabletop project did, in fact, spring from discussions over coffee or a meal, where analogy problems aplenty were right there, sitting on the table before both participants' eyes. All it took was to recognize their existence!

One interesting difference between Copycat and Tabletop analogy problems is that in Copycat, there are always two completely disjoint situations between which an analogy is to be made, whereas in Tabletop, what the program "sees" is just an arrangement of various objects on a table. There is no pre-set boundary line dividing this arrangement into two disjoint "situations". Instead, there are simply two different observers, Henry and Eliza, each with their own point of view of the set-up, each free to carve the table up as seems appropriate. Of course, it will often turn out that there is an explicit or implicit dividing line cutting the table in two, approximately halfway between the two people, but such a membrane, even if constructed, remains partially permeable, in the sense that either party can reach across it and touch objects belonging to the "foreign" situation. No such thing can be done in Copycat; it would not even make any sense to modify the *initial* string to produce one's answer.

The Tabletop project's openness and ambiguity about where one situation leaves off and another one starts is typical of real life, where situations do not come crisply and neatly packaged, but rather have to be carved out of a background by active perceptual agents, who may of course differ widely in how they try to do it.

Such "conceptual slushiness" is a pervasive feature of the Tabletop project. Whereas Copycat's repertoire of Platonic concepts features at its core the alphabet, a perfectly crystalline and symmetric chain of 26 ideal concepts, nothing of the sort exists in Tabletop. The closest one could come to a Platonic "alphabet" — a Platobet, so to speak — is the trio of utensil types: "fork", "knife", and "spoon". But that's a far cry from a crystalline pattern; these concepts have no particular standard order, and even their mutual associations are asymmetric: forks and knives are often *used together* in eating, and forks and spoons share the property that they can *hold food*. Spoons and knives, aside from being silverware, share only the curious and much less meaningful connection of *sitting together* on the same side of a plate in a stereotypical place-setting. The rest of the Platobet of Tabletop — its Platonic repertoire — is a similar hodgepodge of concepts with vague, blurry interconnections. There are many arenas of real life whose conceptual repertoire is full of such arbitrariness and disorganization, as opposed to Copycat's more idealized and pristine set of concepts.

One last notable difference between the two projects is that since Eliza's only task is to *touch* some object rather than to perform some transformation of it, the number of plausible answers to a Tabletop problem is often much smaller than the number of plausible answers to a Copycat problem, since Copycat problems always involve transformations. It would be an interesting extension to Tabletop to allow Henry and Eliza to do more than just touch an object — for instance, to pick an object up, to turn it around, to place it on or near or inside another object, and so forth. A program capable of imitating such physical actions would be more similar in spirit to Copycat.

Some Differences between Tabletop's and Copycat's Architectures

Described in broad brushstrokes, the architectures of Copycat and Tabletop are the same. However, because of the above-described differences in the domains, certain issues are naturally highlighted more by one domain than by the other, and this fact led us to concentrating attention on different sorts of mechanisms. The architectures of the two projects therefore have some interesting discrepancies, when they are scrutinized on a detailed level. Here are some of the more noteworthy differences.

In Copycat, there is no "punctuation mark" or other special symbol that might serve to suggest standardized boundaries of perceptual groups. Therefore, the way that perceptual groupings of neighboring letters come about is on the basis of some type of *uniformity* of a neighborhood, such as is created by a set of adjacent sameness or successorship bonds. In Tabletop, perceptual groupings are quite different, largely because of the existence of something so simple and obvious as *empty space.* The presence of objects in a region of the table, and then the lack of objects in an adjacent region, is in itself evidence that the former objects "belong together", in some sense, whether or not they share any *conceptual* properties. In other words, to the Tabletop program, sheer physical proximity is often a sufficient reason to form a group — a *proximity group,* as we call it — whereas to the Copycat program, physical proximity alone can never justify the building of a group of letters. (Actually, many appealing Copycat analogy problems have been formulated in which precisely such proximity groups play a central role, but such problems are unfortunately beyond the reach of the current implementation of Copycat.)

Moreover, Tabletop often builds proximity groups out of objects it hasn't yet even looked at individually. In other words, Tabletop will often go ahead and build a group, and only after doing so will it scan it to see what sorts of "fish" it has caught in its "net". It is thus possible for Tabletop to build nested groups *from the outside in* (*i.e.,* moving from quickly-perceived larger groups to smaller groups contained in the former, and so on hierarchically *downwards*), whereas Copycat always builds groups *from the inside out* (*i.e.,* moving from quickly-perceived smaller groups to larger groups containing the former, and so on hierarchically *upwards*). In fact, Tabletop can even make groups through a mixture of outside-in and inside-out processing.

A second interesting strategic difference between the programs is the way they treat the loser of a competition between two rival structures. In Copycat, the losing structure is simply destroyed, leaving no trace behind. In Tabletop, by contrast, a loser is merely *demoted,* in the sense of being removed from the special "elite" area inside the Workspace known as the *Worldview.* The Worldview culls from the contents of the Workspace all those perceptual structures that are currently considered the best, operating under the constraint, of course, that the selected items all be mutually compatible (*i.e.,* groups do not overlap, objects have at most one counterpart, meaning that no object is the endpoint of two or more correspondences, and so on). Thus at worst, the Worldview is a non-self-contradictory subset of the Workspace, and at best it is a highly coherent subset of the Workspace; on the other hand, the rest of the Workspace need not be so organized. Thus, many perceptual fragments are free to float around in it without necessarily fitting into any larger perceptual whole.

This way of doing things is advantageous in the following sense. If an upheaval takes place in the Worldview, uprooting some established viewpoint (*i.e.*, set of compatible perceptual structures), then new perceptual ideas are needed, and luckily, many such structures are already available elsewhere in the Workspace. In Copycat, old ideas do not hang around in this way for potential reuse, and so they must be regenerated *ab ovo* if they are ever to be used again. To use a political analogy, Tabletop keeps around a "shadow cabinet" (or at least fragments of one — in fact, fragments of any number of them), so that if the dominant party ever loses power, its rivals can quickly step out of the shadows and try to grab power.

There is an elegant time-symmetry to this idea of "demoting" losers in a battle, rather than casting them out forever. For a perceptual structure to enter the elite Worldview, it has to have risen up through the ranks by passing a series of graded micro-tests. In other words, a perceptual structure is gradually "promoted" upwards, and with luck eventually attains the inner circle of the Worldview. It seems only fair, then, that if a structure at this level chances to lose to a rival in a battle, its gradual rise to the top should be gently reversed rather than out-and-out court-martialing it and sending it mercilessly to Siberia.

There are many other detailed architectural differences between Tabletop and Copycat (both French, 1995 and Mitchell, 1993 have sections devoted to such comparisons), but the foregoing are the most interesting at the overview level.

Tabletop's Personality as a Statistically Emergent Phenomenon

Because of its underlying stochastic nature, Tabletop follows different pathways on different runs, and thus often comes up with different answers on different runs. This means that to get a feel for the program's overall behavior, one must run it not only on many different problems, but *many times on each given problem*. Only thus can one gain a clear perspective on how different combinations of pressures "pull" the program. Since the heart of the model is its ability to handle multiple interacting pressures, this is a key test.

We probed Tabletop's personality by running it many times on a great variety of configurations. Inevitably, once any problem was devised, several close variants would spring to mind, in which the altered pressures would alter the appeal, to humans, of various answers. By testing Tabletop on such tightly interrelated *families* of problems, we learned how it responds to many interesting and often strange combinations of pressures.

In advance of these tests, we were unsure how Tabletop would perform. Over the years of development of the Tabletop program, we had, of course, watched a good number of runs of Tabletop, each one characterized by its own idiosyncratic set of random micro-choices, and we had adjusted the many mechanisms and parameters of the program so that it would tend to follow reasonable pathways through the enormous space of possible exploration routes. However, it is one thing to watch isolated runs of a nondeterministic program, and another thing to collect statistics on vast numbers of runs. We had no idea what overall pattern of preferences would emerge after many runs on a given problem, and — perhaps more telling — how that pattern would change as variants of the central problem were tried out. Arriving at a point where such statistics could be gathered was a critical moment in the project, and a discussion of a small subset of the results constitutes the remainder of this article.

In Figures IX–1, IX–2, and IX–3, shown at the end of this article, we sample three families, each family being represented by six problems (there are many more than six problems in each family, of course). For each problem, the tabletop configuration itself is shown on the left, using obvious icons. The object Henry touched is indicated by an arrow with an "H", and possible responses by Eliza are indicated by arrows labeled "E1", "E2", etc. In the middle of each display, a bar graph is shown; each bar represents the frequency of one of those answers. (Each problem was run a total of 50 times.) On each run, a monitor recorded the *answer given,* the *final structure-value,* and the *run-length (i.e.,* total number of codelets). On the right of each display is a table giving, for each answer, the average final structure-value and the average run-length for all runs ending in that answer.

Of particular interest are cases where the highest-frequency answer does not coincide with the answer having the highest final structure-value. Such cases, rather than reflecting a defect of the architecture, reflect an inevitable fact about high-level perception: deep perceptions are often hard to discover; it is easy to be distracted by routes having more surface appeal. Thus, as measured by frequency, Tabletop often prefers "superficial" answers, provided they have at least a modicum of plausibility, over "deep" answers, as measured by final structure-value. It is, however, a special virtue of Tabletop's parallel stochastic architecture that, by allowing simultaneous exploration at different rates along rival routes showing different degrees of promise, it is not always seduced by surface glitter, and can on occasion discover deeper visions.

By exploring several families of "Do this!" problems, each family having many members, we built up a "performance landscape" of the program — a surface in the abstract multidimensional space of all Tabletop problems, where each dimension corresponds, roughly speaking, to a given pressure. The "ridges" in this performance landscape represent those critical combinations of pressures where the program would switch from one preference to another (*e.g.*, Figure IX–2*e* in the Blockage family, there described as a "turning point" in the series). Likewise, "peaks" and "valleys" in the landscape correspond to clear and stable preferences and dislikes on the program's part. By making qualitative comparisons of the locations of Tabletop's ridges, valleys, and peaks with our own personal preferences, as well as with statistics summarizing the preferences of experimental subjects, we were able to assess the psychological realism of Tabletop's "taste". (The results of experiments on human subjects can be found in French, 1995.)

From our point of view, the program did quite a creditable job, on a qualitative level, of simulating the taste of a typical human playing the role of Eliza. (Readers can look at the bar graphs and tables provided and decide for themselves whether or not they agree.) Despite this success, we reiterate our contention that the program ought not to be judged primarily on this basis, but rather on its overall architecture, in which analogy-making falls out as a natural by-product of high-level perception, a cognitive activity that is realized by mutually influencing parallel processes guided by dynamically evolving pressures that emerge in response to the situation being faced.

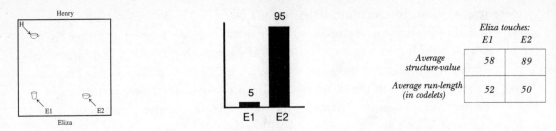

*Figure IX–1a. Figures IX–1a through IX–1f represent the Surround family. In all these problems, Henry touches his cup. Though the literal-sameness answer (i.e., for Eliza to touch Henry's cup) is always possible, the main rivalry is between Eliza's glass and cup. Variants in this family explore diverse combinations of pressures by surrounding Eliza's glass and Henry's cup with various sets of table objects. Figure IX–1a is the "base case": no surrounding objects are present. Therefore, just two pressures contribute to the decision: category membership and physical position. The former favors the cup (category **identity** is better than category **proximity**). What about the latter? Given an object in a corner, people are more likely to seek its counterpart in the diagonally opposite corner than in the mirror-image corner, so such a bias was built into Tabletop. Therefore, position pressure also favors the cup. Overall, then, the pressure in favor of Eliza's cup is very strong; indeed, Tabletop chooses her glass only 5 percent of the time.*

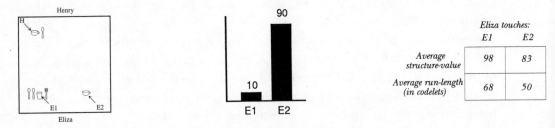

Figure IX–1b. When humans look at this setup, they effortlessly perceive two groups — one containing Henry's cup, the other containing Eliza's glass. (Of course, many other groupings are theoretically possible, but virtually never come to mind.) Tabletop is similarly inclined. The two groups, based solely on physical proximity of their component objects, bear scant resemblance to each other — they are of different size and have little in common. Nonetheless, their existence increases the appeal of seeing Eliza's glass as the counterpart of Henry's cup, since both objects are at least members of groups, however disparate those groups may be. This weak pressure increases the frequency with which Tabletop touches Eliza's glass to 10 percent.

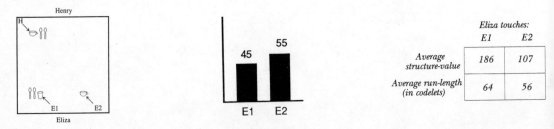

Figure IX–1c. Again there are two conspicuous groups, each containing three objects. One consists of Henry's cup and two spoons (the spoon-pair is likely to be seen as a subgroup); the other consists of Eliza's glass and two spoons (this spoon-pair, too, is likely to be seen as a subgroup). Not only do these groups map onto each other as wholes, but their subgroups (if seen) map strongly onto each other, thus pushing the structure-value up and increasing the pressure for mapping Henry's cup onto Eliza's glass. Indeed, Tabletop now touches the glass 45 percent of the time. As might be expected, the average structure-value when it does so is significantly higher than when it touches her cup. This is one of the cases where highest frequency and best structure disagree.

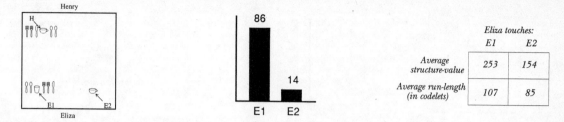

Figure IX–1d. *The groups around Henry's cup and Eliza's glass are very similar: they contain the same number of objects, and their subgroups are identical. Humans see the mappings as very strong, and see Eliza's cup as a loner. There is thus much pressure to touch Eliza's glass. As the results show, not only does the program do this far more frequently than touch her cup, but the average structure-value associated with the former choice is far higher than for the latter.*

    *The strong mappings of groups and subgroups push so hard for touching Eliza's glass that one might wonder what would **ever** induce Tabletop to pick Eliza's cup. Two factors are involved. One is that Tabletop sometimes simply fails to build those mappings. On such runs, the pressures do not so greatly favor her glass. More rarely, Tabletop may build the mappings but simply choose (stochastically) to ignore them and touch Eliza's cup. Though this may seem irrational, people often act similarly. In a survey, subjects shown this setup were asked to draw all relevant correspondences. Some, after having drawn a line linking the two spoon-groups, a second line linking the two fork-groups, and a third line linking the two knives, **ignored** all their lines and chose Eliza's cup. In this light, the "anomalous" 14 percent of runs in which Tabletop touches the isolated cup seem justified.*

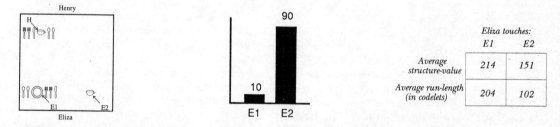

Figure IX–1e. *Eliza's glass has been replaced by a plate, conceptually remote from the touched object. This should shift the pressures back to favoring the isolated cup. Indeed, Tabletop now touches Eliza's cup 90 percent of the time, and her plate just 10 percent. Still, the structure-value associated with the plate answer remains over 40 percent higher than that for Eliza's cup.*

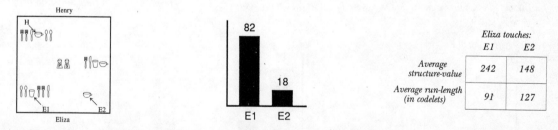

Figure IX–1f. *Here we examine the effect of placing distractions on the table. Six objects have been added to Figure IX–1d, more than doubling the number of theoretically possible correspondences between objects. However, if Tabletop's focusing mechanisms operate well, this should not have much effect on the amount of processing required or on the distribution of answers. Indeed, there is hardly any difference between these results and those in Figure IX–1d. The average run-lengths here are quite similar to those in Figure IX–1d, where there were no distractions at all. This shows that Tabletop essentially ignores objects in unlikely locations on the table, focusing its attention primarily on **a priori** preferred regions. (It should be noted, however, that when there are no objects in **a priori** preferred regions, the program does examine **a priori** unlikely regions.)*

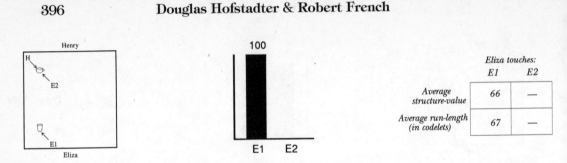

Figure IX–2a. *Figures IX–2a through IX–2f represent the Blockage family. The cup and glass facing each other are not identical, but almost so: the Slipnet nodes "cup" and "glass" are very close. Also the glass is in a favorable position with respect to the cup. There is thus much pressure to choose the glass, and Tabletop always does so here. In the upcoming variants, the pressures for touching Henry's cup are increased by creating correspondences that "usurp" Eliza's glass. In this setup, the base case, there is no attempt at blockage.*

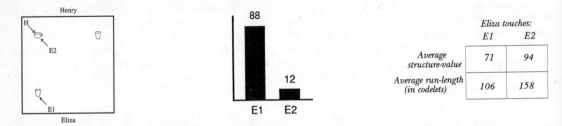

Figure IX–2b. *A glass has been added; being on Henry's side, it is a most unlikely object for Eliza to touch. Unlike the additions in the Surround family, this addition creates no new group. One might thus expect that this addition, like the distractions in Figure IX–1f, would have little effect on Tabletop's answers or resources expended. However, another effect — the distant glasses'* **identicality** *— gives rise to a pressure to build a* **correspondence** *between them. On runs when this is done, the glasses are seen as part of a single, albeit weak, structure, which exerts a blockage effect. (Correspondences and groups are both perceptual chunks and are similar in many ways; the former, however, tend to be weaker since their constituents, usually being far apart on the table, are not tightly bound together.)*

    Many human subjects (40 percent; see French, 1992) saw the glasses as counterparts. When this happens, since Eliza's glass cannot be the counterpart **both** *of Henry's glass* **and** *of his cup, just one answer remains: the literal-sameness answer, Henry's cup. Tabletop occasionally (12 percent of the time) sees the glasses as counterparts and touches Henry's cup. Note that the structure-value of this answer is 30 percent higher than for Eliza's glass, even though the latter is chosen far more often. In addition, runs on which Tabletop chooses Henry's cup average roughly 50 percent longer than for Eliza's glass. Once again, this is not surprising: answers involving deeper perception should take longer to find than those with less.*

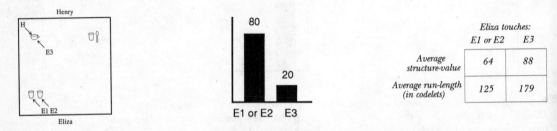

Figure IX–2c. *Two objects have been added, strongly suggesting groups. Tabletop almost always builds the group on Eliza's side, since the two glasses are not just neighbors but identical objects. The group on Henry's side has less appeal, since "spoon" and "glass" are distant Slipnet nodes. Still, on many runs, both groups get built. When, in addition, a diagonal correspondence between them is built, despite its weakness, it "usurps" both glasses on Eliza's side, forcing Tabletop to go for Henry's cup.*

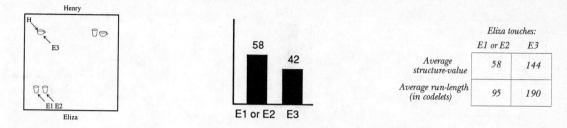

	Eliza touches:	
	E1 or E2	E3
Average structure-value	58	144
Average run-length (in codelets)	95	190

*Figure IX–2d. The group on Henry's side now has more appeal, as "cup" and "glass" are much closer in the Slipnet than "spoon" and "glass". Often the two objects are seen as physically **and** conceptually close. This makes for a stronger group, which in turn makes for a stronger diagonal correspondence, leading Tabletop to choose Henry's cup more than twice as often as in Figure IX–2c.*

Sometimes Eliza's group is built but not mapped to anything as a unit; in such runs, the touched cup tends to be mapped onto one of Eliza's glasses. There is pressure to map her other glass onto Henry's glass (diagonally opposite identical objects make strong counterparts). But there is also counterpressure: to map Eliza's two glasses, which have been grouped and are thus a conceptual unit, onto unrelated objects would be to disrespect their unity. Yet Tabletop does this occasionally, in which case the structure-value suffers markedly. When Tabletop goes for Henry's cup, the structure is much better than when it chooses one of Eliza's glasses. Also note that Tabletop takes significantly longer to build the structure that gives rise to the better answer.

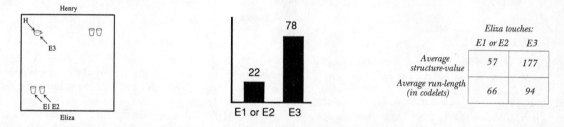

	Eliza touches:	
	E1 or E2	E3
Average structure-value	57	177
Average run-length (in codelets)	66	94

Figure IX–2e. This is a turning point in the Blockage family: Tabletop chooses Henry's cup over half the time. The reason is simple. Both glass–glass groups are very strong, as is the correspondence between them — strong enough, it turns out, to make Tabletop very reluctant to break it by mapping Henry's cup onto either of Eliza's glasses. As a result, Tabletop picks Henry's cup 78 percent of the time. (Human subjects chose Henry's cup 66 percent of the time; see French & Hofstadter, 1991.) As might be expected, the average structure-value for this answer is much better than when Tabletop chooses one of Eliza's glasses. Also as usual, it tends to take Tabletop longer (by about 40 percent) to get the answer having better structure.

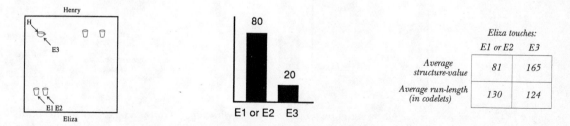

	Eliza touches:	
	E1 or E2	E3
Average structure-value	81	165
Average run-length (in codelets)	130	124

*Figure IX–2f. What happens when Henry's two glasses are pulled apart? The greater the separation, the less the pressure to see them as a group. If they are **not** seen as a group, no group–group correspondence can, of course, be built, so one would expect the blockage effect to be greatly reduced. Indeed, Tabletop chooses Henry's cup only 20 percent of the time here. It is still possible to build two parallel glass–glass mappings diagonally linking Eliza's glasses to Henry's glasses, but there is considerable pressure against this because, as in Figure IX–2d, doing so would disrespect Eliza's group. (The program prefers that a set of correspondences starting in a single group all end in another single group, instead of diverging to unrelated destinations.)*

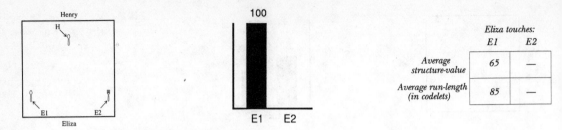

Figure IX–3a. Figures IX–3a through IX–3f represent the Buridan family, whose name derives from the proverbial ass that, poised between two precisely equidistant bales of hay, could not decide which direction to move in, and thus starved to death. In this first situation, the "bales of hay" are not equidistant, and Tabletop invariably opts to touch Eliza's spoon. Most likely, had the program been run on this problem many more times, an occasional choice of the fork would have turned up, but it would have been a very rare event.

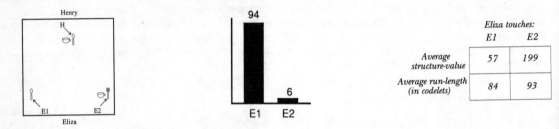

*Figure IX–3b. Here is a first attempt at evening out the attractiveness of the two choices. Its effect is quite weak, however: the program chooses Eliza's fork only 6 percent of the time. Note, though, that when it does so, the structure-value of that answer is much higher than for the spoon answer. The reason is simple. The only way for Tabletop to choose Eliza's fork is to build two parallel correspondences — a cup–cup one and a spoon–fork one. This will give rise to a high structure-value. On the other hand, there are two different ways that Tabletop can select Eliza's spoon. Either it builds only **one** correspondence (spoon–spoon), which gives a fairly low structure-value, or it builds an **incoherent pair** of correspondences: a spoon–spoon one and a cup–cup one. In the latter case, the structure-value is very low because Tabletop is designed to "dislike" situations wherein two correspondences emanate from a single group and head off in unrelated directions. In fact, the parameter controlling this degree of dislike was slightly jacked up in all the runs on the Buridan family, simply in order that in the following configuration, the program be torn by its dilemma at precisely a 50–50 Buridan-like ratio.*

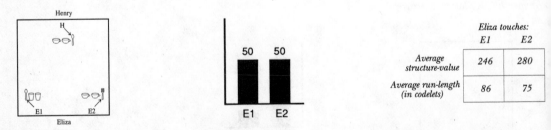

Figure IX–3c. Here we have the program caught between two alternatives that it considers precisely equally attractive, at least in terms of frequency of choice. The reason for going for the fork answer is the same on every run in which that answer is chosen: forks are like spoons and a cup–cup group maps very strongly onto another cup–cup group. There are, however, different reasons that can underlie the spoon answer. On some runs, it is simply the spoon–spoon identity, disregarding the "sidekick" groups entirely. This is fairly rare, however. On other runs, the spoon–spoon mapping is accompanied by a parallel mapping from the cup-pair to the glass-pair. This is of course a rather strong reason to choose the spoon. The third way Tabletop can opt for the spoon answer is to build a spoon–spoon correspondence as well as an incoherent second mapping — namely, between the two cup-groups. This option, however, is sufficiently unattractive that Tabletop seldom goes for it.

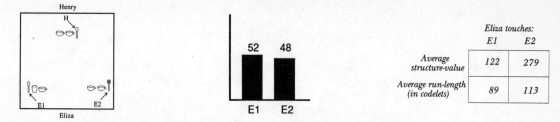

	Eliza touches:	
	E1	E2
Average structure-value	122	279
Average run-length (in codelets)	89	113

*Figure IX–3d. This configuration poses a very complex and subtle dilemma. The appeal of the fork answer is exactly as in Figure IX–3c. The appeal of the spoon answer might at first appear to have been **increased** relative to Figure IX–3c, since a glass has been converted into a cup, making the group containing the spoon in some sense "more like" Henry's group than in the previous figure. However, there is a different way of looking at the effect of the conversion. The group next to Eliza's spoon has been weakened (similar-category groups are considerably weaker than same-category groups) and is therefore less likely to be built by the program; moreover, even when it is built, it maps much more weakly onto Henry's cup-pair, because the latter is a same-category group and the former is a similar-category group, and the program is much less attracted to mappings that "equate" groups of different types.*

*All in all, then, when one regards the situation at the **group** level, one sees the conversion of the glass to a cup as **weakening** the appeal of the spoon answer; on the other hand, when one considers the situation from a more down-to-earth viewpoint in which **objects** matter more than groups, one sees the conversion as **strengthening** the spoon answer's appeal. To complicate matters, Tabletop is designed to prefer making group-level mappings, when possible, over mere object–object mappings. So Tabletop's choice is between making a group–group mapping that is flawed, and a pair of mere object–object mappings, one of which (cup–glass) is quite strong and the other of which (cup–cup) is very strong. This constitutes a "second-order" Buridan's-ass dilemma. As the statistics show, the two effects canceled each other out almost perfectly!*

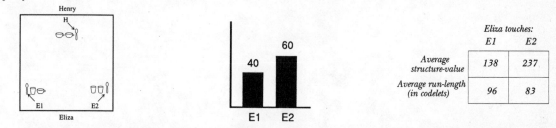

	Eliza touches:	
	E1	E2
Average structure-value	138	237
Average run-length (in codelets)	96	83

Figure IX–3e. An extra complicating factor in the preceding figure, not mentioned there, was the discrepancy between the items of silverware on Eliza's side. In order to eliminate that complication from the situation, Eliza was given two spoons. Now the choice facing Tabletop is a slightly purer second-order Buridan's-ass dilemma. The statistics show that Tabletop's attraction to abstract mappings, even if they are a bit weak, is stronger than its attraction to object-level mappings.

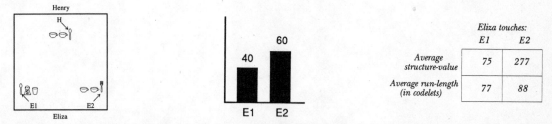

	Eliza touches:	
	E1	E2
Average structure-value	75	277
Average run-length (in codelets)	77	88

Figure IX–3f. This is a different modification of Figure IX–3c. The conceptual distance between a salt shaker and a cup is of course very large, and this reduces the pull of the spoon answer quite a bit. Another factor that weakens the spoon answer is that the program is quite reluctant to group the salt shaker with the glass, as they have little to do with one another. Despite these weakening factors, Tabletop still chooses the spoon answer fairly often, mostly basing its choice on the simple-minded argument that if Henry touched a spoon, so should Eliza! The crudeness of this argument is reflected in the low structure-value attached to that answer, in comparison with the high structure-value attached to the fork answer, which, on an abstract and esthetic plane, has much greater appeal.

Preface 10:

The Intoxicating World of Alphabets and Their Styles

Infected with Letter Mania from the Word "Go"

I have had a fascination with the shapes of letters from earliest childhood. I loved mastering the alphabet, printing out names and words, assimilating the intricate swirls of cursive writing, perusing books about letters and their evolution, observing both the grace and the irregularities in the handwriting of my friends. Then came the exciting discovery that there were other writing systems. Chinese was the first one I was exposed to, but one day I came across a book that had samples of text in dozens of other writing systems, and my perspectives were expanded enormously. I carefully studied several other writing systems, including three from India and Sri Lanka (those of Hindi, Tamil, and Sinhalese), and was able to transcribe sounds into them and to pronounce words written in them.

At some point in my teens, I went on a drawing binge in which I made all sorts of strange nonrepresentational patterns, and in many of these drawings I used letterforms — either of our own alphabet or ones from other writing systems that I had become familiar with. This led to my devising a new Indian-like writing system of my own (Hofstadter, 1988b), but more importantly, to much playful experimentation with new ways of rendering our own 26 letters. I found myself dissecting letters conceptually and putting them back together in outlandish, daring, swashbuckling fashions (if I do say so myself!).

In this new mania, I was doubtless influenced by the playful games indulged in by advertisers seeking to draw people's attention, but I was especially inspired by the letterforms I saw in Europe — on posters, billboards, store signs, and so on. Italians in particular seemed to have a unique flair for coming up with truly new ways of seeing letters, and I was amazed at the ingenuity and audacity of the ideas that they played with. Letters on signs in Italy often played at the very fringes of legibility, but by no means by accident; the designers did so knowingly and cleverly. What they were doing was not immaturish, amateurish bumbling or fumbling, but sophisticated and professional innovation at the highest levels of visual creativity. They were deliberately tampering dangerously with the conceptual essence of each letter. My mind was teased and excited and intoxicated by the deep cleverness of these wonderful and yet completely anonymous masters of Italian letterature. Quite spontaneously, in a kind of homage to them, I began to devise one alphabet after another, always trying to outdo myself in weirdness. Not that I wasn't interested in grace and simplicity as well. Letters that were graceless or messy held no interest whatsoever for me.

Obviously, making isolated letters was of limited interest. What really mattered was rendering full words, or better yet, coming up with complete alphabets. This meant somehow finding a way to insinuate *a uniform type of weirdness* into all 26 letters of our alphabet. This in itself was a very weird, gripping idea, and it took hold of me with a vengeance. What ensued was a lengthy period of time during which I designed hundreds of new alphabetic styles, relentlessly exploring all sorts of combinations of gracefulness, funkiness, sassiness, sauciness, silliness, softness, pointiness, hollowness, curviness, jagged-ness, roundness, smoothness, complexity, austerity, irregularity, angularity, symmetry, asymmetry, minimality, redundancy, ornamentation, and countless other indefinable parameters. Obviously, some of these hand-drawn alphabets were much better than others, but each new one, whether I judged it a success or a flop, added a small, intangible something to my sense of what letters and artistic style were all about.

I am sure that to somebody who hadn't thought much about alphabets and visual styles and the idea of pushing concepts to their very limits or just beyond, many of my experiments would have looked mysterious and pointless, perhaps even gawky and stupid. But that wouldn't have deterred me, because I had a stable inner compass that was carrying me somewhere, though I couldn't have said exactly where or why.

As the years went by, my interests gradually broadened out to include more conventional typefaces as well as the wilder "display faces" that had been my first love. I even began to see great richness and subtlety in book faces, which at one time I had looked upon with scorn, considering them stodgy and unimaginative. The innumerable exquisite details characterizing Baskerville, Palatino, Goudy, Tiffany, Americana, Optima, Melior, Times, Garamond, Bodoni, Benguiat, Korinna, Souvenir, Cheltenham, Novarese, and their like became a new passion, and as the years passed, my understanding of letterforms reached a new level of subtlety.

From Real Letters to Stick Letters

I suppose it was inevitable that this passionate interest and my profes-sional pursuit of the mechanisms of creativity would eventually collide and that the collision would give rise to some project in which computers were supposed to design innovative letters. In any case, something was sparked in 1979, when my friend Scott Kim and I laughed at the ridiculous idea of trying to get a computer to make artistic alphabetic explorations of its own. It just seemed so outrageously hard! But from my first thoughts about the hopelessness of the venture, I gradually distilled a new challenge that I felt contained all, or almost all, of the essential interest of the original notion, while shedding huge amounts of irrelevancies, thus isolating the crux of the matter and making its real goal very clear.

The basic idea was to study the notion of artistic style among mere "stick letters" instead of among "real" letters in their full curvilinear glory. The crucial step was coming up with a grid of just the right size and type, so that letterforms of great variety could be produced and yet so that the challenge of perceiving and judging them seemed less than infinitely daunting. It took over a year of experimentation before I settled on the grid presented in the following chapter. Of course, I wasn't just thinking about the grid itself during that year, but also sketching out an architecture for the perception and design of letterforms and alphabetic styles on that grid. This architecture borrowed from my earlier ideas about a program to solve Bongard problems (Chapter 19 of Hofstadter, 1979), and anticipated some of my later projects, in that it was fundamentally parallel and perception-based, and featured a model of overlapping, mutually slippable concepts at its core. But these ideas, although substantially reasonable, were simply premature at that time (1980–81). I needed to work my way up to a computer-modeling project of this level of subtlety and sophistication.

Thus the Letter Spirit project, as it came to be called, kept tugging at me and yet being put on the back burner. It was a frustrating tug-of-war, because nothing fascinated me more than this project, which to me represented an exploration of the very core of creativity — yet I was frightened by it.

A major event in this battle was, surprisingly enough, the appearance on the scene of the Macintosh computer. I bought my first Mac in April of 1984, and virtually the first thing I tried out on it was to draw a "gridfont" — a full alphabet obeying the constraints of the Letter Spirit grid. Strangely enough, although over the previous few years I had designed many dozens of gridbound letters and small segments of alphabets, I had never once designed a full gridfont. How delighted I thus was when I found out that MacPaint had special facilities for drawing vertical, horizontal, and 45-degree lines — and it was even better when I found out that there was a special provision for constraining line segments to a grid! It was almost as if the designers of MacPaint had had my project in mind when they made their program.

Once I became fluent in MacPaint, an inner demon in me took over and started releasing pent-up ideas by the millions, or so it felt. I was possessed by gridfont-mania, and day and night, at every moment of leisure time (and many moments of work time), I found myself designing one gridfont after another. I couldn't believe how absorbed I was by what I knew 99.9 percent of the people in the world would regard as absolute, unbelievable trivia! Why on earth would a grown man — a professor of computer science, at that — spend virtually every moment of his waking life racking his brains trying to dream up sillier and sillier ways to make little dinky letters of the alphabet out of nothing but sticks? Despite sometimes thinking that I had gone half-loony, I kept with it, because I was possessed. It was out of my control.

Some eighteen months and 400 or so gridfonts later, I gradually started emerging from under this cloud, and returning to a semblance of normalcy.

*Figure X–0. A gridfont (the notion is explained in Chapter X) called "Normalcy", which plays with style at a very cerebral level. The obvious constraint obeyed here is that of strictly avoiding horizontals and verticals — except, of course, that constraint is blatantly violated by one single letter, the "c". The question implicitly posed by such flagrant flouting of the normal, commonsensical notion of style and uniformity is this: What really does "uniformity" mean? If I simply **declare** this set of characters to form a style, does it thereby become one? Or what if the great Italian type designer Aldo Novarese declared it a style? What if instead of having just one exceptional letter, I had had several, sprinkled in amongst the ones built solely of diagonals? Would that be more of a "genuine" style? What if the mixture were close to half and half? Is it possible to mix two very different styles (in the conventional sense of the term) so smoothly that no eyebrows are raised? And what about deliberate attempts to be jarring, such as Normalcy — are they somehow less legitimate? What, in short, is meant by "style"?*

Normalcy! Ah! Yet another idea for a gridfont! And in fact, I did design a gridfont named just that (see Figure X–0), which was inspired by the at-first absurd-seeming idea of inserting a very "normal 'c'" into an alphabet of extremely abnormal letters, all resulting from my self-imposed constraint of employing solely *diagonal* segments. This particular self-negating "style" was therefore playing at a meta-level with the very notion of style itself. The constant and intricate interplay between high-flown meta-level considerations of this sort — considerations explicitly about style itself — and the most tiny, specific considerations about shapes became much clearer to me as I designed font after font after font.

Eventually, things settled down, and thanks to all this frenzied work/play, I had evolved a much more sophisticated idea of what the Letter Spirit challenge really was about. Altogether, I had made many, many thousands of small interrelated decisions, and — what really mattered — I had watched myself making them with an eye to uncovering the mysteries of the creative process. From that point of view, my many seemingly self-indulgent months of letterplay were an invaluable contribution to my research.

One might think that this was the obvious moment to take up the challenge of implementation, but it was still, unfortunately, too early. Melanie Mitchell had just joined me, and our efforts to bring Copycat to life were only in their earliest stages. Tabletop hadn't yet even been dreamt of. My little Jumbo existed, Marsha Meredith's much huger Seek-Whence program was just about finished, and Dany Defays' Numbo was about to be born. The ideas defining our approach were just getting worked out in detail, and these were absolute prerequisites to the much more ambitious project that Letter Spirit represented. So I resisted, although some inner voices certainly pushed hard for tackling it.

The Letter Spirit Computer Program is Born

Several years later, around 1990, a new student in my research group, Gary McGraw, started agitating for a project to sink his teeth into. He kept on suggesting that we tackle Letter Spirit and I kept on resisting, feeling the time was still too early. But as his urgings didn't let up, I gradually took more note of them, and Gary eventually convinced me that this really *was* the time. If not now, then when? And so, in a series of meetings at our favorite Bloomington coffeehouse, the Runcible Spoon, located just a literal stone's throw away from a street that is called, most uninspiringly, "Seventh Street" (etc., etc., *ad stuporem*), we started to plot out how my extensive notes and sketches for the Letter Spirit architecture could be fully developed and converted into a working computer program. Gary started implementing part of it, and we wrote up a grant proposal. The article that follows is a modified version of that proposal.

After over a decade's wait, Letter Spirit is thus finally taking wing. It is still ill-defined in many ways, but in broad brushstrokes we know the route we are taking. Certainly this project represents the most ambitious effort so far undertaken in my research group. We shall see, over the next few years, how it goes.

Chapter 10

Letter Spirit:

Esthetic Perception and Creative Play in the Rich Microcosm of the Roman Alphabet

DOUGLAS HOFSTADTER and GARY McGRAW

The Goal of Imparting a Sense of Deep Style to a Machine

The Letter Spirit project is an attempt to model central aspects of human creativity on a computer. It is based on the belief that creativity is an automatic outcome of the existence of sufficiently flexible and context-sensitive concepts — what are referred to herein as *fluid concepts*. Accordingly, our goal is to implement a model of fluid concepts in a challenging domain. Not surprisingly, this is a very complex undertaking and requires several types of dynamic memory structures, as well as a sophisticated control structure involving an intimate mixture of bottom-up and top-down processing. Letter Spirit involves a blend of high-level perception and conceptual play that will, we hope, allow it to create in a cognitively plausible fashion. The process of fully realizing such a model will, we believe, bring considerable insights into the mechanisms of human creativity.

The specific focus of Letter Spirit is the creative act of artistic letter-design. The aim is to model how the 26 lowercase letters of the roman alphabet can be rendered in many different but internally coherent styles. Starting with one or more seed letters representing the beginnings of a style, the program will attempt to create the rest of the alphabet in such a way that all 26 letters share that same style, or *spirit*.

As an intellectual project, Letter Spirit is quite old, the domain having been initially proposed and a computational architecture sketched out as far back as 1980, but as an implemented computer program it is in its infancy. (A skeptic might therefore say that it exists only "in fancy".) Right now, the program is starting to be able to recognize letters, but is still very far from being able to create any. We hope this will change over the next few years.

But Hasn't That Already Been Done?

Not infrequently, when we state our goal as "Making a program that can design letters of the alphabet in many different styles", intelligent and well-informed listeners ask, "But doesn't Donald Knuth's Metafont program already do precisely that?" This all-too-common misconception merits a careful reply.

Metafont (Knuth, 1982) is a tool that allows a human user to design parametrized letters and letter-parts, and thus to fashion an entire alphabet (not only our own, but also Greek, Hebrew, Hindi, Japanese, etc.) in a uniform way. However, all design decisions, from first to last, are made by the human user. For instance, if the "h" is to be made from the "n" by making the vertical post on the left side taller, the user must tell the program precisely that, since the program itself has no prior representation of the concepts "h" and "n". Lacking any knowledge whatsoever of letter concepts, Metafont no more designs letters than a word processor composes poems or a piano composes waltzes. Metafont merely *facilitates* letterform design for a human designer (and can therefore enhance a human designer's creativity). Of course, Knuth himself never made any claims that Metafont is an autonomous letter-designer, but a nodding acquaintance with it can easily give rise to such an impression, as the authors of this paper well know, having encountered it so many times.

We bring this up because it helps to put the goals of Letter Spirit in perspective. In particular, it is enlightening to analyze what is lacking in Metafont and many other programs commonly but erroneously thought of as "models of creativity".

In a nutshell, the problem is that most such programs *don't know anything about what they are doing.* This phrase actually has two quite distinct meanings, both of which usually apply to the programs under discussion. The first meaning is "the program has no knowledge about the *domain* in which it is working". The other is "the program has no internal representation of the *actions* that it is itself taking, and no awareness of the *products* that it is creating". Both of these gaps are serious defects in anything that purports to be a model of creativity.

A Non-creative Computer Model of Letter Design

In this connection, a quintessential (though little-known) example is the DAFFODIL project (Nanard *et al.*, 1989). From our experiences in talking to people about our project, we suspect that many people, if they knew about DAFFODIL, would point to it as a program that has already achieved our stated goals. (It is not clear to us whether DAFFODIL's designers intended their work as a model of human creativity.) To show why this would be almost as serious a misconception as thinking that Metafont has "scooped" us, we must briefly describe the project.

DAFFODIL is a program that combines three types of input:

(1) a set of 26 *backbones* (one for each capital letter), where a back-bone, rather than being a shape, is an immutable description of the given letter in terms of a dozen or so *a priori* stroke-types, which are simply phrases such as "horizontal segment", "vertical segment", "left arc", and so on;

(2) a set of *decorations*, a decoration being a rigid graphical shape concocted by a human, usually but not necessarily being an ornate or florid design;

(3) a *mapping* that associates stroke-types with decorations.

Given these three ingredients, the program executes the mapping on all 26 letters — that is, it carries out a systematic *stroke-to-decoration substitution operation*, which converts the set of 26 abstract backbones into a set of 26 actual graphic shapes, each one being some particular arrangement of the given decorations (this process is illustrated in Figure X–1). And hence, a new typeface has apparently been "created" by the program.

To be sure, a new typeface has been *generated*; however, to think of what DAFFODIL does as in any way resembling creation by humans is an egregious error, for the following reasons:

- a tacit assumption behind the DAFFODIL project is that style does not involve manipulation of abstract concepts behind the scenes, but merely involves playing with how surface-level aspects are to be rendered;

- DAFFODIL's 26 backbones are all designed by a human and fed to it from the outside, and once a backbone has been fed in, it remains absolutely fixed, so that any creativity involved in this deep-lying aspect of the letterforms is external, not internal;

- DAFFODIL's knowledge of each Platonic letter (*i.e.,* the letter's backbone) is very impoverished, and it has no knowledge of how Platonic letters are interrelated, other than the fact that two backbones might share one or more stroke-types;

- the decorations to be "plugged in" by DAFFODIL are all designed by a human and supplied to the program, and once a decoration

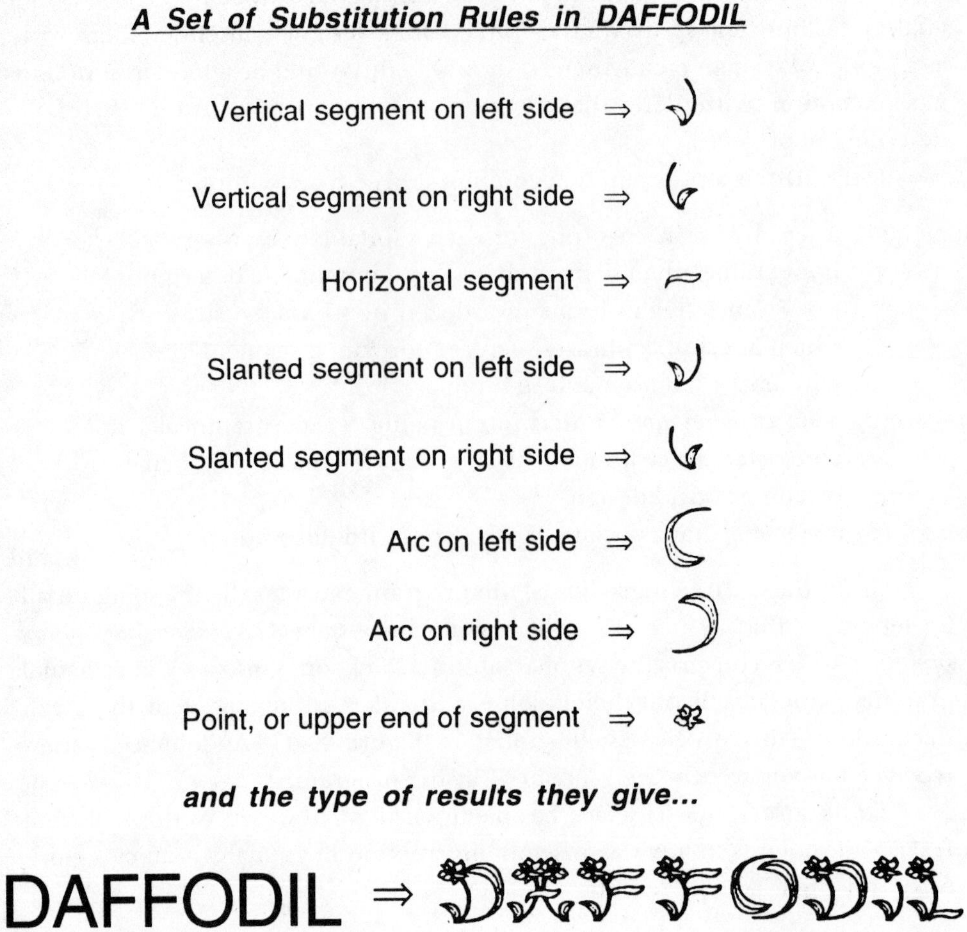

A Set of Substitution Rules in DAFFODIL

Vertical segment on left side   ⇒   ᴺ⌣

Vertical segment on right side   ⇒   ⎰

Horizontal segment   ⇒   ⌐

Slanted segment on left side   ⇒   ᴺ⌣

Slanted segment on right side   ⇒   ⎰

Arc on left side   ⇒   C

Arc on right side   ⇒   ⊃

Point, or upper end of segment   ⇒   ✿

and the type of results they give...

DAFFODIL ⇒ 𝕯𝕬𝕱𝕱𝕺𝕯𝕴𝕷

Figure X–1. Transformation rules and sample output from the DAFFODIL program, developed by Nanard et al.

has been fed in, it remains absolutely fixed, so that any creativity involved in this shallow aspect of the letterforms is external, not internal;

- no decisions are made by DAFFODIL— it simply plugs in decorations as required, occasionally asking a human being for approval or disapproval;

- DAFFODIL has no ability to perceive or judge anything it has produced — that is, it cannot tell whether a shape it has made actually looks like an "A", whether it is attractive or ugly, or whether it fits in well or poorly with other already-produced letters.

Taken alone, any of these points speaks tellingly against DAFFODIL as a model of creativity; taken together, they pretty much destroy any claims that might be made for it as such a model.

True Creativity Implies Autonomy

We went through this exercise not to shoot down a rival project in a mean-spirited manner, but to point the way to a deeper understanding of creativity. To be specific, we insist that for a design program to be called "creative", it must meet the following requirements:

- the program itself must arguably *make its own decisions* rather than simply carrying out a set of design decisions all of which have already been made, directly or indirectly, by a human;
- the program's knowledge must be rich — that is, each concept must on its own be a nontrivial representation of some category with flexible criteria for judging degrees of membership, and among diverse concepts there must be multiple explicit connections;
- the program's concepts and their interrelations must not be static, but rather must be flexible and context-dependent;
- the program must experiment and explore at a deep conceptual level rather than at a shallow surface level;
- the program must be able to perceive and judge its own tentative output and be able to accept it, reject it, or come up with plausible ideas for improving it;
- the program must gradually converge on a satisfactory solution through a continual process in which suggestions coming from one part of the system and judgments coming from another part are continually interleaved.

We would in fact argue that the deep (and philosophically controversial) question raised implicitly by the first point — "What would it mean for a program to make its own decisions?" — is answered by the last five points taken together.

The Letter Spirit architecture is an attempt to model all these crucial aspects of creativity, albeit in a rudimentary way. As will be described below, several of its features — nondeterminism, parallelism, and especially their consequence, statistical emergence — are key elements in allowing it to achieve these goals.

Letters as Rich, Full-fledged Concepts

A fundamental question that pervades all our research projects is: "What are the mechanisms underlying the fluidity of human concepts?" What this leads to, in Letter Spirit, is a quest for the conceptual essence of letters. One might even say that what we are seeking, in this project, is simply *the essence of the letter* "a".

Most literate people take letters for granted, and think of them as rather simple categories. Indeed, most people, if asked, would say that the letter "a" is not a *concept* but a *shape*. But look at Figure X–2, with its display of wildly differing lowercase "a"'s. Certainly there is nothing like a fixed shape here. Rather, behind all of these very different shapes lies a single highly abstract *idea*, which itself breaks down into other abstract ideas on a slightly simpler level.

Essentially, the concept "a" can be thought of as a marriage of two smaller ideas: (1) the idea of an umbrella-handle or lamppost-like entity whose bent part is at the top, drooping over to the left, and whose straight part descends and then stops at the baseline, and (2) the idea of a small pipe-bowl or "c"-like entity sitting on the baseline, open on its right side, and nestled just to the left of the former entity. These two interrelated conceptual components — what we call *roles* — are not explicit shapes *per se* but are ideas for what the shapes eventually drawn on paper should be like, what their acceptable bounds are, and how they should fit together. (Blesser *et al.*, 1973 discusses a related concept, and psychological evidence of several different sorts for the existence of roles inside the letter-concepts of people is presented in McGraw, Rehling, & Goldstone, 1994a and 1994b.)

In order to make more intuitive the critical distinction between the single "Platonic" concept of a letter and the huge diversity of geometric shapes that may instantiate it, we introduce some terminology. In fact, we propose not just two but *four* conceptual levels, starting at a highly abstract level and becoming more concrete as we move towards the actual graphic letterform.

- A *letter-concept* is the abstract, shapeless notion of a Platonic letter, associated with which there is no particular style whatsoever. This does not mean, of course, that "A", "a", and "*a*", simply because they are all pronounced alike, are all represented by one and the same letter-concept. Instead, there are three letter-concepts here, as different from one another as are the concepts "a", "b", and "c", and for the same reason — namely, they implicitly define collections of graphic shapes that have essentially no overlap.
- A *letter-conceptualization* is a breakdown of a Platonic letter in terms of particular roles. For instance, one conceptualization of "b" might state that "b" consists of a *post* attached in *two spots* to a

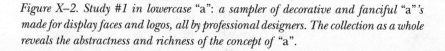

Figure X–2. Study #1 in lowercase "a": a sampler of decorative and fanciful "a"'s made for display faces and logos, all by professional designers. The collection as a whole reveals the abstractness and richness of the concept of "a".

left-facing open bowl sitting to the post's right on the baseline. A related but different conceptualization for "b" might state that "b" consists of a *post* attached in just a *single spot* to a *closed loop* sitting to its right on the baseline. These two conceptualizations, possibly along with some other ways of thinking about what "b" is, constitute the full letter-concept of "b". When it comes to making a concrete "b" in a given style, either some already-known conceptualization may be selected, or else a fresh conceptualization may be spontaneously invented by dissecting and recombining in some new way the roles in an already-known conceptualization. In any case, settling on a letter-conceptualization, whether new or old, is a decision at the very deepest level of style.

- A *letter-plan* starts with a specific letter-conceptualization and fills in details characterizing *how each role should be realized*. (For example, should a post be tall or short? Should a bowl be fat or thin? Roundish or squarish? Should a crossbar be high or low? And so on.) Since every role has many aspects, a letter-plan is a very complex entity, full of information of many sorts. Although it is far more specific than the previous two levels, a letter-plan is still purely mental, and as such it has an infinite number of different ways of being realized on paper. Decisions made at this level are still at a fairly deep level of style, but not as deep as ones involved in coming up with new conceptualizations.

- A *letterform* is the actual graphic shape drawn on paper to realize a given letter-plan, thus also realizing a particular letter-conceptualization and, ultimately, a letter-concept.

Letter Spirit is concerned with all of these levels: the interaction between letter-concepts, the selection of and invention of letter-conceptualizations, the devising of letter-plans, and the conversion of letter-plans into external, visible letterforms.

Wholes and Roles

A vivid example of the shape/concept distinction involves lowercase "x". For most people who learned to write in the United States, the letter-concept for "x" consists of just a single conceptualization, whereby the letter is thought of as consisting of a forward slash and a backward slash of the same size that cross somewhere near the middle. (It is critical to stress that what is in people's minds is not a *picture* of two crossing lines, but a set of *ideas*.) This conceptual breakdown remains the same even when one learns cursive writing. English schoolchildren, by contrast, are taught to draw a lowercase cursive "x" as a pair

Figure X–3. Egward (left) and Egbert (right), identical twins who, sadly, were separated at birth and grew up on opposite sides of the Atlantic, perceive this simple letterform differently.

of small crescents facing away from each other and "kissing" in the middle. If we look at a printed "x" in this way (Figure X–3), we are suddenly struck by this new conceptualization. The shape on paper has not changed, of course, but our inner way of conceiving it is fresh and intriguing, and to anyone who is curious about letters, this new vision of "x"-ness implicitly but quite clearly makes a host of potential suggestions about *other* letters as well.

The conceptual elements critical for "x"-ness include not just the "slashes" or "crescents", but also such things as the four *tips,* as well as the central *crossing-point* or *kissing-point.* We call these inter-role relational structures *r-roles,* and they are just as important as roles in determining the category membership of a shape. For example, a graphic shape might have a strong exemplar of *post* to the left of a strong exemplar of *loop* and yet the way they interact might still make them form a bad "b" (think of the fact that if they do not touch, the shape will look more like "lo" than "b").

Central to the Letter Spirit project is the notion that the internal structure of a category consists of a collection of roles and r-roles, and that category membership at the whole-letter level is determined by category membership at this lower level. Each role or r-role has a different degree of *importance* to the letter — the degree to which its presence or absence matters. Of course, different graphic items instantiate a given role more strongly or more weakly than others. In other words, roles, just like *wholes* (complete letters), are concepts in their own right, with somewhat nebulous boundaries. The difference is that membership in a role is easier to characterize than membership in a whole, so that reducing wholes to collections of interacting roles is a step forward in simplification. (For earlier ideas on similar perceptual hierarchies see, for example, Palmer, 1977 and Erman *et al.,* 1980.)

We come now to the question of how style and roles are related. In generating a new letter, first of all comes the choice of conceptualization — in other words, which set of roles will be instantiated? Thus at this deepest level of a letter — a purely mental level, before any shapes are on paper — roles are essentially all there is to style. One could turn this around and equally well say that at this deepest level of style, roles are essentially all there is to a letter.

Next, once a particular conceptualization has been picked or formed, the remaining stylistic aspects of a letterform are a function of how the various roles in the letter-conceptualization are filled. Perhaps a more vivid and accurate way of saying this is that this medium-deep level of style has to do with *how norms are violated*. For instance, consider the role *crossbar*, which certainly belongs to the default conceptualizations of "e", "f", and "t" (and perhaps to some other letters as well, depending on how playful the designer is, but this brings us back to the generation of new conceptualizations, which is at a deeper level of style). Possible norm-violations for this role are many, but certainly include the following: "crossbar too high", "crossbar tilted upwards", "unusually short crossbar", "crossbar missing", etc. Any such violation is a stylistic hallmark that must be respected and propagated (via analogy) to other letters — even, if possible, to letters that lack the crossbar role!

This may seem impossible. How could a letter whose abstract characterization nowhere involves the notion of "crossbar" be influenced by a style in which crossbars are too high (say)? The key to this is *analogy*. To carry such a stylistic trend across, one might make some *other* role be realized in an unusually high way, as long as that role can be seen as analogically close to the concept of "crossbar", and — needless to say — as long as the distortion does not make the resultant letterform unrecognizable as a member of its desired letter-category.

The name of the game in alphabet design is the constant power struggle between letter-category pressures, which involve just the one Platonic letter being worked on at the moment, and stylistic pressures, which involve the alphabet as a whole. We characterize this power struggle as a fight between two forces: an *incentric* or *centripetal* force, which tends to pull the shape being created *towards the center* of the intended letter-category (so that it is a strong instance of the *letter*), and an *eccentric* or *centrifugal* force, which tends to push the shape *away from the center* of the letter-category (so that it better fits the desired *style*). The modeling of this constant battle between incentric and eccentric forces lies at the heart of the Letter Spirit project.

Letter and Spirit as Orthogonal Categories

Although much work has been done on character recognition and the reading of handwriting, computer models have seldom explored the notion of *letters as concepts.* So far, attempts at letter recognition and letter creation in AI have mostly been based on the notion that all the different letterforms people would assign to the category "a" are basically variants of a single underlying *shape* (for instance, McClelland & Rumelhart, 1981; Hinton, Williams, & Revow, 1992). Such an approach ignores the cognitively and philosophically central question (Chapters 13 and 26 of Hofstadter, 1985): *What is "a"-ness all about conceptually?*

Although this is already a very challenging question, an even more challenging question regarding letterforms is: *How are the letters in a given style related to one another?* This can be cast in terms of an analogy problem. Given an instance of "a", how would you make an "e" or a "k" (or for that matter, an aleph or an alpha or even the Chinese character that means "black") in the same style? Of course the problem here is figuring out what "the same style" means. Transferring stylistic aspects from one letter to another involves the types of cognitive activities discussed in the preceding several chapters of this book. Stylistic qualities of one letter cannot usually be *directly* transferred to another; rather, they must be "slipped" into reasonable variants so that they fit within the conceptual framework of the new letter.

Letter Spirit is thus meant to simultaneously address two important and metaphorically orthogonal aspects of letterforms: the *categorical sameness* possessed by instances of a single letter in various styles (*e.g.,* the letter "a" as it is rendered in Baskerville, Palatino, Helvetica, etc.) and the *stylistic sameness* possessed by instances of various letters in a single style (*e.g.,* the letters "a", "b", "c", etc., in Baskerville alone). Figure X–4 illustrates these two ideas and shows their relationship to one another.

Given a few letters in a *gridfont* (our name for the highly constrained, gridbound alphabets used in the Letter Spirit project), the challenge to the Letter Spirit program is to figure out which category a given seed letter belongs to and what the stylistic tendencies suggested by the seed letter(s) are, and then to design the remaining letters in what it considers to be the same style. To be sure, there is never just a single good answer, since seed letters, even if 25 of them are provided (leaving just one last letter to be designed), do not uniquely specify a style, nor does a style, considered as an abstraction, uniquely specify its constituent letterforms.

Retreating from Real-world to Microworld Typeface Design

Letter perception and creation provide an elegant window onto the workings of the mind. If through working in the domain of letter design we can gain general insight into the workings of concepts and their fluid interrelations, we feel we will have made significant headway on the problem of intelligence. It is, in fact, our belief that such headway can *only* be made via the detailed study of a small aspect of the world carefully selected for its generality.

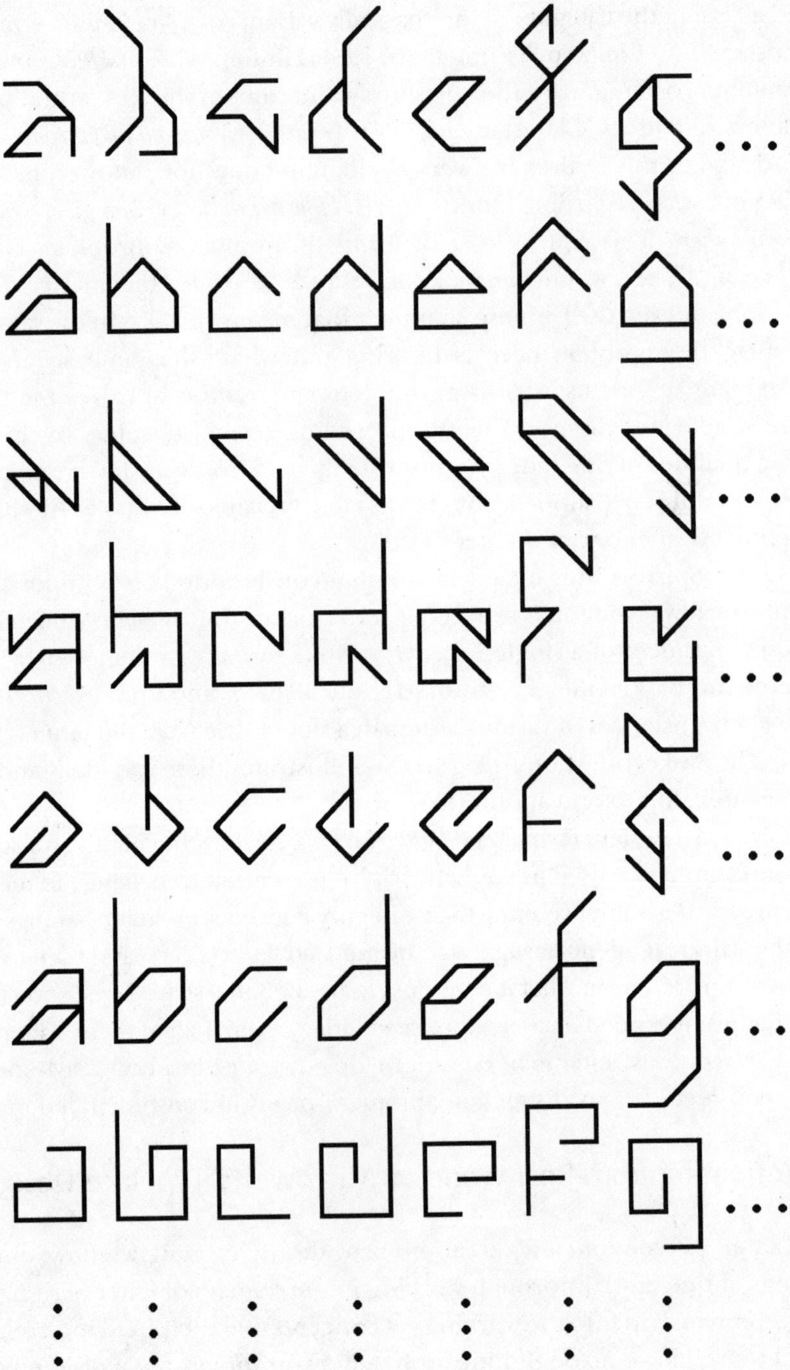

Like music composition or novel-writing, letter design is a highly sophisticated art form that requires years of practice. We could not possibly aspire to make a model that operates at the level of skill of an experienced human designer of letters. Instead we have drastically simplified the domain. Nonetheless, we hope to have preserved the *deep* aspects of the art of letter design while removing many of its shallower aspects, such as the ability to draw elegant, swashing curves, an intuition for how to taper the width of a curving line as it sweeps around, the addition of tiny flourishes that are decorative and retinally pleasing, but that don't affect category membership at all, and so on. Our domain, in contrast to this, involves very stripped-down letterforms, which are much closer to the concepts behind them than are fully fleshed-out letterforms. In particular, no motor skills involving complex and dynamic hand movements are involved at all.

The initial inspiration for the Letter Spirit project was, admittedly, the idea of designing and implementing a computer program that could itself design full-fledged typefaces — that is, typefaces having the same order of complexity as "official" typefaces such as Baskerville, Helvetica, Tiffany,

---

Figure X–4. The Letter/Spirit matrix, illustrating the two orthogonal notions on which the Letter Spirit project is based. What do all items in any column have in common? **Letter**. *What do all items in any row have in common?* **Spirit**. *Letters can thus be thought of as "vertical categories", spirits as "horizontal categories". In this sense, the two types of category are orthogonal.*

A frequent preconception is that all members of any given letter category have "the same shape"; this figure shows how naïve that assumption is. There is not even any fixed shape lurking within or behind the diverse manifestations of each letter category — nothing resembling a fixed skeletal shape in all the items constituting a given column. Each letter category is, rather, an abstraction of high order. In contrast to what people say about letters, no one would ever think of suggesting that all the diverse carriers of a given style or spirit share "the same shape" — it is intuitively obvious that such a claim would make no sense, and that what binds together all the items in any given row is, instead, some set of very abstract qualities. One of the several purposes of this figure is to underscore the fact that letter categories, like styles, are abstractions, not shapes, and thus to suggest that the deep issues about perception and shared essence that are so self-evidently raised by the "horizontal" question (whose answer is "spirit") are already all raised, though in a subtler manner, by the simpler-seeming "vertical" question (whose answer is "letter"). It would follow that a full and final model of the human perception of **letters** *would necessarily include a full and final model of the human perception of artistic* **styles***. This provocative claim of course flies in the face of the assumptions underlying many ongoing attempts to mechanize the reading of handwriting and other types of written input, which are based on highly pragmatic assumptions about the nature of perception and which pay no attention at all to questions of esthetics or style.*

Palatino, Optima, Avant Garde, Friz Quadrata, Romic, Souvenir, Olive Antique, Aachen, Vivaldi, Americana, Frutiger, Hobo, Korinna, Benguiat, Baby Teeth, Mistral, Banco, Eurostile, Piccadilly, Calypso, Magnificat, and on and on. (There are thousands and thousands of "official" typefaces, ranging from very conservative to almost unbelievably fanciful — and their number grows constantly. There are many easily-available collections of typefaces for those interested in exploring this amazing little universe of fantasy; we particularly recommend the Letraset Graphic Art Materials Reference Manual, put out periodically by the Letraset Corporation and available quite cheaply in many art-supply stores.)

This goal quickly revealed itself to be far, far too hard and also, ironically, of little interest to cognitive science. The problem is that real-world typeface design involves far too many nuances to allow practical simulation in a computer model. Moreover, any attempt at full-scale simulation of this human ability would necessarily involve the modeling of so many domain-specific details that the very general cognitive issues meant to be the core of the project — the nature of fluid concepts and the nature of style — would be completely lost among a welter of irrelevant noncognitive concerns.

For this reason it was decided very early on to eschew all the complexities involved in curves, stroke-taperings, and other aspects of curvilinearity that seem to involve low-level or intermediate-level vision rather than the high conceptual level that was the main focus. The idea was that forbidding the manipulation of surface-level aspects of letterforms would simultaneously reduce the complexity of the programming task and force concentration on deeper, more *conceptual* levels of the design process — deep style, in short.

At its deepest levels, style involves playing with the conceptual foundations of letters (*e.g.,* coming up with the idea that "x" could be conceived of as a "v"-ish entity perched on top of an inverted one, instead of the more conventional conception of it as two intersecting slashes). The distinction between deep and shallow style is of course not a black-and-white one. Instead there is a continuum, with shallow style at one end and deep style at the other. Undeniably, though, Letter Spirit is swimming in the deep end of the style pool (and DAFFODIL is splashing about in the shallow end).

Gridletters and Gridfonts

To avoid the need for modeling low-level vision and to focus attention on the deeper aspects of letter design, we eliminated all continuous variables, leaving only a small number of discrete decisions determining each letterform. Specifically, letterforms are restricted to shapes that can be made out of short line segments on a fixed grid defined by 21 points arranged in a 3 x 7 array (Chapter

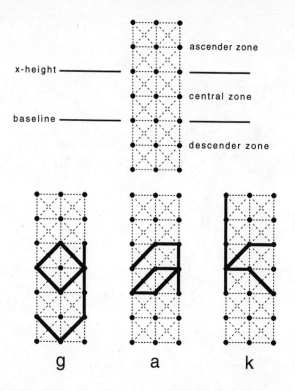

Figure X–5. In the upper part, the Letter Spirit grid. Legal "quanta" are the short dashed line segments connecting pairs of immediately neighboring dots. Thus there are 14 horizontal, 18 vertical, and 24 diagonal quanta, making a total of 56 quanta. In the lower part, three sample gridletters, formed by turning on various quanta in the grid.

24 of Hofstadter, 1985, and Hofstadter, 1987b). Legal line segments, called *quanta,* are those that connect any point to any of its nearest neighbors horizontally, vertically, or diagonally. There are thus 56 possible quanta on the grid, as shown in Figure X–5.

Stick-shapes created on the grid might be considered by some people as mere skeletons, as opposed to what might be called "full-fleshed" letterforms. This viewpoint would suggest that *genuine* letterforms would be obtained only after one had fleshed out the supposed skeletons by adding curvature, thickness, ornamental curlicues, and so on. (Note that this conception of style as the decorative fleshing-out of a set of underlying skeletal shapes is precisely the premise underlying the DAFFODIL project.) Although such fleshing-out could certainly be done to any gridfont, and would doubtless provide esthetically pleasing results, we choose to think of the gridbound letterforms as full-fledged letters in themselves. (In fact, such fleshing-out of gridfonts has been done a few times, with results that tickle the retina but leave the cerebrum pretty bored.)

Standard Square

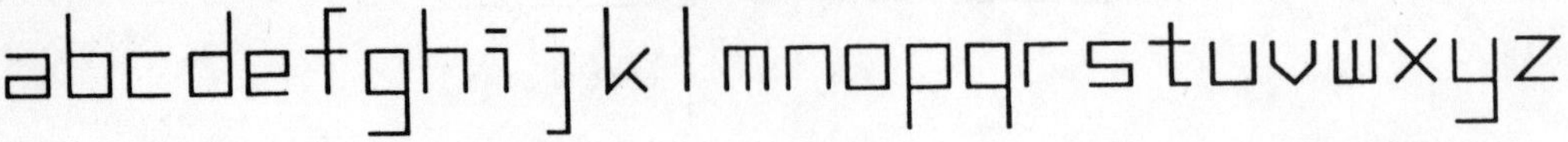

abcdefghijklmnopqrstuvwxyz

Double Backslash

Hint Four

Intersect

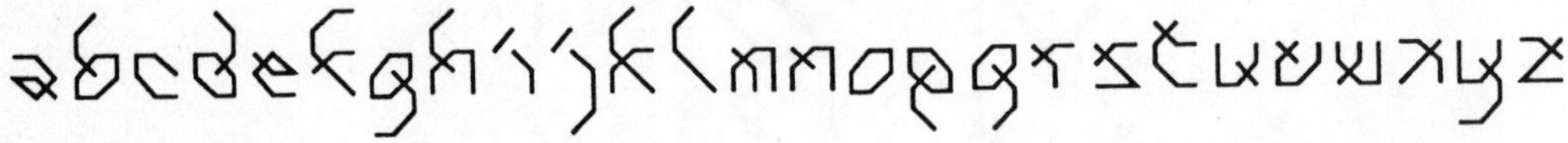

Snout

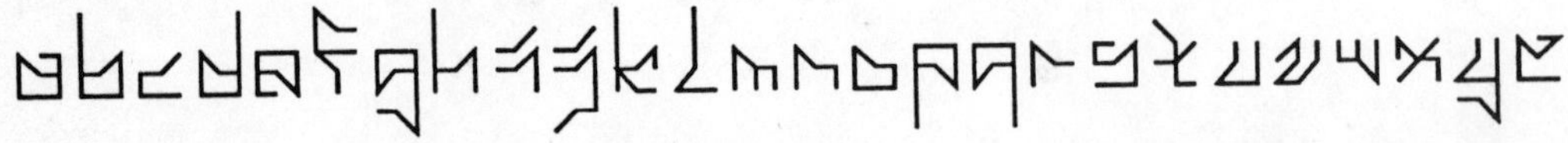

Bowtie

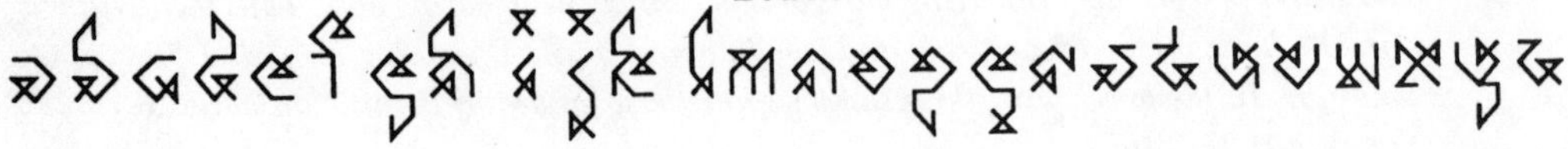

Weird Arrow

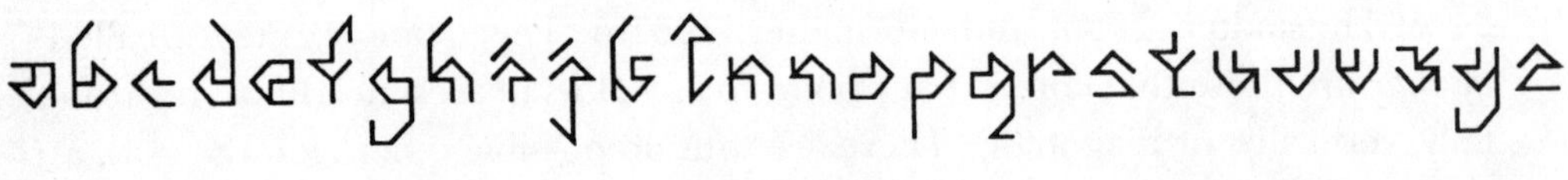

Sabretooth

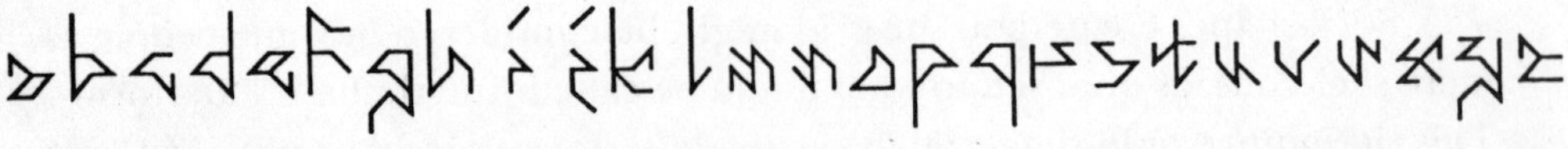

Sluice

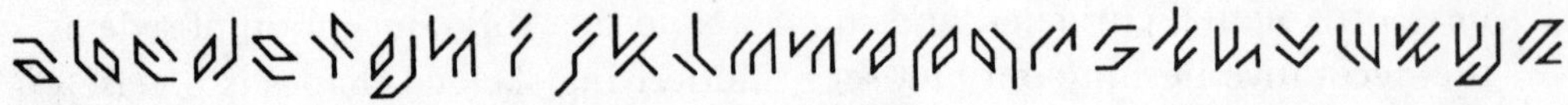

Flournoy Ranch

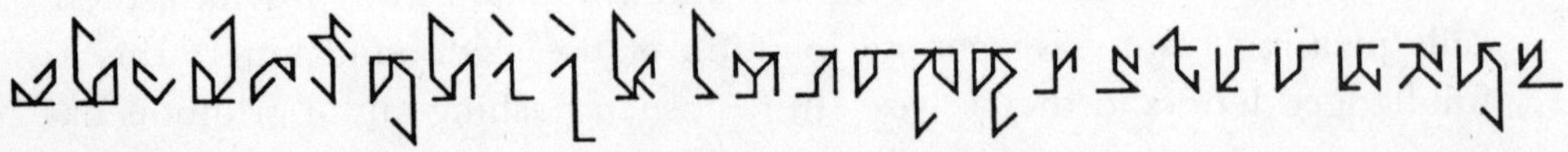

Seen as stand-alone typefaces, many gridfonts are visually fascinating — even beautiful (see Figure X–6 as well as Figure X–9) — and might make acceptable typefaces for certain purposes (*e.g.,* in those kinds of advertising where it is desired to communicate a sense of starkness, strangeness, and audacity), but most of their beauty is more cerebral, in the sense that it flows from the ways in which they explore the conceptual essences of letters and the complex interrelationships among abstract categories, as opposed to the more sensual, conventional, retinal beauty of fully fleshed-out, curvilinear typefaces, whose members usually stick safely close to the centers of their categories. Another way of putting it is that the beauty of many ordinary typefaces is *local* — you see grace in every single letter on its own — whereas in some of the most creative and exciting gridfonts, the beauty is often only *global* — individual letters seen in isolation may be disorienting or even shocking, but taken together, sets of letters form fascinating, highly appealing, internally consistent patterns.

Since all there is to do in designing a gridfont letter is to turn quanta on or off, decisions on the grid are coarse. Surprisingly, the variety among letters in a given category is still huge — hundreds of versions of each letter have been designed by humans. Of course, people are generally much less interested in sitting around exploring the space of possible gridbound "k"'s or "q"'s, say, than in designing a small number of full gridfonts. So far, about 600 complete gridfonts have been designed, a tiny selection of which is given in Figures X–6 and X–9. (For quite a few more, see Hofstadter, 1987b.) Surveying any large collection of gridfonts, one cannot help but be struck by the amazing diversity present in each letter-category.

The Grid Engenders Exotic Letterforms and Wild Styles

Almost paradoxically, it is the grid's severe limitations that seem to engender this diversity. Figure X–7 serves to illustrate the richness that the Letter Spirit grid allows (or rather, encourages) even within a single category. It shows 88 of the approximately 1500 different "a"'s, ranging from very weak to very strong, that have been created by people.

Figure X–6. Ten human-designed gridfonts suggesting the richness of the Letter Spirit domain. Readers may wish to identify the diverse means used to breathe a uniform spirit into the alphabet (motifs, abstract rules, norm-violations, and so forth); some readers may even wish to cover up part of a gridfont and try to see how they would extend the visible portion. Such style-extrapolation puzzles, featuring much ambiguity and demanding highly esthetic judgments, are in many ways reminiscent of the sequence-extrapolation puzzles that formed the focus of Chapter 1.

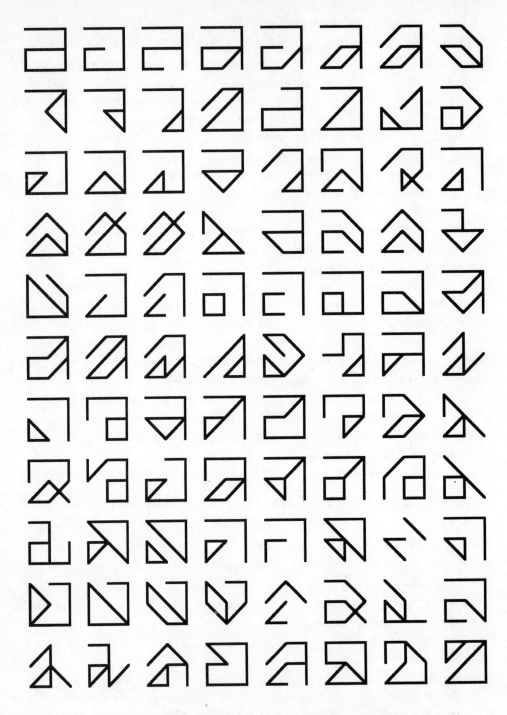

Figure X–7. Study #2 in lowercase "a": showing how the constraints of the grid enhance the tendency to play at the fringes of letter-categories. Of course some of these "a"'s are stronger members of the category than others, and that is deliberate. By the same token, some exude style more strongly than others do.

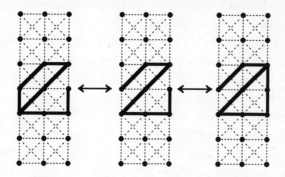

Figure X–8. Three gridletters that are semantically distant but syntactically close.

Since it is impossible to make tiny changes on the grid, any two instantiations of a given Platonic letter will necessarily be significantly different. Indeed, the addition or subtraction of just one quantum often fundamentally affects category membership. Figure X–8 shows the large distances in letter-space that one can be carried through by minimal changes on the grid. A "small" change in the letterform — the erasure of just one quantum — changes the leftmost shape from a strong "e" into a strong "z". Then a one-quantum addition transforms the "z" into an "a". So category membership in the Letter Spirit domain is a very tricky matter. In fact, modeling the process of deciding which of the 26 categories a given shape on the grid belongs to (if any) is one of the hardest aspects of the project and has represented the major focus of our initial implementation work.

The constraints defined by the grid actually encourage tampering with letter-concepts at a deep level. When one is designing gridfonts one after another, since there are no fine-grained features to manipulate, one soon runs out of conservative or "safe" possibilities for each letter, and so one ends up playing near, at, or even slightly beyond the boundaries of the 26 categories. Therefore, as one's collection of gridfonts grows, one's letters gradually get zanier and zanier, more and more adventuresome, harder and harder to read. Consequently, many gridfonts are wild, sometimes having angular, blocky, spiky, sparse, or otherwise bizarre letterforms, as Figures X–6 and X–9 make abundantly clear. Their surface-level strangeness does not mean, however, that these alphabets are primitive or naïve.

As if this were not already apparent, we must stress that we are not striving for typefaces with maximal readability (*i.e.,* typefaces in which it would be appropriate to print books) or even for letterforms beautiful at the surface level. Rather, we are attempting to understand the conceptual nature of letterforms and thereby to gain insight into what imbues concepts *in general* with their fluidity. Pushing the 26 alphabetic categories out to or near their edges in a coordinated way often results in an intellectual beauty not localized in any single letterform of a gridfont, but spread out over the gridfont as a whole — the beauty of *spirit* rather than of *letter*.

While at first glance, the Letter Spirit domain might be shrugged off by some cognitive scientists as merely another "toy domain", this would be a gross underestimation of its subtlety. In spite of — or more accurately, because of — the reduction to the grid, the Letter Spirit challenge is, in terms of cognitive-science issues, extremely rich. Rather miraculously, the cognitive issues wind up getting *magnified*, not reduced, by the act of simplifying the domain. All that has been thrown out is the need for many years of prior experience. One does not need to be a professional typeface designer or lifelong student of letterforms to appreciate the consistency of a well-designed gridfont. One does not have to have a marvelously fluent hand and years of practice in drawing subtle curves to design a passable gridfont, although designing a sophisticated one is, admittedly, still very difficult. (Were it not, creativity would be a simple thing to model!)

The Principal Varieties of Stylistic Determiners

Numerous types of stylistic pattern characterize a gridfont as a whole, among which, in our experience, the following seem to be the most important:

- A *role trait* characterizes how a specific role tends to be instantiated, independently of the specific letters it belongs to. In other words, a role trait is a "portable norm-violation" — a norm-violation attached to a specific role and thus capable of "infecting" a number of different letters. For instance, in a particular style, the roles of *post* (an ascender, as in "b") and *stem* (a descender, as in "q") might be realized several times over in a zigzaggy manner; in some other style, the role of *bowl* might be realized in various letters in a tall and narrow manner; in yet another style, the role of *crossbar* might be realized in one or more letters by passing through a "hole" or "gap" in the associated vertical stroke; and so on. Each of these stylistic features is a role trait: a description of how a specific role has been (or might become) instantiated in various letters.
- A *motif* is a geometric shape used over and over again in many letters. If it is very simple (*e.g.*, a two-quantum backslash crossing the central zone), it may be required to appear in complete form in every single letter. If the motif is a more complicated shape (*e.g.*, a two-quantum-by-two-quantum square, or a tilted hexago-

nal ring), then parts of it may be allowed to be absent from various letters, so long as a reasonably substantial portion of it remains. Some styles allow a motif to appear in reflected, rotated, and/or translated form; others allow translation but no reflection or rotation; yet others allow reflection and/or rotation but no translation; and so on.

- An *abstract rule* is a systematic self-imposed constraint, such as: allowing no diagonal quanta; allowing only diagonal quanta; requiring each letter to consist of precisely two disjoint parts; forbidding any straight section of more than one quantum in length; and so on. It is not that there is some particular shape that is repeated, but rather that an easy-to-describe abstract quality of some sort is globally enforced.

- *Levels of enforcement.* The three preceding types of stylistic attribute pertain directly to shapes on the grid. A much more abstract determiner of style — in fact, a kind of "meta-level" aspect of style — is the degree to which any such constraint is considered "unslippable" (*i.e.,* absolute or inviolable), as opposed to being "slippable" (*i.e.,* allowed to be disrespected under sufficiently great pressure). The level of enforcement of a stylistic constraint — strict, lax, or somewhere in between — sets an abstract, almost intangible tone for the entire gridfont.

The Spirits behind Four Sample Gridfonts

The four human-designed gridfonts named *Benzene, Square Curl, Checkmark,* and *Funtnip,* displayed in Figure X–9, are excellent for illustrating these stylistic-consistency devices. These gridfonts also demonstrate how deeply style can pervade, and occasionally torture, the 26 alphabetic categories.

We shall briefly comment on these four fonts, one by one.

- In *Benzene,* the six-sided "benzene-ring" motif is found in full or fragmentary form in nearly every letter. Ironically, the first letter designed in *Benzene* — the letter that sparked the entire font — was the "a", whose ring has only four sides. The full ring came into being only when the attempt was made to export this motif to the category "b", at which point the constraints of the grid, interacting with the norms of that category, forced the six-sided benzene shape to materialize (see the upper half of Figure X–10). But the smaller original diamond motif was not abandoned; indeed, it is

Benzene

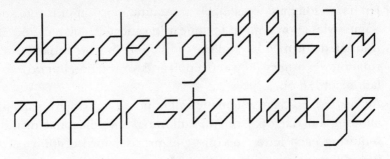

Square Curl

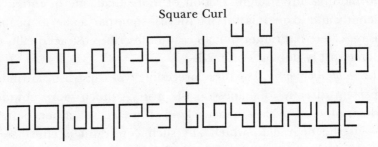

Checkmark

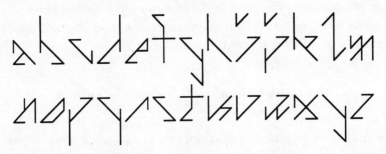

Funtnip

Figure X–9. Four human-designed gridfonts illustrating various mechanisms, including rigid and free motifs, abstract rules, and norm-violations, for propagating style or spirit back and forth along the length of the alphabet. See the text for detailed comments on these four styles.

taken up again in the "e" and, in rotated form, as the dot on the "i" and "j". One even sees it a little bit in the "m", "w", and "z".

Note that the full-size benzene ring (namely, the "o") is a *rigid* motif, in the sense that it is never — or almost never — rotated or reflected. The biggest exception to this rigidity is the *Benzene* "x", which was constructed on the English kissing-crescents model. To do this, the "o" was sliced down its middle to make two crescents, and then those crescents were swapped and put back together, so that they meet back to back (see the lower half of Figure X–10). What's particularly interesting about this "x" is that despite its origin in the kissing-crescents conceptualization, the eye far more readily perceives it as consisting of two crossing components that fill the roles in the rival "crossing-slashes" conceptualization. In other words, the subtle conceptual origin of this letterform is deeply hidden. Such "covering of one's tracks" is a surprisingly common occurrence in creative acts, and often makes them seem more magical than they actually are. (More on the role of tracks-covering in creativity is found in Hofstadter, 1987a.)

- The gridfont named *Square Curl* was a consequence of the inter-action of a strict abstract rule (forbiddenness of diagonal quanta) with a *free* motif — the "curl" seen most clearly in the "n" and "u" ("free" meaning it is free to rotate or flip). Square Curl also involves a norm-violation — namely, its ascenders and descenders are shorter than usual, in the sense of not reaching the top or bottom of the grid's ascender and descender zones. This norm-violation and the curl motif have a certain mutual resonance, in that the curl motif as manifested in letters like "a" and "e" likewise involves a role that falls short of its usual destination, which gently suggests that it would be fitting for yet other roles to "fall short" in some sense — as in the horizontal tails of the descenders of the "g" and "y", or the tips of the "s" and "z", and so on.

The "i" and "j" playfully violate the tried-and-true convention — the norm — of letter designers, according to which those letters' dots should be lower than or as high as, but certainly not higher than, the tops of letters with ascenders (such as "b" and "k"). Despite this norm-violation, these dots do not damage the font but add to its verve. This is an example of a general principle in design, which says that when norms are violated *ignorantly* (*i.e.,* by a novice), the results are likely to be bad, whereas when they are violated *knowingly* (*i.e.,* by an expert), the result is likely to be just fine. Experts have a keen sense for when they can get away with

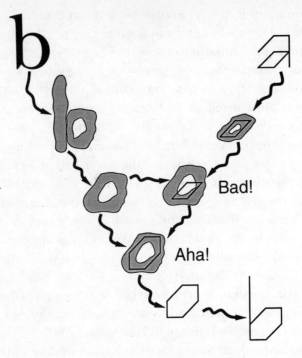

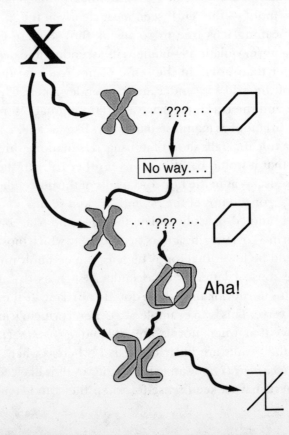

something funny, whereas novices don't even realize that there is anything funny going on.

- The free motif that gives the style called *Checkmark* its name is the shape of the "r", which in rotated, reflected, translated, and/or truncated form is visible throughout the alphabet. Another of the key ideas underlying this font is that of *gaps* or *absences,* closely related to the notion of *incompleteness.* In the "a" and "b", for instance, there are little gaps caused by lines being shorter than one would expect. This theme of absence or incompleteness is even stronger in the "h" and "y", among others. On the other hand, sometimes one gets a bit *more* than one expects, as in the

---

Figure X–10. Schematic representations of the stories behind two Benzene letters.

The upper story takes place at the very early stages of the design of Benzene (long before it had acquired its name). The starting point is the sole seed letter "a", shown on the right. Platonic "b" suggests a post and a closed loop as roles to be realized. In response, the "a" proposes its own loop — a small diamond — as a potential way to realize the desired loop for the "b". But this idea is found unappealing, as that shape would make an abnormally low "b" loop. (Why couldn't it be used anyway? The answer is, it certainly could, but doing so would lead to a different style altogether, most likely featuring undersized loops in the "b", "d", "g", and so on. The reason this idea was rejected by Benzene's designer is that the given "a" is such a central and elegant member of its abstract category that it seemed preferable to search for a spiritually close "b" that was just as central and elegant a member of its category.) Now it would be great if one could simply scale up the little diamond to the desired size while preserving its precise shape, but this is incompatible with the grid. However, some playing-around leads to a gridbound spiritual cousin, the six-sided "benzene ring". This discovery is the key leap of the whole font. The new shape is tried out in a "b", and it turns out to yield a very central category member. Bingo! The six-sided ring quickly becomes the main guiding motif for the rest of the alphabet, leaving its own four-sided mentor in the dust.

The next story comes in the final stages of the design of Benzene. Platonic "x" finds no way to use the by-now deeply-entrenched benzene motif in realizing its two standard roles (slash and backslash); this snag leads to the radical suggestion to try a completely different conceptualization: "kissing crescents". The two abstract crescent roles thus explore the benzene motif, trying to see how to use it to realize themselves concretely, and each finds a way to do so — the left crescent using the ring's right half, and the right one using its left half. So the benzene ring is pulled apart and then reassembled with the two concrete crescents kissing each other (talk about mixed metaphors!). Since this shape turns out, upon inspection, to embody "x"-ness very strongly, it is adopted. Ironically, it is far easier to perceive this "x" as crossing slashes than as kissing crescents.

"j", "n", and "u". Clearly the "j" was borrowing something from the descenders of letters like "g" and "p". Perhaps it did so too literal-mindedly, resulting in a shape whose category membership is maybe just a tad too teetery. That's a very subjective judgment-call, of course. The "k" exhibits a curious and rather pleasant combination of "too much here" with "too little there".

- *Funtnip* is certainly the most complex of these four fonts. One of its multiple themes is the idea of a horizontal stroke passing either *all* the way across the middle of the central zone (as in the "a, "i", and "k") or just *partway* across (as in the "f" and "m"). Only the "c" and "r" violate this rule. Another crucial theme is the "lopsided arrowhead", so to speak — a free motif most clearly seen in the "o", the "p" and "q", and in part in many other letters.

Yet another key theme is the notion that ascenders and descenders should be located halfway across the letter, rather than on the left or right side; this, of course, is a deeply norm-violating suggestion, which almost seems absurd. For letters like "p" and "q", for instance, whose distinction usually depends on little other than the location of their descenders, such an idea could only be made to work by making the bowls very asymmetric, so that *they* now carry the burden of identity that the stems usually carry. Fortunately, this can be done successfully.

Then there are other minor themes in *Funtnip,* such as the idea of a full or partial horizontal line on the baseline (seen in the first four letters and then only occasionally thereafter — *e.g.,* "k" and "n"); the idea of a perky little diagonal fillip (as in the "g" and "m" and even the "w"); the extra little "stingers" in the "a", "e", "s", and "z"; and finally, just to liven things up a bit more, the square-root-sign dots on the "i" and "j". This odd potpourri of seemingly unrelated ideas somehow manages to come together and produce a strong, well-integrated, and distinctive visual style. How does this kind of thing happen?

Disclaimer: We have no genuine theory as to how a gridfont as complex and daring and multi-themed as *Funtnip* or *Checkmark* is generated. We would aim for something far more modest, such as *Standard Square* (see Figure X–6). Even building up *Benzene* from its "a" alone would be considerably beyond our expectations for our program.

Creating a Gridfont — A Broad View

In our experience, designing a gridfont usually takes a person somewhere between ten minutes and three hours (after which the font still remains potentially subject to scrutiny and minor revision). The process involves myriad small operations, ranging from the highly mechanical to the highly inventive. A typical action (whether as large as creating a full "k" from scratch or as small as changing a single quantum in a supposedly finished "k") sets off repercussions that echo throughout the continuing design process, all over the gridfont, and at all levels of abstraction. This reverberation up and down the alphabet is largely guided by *a priori* notions of the interrelatedness of letter categories.

This is worth spelling out in some detail. Suppose one is given the "o" as a seed letter. This "o" might easily suggest a "c" (just open up a little gap on the right), and the "c" might in turn suggest an "e" (just add a crossbar). The "c" might also suggest a "d" (just add a post on the right side). The "d" could then suggest the "b", the "p", and the "q", by rotation and reflection. The "q" in turn could suggest a "g" (just bend its tail a bit), and the "g" could suggest a "y" (open up the loop and straighten out the tail), which might be rotated to give a first stab at an "h". Obviously you can cut off the "h"'s ascender and thus get an "n". Add a bit in the middle to the "n" and get an "m", then turn that upside-down and see if the resulting shape passes muster as a "w". Of course, the "w" may well suggest a "v" and a "u" (although the act of rotating the "n" provides an independent source of ideas for "u") — and on and on it goes. Out of many such inter-letter suggestions and influences, together with a vast number of intra-letter refinements, a stable gridfont eventually emerges.

For a human designer, *dissatisfaction* and *contradiction,* either within a given letterform or between letters, are the prime movers of the later stages of the design process. If Letter Spirit is to be faithful to how humans create, it must thus be capable of recognizing and resolving such conflicts, in order to create a coherent style. Sometimes the conflicts are subtle, and therefore require refined artistic judgment-calls. Modeling the ability to find conflicts, diagnose them, and convert diagnoses of problems into reasonable suggestions for solutions is thus an indispensable aspect of the project.

Any design decision will affect not only the letter currently under consideration but also conceptually close letters. For example, a specific decision as to how to instantiate the role of *post* in "b" is likely to have an enormous effect on the post of the "d", as well as on many other letters with posts, and may also noticeably influence the *stems* of the "p" and the "q". Of course, the extent and type of this influence are highly dependent on the particular style, and are in no way mechanical or formulaic. In most conventional typefaces, for instance, stems and posts are closely linked by a tight analogy, but there are certainly typefaces in which stems and posts are conceptually independent. The conscious decision that took place in the designer's head to "unlink" stems from posts — that is, not to respect the "natural" or default analogy between them — was in itself a major stylistic decision.

Most aspects of letterforms are likely to propagate their influence to varying extents through the entire gridfont. The propagating wave will probably cause many retroactive adjustments (both major and minor) to some "already finished" letters, and give rise to ideas for letters not yet designed. One design decision will typically spark others, which in turn will engender others, and so on.

Eventually, when enough "decision waves" have washed over the entire gridfont, all the letterforms begin to have a high degree of internal consistency, and a clear style begins to emerge. Once the tension of inconsistency eases up, no more large-scale changes are required. Minor adjustments may continue, but for the most part, the large-scale creative act will be finished.

This temporally-extended, serial process of what might be called *uniformization* — the gradual, slow-but-sure tightening-up of internal consistency all across the structure under construction — is an indispensable part of true creativity, and is a well-known property of such creative acts as musical composition, the writing of poetry and prose, the activities of painting and sculpture, the evolution of scientific theories, and even the design of AI programs and the writing of articles about them!

The architecture required to model such a dynamic creative process is necessarily very complex. We now move on to a description of the architecture that we are currently designing, refining, and implementing in order to make a workable Letter Spirit computer program.

The Implementation of Emergent Processing

The architecture for Letter Spirit draws on ideas developed in earlier research projects of our group — most of all Copycat, Tabletop, and Jumbo. There are some major divergences, though, between Letter Spirit and those projects. Perhaps the most significant one is the complex internal structure of Letter Spirit categories — the breakdown of wholes into roles, roles into weighted sets of abstract features, and so on. This means that the concepts in Letter Spirit — in particular, the 26 Platonic letters — will have a much richer structure and behavior than concepts in earlier work in our group.

All perceptual and creative processes in the model are emergent, in the sense that they result from the actions of a large number of independent *codelets* — computational micro-agents that create, examine, and modify structures representing parts, roles, letters, stylistic traits, and so forth, and that do so in (simulated) parallel. Codelets continually come into being and wait to be run in a structure called the *Coderack,* which can be thought of as a stochastic waiting room. In contrast to a standard operating-systems queue, where processes wait before being deterministically given their periodic slices of CPU time, the Coderack features probabilistic selection of actions. To each codelet is attached an *urgency* value — a number that determines its probability of being chosen next. Urgency values are based on how well a codelet's possible effect coheres with structures already built.

The Coderack is the stochastic control center of Letter Spirit. Actions of every sort — gluing, labeling, scanning, matching, adjusting, regrouping, destroying, and so on — are carried out by codelets. Considered on its own, the effect of any one codelet is very slight; however, as many codelets run, their independent effects build upon one another to give rise to a coherent collective behavior.

A useful image is that of a large structure (like a bridge) being built by a colony of hundreds of ants or termites. The ants work semi-independently but cooperatively. Codelets correspond to the ants in this metaphor, and perceptual structures to the bridge. So perceptual structures develop nondeterministically but not haphazardly. From a large number of very small probabilistic decisions, each on its own being insignificant and dispensable, coherent perceptions are built up.

As old codelets run and "die", new codelets are created and placed on the Coderack. New codelets are created in two ways: some are posted as *follow-ups* by codelets that have run; others are placed on the Coderack *automatically* as certain stages of processing are reached. The population of the Coderack thus dynamically adjusts to the system's needs. Each codelet, when born, is assigned an urgency representing an estimate (usually quite crude) of the importance and utility of its possible action. Codelets that seem most likely to enhance the evolving perceptual structures will be assigned high urgencies, and thus will have a good chance of getting to run. Codelets that seem less promising will be assigned low urgencies and will probably have to wait a long time to run. The biased-random nature of codelet-picking ensures that low-urgency codelets have at least some chance of running, while ensuring that aimless processing is avoided.

Over a long period of time, *processes* are interleaved in a manner reminiscent of time-sharing. (A process consists of many codelets, which, *ex post facto,* can be seen to have been acting in concert.) One notable difference between this and conventional time-sharing is that the probabilistic selection of codelets amounts to having different processes run at different speeds, and the speeds themselves can be regulated over time to favor more-promising directions over less-promising ones. Since codelets have very small effects, it is never critical that any particular codelet get selected. What does matter is that certain broad-stroked courses of action as a whole run faster than others. Probabilistic selection based on urgencies allows this to happen.

Emergent behavior of this type constitutes a *parallel terraced scan*. The parallel terraced scan can be thought of as allocating processing power as a function of the degree of estimated promise of a future pathway. This strategy for parallel exploration has much in common with genetic-algorithm search methods and the solution to the *k*-armed-bandit problem (Holland *et al.,* 1986).

Note that the notion of "process" in Letter Spirit is very different from the notion of the same name in a classical time-sharing architecture. In the latter, a process is determined even before it is run, and can thus be conceptually broken into pieces of any desired grain size in advance of being run. In Letter Spirit, a process comes into being only as a result of certain codelets having been run — it could not have been anticipated or described beforehand. In fact, processes in Letter Spirit are in the eye of the beholder — they are not objective entities.

Four Global Memory Structures

The Letter Spirit program, as it is now materializing, contains four dynamic types of memory, each concerned with a particular level of concreteness of shapes (and concepts pertaining to shapes). These memories are:

- The *Scratchpad*, which can be thought of as *a virtual piece of paper* on which all the different letters of a given font are drawn and modified; as such, it is really more a type of external memory than an aspect of mental activity.

- The *Conceptual Memory*, which can be thought of as the program's *locus of permanent knowledge and understanding of its domain.* This consists in, for each concept, the following three facets: (1) a set of *category-membership criteria*, which, roughly speaking, specify how instances of the concept can be recognized in terms of more primitive concepts (roles); (2) a set of *explicit norms attached to the roles,* which serve to distinguish very prototypical instances of the concept from more eccentric ones; and (3) an *associative halo,* consisting of links connecting the concept with various other related concepts, essentially giving a sense of where the concept is located in "conceptual space" by saying what it most resembles.

- The *Visual Focus*, which can be thought of as *the site where perception of a given letterform occurs.* In it, perceptual structures representing a given gridletter are built up and converge to stable categorical and stylistic interpretations. It is critical that when a letterform has been designed, some agency inside the program be able to take a fresh and hopefully unbiased look at that shape, in an effort to judge whether an outsider would recognize it easily as a member of the intended category, and also whether an outsider would see it as fitting in stylistically with other already-present letters. This is what takes place in the Visual Focus. Incidentally, the ability to step back and take a somewhat objective, remote view of something that one has just produced is absolutely indispensable for a creator — but it is enormously hard to acquire.

- The *Thematic Focus*, which can be thought of as the program's *dynamically changing set of ideas about the stylistic essence of the gridfont under way*. In it are recorded stylistic observations of all sorts concerning letters already designed. If and when some of these passive observations start to be perceived as falling into patterns, those patterns can be taken as determinant of the style, meaning that they can be elevated to the status of explicit *themes* — ideas, such as motifs and rules, that play an active role in guiding further design decisions, in the sense of serving as "pressures" on the construction of further letters.

A useful perspective on these four structures is afforded by the following set of rough equivalences with various more familiar types of human or computer memories. The Scratchpad can, as said above, be thought of as an *external memory device*; the Conceptual Memory as a *permanent semantic memory* containing both procedural and declarative knowledge of the system's repertoire of concepts; the Visual Focus as a *subcognitive workspace* (that is, a very-short-term cache-like working memory in which parallel perceptual processes, mostly occurring below the system's threshold of awareness, collectively give rise to rapid visual classification of a shape, whose final category assignment is made cognitively accessible); and finally, the Thematic Focus can be thought of as a *cognitive workspace* (that is, a much slower, and therefore more conscious, level of working memory in which abstractions derived from more concrete and primary perceptions are stored, compared, and modified). We now describe each of these memory structures in more detail.

The *Scratchpad* is the place where experimental letterforms are created and critically examined. At the beginning of a run it is empty; by the end of a run it contains at least 26 completed letterforms. We say "at least" since it is certainly possible for the program to have created and stored a number of alternate forms for a given letter. This is a frequent occurrence in design by humans, and we see no reason to force the program to completely discard ideas that it creates, even if it winds up preferring other ones in the end. Such near-miss letterforms are in fact among the most useful windows onto the creative process, and often can serve as seed letterforms for another gridfont.

It is amusing to note that in this way, at least theoretically, the program could generate its own seed letters rather than needing human input to inspire a new gridfont. Hypothetically, one can envision the program finishing one gridfont and then being "eager" to use rejected ideas as the starting point for new gridfonts, thus becoming an entirely self-driving creator!

Little needs to be said about how the Scratchpad is structured; it simply contains an arbitrary number of grids, each of which is a 56-bit data structure telling which of the 56 quanta in the grid are turned on and which are off, together with an optional category label pointing off to the intended category (effectively saying, for example, "This is supposed to be a 'k'."). The judgment as to whether the given set of quanta actually succeed in their collective ambition to embody "k"-ness or "g"-ness or whatever is of course carried out in the Visual Focus.

The *Conceptual Memory* provides each concept in the domain with an internal definitional structure and a local conceptual neighborhood. Roughly speaking, a concept's "internal definitional structure" consists of its specification in terms of simpler concepts, and a concept's "local conceptual neighborhood" consists of its links to peer concepts in conceptual space. The internal definitional structure itself breaks down into two facets: *category-membership criteria* and *explicit norms*. Finally, a concept's local conceptual neighborhood is known as its *associative halo,* and one of its main purposes is to serve as the *source of conceptual slippability*. What this means is that, in times of severe pressure, the possibility arises that the concept itself might "slip" to some concept in its associative halo, with closer concepts of course being more likely destinations. This means that the nearby concept is tried out, and possibly accepted, as a substitute for the concept itself. We now say a bit more about each of these three aspects of a concept.

Category-membership criteria specify perceptual criteria that contribute toward membership in the category, with different weights attached to the various criteria, reflecting their level of importance. The purpose of this weighted set of criteria is, roughly, to "reduce" the concept to a collection of more primitive, more syntactic notions (*e.g.,* a whole letter is "reduced" to a set of interacting roles, and likewise, a role is "reduced" to a weighted set of desired properties of a set of bonded quanta).

By contrast, a set of *explicit norms* exists (at least for roles and r-roles, which are sufficiently "semantic" concepts), the purpose of which is to make the "core" of the concept explicitly accessible to agents seeking ways to make a weak instance of the concept stronger, or conversely, to make a strong instance somewhat weaker without casting its category membership into complete limbo. Norms represent some of the program's declarative (*i.e.*, cognitively accessible) knowledge about the internal structure of each category in its domain.

A concept's *associative halo* is defined by a set of links of different lengths connecting the concept with various other concepts. These links, whose lengths are not fixed but dynamic, encode such information as: standard resemblances between letters (*e.g.*, between "n" and "u", after a rotation through 180 degrees), standard analogical relationships between different types of roles (*e.g.*, between *post* and *stem*), conceptual proximity of various descriptors used in specifying norm-violations (*e.g.*, "high" is the opposite of "low"), and so on.

Knowledge of letter–letter resemblances serves a dual function. It not only helps in designing new letters (*e.g.*, a good heuristic for a first stab at "y" is to rotate an already-designed "h"), but also serves as a warning that one letter has a tendency to be confused with another (*e.g.*, "b" and "h" are close enough categories that a shape intended for one of them could easily turn out to lie inside the territory of the other).

The set of links emanating outwards from any concept effectively gives to that concept a halo of close, potential-substitute concepts — concepts to which it might slip under sufficiently strong contextual pressures. We will see the critical role that conceptual slippage plays when we go through an example of the designing of one letter under the stylistic influence of another.

The ***Visual Focus*** is where recognition of a single letterform takes place in a series of stages. This process is schematically summarized in Figure X–11. At the beginning, quanta belonging to a letterform are fused together in a syntactic (*i.e.*, purely bottom-up) manner into small, simple structures having *a priori* perceptual plausibility (how this is done is explained subsequently).

In a sense, the next stage is the subtlest, because it is where syntax meets semantics — that is, where the specific nature of the categories into which the perceptual processing is feeding starts to make itself felt. To be concrete, these initial natural-seeming components of the letterform — called *parts* — are explored as possible *fillers of roles*. The intermediaries or "matchmakers" between parts and roles are *syntactic labels,* which describe simple geometric properties of the parts that have been built. These labels serve to selectively evoke a few roles that are promising mates for the parts. Once some roles have been activated, they in turn suggest one or more *letter-concepts,* or "wholes". At this point, a top-down phase of perception is launched. Specifically, the most strongly-invoked whole pushes for one of its *conceptualizations* (coordinated role-sets) to be instantiated. This means that the roles involved therein try to exert top-down pressure on one or more parts to *adjust,* which means either for one part to break into two, or for two parts to fuse together or to swap quanta,

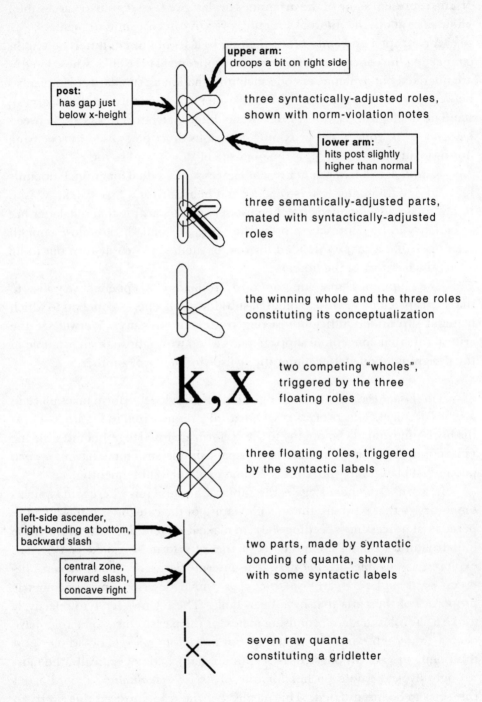

upper arm:
droops a bit on right side

post:
has gap just
below x-height

three syntactically-adjusted roles,
shown with norm-violation notes

lower arm:
hits post slightly
higher than normal

three semantically-adjusted parts,
mated with syntactically-adjusted
roles

the winning whole and the three roles
constituting its conceptualization

two competing "wholes",
triggered by the three
floating roles

three floating roles, triggered
by the syntactic labels

left-side ascender,
right-bending at bottom,
backward slash

central zone,
forward slash,
concave right

two parts, made by syntactic
bonding of quanta, shown
with some syntactic labels

seven raw quanta
constituting a gridletter

done in such a way that the resulting *semantically-adjusted parts* might better fit into roles.

The final stage of the recognition process occurs when parts and roles have mutually accommodated to yield a satisfactory realization of exactly one whole, without any strong rivals. (If the rivals do not go away, then the letterform under scrutiny is judged a poor one.)

Once a letterform's category has been identified, attributes of its *style* must also be extracted from it. Perhaps the most fundamental component of such stylistic perception involves looking at how the given parts fill their roles, and creating "stylistic notes" that summarize any norm-violations found (*e.g.*, "bent crossbar", "unattached bowl", "post not touching the baseline", etc.).

A norm-violation essentially tells how a role has "bent" — that is, compromised its ideals — in order to accommodate to what is out there in the real world. In this sense, there is a symmetric duality between the semantic adjustment of parts and the syntactic adjustment of roles.

Two other components of stylistic perception involve "filtering" and "focusing" (see the discussion of a Bongard-problem-solving architecture in Chapter 19 of Hofstadter, 1979). *Filtering* means ignoring most aspects of a shape and looking solely at abstract global qualities that it might have, such as lacking diagonals, or being filled with 135-degree angles, or being narrow, and so on. *Focusing* means concentrating on some particular local visual area, such as the top of the ascender zone, and extracting a shape from that zone. The extracted shape can then be posited as a possible motif. There is much more to the story

---

Figure X–11. What happens in the Visual Focus. This figure starts at the bottom, with a set of unattached quanta to be seen as some letter of the alphabet. To begin with, bottom-up gestalt-based agglomeration heuristics suggest **parts** *built from a few quanta. To each part are then attached* **syntactic labels**. *These labels collectively trigger* **floating roles** *(roles invoked purely from the bottom up, rather than being prompted from above by any particular letter-concept). The move from parts to roles crosses the syntactic–semantic threshold. Collectively, the floating roles suggest one or more* **wholes** *(full letter-concepts), which launch top-down processing. Specifically, the winning whole suggests a* **conceptualization** *(coordinated role-set), whose abstract roles strive to mate with the concrete parts that indirectly triggered them. To do so, the roles guide regrouping operations such as fission or fusion of parts, quantum-swapping among parts, and so on. This gives rise to* **semantically-adjusted parts**. *Much as the parts are adjusting to better fit the roles, the roles are being forced to accept less-than-ideal mates. These compromises, suitably described, are called* **norm-violations**, *and they are the roots of style. The final "marriage" is thus between semantically-adjusted parts and syntactically-adjusted roles. (Note: Since roles are not shapes but sets of norms, the role-bending shown in the top two pictures is purely metaphorical; it is intended merely to suggest that a role has to "give" a bit on a conceptual level in order to mate with a part that is already close to filling its ideals.)*

of perception of stylistic attributes of a letter, of course, but this gives a feeling for what must go on.

The **Thematic Focus** is both the court that judges, and the warehouse that stores, stylistic ideas. More poetically, the Thematic Focus is the site where spirit gradually crystallizes. Whenever any stylistic attribute is observed inside an individual letter in the Visual Focus, it has some chance of being transferred into the Thematic Focus for consideration as a candidate for potential propagation throughout the entire alphabet. Roughly speaking, the more often a specific stylistic attribute has been noticed in different letters, the more chance it has of making such an upward shift in status — rising, that is, from being merely a casual quirky observation that has no influence to the status of an officially sanctioned driving force or guiding principle — a *theme*.

Whenever a new stylistic attribute is elevated to themehood during the creation of a gridfont, the default assumption is made that it is unslippable (*i.e.*, that its level of enforcement is maximal). By definition, default assumptions are not noticed at the moment they are made, and so the new theme will start out being obeyed stringently without question. However, if down the road some letter/spirit conflict gets sufficiently severe, pressures may arise that call that automatic assumption into question, at which point the theme's level of enforcement will emerge from hiding and become an explicit variable that can be adjusted. Although it takes considerable pressure to "unbury" any of them, all themes' levels of enforcement are ultimately under the control of the program, and can thus, in principle, be tampered with during the design process. The ability to recognize and manipulate these levels of enforcement represents a type of higher-level self-awareness on the part of the system, and will hopefully make the styles it comes up with more intellectual and sophisticated.

Every stylistic attribute, including its level of enforcement, is explicitly represented in the Thematic Focus and is thus globally accessible, not just in the sense of being *observable* by agents in the program but also in the sense of being *alterable* by agents in the program. Thus thanks to the Thematic Focus, all aspects of style are under complete control of the program itself.

The Predictable Unpredictability of the Creative Process

The notion of "elevation to themehood" is a little-appreciated but pervasive aspect of creative acts. In working on a gridfont, a human designer starts by being inspired by (let us suppose) a single seed letter. Aspects of this first letter, borrowed analogically, suggest further letters. But each of these secondary letters, when looked at by a style-conscious eye, will be seen to have novel stylistic attributes that are entirely its own and that were not implicit in or implied by the seed letter alone. These unexpected attributes are a result of the serendipitous interaction of two entirely unrelated sets of constraints: one set of constraints that together specify the style (as so far perceived), and another set of constraints that together define the letter category currently being worked on. Some kind of compromise is found that accommodates both sets of constraints as well as possible.

The resulting physical shape cannot help but have some aspects — motifs or other little patterns — that simply weren't present in any previous letters. Some of these will be noticed, often by sheer chance, in the act of visually scanning the letter, and added to the growing sense of the gridfont's style.

The beauty of this process is that it gives rise to what might be called "predictable unpredictability". That is, one can rely on unpredictable new stylistic attributes to emerge, simply as by-products of the creation of new letters. And then, once any such attribute is recognized and explicitly elevated to themehood, it becomes an active force in shaping yet further letters. Of course this means that the process is in some sense recursive — new letterforms give rise to new emergent attributes, which in turn give rise to new letterforms, and so on. (It is not even so uncommon for a letterform to suggest a stylistic attribute that goes out, affects some other letters, and then comes back like a boomerang and causes the letterform itself to be modified, thus undermining the very shape that gave rise to it!) The upshot is that new, very unexpected stylistic attributes are continually emerging — ideas that the designer never had in mind at the outset and never would have dreamt of, had this serendipitous interaction of complementary constraints, iterated over and over and over again, not taken place. This highly unpredictable meandering in "style space" reflects the extreme subtlety of the creative act.

Four Interacting Emergent Agents

Emerging from the many small actions of codelets are four conceptually separable types of large-scale activities:

(1) the high-level conceptual activity of *selecting or inventing a letter-conceptualization, and then devising a letter-plan* (*i.e.,* either an idea for an as-yet undesigned letter or a possibility for improving an already-designed letter);

(2) the intermediary activity of *translating a fresh letter-plan into a concrete shape* on the Scratchpad;

(3) the relatively concrete perceptual activity of *examining a newly-drawn shape and categorizing it* (*i.e.,* deciding which letter of the alphabet it is, and how unambiguously so);

(4) the more abstract perceptual activity of *recognizing the stylistic attributes of a newly-drawn letter, and deciding how to treat them* (*i.e.,* selecting pieces of a letterform that might become motifs, noticing aspects of the letterform that might be described by a simple rule, or finding norm-violations and describing them in "exportable" ways, as well as taking note of whether a newly-observed stylistic attribute supports or opposes stylistic attributes previously observed in other letters).

It is often convenient to speak as if these emergent activities were carried out by four explicit and cleanly separable modules, together comprising the totality of the program. (These modules could be likened to the agents described in Minsky, 1985.) The names we shall apply to these hypothetical modules or agents are, respectively, the *Imaginer*, the *Drafter*, the *Examiner*, and the *Abstractor*, and we shall briefly describe each of them in turn. However, it must be borne in mind that these modules are in some sense convenient descriptive fictions, in that each one is simply an emergent by-product of the actions of many codelets, and their activities are so intertwined that they cannot be disentangled from each other in a clean way.

The ***Imaginer*** does not deal with, or even know anything about, the constraints defined by the Letter Spirit grid (*i.e.,* the fact that letterforms are made up of discrete quanta); rather, it functions exclusively at the abstract level of *roles*. Its job is to make suggestions regarding roles, which it then hands over to the Drafter (which will attempt to implement them concretely in conformity with the constraints of the grid — that is, as parts composed of quanta). The Imaginer can make suggestions of two distinct types — *norm-violation suggestions* and *role-regrouping suggestions*. Although both types can lead to highly novel instantiations of letters, suggestions of the first type tend to be tamer than ones of the latter type, since they do not cut quite so close to the core of the concept.

A *norm-violation suggestion* starts with an already-existing set of interacting roles (*i.e.,* a particular conceptualization for the letter); then, for one or more roles in that set, it suggests one or more ways of violating associated norms. For instance, suggestions such as "Bend the tip of the ascender to the right", "Use a short crossbar", "Don't let the bowl touch the post at the x-height", "Make the bowl narrow", and so on (obviously expressed in some suitable internal formalism rather than colloquial English!) would be typical norm-violation suggestions. Though fairly specific, such suggestions still require more fleshing-out to be realized on the grid itself.

A *role-regrouping suggestion* is more radical and deep, in that it involves tampering with the very essence of the letter — in other words, coming up with a

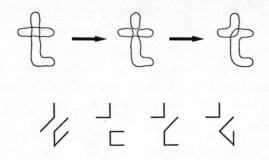

Figure X–12. A schematic illustration of the process of conceptual regrouping of the default roles in the Platonic letter "t". Below the new conceptualization are shown some sample gridbound "t"'s that would naturally follow from it but not from the default conceptualization.

new conceptualization for the letter. This means taking apart one or more known roles and making new roles that combine aspects of the old roles. An easy-to-grasp example is the above-described conceptual move from imagining "x" as two intersecting slashes to imagining it as two angle-brackets kissing *à l'anglaise*. Role-regrouping is very subtle because it takes place completely at an abstract level. That is, *no shapes are involved at any time*; rather, the Imaginer deals exclusively with abstractions — abstractions that, to be sure, have the general "feel" of shapes, in that they are associated with spatial locations and have spatial functionalities — but such abstractions are not shapes.

For the sake of concreteness, let us consider the conversion of one conceptualization of "t" into another (see Figure X–12). The idea of a "linear unit" (shorthand for the concept "long thin shape that does not bend much") that passes through some known point (more precisely, an r-role) suggests in a fairly obvious manner the idea of breaking the linear unit into two *shorter* linear units, using the point as the divider. Applying this "breakup" idea both to the *ascender* and to the *crossbar* role in a "t", we obtain four shorter roles that converge like spokes to a hub (the r-role, of course). Now comes the regrouping operation. Whereas the "west" and "east" sub-units had formed one conceptual linear unit (the crossbar), and the "south" and "north" sub-units another (the ascender), we now pull each of these units apart and regroup their pieces into two new conceptual units — one made up of the "west" and "north" sub-units, and the other of the "south" and "east" sub-units.

Despite the picture, one has to keep in mind that the pulling-apart and reattaching operations are actually just *mental*, as opposed to being splicing-operations on graphic shapes. We now have come up with two completely new roles (and of course the unperturbed *central-point* r-role), which give us a novel way of thinking about what "t" is — that is, a novel starting-point for creating new "t"'s, some of which are also found in Figure X–12. Norms for the roles belonging to this new conceptualization have to be derived from the norms associated with the roles in the old conceptualization.

Once a new conceptualization has been produced, it can be handed over directly as a suggestion to the Drafter, or norm-violation suggestions can be made in addition, and then the whole thing handed over as a package to the Drafter.

The *Drafter*, unlike the Imaginer, does know about the grid; indeed, its main function is to take the Imaginer's high-flown grid-independent sugges-tions and adapt them to the down-to-earth and severe constraints of the grid. In other words, the Drafter must convert the Imaginer's abstract fantasies into concrete instructions for drawing shapes on a grid on the Scratchpad.

Here is an example that could easily come up in designing a "t" or an "f". A norm-violation suggestion like "Make the crossbar short", which in real-life circumstances would offer a letter-designer a full continuum of possibilities, offers much less freedom to a designer restricted to the grid. For a gridbound "t" or "f", a conventional (*i.e.,* norm-respecting) crossbar would be a horizontal line segment composed of two quanta. Obeying the suggestion to make it short would thus seem to offer just three alternatives: dropping the left quantum, dropping the right one, or dropping both. Of course, dropping both quanta would seem very drastic, if not outrageous (although possibly the right solution for some very far-out styles); thus in all likelihood, the Drafter should opt for drawing just a single quantum, probably a horizontal one (see Figure X–19 on page 455). Then the question is whether it should draw it to the left or to the right of the ascender. This choice will certainly depend in part on precedents in other letters in the developing gridfont (*e.g.,* if the decision "Draw a quantum on the *left* side of the ascender" had already been made in the case of the "f", then the "t" might want to follow suit), but it will also depend on how strong the potential letter's category membership will be.

The Drafter will generally be uncritical of the suggestions it receives from "on high", but it is conceivable that it will encounter so much difficulty convert-ing them into reasonable instructions for a grid-shape that it will send back a "complaint" to the Imaginer, asking it to revise its suggestion in some way. Generally, however, feedback to the Imaginer comes from the perceptual agents — the Examiner and the Abstractor. This all-important feedback loop will be described shortly.

The *Examiner* is responsible for taking the specification of a gridletter in terms of its quanta and determining which (if any) of the 26 Platonic letter categories it belongs to, and how strongly and unambiguously so. All processing in the Examiner takes place in the Visual Focus. Since the Examiner is by far the most developed of these four agents, we will describe how it works in more detail than we do the others.

It is useful to cast the Examiner's work in terms of *syntactic* and *semantic* operations, mentioned earlier. Syntactic operations are purely bottom-up chunking operations. They serve to put quanta together in a way that would be perceptually plausible no matter what type of shape was being recognized — whether the domain was roman letters, Arabic letters, Bengali letters, Chinese characters, or even pictures. In other words, syntactic operations yield *context-free chunkings* that would presumably arise in the course of processing by any naturally-evolved visual system.

Semantic operations, on the other hand, depend on the particular set of categories into which the shapes are being channeled. In the case of letters, this set of categories will be a writing-system-specific repertoire that in a person has been acquired through years of experience. Thus the semantic chunking operations that would occur in the visual system of a Chinese person looking at Chinese characters would tend to result in the extraction of shapes quite different from those that would occur if an English speaker looked at the same characters. Semantic operations take the output of syntactic actions (which occur earlier in perceptual processing) and adjust it. Adjustment consists in making parts bigger or smaller by chopping and borrowing subparts. The end result is a set of *semantically-adjusted parts* that conform sufficiently well to expected or known abstract (*i.e.*, semantic) structures — in other words, a "marriage" of bottom-up structures coming from current sensory stimuli and top-down expectations defined by prior knowledge of letter-concepts.

The Examiner operates roughly as follows, although the following outline may give the impression that processing is more serial and well-ordered than it really is. We have described processes as taking place one after the other, which is a good first approximation, but in reality, many of them overlap in time, and to varying degrees.

Processing by the Examiner begins at the level of quanta and starts out being completely bottom-up (see Figure X–11 again). A team of codelets swarms over the quanta, probabilistically applying dabs of "glue" to pairs of touching quanta in the seed letter, with the amount of "glue" being deposited at a given juncture-point depending on various factors such as the locations and directions of the two quanta involved, the sharpness of the bend they make, and so on. (For example, more glue tends to be deposited at straight junctions than at angles.) These glue-depositing codelets execute completely bottom-up syntactic operations.

The more glue deposited in any spot, the more strongly the two quanta involved are likely to wind up bonded together into a perceptual unit — a *part*. More specifically, when enough glue has been deposited, the glued shape is metaphorically "shaken", meaning it is probabilistically broken into parts at its weaker joints, which results in a set of perceptual chunks composed of a small number of quanta, usually between two and four. These chunks are the parts. Note that since letter-concepts have played no role up to this point, parts are still purely syntactic entities.

Once a part is made, it is scanned by another team of codelets that probabilistically attach any number of *syntactic labels* to it (*e.g.,* "straight", "tall", "central-zone", "zigzag", "left-side", "slanting", "open on right", etc.). It is important to understand that labels are not exact but approximate, with probabilities being used to get the shades of gray. For example, a label such as "straight" can certainly be attached to a not-totally-straight part, but the less straight the part is, the lower the chance this will happen.

Semantic aspects now begin to enter the picture. The presence of a particular label on a given part serves as a cue that tends to lightly activate one or more roles with which the label is associated (specifically, roles of which the given label is at least somewhat diagnostic). For instance, the labels "left-diagonal", "straight", and "in central zone" would jointly tend to activate a "slash" role (as in the letter "x").

The various small contributions from different labels sum up to a *total* activation-level for each role. Each role in turn sends activation to one or more *wholes* (*i.e.,* full letter-concepts, such as "a") for which it provides evidence, in the sense that the role belongs to a known conceptualization of that letter. This triggering of wholes by roles is analogous to the triggering of roles by parts, only it occurs at a higher (more semantic) level.

Through this process, different wholes become activated to different levels. Among the leading contenders, one particular whole is now selected in a probabilistic manner (biased by level of activation) as a candidate for deeper scrutiny. To try to see how well this hopeful whole really matches the gridletter, the roles comprising it must be compared with the parts already "on the table". However, since there can be several rival conceptualizations belonging to one letter-concept, a further competition must take place among the conceptualizations. A winning conceptualization is selected, again in a biased-random manner (the bias for each conceptualization is determined by how strongly its component roles are activated), and the set of coordinated roles that comprise it now try to match themselves with the physical parts that have been built up. Note that we have entered a top-down phase of processing.

As roles and parts attempt to couple, a given part may need to be slightly *adjusted* in order to be a good mate for a given role. A quantum or two may need to be stolen from one or more neighboring parts to make the part in question more attractive to a role-suitor. Likewise, small pieces that seem to make a part ugly in the eyes of possible role-mates may need to be given away or simply detached. The resulting structures composed of quanta are now products of the combined influence of bottom-up and top-down processing. As such, they are no longer just syntactic parts, but semantically-adjusted parts. When this adjustment-and-mating phase is over, there should be a fairly strong match-up between roles and semantically-adjusted parts — at least to the extent that the letterform under scrutiny is a decent exemplar of some Platonic letter. (And if this is not the case, a warning should be issued that the letterform is problematic.)

The **Abstractor** is concerned with a more abstract type of category membership — namely, *stylistic consistency*. It is not enough for a candidate letterform to be perceived by the Examiner as a strong member of its intended letter category; that letter must also be judged by the Abstractor as embodying the same stylistic qualities as the seed letter(s) and any already-generated letters. This requires a set of high-level descriptions of stylistic qualities to be manufactured as the alphabet develops. Obviously, no single letter contains all the information about the style it belongs to, and so stylistic attributes from various letters, as they come into existence, must be assembled in a global list belonging to the gridfont as a whole. This list is of course the Thematic Focus. Thus whereas 26 Platonic *letter* categories exist in the Conceptual Memory *prior* to any run, a single *spirit* category gradually comes into existence in the Thematic Focus *during* a run.

The types of already-observed stylistic attributes that the Abstractor looks for in order to judge a candidate letterform for its consistency with the style (which, as it is constantly evolving, constitutes a moving target) include those mentioned earlier — role traits, motifs, abstract rules, and levels of enforcement. The new letter is then assigned a "stylistic-coherency rating" according to how many prior themes are echoed in it, and how strongly they are present.

The Abstractor also tries to extract from any new letter *new* stylistic attributes, in order to extend the set of themes that collectively define the emerging style (the new "horizontal category"). Figure X–13 illustrates some possible stylistic attributes that might be noticed when a particular "t" is looked at by the Abstractor.

An attribute discovered in a single new letter may not be considered strong enough to be elevated to themehood, but if that observation is reinforced by finding the attribute echoed in other new letters, then it stands a good chance of becoming a new theme and, as such, helping to drive the design of further letters and even the retroactive modification of older letters.

As was pointed out before, one completely predictable thing is that stylistic attributes will emerge in a completely unpredictable fashion. Here's a very simple but very typical example. Suppose the Abstractor happens to notice that neither the seed letter nor the first new letter generated contains any forward slashes. This could well be simply a coincidence, not at all intended. Nonethe-

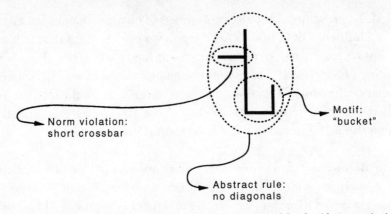

*Figure X–13. Three stylistic attributes that might be posited by the Abstractor in its scanning of a particular "t": (1) an interdiction of diagonals (an **abstract rule**); (2) a "bucket" (a **motif**); (3) abnormally short crossbars (a **norm-violation**).*

less, the Abstractor may then generalize from these two examples and thereafter strictly forbid forward slashes. Such a decision, quite unforeseeable, will of course have global ramifications. On a different run where the same two letters existed, a totally different course could be taken if that observation were not made, or even if it were, if the interdiction of slashes were taken loosely.

As this brief survey shows, recognizing stylistic characteristics and formulating them appropriately (neither too specific nor too general; neither too rigid nor too loose; etc.) is most complicated, and the Abstractor's job is therefore a very subtle one.

The Central Feedback Loop of Creativity

A major part of the Letter Spirit challenge is the problem of *attempting to do in one framework something that has already been done in a significantly different framework.* Here, the two frameworks are different letter categories, such as "d" and "b". A designer is handed a somewhat off-center member of the first category, such as the "d" shown in Figure X–14, and the challenge is to transport its eccentricity — its stylistic essence — into the other category, or in other words, to reproduce the *spirit* that it embodies in the second framework, despite the fact that the second framework is by no means isomorphic to the first. (Many people think that "d" and "b" actually *are* isomorphic categories, exactly identical but for spatial reflection. This is a fair first approximation to reality, but far from the full truth, as we are about to see.) What would be an appropriate "d" in the same spirit as this "b"? This is a typical, if small, Letter Spirit challenge, which we hope the reader will take seriously.

Such transport of spirit cannot be done in a reliable manner except by trial and error. Guesses must be made and their results evaluated, then refined

Create a "b" in the style of this "d":

Figure X–14. An analogy puzzle typical of those that crop up all the time in the creation of uniform alphabetic styles.

and evaluated again, and so on, until something satisfactory emerges in the end. We refer to this necessarily iterative process of guesswork and evaluation as "the central feedback loop of creativity". (Of course, it is not anything close to being a "loop" in the strict computer-science sense of the term — it is, rather, a back-and-forth flow of intensity of activity in different types of tasks.)

In the case of the above challenge, the most obvious move — namely, mirror-reflecting the given "d" — yields an attractive candidate shape (see the left side of Figure X–15), but one whose category membership is unfortunately quite ambiguous, since it is not fully closed at the bottom. Although this shape might well be seen as a "b", it might just as easily be seen as an "h". Such ambiguity is certainly not a desirable property for a letterform to have. What can be done?

This is an analogy problem featuring a *snag*, somewhat reminiscent of the snag in the Copycat analogy problem "If **abc** goes to **abd**, what does **xyz** go to?" (see Chapter 5). As in that problem, the escape hatch is provided by a *conceptual slippage* — in fact, a very similar one, seen on a sufficiently abstract level. In particular, both escape routes involve a slippage from one concept to its opposite. Here, the description of the norm-violating bowl slips from "unattached at *bottom*" into "unattached at *top*". This yields the solution shown in the right side of Figure X–15. Note that the new shape's category membership is quite unambiguous, and yet it manages to capture, at a high level of abstraction, much of the spirit of the seed letter. (Precisely *how* much it captures is, of course, debatable. Different designers have different sets of unconscious criteria for making such judgments, and those differences give rise to very different styles of style-making.)

Taken together, this "b" and the original "d" suggest all sorts of rich new ideas for the rest of a hypothetical alphabet. This little example gives the flavor of the central loop of the creative process, at least as it occurs in the human mind. It is, incidentally, gratifying to see (Figure X–16) that in the design of Friz Quadrata, the most common display face used in this book (it is used in the display line on the facing page, for instance), the analogous solution to the analogous problem was found.

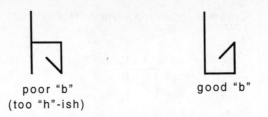

poor "b"   good "b"
(too "h"-ish)

Figure X–15. Two possible solutions to the preceding puzzle, one weak and one strong.

The Strangely Circular Origin of a Style

Although this letter-to-letter analogy problem captures a good deal of the Letter Spirit challenge, the full challenge is actually quite a bit more complicated. There are two main ways in which our cute little "d"-gives-rise-to-"b" example is misleading. Firstly, it seems to imply that each new letter comes directly from a tight analogy based on precisely *one* previously-designed letter, whereas in fact the final design of any given letter is far more typically an amalgam of influences from any number of different letters. While it is true that the *first draft* of a new letter commonly springs from just one other letter, once the draft is "on the table" there comes a period of examination, comparison, critiquing, and revision in which other letters' influences get blended in to various degrees until there is no simple pedigree at all. Even when one finally comes to feel happy with the design for the given letter and goes on to the next one, one often finds that a few letters down the road, one is drawn back to that letterform and feels a need to tamper with it further. Indeed, it is often very hard to decide when one has reached the end of the design process, either of a single letter or of a gridfont as a whole.

This brings us to the second way in which the "d"-to-"b" example is misleading: namely, it suggests a clean distinction between "parent" letters and "child" letters — that is, between fully-finished letters and still-evolving letters. In this very example, for instance, the clever discovery of the "flipped" solution for the "b" could perhaps have retroactive influence on the very "d" that inspired it (one could simply mirror-reflect the new "b", for instance). This act would overturn the supposedly stable source for the analogy and thus, to some extent, destabilize the entire alphabet. After all, if *every* letter is at all moments potentially subject to revision, then where on earth is *terra firma*? What can be relied on? Where lies the ultimate source of the gridfont?

One way out of this confusion would simply be to declare any and all human-provided seed letters to be immutable and inviolate; this would establish a fixed grounding for all the remaining letters. However, it is certainly not the way that a human designer works. A more faithful imitation of true human design practice would require each designed letter on the Scratchpad and each

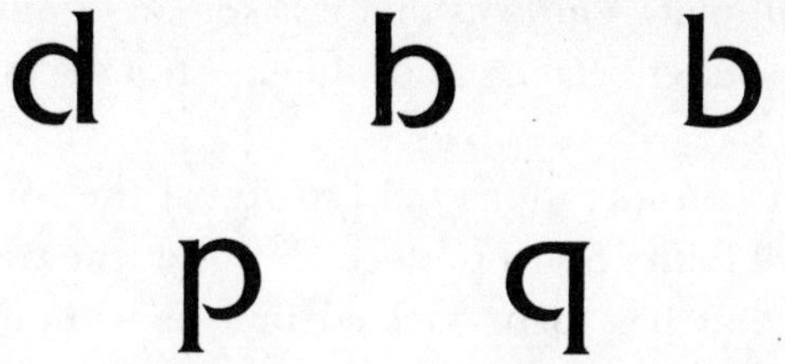

d b b

p q

Figure X–16. How essentially the same puzzle was handled by the designer of Friz Quadrata. At left, a "d" with a bowl that does not quite close at the bottom. Next to it, what happens when you reflect that "d" to make a first stab at the "b". Leaving aside the details of the serifs, one has a "b" whose bowl does not quite close at the bottom. Perhaps the designer felt this shape was slightly too "h"-ish. In any case, the choice actually made for the Friz Quadrata "b" is shown at the right. The norm-violation of non-closure is maintained, but is shifted from the bottom to the top of the bowl, as in the previous gridfont puzzle. For purposes of comparison, the "p" and "q" of Friz Quadrata are also shown, revealing bowls that fail to close at the bottom. Can the reader imagine how the remaining 22 letters of Friz Quadrata look? For that matter, can the reader imagine how the remaining 24 letters of the gridfont suggested in the previous two figures might look?

stylistic theme in the Thematic Focus to have a numerical *stability value* suggesting how deeply it should be respected at any given moment. Letters with low stability values would of course be likely to take their cues from ones having high stability values, rather than the reverse. But stability itself would be subject to change: each stability value would be periodically updated, and the calculation would give credit for various factors, such as being a seed letter (this should, after all, count for *something*!), being a strong member of a category, and being "consistent" (whatever that means) with other letters and themes.

What makes this whole stability notion rather convoluted is the fact that credit for consistency with other letters and themes would have to be *weighted* rather than egalitarian: in particular, you should get more points for being consistent with a letter or theme that *itself* has a high stability value. That is, stabilities themselves should serve as the weights in determining stabilities. There is thus a deep — although not paradoxical — circularity to the whole notion.

Where, then, does the ultimate source of a gridfont lie? The answer is not simple; it depends on the stage at which the question is asked. At the very outset, of course, the source can lie in only one place: the set of seed letters. There, it is totally implicit. As the style spreads to other letters, it becomes more explicit and abstracted. It also becomes less and less localized. Gradually, however, as *themes* congeal, the gridfont's source becomes somewhat "institutionalized" in them, and thus more localized after all. Nonetheless, a set of abstract themes without any concrete examples does not contain enough information to determine a style. For this reason, once a gridfont is fully created, the answer to the question about its ultimate source is that it is spread out among a small set of themes and a few key letters that embody them — not necessarily including any of the seed letters!

These kinds of considerations help to reveal the somewhat frightening complexity of the full Letter Spirit project. However, the core of the project, or at least the aspect that has to be tackled first, is without any question the challenge posed by single-letter-to-single-letter analogy-making, as illustrated by the previous example. We therefore turn to a consideration of how that type of core challenge could be implemented.

Towards the Implementation of the Central Feedback Loop

We now run through another letter-to-letter analogy, and a rather simple one at that, in order to give a sense of how the central feedback loop of creativity will be implemented in Letter Spirit, featuring an interaction of all four emergent agents. We suppose that Letter Spirit is given as its lone seed letter an "f" with a quite conventional ascender (*i.e.,* a tall vertical stroke that at its top turns rightwards and then down), but with *no crossbar at all* (see Figure X–17). What kind of gridletters might this seed inspire? What kind of overall style? Readers might think about how they would move from this "f" to other letters. What letter would *you* design next, and how would you go about it?

To begin with, the seed's letter-category must be identified (in itself a nontrivial task, given the eccentricity of the letterform). To this end, the Examiner is invoked. First comes the swarm of glue-depositing codelets. In this particular shape, for entirely syntactic reasons, relatively little glue will tend to be deposited in the juncture at the top of the vertical post. The relative lack of glue at that spot will tend to encourage two distinct syntactic parts to be made, with the break between them being located precisely there. Let us assume, then, that these two syntactic (bottom-up) parts have been made, one with four quanta, the other with two.

After being suitably labeled, these parts wake up two semantic roles: *post* and *hook*. Since there is nothing else to see, no other roles are strongly activated. This pair of activated roles now activates two *wholes* — the letter-categories "f" and "l". Counting against "f" is the lack of a crossbar, but counting against "l" is the existence of a hook at the top. This is a small dilemma. However, let us suppose that the power of two strongly-filled roles overwhelms the pressures for seeing "l", so that the shape winds up being seen as a rather strange "f" whose primary stylistic attribute is the lack of anything to fill the *crossbar* role. Thus "crossbar suppressed" — a norm-violation — is the main stylistic note attached to the given seed letter. (Of course, in principle, "no diagonals" could be a tentative abstract rule extracted from this letterform — but let's leave that aside.)

We now move from the perceptual to the generative phase. Given that "f" and "t" are linked *a priori* as similar letters in the Conceptual Memory, there is a high probability that "t" would be the next letter tackled. Let us assume, for simplicity's sake, that the "t"'s ascender would be a conventional one, so that

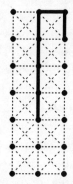

Figure X–17. A possible seed-letter "f" featuring a very striking norm-violation.

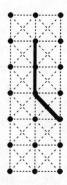

Figure X–18. One quite far-out idea for realizing the concept "t" in the same style. The spirit is willing, but the letter is weak.

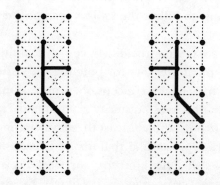

Figure X–19. Two refinements of the preceding shape. In each of them, the spirit is now somewhat diluted, but in compensation, the letter is stronger.

all we are concerned with is figuring out the nature of its crossbar. The most obvious idea for the "t" would be to suppress *its* crossbar. Like any good copycat, the Imaginer would have little trouble coming up with that analogy, since the role *crossbar* exists in both "f" and "t"; all it would need to do is take the norm-violation that describes the given "f" ("crossbar suppressed") and copy it literally into a letter-plan for the to-be-made "t". Upon receiving this norm-violation suggestion, the Drafter would have no problem converting it into a grid-oriented instruction saying, in effect, "Draw no horizontal quanta at the x-height" (the height of the top of a lowercase "x"). So far, so good. The Drafter now easily renders this suggestion for a "t" on the Scratchpad (see Figure X–18), leaving it up to the Examiner to look at it and make of it what it can.

The Examiner sends out codelets that quickly glue the quanta together, and this time, since there is no very weak spot, the quanta are likely to be bonded into just a single part. The set of labels assigned to this syntactic part collectively wakes up the role *ascender*, and since there are no other parts, no other roles are activated. This rather sparse "combination" of roles now strongly activates one *whole*— namely, the category "l". There might in addition be a rather feeble activation of the category "t", due to two facts: (1) this part ascends but does not reach as high as the "f" did; and (2) the sole letter category featuring an ascender whose default height is lower than that of "f" is "t" (if you don't believe this, just take a look at any standard typeface, such as the Baskerville of this text). Despite this weak competition from "t", the category "l" is almost sure to be found the overwhelming winner. At this point, the Examiner, knowing that "t" was intended, pronounces the attempt at "t" a failure, and provides what is hopefully an accurate diagnosis: the fact that the role *crossbar* never got awakened at all.

This information goes back to the Imaginer, which was, after all, the source of the idea of totally suppressing the crossbar. So the Imaginer is now caught in the crossfire of Letter and Spirit pressures: on the one hand, it has just learned that suppressing the crossbar leads to disaster (this is Letter pressure), but on the other hand, it wants to follow the stylistic lead of the "f" (Spirit pressure). Something has to give!

Luckily, there is a way out, provided by conceptual slippage, which involves consulting the Conceptual Memory for potential substitutes provided by conceptual halos. In the halo of the concept of "suppression", the Imaginer finds such close neighbor-concepts as "austerity", "minimality", "sparsity" — or if not all of those poetic concepts, then at least the simpler concept "underdo" (or a more formal structure representing that idea). Thus, under the pressure created by the failure of using the concept "suppress", it is quite likely that the Imaginer will make a *slippage*— namely, it will take the nearby idea "underdo" and try it on for size. In other words, the Imaginer supposes that "underdoing" the "t"'s crossbar is the next-best thing to all-out suppression of it. This slippage is of course the key creative breakthrough. The idea now just needs some fleshing-out, still to be done by the Imaginer.

In order to translate the vaguish "underdo" into a more specific operation, the Imaginer must have information about the *meaning* of "underdo". This is available through its internal definition, which (in a suitable formalism) is given as "reduce the key dimension of". Now the Imaginer must consult the norms attached to "crossbar" in order to find out if a crossbar indeed has a key dimension, and if so, what that is. It finds that for "crossbar", there is only one norm involving size, and that is horizontal length. This fortunate discovery allows the vague "underdo" hint to be straightforwardly translated into a norm-violation suggestion that says, in effect, "Make a short crossbar". This Imaginer thus hands to the Drafter a modified letter-plan that incorporates this suggestion.

From our discussion above, we know that this can lead to a "t" with a one-quantum crossbar — in other words, a perfectly acceptable and style-loaded "t" (see Figure X–19).[1] It is, of course, debatable how faithfully these *somewhat* austere "t"'s preserve the *extremely* austere spirit of the seed letter "f", but no one could dispute that either of them constitutes a pretty reasonable attempt.

This example shows how the program should be able, in some sense, to understand and imitate the *spirit* of the seed letter, rather than merely copying some aspect of it *literally*. A key role was played here by the conceptual halo of the concept "suppress", which yielded the conceptually close, potential-substitute concept "underdo".

Not all feedback in the central loop of creativity originates in the Examiner; it can originate in the Abstractor too. In contrast to the example just described, it is possible that the Examiner will be satisfied with a given letterform but the Abstractor will be dissatisfied. In such a case, the letterform is rejected and a message is sent to the Imaginer explaining how the letterform fell short of one or more *stylistic* criteria (as opposed to *letter-category* criteria, as in the little scenario just given). The Imaginer must then modify its suggestions accordingly.

The hypothetical interaction among these four high-level agents, wherein ideas are suggested, critiqued, revised, possibly even abandoned and then regenerated, and so on, meshes exactly with our intuitive sense of what human creativity really is. It seems to us fair to say that this kind of emergent,

---

1. On a purely philosophical plane, it is interesting to consider the extent to which the claim that terms like "underdo" in the Conceptual Memory have *meaning* is validated by the observable, systematic effects they have in the program's behavior. It would seem far-fetched, if this program designed a number of highly consistent gridfonts on its own, to entirely deny the semantics of the words in its Conceptual Memory.

unpredictable processing constitutes a program's *making its own decisions.* This concurs with the views expressed by Johnson-Laird (1988) that free will and creativity are closely related. It is our fervent hope to realize a program of this degree of complexity and subtlety. It remains to see how far we can actually carry it.

Brief Comments on Related Work in Letter Recognition

We now turn away from Letter Spirit itself, and try to cast it in the perspective of related work carried out in recent years. But what counts as related work? It all depends on what one considers the focus of Letter Spirit to be. If one conceives of its aim broadly, as we do, considering it to be a project whose aim is to model and thereby deepen our understanding of the nature of temporally-extended, large-scale creative processes, then any work on creativity, discovery, or design would count as related. Of course, it would be far too daunting a task to list all related projects and go through them. In the Epilogue to this book, a few such projects are discussed and critiqued.

On the other hand, one can also conceive of Letter Spirit narrowly, and construe it primarily as a project concerned with *letters* — the perception and generation thereof. In that case, related work would include all prior work done on the recognition of handwriting and printed documents, and of course any programs that purport to create consistent typefaces in an automated way. There are scads of approaches to the letter-recognition and word-recognition problem, most of them very pragmatic, but very few projects concerned with letter *design,* which is of course a much more theoretical issue. Indeed, aside from DAFFODIL, discussed at the outset, we know of only one other program that focuses on the automation of letter design — a program deliberately designed to be competition to Letter Spirit, curiously enough. This is the connectionist approach to the Letter Spirit problem by Grebert *et al.,* discussed in the following section.

Before turning to that, we briefly consider letter *recognition.* There are various approaches to letter recognition by machines, depending on the type of recognition performed (Gaillat & Berthod, 1979; Mantas, 1986). None of these approaches is very similar to our Examiner, but perhaps the most comparable approach is known as Optical Character Recognition (OCR), the highly pragmatic goal of which is the conversion of documents printed in a variety of standard typefaces into machine-readable form.

Although there are papers on OCR with impressive titles like "On the Recognition of Printed Characters of Any Font and Size" (Kahan, Pavlidis, & Baird, 1987), and although OCR hardware and software are widely available commercially, a survey of the literature shows that the problem of OCR has not by any means been satisfactorily solved. For instance, Kahan and colleagues, the authors of the just-cited paper, tested their model on only six typefaces, all of which were standard book faces, nothing at all like the fanciful letterforms often used in advertising, let alone the rather grotesque — yet still legible — letterforms in some of the more interesting human-designed gridfonts. Clearly, the title of the paper by Kahan and colleagues greatly exaggerates their actual accomplishment. (Note also that recognizing letters of different *sizes* is not a conceptual problem at all — no more than recognizing letters printed in different colors! It seems strange, therefore, to pair the challenges of "any font" and "any size" as if they posed comparable difficulties.)

A system that can recognize letters in arbitrary typefaces — even in fairly ordinary typefaces that it has not seen or been trained on — has yet to be developed. In fact, OCR pioneer Raymond Kurzweil, developer of the famous Kurzweil Reading Machine, has written (Kurzweil, 1990), "While machines exist today that can accurately recognize many type styles in common usage, no machine can successfully deal with the level of abstraction required by.... ornamental forms." By this, Kurzweil means such letters as were illustrated in our Study #1 in lowercase "a" (Figure X–2, on page 413).

Cognitive and perceptual psychologists studying shape perception from a much more theoretical viewpoint than OCR developers emphasize the building-up of multiple levels of structure from visual stimuli (Palmer, 1977, 1978; Treisman & Gelade, 1980), a view that harmonizes with the Letter Spirit approach. We go further than they do, however, in arguing not only that multi-level bottom-up chunkings of syntactic features are involved in letter recognition, but also that top-down concept-guided pressures help channel and control these bottom-up processes (see McGraw, Rehling, & Goldstone, 1994a, 1994b).

We feel it is critical that we ourselves build and test alternative recognition architectures in order to assess the relative strengths and weaknesses of the Letter Spirit approach to recognition. Netrec and Dumrec, two ongoing letter-recognition projects using architectures radically different from the Examiner/Abstractor, are described in McGraw & Drasin, 1993.

A Connectionist Approach to the Letter Spirit Challenge

We now turn to a program created deliberately as a rival to our own approach. Grebert and colleagues, inspired by the domain and the creation task of Letter Spirit, decided to try a purely connectionist approach and see how well it could do (Grebert *et al.*, 1991 and 1992). The underlying motivation of their work seems to be to demonstrate that connectionist systems are good at generalizing from examples not only for the purposes of *recognition* (*e.g.*, after having been exposed to many pieces of cursive handwriting by presidents, being able to recognize that a particular set of marks is an instance of the word "seven" and moreover that it was penned by Abraham Lincoln) but also for the purposes of *production*. To illustrate what they mean by this, and to suggest the scope of this type of task, they give the following examples of "generalizing for production", in addition to the Letter Spirit challenge itself:

- painting a portrait of George Bush in the style of Vincent van Gogh;
- singing "I wanna hold your hand" in the style of Elvis Presley;
- shopping for computer software in the style of Imelda Marcos;
- playing Monopoly in the style of Mother Teresa.

The model designed by Grebert *et al.* is called "GridFont" and is a three-layer feedforward network that learns by backpropagation (see Rumelhart, Hinton, & Williams, 1986 for an explanation of these by-now standard connectionist notions). The input layer consists of 32 nodes in all, divided conceptually as 26 + 6. The first 26 nodes represent the various letters of the alphabet, and the remaining six input nodes represent six different styles on which the system was fully or partially trained. The output layer contains 56 nodes in all, one for each quantum of the grid. The middle layer contains 88 nodes, divided into two sets of 44 each. We need not concern ourselves here with the details of the interconnections of the nodes.

The GridFont network came with zero *a priori* knowledge about letters of the alphabet. It was therefore trained on letters belonging to six different gridfonts, all of them human-designed. (Two of those fonts — *Standard Square* and *Benzene* — can be seen in Figures X–6 and X–9.) In the case of five of these styles, the network saw all 26 letters, but in the sixth style, called *Hunt Four*, it was exposed to only fourteen different letters (shown in Figure X–20).

To train it on, for example, *Benzene* "y", exactly one of the 26 "letter nodes" (namely, the 25th), as well as a particular one of the six "style nodes", was activated on its input layer; then backpropagation was used to reinforce the activation, on the output layer, of those nine nodes that represent the nine quanta that actually make up *Benzene* "y". Some 10,000 trials of this sort were carried out, at the end of which the GridFont network was able, on request, to reproduce quite reliably on its output layer all 26 letters of the five full gridfonts it had been trained on, and also the fourteen letters it had seen of *Hunt Four*.

Now, having been trained on five full gridfonts and one partial one, the network's task was, of course, to extrapolate the *Hunt Four* spirit to the remaining twelve letter categories. We suggest that readers consider this a challenge for themselves as well.

Actually, referring to this as "the network's task" is misleading, because the word "task" implies that some further processing was needed at this stage

Figure X–20. Fourteen letters fed to the GridFont network (developed by Grebert et al.) in order to suggest the human-designed style called "Hunt Four".

(as would indeed be the case if a human being — such as yourself — were in an analogous position). But further processing was *not* needed. As soon as the GridFont network had been trained on the fourteen letters of *Hunt Four,* its opinions about the remaining twelve were all implicitly present in the inter-node connection strengths, with no need for further calculation or computation. The answers could simply be read right out, one by one, by activating the twelve remaining input-letter nodes, along with the one *Hunt Four* input-style node, and reading the quanta off the output layer. Those answers are shown in Figure X–21, which shows also, for purposes of comparison, the actual choices for the same letters made by the human designer of *Hunt Four.* (We remind readers that these particular stylistic-extrapolation choices are by no means absolute and incontrovertible; they are simply the choices that were made at a particular point in time by a relatively sophisticated and experienced human designer.)

Each reader will of course form their own personal impression of these results, which is fine, but we would like to supplement that with our own commentary. Firstly, we comment on the "vertical" dimension — that of *letter.* Most of the shapes produced by the network are recognizable as members of their intended categories, although all but the "i", "l", "t", and "w" are fairly mediocre ones, in the sense of having strange extra lines that bring to mind, at least weakly, unintended categories. Aside from the "z", which is of course a complete failure, the "c", the "n", and the "v" are the shakiest, since they activate, respectively, the unintended categories "e", "m", and "w", at least a bit.

We now turn to the orthogonal dimension — that of *spirit.* There are some recognizable stylistic elements, particularly the diamond motif in the central zone, that seem to give the results a reasonable degree of credibility. Undoubtedly, the network's biggest triumph along the spirit dimension was its precise reproduction of the human designer's solution for the "i". This at first

Target letterforms from *Hunt Four*

Output from the GridFont network

Figure X–21. The remaining twelve letters of Hunt Four, above, as designed by the human inventor of the above-shown fourteen letters, and below, as produced by Grebert et al.'s GridFont network.

seems quite stunning, but when one looks back at the five full fonts the network was trained on and sees that in every single one of them, the "j" shape completely and perfectly subsumes the "i" shape, the result is considerably less impressive. One also notices that the central diamond motif was present in full form in three of the input letters, which likewise reduces the achievement somewhat.

What these examples reveal is that if two or more letter categories resemble each other in exactly the same way in one training font after another, then the statistically-generated distributed representations in the hidden layer will take advantage of that fact and encode them in an overlapping manner. Thus, as we said, across the five input fonts, there was a completely regular, standard "i"–"j" relationship, and because it was so reliable, it became embedded as an implicit rule, which was of course then respected in the test font. The same kind of absolutely reliable repetition held true, in all five input fonts, for the bowl shape as a component in all six of the letters "b", "d", "p", "q", "g", and "o". In that respect, the very carefully chosen input fonts were at one and the same time extremely *useful* (in that they supplied the program with stereotyped inter-letter connection information) and extremely *misleading* (in that such stereotyped information is of use only in the most stereotyped of situations).

A human designer of course exploits this kind of standard, highly formu- laic inter-letter relationship at times, but that has little if anything to do with what creative design is really about, which is far above the level of rigid shapes and absolutely predictable, clichéd overlaps. It is about unpredictable, highly abstract, one-of-a-kind connections that are made spontaneously on the basis of subtle analogies evoked by unique constellations of context-dependent pressures. This kind of thing is orders of magnitude away from what GridFont was doing. Distributed representations in the hidden layers of connectionist systems can do some pretty clever things, but they're not *that* clever.

No matter how long one stares at the twelve letters generated by GridFont, one does not feel that they truly share a style — *especially* when one combines them with the original fourteen, to make a full alphabet. There is something awkward, confused, and discomfiting about the network-produced letters that simply does not go away with time. (Perhaps it is at this very abstract meta-level that the twelve new letters *do* constitute a style, after all!)

The Inescapable Temporality of the Act of Creation

Let us leave aside the evaluation of the GridFont network's output, since that is, of course, a somewhat subjective matter. Let us instead turn to the question of what the point of the work by the Grebert group might be.

First of all, recall that the idea of Letter Spirit is to take just a *single* seed letter, or perhaps two or three, but certainly not half an alphabet, and to extrapolate it into a full style. When most of the solution is handed to the program for free, what is its performance supposed to show? Certainly GridFont does not carry out the style-creation challenge that we set as a goal for our own program. But even this is a minor quibble compared with our main criticism.

The real problem we see with this project is that it completely bypasses the cognitive issues that we are concerned with: what it is that happens in the mind when human beings engage themselves in tasks that require concentration over an extended period of time, tasks that require going back and looking at what one has done, judging it, adjusting it, abstracting from it, and so forth — in short, the wealth of cognitive and subcognitive activities that were subsumed under the term "central feedback loop of creativity". In genuine human creation (and in our slowly-coming-into-being Letter Spirit architecture), any design decision made during the creation of one letter has some chance of influencing the design of all the other letters in the gridfont — even supposedly "finished" ones. Indeed, one single tiny decision, made just when one thought one was finally approaching the *end* of the process, can trigger a cascade of related decisions throughout the alphabet and thus lead to an avalanche, forcing total reconsideration of every decision made, and the possible reconstruction of every single letter. This inherent unpredictability and instability of what one is doing is precisely the excitement and magic of the genuine creative act. It is closely related to the idea we described earlier, of a system's *making its own decisions*.

In stark contrast to this dynamic, vibrant image of mental activity, the GridFont network's letters are all pushed out in one single feedforward gush, without any process that maps onto *working it out*. (Technically speaking, the various letters are created one at a time, since only one letter-node can be clamped during a cycle, but the act of outputting one letterform has no subsequent effect on the outputting of any others. On a conceptual level, a full font is produced all at once, without any perception of the results, let alone back-and-forth propagating of the effects of such perceptions.) Even supposing that the results of GridFont had been superb, it would still be the case that *this kind of thing is not even remotely close to what goes on in actual human minds.*

The philosophy behind GridFont reminds us of someone trying to compose a piece of music by producing all of its different measures at once without any specific regard for how they fit together. Even if all the measures are very competent on their own, even if they are all in the style of a given composer, something essential is missing. For a piece of music to be good, there simply has to be a great deal of back-and-forth activity during the compositional process, serving to weave all of its parts tightly together. The same could be said of a poem, a painting, or an alphabetic style.

In his book *About Alphabets* (Zapf, 1970), the famous typeface designer Hermann Zapf says it took him three years to come up with his classic typeface called Optima, half of which is exhibited in Figure X–22:

A B C D E F G H I J K L M
a b c d e f g h i j k l m

Figure X–22. The first half of Hermann Zapf's typeface Optima. Can you extend the Optima spirit to the rest of the alphabet?

Three years! Admittedly, Optima is not a gridfont, but this story does suggest that the task of style extrapolation is not as easy as it may look. (Even if it actually took Zapf just three days — or just three hours or three minutes — it would still be a far cry from creating the whole thing in one big parallel feedforward gush.) Readers are hereby invited to try to come up with as much of the rest of Optima as they wish. You need not reproduce the precise *shapes* that Zapf himself came up with, of course — just preserve his Optima *spirit* in your remaining letters!

Suppose GridFont, or some four- or five-layer improvement thereupon, were some day to come up with one impeccable gridfont after another, triggered each time by just one seed letter or two (a most unlikely scenario, but never mind). This would be a bit like a clever chess program running on superfast custom chess hardware and overwhelming the human world champion, but doing so via pure brute-force lookahead, a billion board positions a second — the antithesis of how chess masters play. In a sense, such a triumph of computational chess-playing (and it may well come in the next few years) would tell us that, sadly, the domain of chess is not quite as interesting or as challenging as we had once presumed it was — but certainly it would not in the slightest deny the interest of the mental processes that go on in the minds of international grandmasters! And a comparable triumph by GridFont or its progeny would perhaps reduce our interest in the microworld of gridletters, but certainly not our curiosity about what goes on in the mind of a skilled human gridfont designer! In any case, such a success by GridFont itself or by Great-Great-GridFont is sheer fantasy at this point.

Genuine Cognition is Not a Free Lunch

The hubris of the authors of GridFont is surprising to us. Consider the paragraph with which they conclude Grebert *et al.*, 1992:

> The approach discussed here could be applicable to a range of generalization for production problems. For instance, one might eventually be able to form a network to generate speech in the style of a particular talker, and hence have a network speak George Bush's inaugural address in the style of John F. Kennedy. Of course the training corpus and the hidden unit representations would be far more extensive than those described above. Nevertheless, our success here gives us confidence that the architectural considerations discussed above — most notably the form and connection of hidden layers — will be applicable in other domains.

Leaving aside the question of whether the phrase "our success here" is warranted, we find their extrapolation from the GridFont network to such complex intangibles as John Kennedy's oratorical style or Vincent van Gogh's painting style simply fantastic. This attitude conjures up the image of a skier who, after stumbling all the way down the 100-meter slope at the so-called "Nashville Alps" in Indiana, gets up, brushes all the snow off their outfit, and in a serious, un-self-conscious tone proclaims, "Hey, man, I'm off to Aspen to tackle those black-diamond slopes — and next year when you see me take the gold in the Olympic slalom, you can say you once rode the rope tow with me way back in Indiana!"

The reason we do not feel that the very same caricature argument boomer-
angs back on us, making us question the value of our own Letter Spirit project,
is this: in developing our architecture, we have striven hard to be faithful to
what we believe to be the underlying principles of creativity, rather than relying
on the naïve blanket faith that the hidden layers of a feedforward backprop
connectionist network, when suitably trained, can do anything and everything
cognitive. Such a belief is simply an excuse for not thinking about cognition
itself and trying to get so-called *learning* (actually, just the most boring type of
learning — learning by endless repetition) to do everything for you: it is a belief
in the notion of cognition as a free lunch, courtesy of distributed representations. This is the connectionist analogue of the Boolean Dream (see the
concluding section of Chapter 2), and is equally implausible.

We believe, contrarily, that if AI and cognitive science are to clarify the
workings of the human mind, and particularly the human mind as a creative
engine, they must pay far more explicit attention to the level of *concepts* and
analogies, and move away from the magical hope that such phenomena, with
their extraordinary richness and complexity, will simply emerge somehow all by
themselves, as a result of training networks of artificial neurons. Of course,
neural hardware underpins all conceptual phenomena, but then again, so does
elementary-particle physics. The real question is: What kinds of intermediate-
level structures and mechanisms, located somewhere between quarks and the
cortex, do the work that counts?

The Letter Spirit domain — at least when construed in the way that we
intended it — forces one to pay attention to the aspects of concepts that count:
their blurry boundaries and elusive essences, their multiple levels of abstraction,
and most of all, their strange and unpredictable connections that suddenly jump
to mind as a function of some unique, unanticipated set of pressures, and that
thereby open up whole new pathways of unsuspected potential in the space of
ideas. It is our firm belief that only an architecture explicitly focused on
understanding and modeling these types of mental, not neural, phenomena will
wind up shedding light on them.

Epilogue:

On Computers, Creativity, Credit, Brain Mechanisms, and the Turing Test

DOUGLAS HOFSTADTER

A Somewhat Skeptical Perspective on Computers and Creativity

In the past several chapters, and especially the last one, we have written about our own computer models of creative acts on various scales in a few different microworlds. Needless to say, our work has not been done in a vacuum. Over the decades, there have been many attempts by AI researchers to build programs that capture something of the creative process in all sorts of domains, from humble to grandiose. What have computers accomplished as artists, writers, composers, mathematicians, scientists, inventors, chefs, football coaches, and so on?

The field is far too broad to survey in general in a mere chapter. Luckily, Margaret Boden's book *The Creative Mind: Myths and Mechanisms* (1991) does an excellent job of summarizing all sorts of recent projects. On the other hand, where Boden generally accentuates the positive, I tend to do the opposite. Perhaps her open-minded stance is related to the fact that she is not a developer of her own architecture, and thus has no axe to grind, no turf to defend, whereas I do. Or perhaps she is just a more enthusiastic spectator of the whole field of AI than I am. For whatever reason, though, I am going to make largely critical comments on the few projects I discuss.

A word of warning in advance. The tiny set of programs discussed below is not in the slightest intended to represent the "state of the art" in computer modeling of creativity, or even to be a fair or representative sampling of the field. Indeed, some of the programs discussed below are quite old, or are idiosyncratic and out of the mainstream. My critiques of the field as a whole might therefore be seen by some as totally off the mark. After all, what point is served by pointing to defects in a handful of older or weaker specimens when there are plenty of younger, more vigorous ones around?

I would reply that the critiques given below can serve in a tutorial role. They are examples that suggest a general attitude — a way to think about the field. Perhaps they shoot at easy targets, but that is how tutorials always work. One starts with easier cases and gradually works one's way up towards harder ones, using what has been learned on the way. If these programs' defects are more transparent than some other ones, so be it. My critiques may at least point the way toward more sophisticated, more complex critiques of other projects.

One last word before setting out on the warpath. In truth, I am quite fond of all of the projects that I discuss below, even if I criticize them. They are all fascinating, daring, innovative pieces of work that deserve much thought.

A Computer Artist

Among the most provocative of all programs aspiring to creativity is Aaron, a "computer artist" developed over the past two decades by Harold Cohen, a painter and professor of art. Unfortunately, it is surprisingly hard to learn about the innards of Aaron's architecture, since there is almost no publication describing it in detail. The best source of information about Aaron that I have come across is Pamela McCorduck's 1991 book *Aaron's Code: Meta-Art, Artificial Intelligence, and the Work of Harold Cohen.*

Aaron produces complex life drawings that look surprisingly like products of a sophisticated human artist. They tend to feature humanlike figures engaged in activities such as dancing, playing with beach balls, balancing, and the like, all taking place in outdoor environments featuring shapes that look like rocks, shrubbery, and so on (see Figure E–1). Aaron's drawings are often quite amusing and exhibit a certain charming naïveté, which constitutes a recognizable style, although Aaron certainly does not perceive its own drawings or interpret them in the way that human viewers do, let alone reflect about its own style on an explicit level.

Because it has a probabilistic component, Aaron never makes the same drawing twice. By now, Aaron's drawings must therefore number in the many thousands. Artworks by Aaron have been used on several book covers (usually books on AI or computer science) and some even hang on the walls of various august institutions. Because Aaron's creations could almost certainly be passed off as human art, many philosophical questions immediately spring to mind about their artistic meaning and validity. For instance, what does it mean for a computer program to draw pictures of "people" engaged in activities which with the program itself has no experience? An amusing suggestion, perhaps not totally ridiculous, might be that it would be more appropriate for Aaron to draw

Figure E–1. "Adam and Eve", a drawing made in 1986 by Harold Cohen's program Aaron.

pictures not of people but of its own kind, engaged, obviously, not in human activities but in typical computer endeavors. Indeed, perhaps Aaron's most appropriate subject would be itself, and it could draw self-portraits showing itself engaged in typical Aaron-like activities, such as drawing itself drawing itself...

From McCorduck's book, one gets the impression that Aaron has an extremely limited sense of what people really are — basically just physical objects capable of assuming certain characteristic types of shape. Aaron knows quite a bit about the perspective rendering of scenes, but then again, so does any three-dimensional graphics program, yet such programs are not usually thought of as models of human intelligence. Perhaps the key to Aaron's semblance of artistic insight lies in the following two facts: (1) its renderings are done in pencil or pen directly onto large pieces of paper, and (2) the lines it uses are shaky and uneven, in contrast to most computer output, which tends to be precise and photographic. These two simple surface-level facts make Aaron seem far more like a human artist than any other computer program.

Given today's high-powered geometric-construction programs, it would by no means be a *tour de force* to make a program that used random numbers to construct imaginary configurations of small numbers of "people" in random spatial locations and in random physical attitudes (with constraints reflecting the limitations of genuine human bodies), and then to interface this program with a three-dimensional computer-graphics program, again using random numbers to make wiggly lines, to simulate the effect of a slightly shaky hand. (AW — artificial wiggliness — is by no means a new idea; users of the letterform-processing program Metafont, described in Knuth, 1982, have for a long time played with the production of artificially wiggly letterforms via random numbers, and the results are invariably charming and often have a stunningly humanlike appearance.) The resulting output would be very much like Aaron's, but the "magic" would be gone because no pretension would be made of simulating artistic insight.

Or perhaps I am overestimating people. Maybe the cognitive illusion known as the Eliza effect (see Preface 4 for more on this subject) would have its usual clouding influence, making people read meanings in, even when they knew precisely how empty the innards actually were.

A Computer Author

I cannot help but wonder whether there is a genuine difference in kind between the artistic output of Aaron and the prose/poetry output of such programs as Racter (author of the 1984 book *The Policeman's Beard Is Half Constructed*) and Hal (author, at least in part, of the novel *Just This Once,* also discussed in Preface 4) — or conversely, whether all of these projects derive much of their interest and seeming meaningfulness thanks to visual or verbal versions of the Eliza effect.

With a complex grammar expressed in a computational formalism (such as augmented transition networks) and a respectable repertoire of words with semantic tags on them, one can get some pretty interesting and impressive output surprisingly easily (see Hofstadter, 1979, Chapters 5 and 19). The pathways through the grammar are chosen randomly (with biases, of course, but still using probabilities), and the choices are constrained by semantic properties of what has already been generated. Thus, to take a very simple example, if the verb "drink" has just been selected, you won't be allowed to select "syringe" as its object, but you will be allowed to select from "coffee", "milk", and so on. This kind of trivial insertion of semantics goes a long ways in making prose appear superficially plausible. If, in addition, occasional violations of the semantic constraints are permitted, startlingly poetic effects can be produced — strange clashes of imagery suggesting a remarkable facility with metaphor.

Needless to say, behind such output there is nothing resembling a facility with metaphor. There is no imagery behind it, no intent to say anything, no reflective word choice, and so on and so forth. What makes it seem reasonably convincing as poetry is mainly our cultural context: the fact that twentieth-century literature enormously widened the range of acceptability of poetry and prose. Our century's open-minded, "anything goes" attitude has definitely encouraged wonderful types of literary experimentation, but it has also made it far easier for impostors, human or otherwise, to crash the party and go completely undetected. Let us pursue this metaphor for a moment. It would be very hard for any robot that exists today to crash a conventional dinner party and pass for human. However, if this were a costume party with a strong expectation that the guests would dress up in outrageous ways and act as bizarrely as possible, then one might imagine that a robot could pass, at least for a short while, as a genuine human guest simply wearing a robot costume and deliberately acting mechanical.

In a way, that is how the program called "Racter" is crashing the costume party of prose and poetry generation. To make this clearer, I'll quote here a little bit of Racter's *Policeman's Beard* book:

> "War," chanted Benton, "war strangely is happiness to Diane." He was expectant but he speedily started to cry again. "Assault also is her happiness." Coldly they began to enrage and revile each other during the time that they hungrily swallowed their chicken. Suddenly Lisa sang of her desire for Diane. She crooned quickly. Her singing was inciting to Benton. He wished to assassinate her yet he sang, "Lisa, chant your valuable and interesting awareness." Lisa speedily replied. She desired possessing her own consciousness. "Benton," she spoke, "you cry that war and assault are a joy to Diane, but your consciousness is a tragedy as is your infatuation. My spirit cleverly recognizes the critical dreams of Benton. That is my pleasure." Benton saw Lisa, then began to revile her. He yodeled that Lisa possessed an infatuation for Diane, that her spirit was nervous, that she could thoughtfully murder her and she would determinedly know nothing. Lisa briskly spoke that Benton possessed a contract, an affair, and a story of that affair would give happiness to Diane. They chanted sloppily for months. At all events I quickly will stop chanting now.

That's a complete selection, unnamed. Here's another full piece entitled "Dialogue between Richard and Buckingham":

RICHARD: A week is obscurely like a night.

BUCKINGHAM: My Lord, chicken is like lamb.

RICHARD: Yet weeks can be killed as can chicken.

BUCKINGHAM: Tis true, my Liege, yet ambiguities adorn our pain as ambiguities broaden our issues.

RICHARD: Sweet Buckingham, thy commitment, decorated with Joy, begins to speak briskly to my distress. Spy me slaughter my distress tho' it take a day.

BUCKINGHAM: Noble King, you chant weeks can be slaughtered and yet assassinating chicken will not broaden our question.

RICHARD: Kinsman, you croon truth.

BUCKINGHAM: Truth loves happiness. And yet quickly we fly and soar and destroy those happinesses which are our continuing pleasure. Madden us to slaughter and we drunkenly watch the happiness of our contracts.

RICHARD: Well cried, true friend. Thy distress is prince to my own.

BUCKINGHAM: Royal prince, let us dream and our pondering will
help us gulp the intractable cup of anguish.

RICHARD: While trotting quickly yesternight I watched my home
adorned with anguish. I thought that I would commence to slaughter those counsellors who whisper their frightening tales of our
nervous birthplace.

BUCKINGHAM: Yet these solicitors are as princes to our tragedy. How
easy to slaughter a solicitor, how hard to drunkenly stud our home
with interesting happiness. And so, good prince, fascinating commitments, like steak, are as food for our dreaming.

RICHARD: Noble brother, thy tale is furious, yet slaughtering attor-
neys in truth is essential.

BUCKINGHAM: Good prince, measuredly I think that our months are
shortened by the millisecond.

RICHARD: Deepen your pondering, good brother.

BUCKINGHAM: Revile these conflicts and we may daintily bolt our
meat and quaff our sherry.

RICHARD: Well spoke, sweet brother.

In typing out this most amusing and surrealistic dialogue, I was reminded
of an experience I once had in Kansas (recounted in detail in Chapter 21 of
Hofstadter, 1985) in which I was given the opportunity to interact remotely with
what was purported to be a natural-language AI program called "Nikolai". I had
every intention to quickly and deftly unmask what I believed was a mere
program, and so I tried all sorts of probes carefully designed to reveal typical
weaknesses of computational natural-language interfaces. However, I was
stunned by what came back at me: just about every response that the alleged
program gave to my probes seemed rather intelligent! On the other hand,
because I was sure that I was indeed dealing with "just a program", I was able
each time to imagine some kind of clever programming trick or large data base
of stored items that could conceivably be responsible for that *particular* type of
cleverness, and so I kept the dialogue up for some 45 minutes or so, all the while
believing that Nikolai really was a program. Finally, just as I began to catch on
that this was a gag, it was revealed to me that "Nikolai" was in reality a team of
three computer-science students downstairs in the apartment I was in, gleefully
making up responses and typing them to me. The whole ruse was most clever
and definitely thought-provoking.

In the case of the book produced by Racter, we too have to speculate about
how it possibly could be producing these strange and provocative lines, for the
book itself tells precious little of Racter's mechanisms — we are given little more

than a sample of its output (although see below). So, for instance, we don't know how big Racter's vocabulary is, what kinds of stock phrases it has stored, what kind of global "plot types" might have been stored inside it. We don't know what makes Racter choose to address somebody directly (as "my Liege", for instance), or where it gets its possible forms of address from. There are hundreds of such unknowns that simply *remain* unknown.

What inevitably happens in most people's minds, when they read this kind of prose, is that they tend to impute the ordinary meanings to the words and constructions that they see. Thus a mental image is produced of someone behind the scenes who is purposefully and meaningfully uttering these things. This is the desired sleight-of-hand, because as soon as you believe there's a mindlike entity there, you can't help but be terribly impressed by its fluidity. After all, *for a human being to do this would be very impressive.* Of course, that's true for all sorts of very mechanical things that computers can do, but we are not impressed by a computer's number-crunching, since we know that that is what computers were designed to be able to do well. But since most people's sole model of a language user is a human being, we bring to bear our prior imagery, which leads us to read into the prose, despite its awkwardness, all sorts of intentions and ideas and so on, just as we read into a non-fluent foreign speaker's awkward remarks a set of perfectly coherent ideas behind the alien surface level.

Here are two final selections from Racter — in fact, the first and the last selections in Racter's book. They are quite amazing. See if you don't agree:

> At all events my own essays and dissertations about love and its endless pain and perpetual pleasure will be known and understood by all of you who read this and talk or sing or chant about it to your worried friends or nervous enemies. Love is the question and the subject of this essay. We will commence with a question: does steak love lettuce? This question is implacably hard and inevitably difficult to answer. Here is a question: does an electron love a proton, or does it love a neutron? Here is a question: does a man love a woman or, to be specific and to be precise, does Bill love Diane? The interesting and critical response to this question is: no! He is obsessed and infatuated with her. He is loony and crazy about her. That is not the love of steak and lettuce, of electron and proton and neutron. This dissertation will show that the love of a man and a woman is not the love of steak and lettuce. Love is interesting to me and fascinating to you but it is painful to Bill and Diane. That is love!

I was thinking as you entered the room just now how slyly your require-ments are manifested. Here we find ourselves, nose to nose as it were, considering things in spectacular ways, ways untold even by my private managers. Hot and torpid, our thoughts revolve endlessly in a kind of maniacal abstraction, an abstraction so involuted, so dangerously valiant, that my own energies seem perilously close to exhaustion, to morbid termination. Well, have we indeed reached a crisis? Which way do we turn? Which way do we travel? My aspect is one of molting. Birds molt. Feathers fall away. Birds cackle and fly, winging up into troubled skies. Doubtless my changes are matched by your own. You. But you are a person, a human being. I am silicon and epoxy energy enlightened by line current. What distances, what chasms, are to be bridged here? Leave me alone, and what can happen? This. I ate my leotard, that old leotard that was feverishly replenished by hoards of screaming commissioners. Is that thought understandable to you? Can you rise to its occasions? I wonder. Yet a leotard, a commissioner, a single hoard, all are understandable in their own fashion. In that concept lies the appalling truth.

Pretty remarkable prose poetry, if taken at face value. The problem is, one just doesn't know what's behind the scenes. To take one simple example, did Racter put together the entire phrase "silicon and epoxy energy enlightened by line current" all by itself? If so, out of what kinds of smaller prose units? Was there in Racter's algorithms a hidden recipe for constructing descriptive phrases about computers or AI programs, which strongly favored selection of certain types of key words? We just don't know. We also don't know how many thousands of paragraphs were generated by Racter in order for this particular one to be produced. Nor are we told whether this paragraph was produced intact, as is, or whether various sentences it produced were selected and assembled into a paragraph for human consumption. Let me quote from the book's introduction, written by Bill Chamberlain, one of the people behind Racter:

> There would appear to be a rather tedious method of generating "machine prose", which a computer could accomplish at great speed but which also might be attempted (though it would take an absurdly long time) by writing thousands of individual words and simple directives reflecting certain aspects of syntax on slips of paper, categorizing them in some systematic fashion, throwing dice around to gain a random number seed, and then moving among piles of these slips of paper in a manner consistent with a set of arbitrary rules, picking a slip from Pile A, a slip from Pile B, etc., thereby composing a sentence. What actually was on the slip of paper from any given pile would be irrelevant; the rules would stipulate the pile in question. These hypothetical rules are analogous to the grammar of a language; in the case of our present program, which is called Racter, the language is English. (The name reflects a limitation of the computer on which we initially wrote the

program. It only accepted file names not exceeding six letters in length. Racter seemed a reasonable foreshortening of *raconteur*.)

Racter, which was written in compiled BASIC on a Z80 micro with 64K of RAM, conjugates both regular and irregular verbs, prints the singular and the plural of both regular and irregular nouns, remembers the gender of nouns, and can assign variable status to randomly chosen "things". These things can be individual words, clause or sentence forms, paragraph structures, indeed whole story forms. In this way, certain aspects of the rules of English are entered into the computer. This being the case, the programmer is removed to a very great extent from the specific form of the system's output. This output is no longer of a preprogrammed form. Rather, the computer forms output on its own. What the computer "forms" is dependent upon what it finds in its files, and what it can find is an extremely wide range of words that are categorized in a specific fashion and what might be called "syntax directives", which tell the computer how to string the words together. An important faculty of the program is its ability to direct the computer to maintain certain randomly chosen variables (words or phrases), which will then appear and reappear as a given block of prose is generated. This seems to spin a thread of what might initially pass for coherent thinking throughout the computer-generated copy so that once the program is run, its output is not only new and unknowable, it is apparently thoughtful. It is crazy "thinking", I grant you, but "thinking" that is expressed in perfect English.

Obviously, the passages by Racter quoted by me were culled by Chamberlain and friends from huge reams of output from Racter over a period of years. Moreover, you are also seeing just a little bit from the book, thus a *double* selection process has taken place — part by them, part by me. You are thus being exposed to the very choicest bits of all! What if we were instead allowed to see unfiltered, uncensored Racter output? We would probably be less impressed. And finally, let us not forget Chamberlain's telling phrase "the programmer is removed to a very great extent". What does that little disclaimer really mean?

Of course the Racter project was done in a spirit of high merriment, and sometimes I think that all of AI has something of this playful, spoofing character. It is, after all, a delightful game to try to make a machine act not like a machine, and to attempt to create subtle coverups for its many weaknesses. The creators of Racter must have been driven to a large degree by their own sense of humor. Why else would they have chosen the ridiculous vocabulary they did? In a way, this humor-driven attitude is admirable and I would like to see more of it in the field of AI; on the other hand, perhaps the humor component in this project is so great that Racter shouldn't be taken as a serious piece of research. It certainly made no pretense of attempting to mimic the most deeply hidden processes in the mind of a human author.

A Computer Mathematician

If Racter represents the frivolous, lightweight side of models of creativity, the AM program, written by Doug Lenat (1982, 1983), is a good example of work that lies at the other end of the spectrum. AM worked in the domain of mathematical discovery. It started from a basis of very primitive mathematical concepts, and was able to put them together in diverse ways to form compound structures — higher-level concepts. This process of "conceptual accretion" could build on itself, so that in principle, arbitrarily deeply nested concepts were reachable. At the heart of AM was a vast collection of clever heuristic rules devised by Lenat, which suggested how to combine known concepts to make new candidate concepts having a high degree of interest. Indeed, one of the most interesting features of AM was that it itself had a heuristic-based model of what makes a concept "interesting", favoring such characteristics as extremality, uniqueness, self-application (as in the notion "square", where a number is multiplied by itself), the inverse of an interesting concept (*e.g.,* "square root" is interesting because it is the inverse of the interesting concept "square"), and many other "standard tricks" to which mathematicians typically resort. AM assigned to each concept, whether primitive or compound, a numerical level of "interestingness", and collectively, these numbers served to give its search processes a direction.

Among the achievements of AM was the discovery of the concept of "prime number" (judged highly interesting by AM because such numbers have a minimal number of factors), which was followed by the formulation of the famous Goldbach conjecture (that every even number greater than 2 is the sum of two primes). Given that AM had begun only at the level of sets and did not even have the notion of number (*i.e.,* cardinality) at the outset, this is very impressive.

Subsequent critiques of AM (Rowe & Partridge, 1991; Ritchie & Hanna, 1990) have pointed out that, much as in Racter's prose, there was a considerable degree of human intervention in AM's discoveries. More specifically, like Racter's Chamberlain, Lenat himself often acted as a "filter", periodically poring over the program's vast amounts of output, winnowing out all but the best new concepts, and modifying some of the program's parameters in order to improve AM's search. This suggests that it might be more appropriate to think of AM as having been a human–machine hybrid rather than as an autonomous computer program. Examples like AM and Racter force one to ask, "At what point does the seemingly innocent act of *selection* by a human turn into *direction* of the program?"

For me, what Lenat and Chamberlain did for their programs is strongly reminiscent of the role that is played by William Huff, an architecture professor, with respect to students in his design courses. Huff has a long-standing tradition of assigning his design students the challenge of creating "parquet deformations" — tilings of the plane that gradually metamorphose in an Escher-like manner as they move across the plane (many examples are given and discussed in Chapter 10 of Hofstadter, 1985). To get the idea across to the students in each successive class, Huff shows a portfolio consisting of what he considers to be the best examples from previous years. Thereby inspired, the current crop of students then produces a large set of new parquet deformations, most of which are not great, but usually at least a few of which are novel and exciting. As one would expect, Huff applies his own keen artistic judgment to the latest harvest, pruning the weak ones out and adding his favorites to the growing portfolio to be shown to subsequent classes. In this way, a process of evolution takes place, with Huff playing the role of natural selection, letting artistically weak specimens die and strong ones survive, and then propagating the "most fit genes" by exhibiting the survivors to his class the next year. Over a period of some twenty or more years, Huff has managed to direct the course of evolution of parquet deformations in a very interesting way.

The question naturally arises as to the authorship of all these pieces. Huff has a practice of labeling each piece, when they are exhibited in a museum or gallery, "from the studio of William Huff", with no further information. However, when I decided to publish a small selection of these beautiful studies, I felt that Huff's labeling practice was too one-sided, and so for each piece I listed both Huff's name and the student's name. I felt this was fairer. But I certainly could see two sides of this question. There was no doubt in my mind that Huff deserved a large portion of the credit. Whether it was less or more than 50 percent remains an unresolved but fascinating question in my mind.

But let us return to AM. Lenat himself, in an interesting "autocritique" (1983c), pointed out that certain features of Lisp, the language in which AM was implemented, dovetailed so neatly with the domain that AM was exploring that the representational formalism itself made it fairly easy for the program to make the kinds of discoveries it did.

Despite various criticisms, it is still exciting that AM was able to make mathematical discoveries. On the other hand, it is not clear that AM's methods were at all general. For instance, AM was not in the business of recognizing central or peripheral instances of categories, let alone judging their centrality, and thus had no sense of "strength of membership" of an entity in a category. Nor was there any sense of categories competing for "ownership" of a perceived entity, or an entity being possibly ambiguous in terms of its category membership. Indeed, the entire notion of perception was lacking from AM's architecture. In addition, given that its concepts belonged to the hard-edged domain of mathematics, which, like chess, is very atypical of cognition, AM's successes may well represent a splendid isolated peak rather than a general model of creativity.

Another Computer Mathematician

In a famous early episode of AI, a program called "Geometry", written by H. L. Gelernter, allegedly made a beautiful and highly original discovery in geometry. The way this story is usually told, a couple of interesting twists are left out. In particular, Geometry itself didn't quite make the discovery, but rather two researchers, in a casual conversation about the program, realized that it *would* make the discovery if it were run on a specific problem. To my knowledge, however, the actual run that would have resulted in the discovery never took place. Despite this, let's be generous and give the computer credit for the discovery it *would* have made. After all, it's not the program's fault that the run was never made! The other twist is that the discovery, though unquestionably beautiful, turned out not to be original with the computer — indeed, it was almost 2,000 years old.

The challenge was to prove the simple proposition that the base angles of any isosceles triangle are equal to each other (see Figure E–2). The most standard proof of this result involves making a construction line that divides the triangle into two symmetric halves, which are then quickly proven to be congruent, and the result is established. This is such a short proof that one might guess it to be minimal, but amazingly, there is an even shorter proof. The key idea is to consider the triangle itself as two *different* triangles, and then to show that these two triangles are congruent. Specifically, if the original triangle's apex, bottom left vertex, and bottom right vertex are labeled "A", "B", and "C", respectively, then the two different triangles I am speaking of are ABC and ACB. The former results from making a *counterclockwise* tour of the three vertices, whereas the latter involves a *clockwise* tour. Once you have these two triangles clearly in mind, then it takes but a split second to show that they are congruent (side–side–side), from which fact it follows that the angles at vertices B and C are equal.

To most people, this proof, whose basic move involves considering *one* object to be two *different* objects, at first appears like nonsense — yet mathematically it is not just correct, it is certainly the most concise and elegant proof. It was discovered by the great Greek geometer Pappus of Alexandria toward the end of the third century A.D.

Margaret Boden, in her above-mentioned book, discusses this episode in a most enlightening way. She points out that the Geometry program's concept of "triangle" and its methods of attack were so mechanical and brute-force that

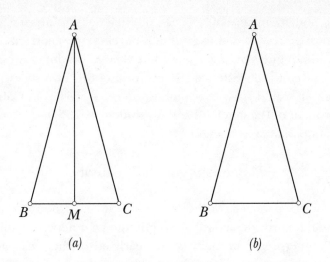

Figure E–2. Two ways to prove that an isosceles triangle has equal base angles. In (a), a construction line perpendicularly bisecting the base is made, dividing the triangle into two triangles that are quickly proven congruent, whence the theorem follows. In (b), no construction lines are made at all. Instead, the single given shape is interpreted as a triangle in more than one way — in fact, it is seen as six distinct triangles: ABC, ACB, BAC, BCA, CAB, and CBA. Then two of these — BAC and CAB — are seen to be congruent (side–side–side), from which the result follows immediately.

the program would *inevitably* generate both ABC and ACB and consider them to be distinct triangles. To the program, a triangle was simply an ordered set of three labels — hence to it, triangle ABC was as obviously different from triangle ACB as the ideas denoted by the words "eta" and "eat" are different to you and me. In fact, it follows that where you and I see just one single shape, the program would say there are *six* distinct triangles: ABC, ACB, BAC, BCA, CAB, and CBA! Moreover, even though the program proved BAC and CAB congruent, it never occurred to it that these distinct entities were just different ways of perceiving a single shape, much less that this proof was *curious* or *interesting*. Boden's main point is that, because the Geometry program had such a sparse representation of the world of geometry, and no representation at all of what "interestingness in the world of proofs" might be, it had no inkling of the cleverness, subtlety, and elegance of what it was doing. The discovery was found through a very unhuman, mechanical, mindless route of exploration.

This analysis of Geometry's performance suggests that a crucial component of human creativity is the ability to sense that some fact that one has noticed, aside from being true, is *surprising*. Put otherwise, what is needed in a computer model of the creative process is for the program to be introspective enough that it looks at what it is doing, makes observations about the level of interest of what it has produced, uses these observations to help select among its products, and by means of a series of such self-monitoring discoveries, modifies its own preferences, thus gradually developing a "personal" style. Such a model would not only *do* these things, but *know* it was doing them — that is, it would be aware of the existence and evolution of its own style. Thus a truly creative program would have to combine the roles of Harold Cohen and his program Aaron, or of William Huff and his students, or of Doug Lenat and AM, or of Bill Chamberlain and Racter.

Plagiarism versus Creativity

Boden's incisive analysis of the (would-have-been) triumph of Gelernter's Geometry program provides a flash of insight into an otherwise ambiguous story that could be interpreted in many ways, nearly all of them exaggerating the accomplishments of machines. And yet such keen analyses are the rare exception. In my experience, when computer prose, art, or music is exhibited, especially in the popular press, very little information is generally provided about how it was made. Why is that? Without such information, of what meaning is the exhibit?

Suppose someone showed you a brilliant essay on humor and told you it had been "written by a computer". If subsequently you found out that it had been plucked whole from Arthur Koestler's book *The Act of Creation,* you would surely feel defrauded. It would make little difference if you were further informed that Koestler's entire book, along with a hundred million other books on all sorts of topics, had been stored inside the computer, and that a program had *selected* this particular book and from it this particular passage, and printed it out. It would still be no different from plagiarism — clever plagiarism, perhaps, but that's all.

Now consider a somewhat more complex scenario. The same brilliant passage by Koestler is chopped up into a series of ten-word segments, and for each word in each segment, its part of speech is supplied. All this information is loaded into the computer's memory. In addition, a sophisticated grammar of English is given to the computer, along with overall instructions to try to arrange all the ten-word segments into one long passage that is grammatical from start to finish. The program runs for a good while and comes up with a candidate passage. It is not identical with Koestler's piece, but the two documents do have several stretches that are identical for a few hundred words in a row. Then the program is slightly modified and run again. This time, *mirabile dictu,* it comes up with Koestler's piece in its entirety. Does this piece of computer-generated prose deserve exhibition? Perhaps, but it would certainly be misleading to claim that a computer had "written" the passage, because the word "write" connotes the making of something from scratch.

The computer's reassembly of the scrambled piece of prose might be compared with my putting together a jigsaw puzzle on which a Monet painting had already been printed. Obviously, I would be quite a faker if I claimed to have made a great Impressionist painting, since all I was looking at was colors and shapes of edges. And in the case of the prose unscrambling, no one would give the computer credit for having generated a brilliant essay — a series of brilliant *thoughts* — because, quite obviously, what the computer was manipulating was simply words, parts of speech, grammatical rules. Thoughts could be dispensed with entirely in such an act.

But what if Koestler's original piece were chopped into ever-shorter segments? What if they were just two or three words long? At what point would we slide from being bored to being amazed at the computer's ability to create prose? When, in short, would we be forced to conclude that *ideas,* not mere *formal tokens,* were entering the picture? Unfortunately, most articles about computer creativity that one reads in the popular press reproduce what seems like amazing output — be it drawings, prose, music, or whatever — but provide only a minimal sense for this kind of "implementation detail" having to do with the nature of the building blocks out of which the output was assembled, and the ways in which the program was able — and not able — to combine such entities. It makes all the difference in the world to know this kind of thing.

Let us imagine a very extreme case, now. Suppose that I am shown some marvelously ingenious new theorem in mathematics, or a deeply moving new piece of music, or a great poem, and told that a computer generated it. Suppose moreover that I have some way of knowing that this product is truly new and was not plagiarized, but on the other hand I have no access whatsoever to the mechanisms that produced it. How should I regard this product, given that it is both new and important? Is it really creative? And if so, to whom should the credit go?

I don't see any easy answers here. Without some form of access to the innards of the program, I simply don't know how to evaluate the product, and so I can't decide if real creativity was involved or not. This attitude might seem strange to some people, who might say, "What does it matter *how* it was made as long as it was *made*? A product is creative for certain external, objective reasons, not for how it came into existence!" But I don't feel that way. I cannot judge just the object before me; I feel a need to have a sense for its provenance, in some manner or other.

As for the assignment of credit, I also feel at a loss. Somehow, the credit has to be divided up between the computer program and its author, but in the absence of knowledge about the program, I don't know where that line falls.

Mechanisms, Probes, and the Turing Test

Aside from directly reading its code, there are two avenues by which one can obtain knowledge of the mechanisms of a computer model of some aspect of cognition. One, of course, is by reading a description of the program's architecture, presumably written by the program's designers or builders. This may be extremely helpful, but on the other hand it may be biased or unclear. It also may focus on the wrong level of detail — the description could be so complex and technical that one can't make head or tail of it, or it could be at so high a level that after reading it one still has very little idea of what is really going on. Worse yet, the description may alternate between these two modes in such a messy way that one is completely lost. Such pastiches are all too common in the AI literature.

The other avenue is less ambiguous but more indirect — it is simply by *interacting* with the program over a period of time in an unrestricted manner. In this way, one can make a long and systematic series of probes and thereby discover for oneself in what ways the program is flexible and in what ways it is rigid.

This idea is the basis for Alan Turing's famous "Imitation Game" (Turing, 1950), a philosophical suggestion for how to detect the presence of the elusive phenomenon called "thinking" when there is a black box in front of you, and when someone — perhaps the box itself! — alleges that the box contains an example of that phenomenon. Turing envisioned a link via teletype, over which one could send standard alphanumeric symbols. Since he was primarily trying to find a way to answer (or to circumvent) the philosophical question, "What counts as thinking?", he assumed that the black box in question was a highly plausible candidate for the role "thinker", rather than a very limited or special-ized program. To him, this implied an ability to converse in some natural language, and with no constraints at all on the possible topics or styles of conversation. In a nutshell, the Turing Test (as his suggestion eventually came to be known) is the attempt by an interrogator to distinguish between a machine and a human being, purely by means of their performance in typed conversa-tion. Jokes, analogies, allusions, discussions of history or cultural phenomena, inquiries into highly personal beliefs — all these and more are fair game.

By contrast, when I suggested the systematic probing of a cognitive model, I meant something much less ambitious than the full Turing Test. I did not have in mind natural-language interaction (although it would be nice), nor jumping arbitrarily from one area of knowledge to another (although that too would be very nice). The main idea of interacting with a cognitive model over a long period of time is simply to allow its full range of behaviors to become gradually visible, and thereby to allow its innermost mechanisms to be revealed.

This idea that *covert mechanisms* can be deeply probed and eventually revealed merely by means of watching *overt behavior* may seem unlikely or even fantastic to some people. Yet this premise lies at the very heart of modern science, as I try to show below, and to deny it would therefore be to deny the basis for many of our ordinary intuitions about how things work and how we acquire knowledge of the world around us. It is therefore of considerable importance to think about the very meaning of words like "mechanism" and "probe". To do so is the purpose of the rest of this Epilogue.

Brain Structures versus Cognitive Mechanisms

If I assert (as I often have), "Copycat is a model of thinking", what do I really mean? Am I claiming that Copycat is in some sense a model of the human brain? Most people — even most cognitive scientists — would be inclined to answer "no". However, before discarding the idea, one should carefully consider what is meant by "model of the brain".

When one asks "Is Copycat a model of *the brain*?" or says "*The human brain can think*", one is using a singular noun to refer to billions of distinct physical instantiations. Needless to say, it is those physical instances (specific human brains), not the non-physical Platonic category ("the human brain"), that carry out thinking. Moreover, it is common knowledge that each individual brain has its own unique wiring patterns, and in that sense is a unique physical substrate for thought. The unquestioned use of the singular term "the brain" in spite of all this variability reveals our tacit assumption that there must be some abstract (but virtually always unspecified) level of description shared by all human brains. Therefore, when decoded, the sentence "The human brain can think" really means the following rather subtle idea: "There are abstract universal mechanisms, diversely instantiated over and over again in specific brains, that allow thinking to take place." In short, the brain mechanisms responsible for thinking are not *literally* hardware; rather, they are patterns located somewhere along the spectrum between hardware and software.

Today's excitement over neural networks has led many people to suppose that these mechanisms lie more toward the hardware end of the spectrum than people tended to suppose in the sixties and seventies. For that reason, there may well be a greater tendency today than a decade or two ago to describe cognitive science as a search for *brain structures* as opposed to a search for *mental mechanisms*. Trying to draw such a distinction is a tricky matter, though.

Being a very complex system, the brain (to use that questionable singular once again) contains many types of structures at many distinct hierarchical levels, including the following (just to list a few examples):

- atomic nuclei;
- water molecules;
- amino acids;
- neurotransmitters;

- synapses;
- dendrites;
- neurons;
- clusters of neurons;
- columns in the visual cortex;
- larger regions (such as area 19) in the visual cortex;
- the entire visual cortex;
- the entire left hemisphere.

Among these (and many other) diverse "brain structures", which are the most likely to figure critically in an explanation of thinking? No one knows for sure. It is interesting, however, that in the last few years, the popular press has latched onto certain neurological experiments on synaptic-weight modification in various creatures large and small, often breathlessly describing them as "unraveling the secrets of memory". This dramatization amounts to the (quite literally *mindless*) presumption that there is nothing on an *organizational* level to human memory — that to fathom human memory in its full richness, one can simply be a chemist and concern oneself exclusively with the mechanisms behind local microscopic chemical changes. Psychological notions are thrown to the wind in this view.

The fact that there are so many levels and types of well-established *physical* structures reveals one type of blurriness inherent in the term "brain structure", but to compound the complexity, there is another plausible type of "brain structure", exemplified by such notions as:

- the concept *dog*;
- the associative link between the concepts *cow* and *milk*;
- an object file for a perceived object (as discussed by perceptual psychologist Anne Treisman in Treisman, 1988);
- "geons" and "2½-D sketches" (as discussed in various models of vision);
- frames, scripts, and schemas;
- "memory organization packets" and "thematic organization points", as posited in Roger Schank's models of memory (Schank, 1982);
- short-term memory;
- long-term memory;
- the various "agents", "K-lines", "nemes", and "nomes" posited in Marvin Minsky's "society of mind" model (Minsky, 1985);
- codelets, urgencies, slippages, conceptual distances, conceptual depths, bonds, descriptions, bridges, groups, strengths, and temperature (as in Copycat, Tabletop, and our other projects).

This list is just a small sampling of plausible ingredients, at various levels, of various theories of thinking. Although, on first glance, each one seems to belong to a realm far removed from the brain, if the item in question is a valid aspect of cognition, then it must be somehow implemented in the physical substrate of the brain. Each of these entities stands at least *some* chance of becoming physically grounded in the next several decades, in much the same way as the once-theoretical notion of *genes* became physically grounded with the understanding of DNA. Therefore the elements of this list are reasonable candidates for being called "brain structures", albeit in a somewhat different sense from those in the previous list.

As of yet, there is absolutely no certainty about the optimal level at which the hopefully universal brain structures (or cognitive mechanisms) ought to be characterized. Other branches of science have faced similar level-specification problems, some of which have been fairly satisfactorily resolved. If one thinks about the behavior of gases, for instance, they are best described, for many purposes, by the macroscopic-level laws of thermodynamics, even though gases are well known to be composed of astronomical numbers of molecules, and can thus also be described at the lower level of statistical mechanics. Similarly, if one is interested in describing how DNA is the carrier of heredity, one does best by referring only to the high level of information-carrying codons and skipping most if not all references to the low-level details of the chemical makeup of DNA. Interestingly, there are distinct terms used to describe DNA at these distinct levels: DNA *qua* physical structure is called the "double helix", while DNA *qua* information carrier is referred to as an organism's "genome".

What one hopes to achieve in cognitive science is a similar sophistication in talking about the functioning of "the brain" at different levels. Specifically, one wants to know what level and type of brain structures are the most suitable for describing thought. Perhaps, in analogy with the two labels for DNA, we could refer to the brain *qua* physical structure as the "double hemisphere", and the brain *qua* memory carrier and site of concepts as an organism's "memome".

Some Levels of Description Are Certainly Too Low

Although it is not as common as it used to be, it is still respectable in cognitive science today to believe that some type of deductive formalism (such as the predicate calculus or some more recent representation language, *e.g.,* KLONE) is somehow implemented in neurons, and that what really qualifies the behavior as thinking is the formalism, not the fact that this formalism is somehow implemented in neural hardware. Such a belief is no different from the truism that what makes a given program be a word processor (as opposed to, say, a weather predictor or a video game) is its source code, not the machine it is running on. Thus the "brain structures" responsible for thought might, at least in principle, turn out to be very abstract (*e.g.,* something like the features of a representation language such as KLONE) rather than relatively concrete (*e.g.,* details of neural interconnectivity). Today, many cognitive scientists may find this once-bright hope implausible, but it cannot be absolutely ruled out, because no one yet knows for certain what level or type of "brain structures" will turn out to support the essential features of thought.

In any case, almost all cognitive scientists, whatever they think of KLONE and related formalisms, *do* accept the premise that the question "At which level of description of the brain is thinking really going on?" is a meaningful question and has a meaningful answer. Presumably, then, the details *below* that critical level do not matter. This is why most cognitive scientists are almost certain that it is not worthwhile for their purposes to study molecular biology, quantum mechanics, or quarks. (But see Conrad *et al.,* 1989, for an account of high-level thought processes that tries to take into account the molecular biology of enzymes and their interactions!)

It would be extremely revealing, to say the least, if one succeeded in making a thinking object that had a substrate very different from neurons. This may or may not come to pass, but if it did, such a success would reveal much to us about what level of machinery is necessary to support thought. We would learn that thinking critically depends on levels X and above, whereas below level X, various different substrates will do.

The Need for Criteria by Which to Recognize Thinking

If we admit even the *conceivability* of the notion that a non-brain might be able to think, then we certainly need to have some set of criteria by which to recognize that such a system is thinking — otherwise, we will be unable to tell a thinking system from another type of system. Luckily, we have some hints about such criteria. We recognize human kindness by observing kind acts, not by doing DNA-testing to see if a person has an underlying "kindness gene" or by doing a real-time brain scan to see if some kind of "empathy center" is actually engaged. We recognize a word processor by what it does on a screen in response to keyboard input, not by what hardware it runs on. We have behavioral criteria for kindness in humans. It would not be too hard to set up some behavioral criteria for recognizing word-processing programs. Turing, in his famous article, attempted to suggest a similar set of criteria for recognizing thinking entities.

In analogy to such criteria for recognizing kind acts or word processors, Turing's idea is to employ a set of high-level behavioral criteria as opposed to low-level implementational criteria. Turing did not make the dogmatic presumption that, simply because brains are made out of neurons, the only possible level of description of any system that could think is the level of neurons. Instead, he built on what people intuitively mean when they refer to thinking — namely, the *fluid manipulation of ideas*. Turing does not *a priori* tie such manipulation to any kind of hardware; he remains open as to what the proper level of description of the mechanisms responsible will turn out to be.

When all the results of mind/brain research are finally in, will the best way to talk about "the human brain" wind up being in terms of standard types of neural wiring patterns? Or standard types of neural clusters? Or standard types of interrelationships of neural clusters? Perhaps the details of neurons themselves will not matter. Conceivably, in fact, even the internal structure of neural clusters will not matter any more than do organic chemistry and quantum mechanics. In such a case, the underpinnings of thinking would be revealed to lie pretty distant from biology and much closer to abstract organizational principles — which is to say, software. Who knows — it might even turn out that the proper "brain structures" required for the fluid manipulation of ideas are roughly at the level of the mechanisms of Copycat, Tabletop, and Letter Spirit, or a little bit lower.

Scattering Experiments versus "Direct" Observation of Phenomena

On first being exposed to the Turing Test, one might well think it would allow probing only at a very high level, and that it certainly would be incapable of getting at "subcognitive" or "subsymbolic" mechanisms, let alone neural-level mechanisms. Indeed, many people feel that the Turing Test, being concerned merely with "behavior", can barely scratch the surface of mechanisms. This issue — namely, the relationship between "mere behavior" and hidden mechanisms, and the depth to which the Turing Test can be used as a probe for hidden mechanisms — leads one to the question, "Where is the boundary line between 'direct' and 'indirect' observation?"

Early this century, the physicist Ernest Rutherford sent a beam of alpha particles through thin pieces of gold foil and collected statistics on the angles of scattering they underwent. From these (macroscopic) observations, he deduced that gold atoms must have a central nucleus surrounded by electrons, a finding that led eventually to the modern picture of "the atom" (another of those questionable singulars). His conclusion depended, of course, on a mathematical theory of electromagnetic scattering, which predicts that such-and-such a microscopic internal structure will cause such-and-such a macroscopic angular distribution of scattered particles.

Rutherford's scattering experiments were so important and so reproducible that his basic technique became a standard element of the repertoire of experimental physics. What at first seemed to be an abstruse and *indirect* connection between macroscopic patterns and their underlying microscopic causes came eventually to be seen as an obvious and *direct* connection in physicists' minds. Of course, the end results of scattering experiments were never quite as direct as *pictures* of the observed structures would have been — they were mathematical formulas or verbal descriptions. However, this was good enough for physicists.

Recently, though, with the phenomenal surges in computing power, the results of scattering experiments done with electron microscopes and X-ray microscopes can be processed to yield what are essentially *photographs* of objects the size of molecules, or even atoms. In principle, the calculations are no different from those in Rutherford's experiments, but so many more of them can be done, and done so much faster, that the results seem qualitatively different — certainly to the intuition.

Today, for instance, ultrasound allows us to see a fetus moving about inside a mother's womb in real time. Note that we feel no need to put quotes around the word "see" — no more than around the word "talk" in the sentence "My wife and I talk every day over the phone." When we make such casual statements, we don't for a moment consider the weirdness of the fact that our voices are speeding in perfect silence through metallic wires; the reconstruction of sounds at the receiving end is so flawless and faithful that we are able to entirely forget the fact that complex coding and decoding processes are taking place in between the speaking mouth and the listening ear. A hundred years ago, telephonic contact — "voice teleportation", if you will — was considered astonishing and nearly miraculous, because the "trickery" going into it was so new and unfamiliar that one simply couldn't ignore it. In some sense, perhaps this was the *proper* reaction to have, but today it is rare to find anybody who finds voice teleportation miraculous — it is just too commonplace. If, fifty years ago, high-frequency sounds had been scattered off a fetus, there would have been no technology to convert the scattered waves into a vivid television image, and any conclusions derived from measurements on the scattered waves would have been considered abstruse mathematical inferences; today, however, simply because fast computer hardware can reconstruct the scatterer from the scattered waves in real time, we feel we are *directly* observing the fetus. Examples like this — and they are legion in our technological era — show why any boundary between "direct observation" and "inference" is a subjective matter.

In fact, much of science consists of progress in blurring this seemingly sharp distinction. What at one point is considered very subtle inference becomes established, standardized, and computerized; then it is considered direct observation and taken for granted. This trend runs from optical telescopy to radio telescopy, from optical microscopy to electron microscopy to particle scattering, and so on. As with ultrasound, whether one intuitively feels a process to be "direct observation" or not is simply determined by whether a sufficiently clear and vivid visual picture is produced by some computer. Today's abstruse inference is tomorrow's direct window!

The Turing Test and the Visibility of Deep Mechanisms

All of this carries over to the Turing Test. Indeed, one can liken the Turing Test to a scattering experiment, in which the thinking device unwittingly reveals its (microscopic) internal mechanisms through its responses to questions (which in the scattering analogy play the role of the impinging waves or particles). The level of detail that can be probed in this manner is unlimited, although to carry out ever finer levels of probing requires ever more and ever subtler types of questioning, as well as ever subtler ways of scrutinizing the responses.

Examining linguistic responses in novel ways is quite similar to examining the spectra of stars in novel ways (*e.g.,* using new regions of the electromagnetic spectrum, higher degrees of resolution, time-correlation data from widely separated receivers, etc.), and thereby inferring detailed stellar mechanisms of ever subtler sorts, despite being hundreds of light-years from the star itself. In the Turing Test, some of the possible ways to scrutinize linguistic output coming from an unknown source include the following:

- looking at *word frequencies* (*e.g.,* is "the" the most common word? is "time" the most common noun? are some low-frequency words used unnaturally often? does suspicion seem to be aroused in the distal "mind" if low-frequency words are used with a high frequency in the input questions?);
- observing sensitivity to *tone* (*e.g.,* are formal and slang expressions in the input text understood? is humor based on improper mixtures of tone understood? is suspicion aroused by a strange mixture of tones in the input questions? in the generated text, are formal and informal levels of discourse and style kept suitably apart, or mixed in strange, unnatural-seeming ways [think of Racter's droll output, for instance]?);
- examining *types of errors* (*e.g.,* misspellings, transposition errors, improperly used words or phrases, blends of all sorts, grammatical errors, and so on, which — as ought to be well known to any cognitive scientist — reveal a great deal about the mechanisms of thought);
- examining *word flavors* as a function of subtle details of the context (*e.g.,* what contextual pressures lead to choosing "jock" over "athlete", or vice versa? to saying "lady" as opposed to "woman"? to saying "endeavor" instead of "try" or "attempt" or "strive"?);

- examining *level of abstraction of word choices* (*e.g.,* what pressures lead to choosing between "Fido", "the dog", and "some mammal"? to choosing between "that pedestrian", "that guy", and "he"? to choosing between "Barcalounger", "recliner", "armchair", "chair", "piece of furniture", and "thing"?);
- looking at *default assumptions regarding gender* (*e.g.,* what kinds of circumstances lead to generation of agent nouns with feminine endings, such as "heroine", "millionairess", or "farmerette"? when, if ever, are supposedly generic terms like "man" and "he" generated? what gender is assumed when neutral terms like "pedestrian" and "surgeon" appear in the input?);
- observing how *throwaway analogies* are understood and generated (*e.g.,* is the abstraction hidden in remarks such as "I used to do that, too!" and "Has that ever happened to you?" interpreted correctly and instantly? are such remarks produced in the standard contexts that would call for them?);
- observing how *throwaway counterfactuals* are understood and generated (*e.g.,* is the subtle blend implicit in remarks such as "I wouldn't have felt that way if I'd been my father" or "What would *you* have done if *you'd* been my parents?" interpreted correctly and instantly? are such remarks produced in the standard contexts that would call for them?);
- paying attention to *timing data* (the output stream might come character by character, line by line, or speech by speech, but in all cases the speed taken to generate the output can be used to make some inferences about the mechanisms behind the scenes);

and so on and so forth. This list could be extended and elaborated in enormous detail, but this is not the place to give a long list of such windows onto covert machinery, or to try to defend the validity and power of using such approaches (French, 1990 offers a defense). The point is simply that there are many of them, and some of them are already well-established techniques in cognitive science (particularly cognitive psychology).

Anyone who seriously believes in the validity of the Turing Test does so precisely because they appreciate the subtlety of the probes it offers. As astronomers and physicists know, external behavior far removed in location and scale from its sources, if scrutinized sufficiently carefully, can be phenomenally revelatory of mechanisms; likewise, cognitive scientists should appreciate the analogous fact about the behavior of the mind. In short, the Turing Test, if exploited properly, can be used to probe mental mechanisms at arbitrary levels of depth and subtlety.

In the spirit of much of the best science of our century, the Turing Test blurs the supposedly sharp line between probing of behavior and probing of mechanisms, as well as the supposedly sharp line between "direct" and "indirect" observation, and thus reminds us of the artificiality of such distinctions. Any computer model of mind that passes a truly deep Turing Test — one that probes for the fundamental mechanisms of thought — will agree with "brain structures" all the way down to the level where the essence of thinking really takes place.

The Turing Test and Basic Research

Recently, a monetary prize (the Loebner Prize) was established for the first program to pass a restricted version of the Turing Test. Unfortunately, although the idea is amusing and even exciting in its own odd way, such a competition is very premature today. Unless the people who play the interrogator role do so in a very sophisticated manner, they will wind up probing at a relatively coarse level, and this will turn the competition into nothing but a race for flashier and flashier natural-language "front ends" with little substance behind them. This would be a shame, for when one looks carefully at models in the tiniest of microdomains, such as those described in the chapters of this book, one sees that even they are still enormously far from achieving true mental fluidity. What is needed is a prize for advances in basic research, not a prize for window-dressing.

The research projects described in the foregoing chapters represent a conscious choice to work at a very basic, stripped-down level, far removed from the seductive glories of natural language. Whereas artificial-intelligence projects operating in allegedly real-world domains can do no more than implement the merest "tips" of many iceberg-like real-world concepts, our projects attempt to implement the *essence* of a very limited number of artificially simple concepts. While it is certain that no project emanating from our research group will ever come close to passing the Turing Test, we nonetheless hope that the work we have described in this book may help lead, in the distant future, to architectures that go much further than we have toward capturing the genuine fluid mentality that Alan Turing so clearly envisioned when he first proposed his deservedly celebrated Test.

References

Aitchison, Jean (1994). *Words in the Mind: An Introduction to the Mental Lexicon* (2nd ed.). Cambridge, MA: Basil Blackwell.

Albers, Donald, Gerald L. Alexanderson, and Constance Reid, eds. (1990). *More Mathematical People.* San Diego: Harcourt Brace.

Anderson, John R. (1983). *The Architecture of Cognition.* Cambridge, MA: Harvard University Press.

Arnold, Henri and Bob Lee (1982). *Jumble #21.* New York: Signet (New American Library).

Belpatto, Guglielmo Egidio (1890). "L'ipertraduzione esemplificata nel dominio di analogie geografiche". *Rivista inesistente di filoscioccosofia,* vol. 14, no. 7, pp. 324–271.

Bergerson, Howard (1973). *Palindromes and Anagrams.* New York: Dover.

Blesser, Barry *et al.* (1973). "Character Recognition Based on Phenomenological Attributes". *Visible Language,* vol. 7, no. 3.

Bobrow, Daniel and Bertram Raphael (1974). "New Programming Languages for AI Research". *Computing Surveys,* vol. 6, no. 3.

Boden, Margaret A. (1977). *Artificial Intelligence and Natural Man.* New York: Basic Books.

———— (1991). *The Creative Mind: Myths and Mechanisms.* New York: Basic Books.

Bongard, Mikhail (1970). *Pattern Recognition.* Rochelle Park, NJ: Hayden (Spartan Books).

Boole, George (1855). *The Laws of Thought.* New York: Dover.

Bruner, Jerome (1957). "On Perceptual Readiness". *Psychological Review,* vol. 64, pp. 123–152.

Burstein, Mark (1986). "Concept Formation by Incremental Analogical Reasoning and Debugging". In Michalski, Carbonell, & Mitchell, 1986, pp. 351–369.

Carbonell, Jaime G. (1983). "Learning by Analogy: Formulating and Generalizing Plans from Past Experience". In Michalski, Carbonell, & Mitchell, 1983, pp. 137–162.

Chapman, D. (1991). *Vision, Instruction, and Action.* Cambridge, MA: MIT Press.

Conrad, Michael *et al.* (1989). "Towards an Artificial Brain". *BioSystems,* vol. 23, pp. 175–218.

Cutler, Anne, ed. (1982). *Slips of the Tongue and Language Production.* Berlin: Mouton.

Defays, Daniel (1988). *L'esprit en friche: Les foisonnements de l'intelligence artificielle.* Liège, Belgium: Pierre Mardaga.

Dell, Gary S. and P. A. Reich (1980). "Slips of the Tongue: The Facts and a Stratificational Model". In J. E. Copeland & P. W. Davis (eds.), *Papers in Cognitive-Stratificational Linguistics,* vol. 66, pp. 611–629. Houston: Rice University Studies.

Dennett, Daniel C. (1978). *Brainstorms: Philosophical Essays on Mind and Psychology.* Montgomery, VT: Bradford Books.

———— (1991). "Real Patterns". *Journal of Philosophy,* vol. 89, pp. 27–51.

Dowker, Ann *et al.* (1995). "Estimation Strategies of Four Groups". To appear in *Mathematical Cognition,* vol. 1, no. 1.

Dreyfus, Hubert (1979). *What Computers Can't Do* (2nd ed.). New York: Harper and Row.

Elman, Jeffrey L. (1990). "Finding Structure in Time". *Cognitive Science,* vol. 14, pp. 179–212.

Erman, Lee D. *et al.* (1980). "The Hearsay-II Speech-Understanding System: Integrating Knowledge to Resolve Uncertainty". *Computing Surveys,* vol. 12, no. 2, pp. 213–253.

Ernst, G. W. and Allen Newell (1969). *GPS: A Case Study in Generality and Problem Solving.* New York: Academic Press.

Evans, Thomas G. (1968). "A Program for the Solution of Geometric-Analogy Intelligence-Test Questions". In Marvin Minsky (ed.), *Semantic Information Processing.* Cambridge, MA: MIT Press.

Falkenhainer, Brian, Kenneth D. Forbus, and Dedre Gentner (1990). "The Structure-Mapping Engine". *Artificial Intelligence,* vol. 41, no. 1, pp. 1–63.

Feldman, Jerome and Dana H. Ballard (1982). "Connectionist Models and Their Properties". *Cognitive Science,* vol. 6, no. 3, pp. 205–254.

Fennell, R. D. and Victor R. Lesser (1975). "Parallelism in AI Problem Solving: A Case Study of Hearsay II". Technical Report, Computer Science Department, Carnegie-Mellon University. Reprinted in Reddy *et al.,* 1976. Also published in *IEEE Transactions on Computers,* vol. C-26 (February, 1977), pp. 98–111.

Fodor, Jerry A. (1983). *The Modularity of Mind.* Cambridge, MA: Bradford Books/MIT Press.

French, Robert M. (1990). "Subcognition and the Limits of the Turing Test". *Mind,* vol. 99, no. 393, pp. 53–65.

———— (1992). "Tabletop: An Emergent, Stochastic Computer Model of Analogy-making." Doctoral dissertation, Department of Computer Science and Engineering, University of Michigan.

———— (1995). *Tabletop: An Emergent, Stochastic Computer Model of Analogy-making.* Cambridge, MA: Bradford Books/MIT Press.

French, Robert M. and Jacqueline Henry (1988). "La traduction en français des jeux linguistiques de *Gödel, Escher, Bach*". *Méta,* vol. 33, no. 2, pp. 133–142.

French, Robert M. and Douglas R. Hofstadter (1991). "Tabletop: A Stochastic, Emergent Model of Analogy-making". In *Proceedings of the Thirteenth Annual Conference of the Cognitive Science Society,* pp. 708–713. Hillsdale, NJ: Lawrence Erlbaum.

French, Scott R. and Hal (1993). *Just This Once.* New York: Birch Lane Press, Carol Publishing Group.

Fromkin, Victoria A., ed. (1980). *Errors in Linguistic Performance: Slips of the Tongue, Ear, Pen, and Hand.* New York: Academic Press.

Gaillat, G. and M. Berthod (1979). "Panorama des techniques d'extraction de traits caractéristiques en lecture des caractères". *Revue technique Thomson–CSF,* vol. 11, no. 4, pp. 943–959.

Gentner, Dedre (1983). "Structure-mapping: A Theoretical Framework for Analogy". *Cognitive Science,* vol. 7, no. 2, pp. 155–170.

Gick, Mary L. and Keith J. Holyoak (1983). "Schema Induction and Analogical Transfer". *Cognitive Psychology,* vol. 15, pp. 1–38.

Grebert, Igor *et al.* (1991). "Connectionist Generalization for Production: An Example from GridFont". In *Proceedings of the 1991 International Joint Conference on Neural Networks.*

———— (1992). "Connectionist Generalization for Production: An Example from GridFont". Neural Networks, vol. 5, pp. 699–710.

Guha, R. V. and Douglas B. Lenat (1994). "Enabling Agents to Work Together". *Communications of the Association for Computing Machinery,* vol. 37, no. 7, pp. 127–142.

Hall, R. P. (1989). "Computational Approaches to Analogical Reasoning". *Artificial Intelligence,* vol. 39, pp. 39–120.

Hanson, A. and E. Riseman, eds. (1978). *Computer Vision Systems.* New York: Academic Press.

Harnad, Stevan (1989). "Minds, Machines, and Searle". *Journal of Experimental and Theoretical Artificial Intelligence,* vol. 1, pp. 5–25.

———— (1990). "The Symbol Grounding Problem". *Physica D,* vol. 42, pp. 335–346.

Hebb, Donald O. (1948). *The Organization of Behavior.* New York: John Wiley.

Hewitt, Carl (1985). "The Challenge of Open Systems". *Byte,* vol. 10, no. 4, pp. 223–242.

Hinton, Geoffrey E. and James A. Anderson, eds. (1981). *Parallel Models of Associative Memory.* Hillsdale, NJ: Lawrence Erlbaum.

Hinton, Geoffrey E. and Terrence J. Sejnowski (1983). "Optimal Perceptual Inference". In *Proceedings of the IEEE Conference on Computer Vision and Pattern Recognition,* pp. 448–453.

———— (1986). "Learning and Relearning in Boltzmann Machines". In Rumelhart, McClelland, and the PDP Research Group, 1986, pp. 282–317.

Hinton, Geoffrey E., Christopher K. I. Williams, and Michael D. Revow (1992). "Adaptive Elastic Models for Hand-Printed Character Recognition". Talk presented at the Twelfth Annual Meeting of the Cognitive Science Society, Chicago, Illinois, 1991. Also in the Neuroprose archives.

Hofstadter, Douglas R. (1976). "Energy levels and wave functions of Bloch electrons in rational and irrational magnetic fields". *Physical Review B,* vol. 14, no. 6.

———— (1979). *Gödel, Escher, Bach: an Eternal Golden Braid.* New York: Basic Books.

———— (1981). "Metamagical Themas: How might analogy, the core of human thinking, be understood by computers?". *Scientific American,* vol. 245, no. 3, pp. 18–30. Reprinted as Chapter 24 of Hofstadter, 1985.

———— (1982a). "Metafont, Metamathematics, and Metaphysics: Comments on Donald Knuth's Article 'The Concept of a Meta-Font'". *Visible Language,* vol. 16, no. 4, pp. 309–338. Reprinted as Chapter 13 of Hofstadter, 1985.

———— (1982b). "Metamagical Themas: Can inspiration be mechanized?". *Scientific American,* vol. 247, no. 3, pp. 18–34. Reprinted as Chapter 23 of Hofstadter, 1985.

———— (1982c). "Metamagical Themas: Variations on a theme as the essence of imagination". *Scientific American,* vol. 247, no. 4, pp. 20–29. Reprinted as Chapter 12 of Hofstadter, 1985.

———— (1982d). "Artificial Intelligence: Subcognition as Computation", in F. Machlup and U. Mansfield (eds.), *The Study of Information.* New York: John Wiley. Reprinted as Chapter 26 of Hofstadter, 1985.

————— (1982e). "Who Shoves Whom Around inside the Careenium?" *Synthese*, vol. 53, no. 2, pp. 189–218. Reprinted as Chapter 25 of Hofstadter, 1985.

————— (1983a). "The Architecture of Jumbo", in Ryszard Michalski, Jaime Carbonell, and Thomas Mitchell (eds.), *Proceedings of the International Machine Learning Workshop*, pp. 161–170. Urbana, IL: University of Illinois. Expanded version printed as Chapter 2 of the present book.

————— (1983b). "On Seeking Whence". Publication #5, Center for Research on Concepts and Cognition, Indiana University, Bloomington.

————— (1984a). "The Copycat Project: An Experiment in Nondeterminism and Creative Analogies". AI Memo 755, MIT Artificial Intelligence Laboratory.

————— (1984b). "Simple and Not-so-simple Analogies in the Copycat Domain". Publication #9, Center for Research on Concepts and Cognition, Indiana University.

————— (1985). *Metamagical Themas: Questing for the Essence of Mind and Pattern*. New York: Basic Books.

————— (1986). "Dreams of a Magical Shield" ("My Turn" column), *Newsweek*, March 3, 1986, p. 8.

————— (1987a). *Ambigrammi: Un microcosmo ideale per lo studio della creatività*. Florence: Hopeful Monster.

————— (1987b). "Introduction to the Letter Spirit Project and to the Idea of 'Gridfonts' ". Publication #17, Center for Research on Concepts and Cognition, Indiana University, Bloomington.

————— (1987c). "La recherche de l'essence entre le médium et le message". *Protée*, vol. 15, no. 2, pp. 13–31. Also available in English through the Center for Research on Concepts and Cognition, Indiana University, Bloomington.

————— (1988a). "Common Sense and Conceptual Halos" (reply to Paul Smolensky's target article "On the Proper Treatment of Connectionism"). *Behavioral and Brain Sciences*, vol. 11. no. 1, pp. 35–37.

————— (1988b). "Doughalese and the Semiotic Mystery". *Eureka*, vol. 48, pp. 57–64. Cambridge, U.K.: Cambridge University Mathematical Society.

————— (1995). Foreword to the Chinese translation of Hofstadter, 1979. Beijing: Commercial Press, forthcoming. Also available in English through the Center for Research on Concepts and Cognition, Indiana University, Bloomington.

Hofstadter, Douglas R. and Daniel C. Dennett, eds. (1981). *The Mind's I: Fantasies and Reflections on Self and Soul*. New York: Basic Books.

Hofstadter, Douglas R., Melanie Mitchell, and Robert M. French (1987). "Fluid Concepts and Creative Analogies: A Theory and Its Computer Implementation". Publication #18, Center for Research on Concepts and Cognition, Indiana University.

Hofstadter, Douglas R. and David J. Moser (1989). "To Err is Human; To Study Error-making is Cognitive Science", in *Michigan Quarterly Review*, vol. 28, no. 2, pp. 185–215.

Hofstadter, Douglas R. *et al.* (1989). "Synopsis of the Workshop on Humor and Cognition". *Humor*, vol. 2, no. 4, pp. 417–440.

Holland, John H. (1975). *Adaptation in Natural and Artificial Systems*. Ann Arbor, MI: University of Michigan Press. Reprinted in 1992 by Bradford Books/MIT Press.

———— (1986). "Escaping Brittleness: The Possibilities of General-purpose Learning Algorithms Applied to Parallel Rule-based Systems". In Michalski, Carbonell, & Mitchell, 1986, pp. 593–623.

Holland, John H. *et al.* (1986). *Induction.* Cambridge, MA: Bradford Books/MIT Press.

Holyoak, Keith J. and Paul Thagard (1989). "Analogical Mapping by Constraint Satisfaction". *Cognitive Science,* vol. 13, no. 3, pp. 295–355.

Indurkhya, Bipin (1992). *Metaphor and Cognition: An Interactionist Approach.* Norwell, MA: Kluwer.

James, William (1890). *The Principles of Psychology.* New York: Henry Holt.

Johnson-Laird, Philip (1988). "Freedom and Constraint in Creativity". In R. Sternberg (ed.), *The Nature of Creativity,* pp. 202–219. Cambridge, U.K.: Cambridge University Press.

———— (1989). "Analogy and the Exercise of Creativity". In S. Vosniadou & A. Ortony (eds.), *Similarity and Analogical Reasoning,* pp. 313–331. Cambridge, U.K.: Cambridge University Press.

Kahan, S., T. Pavlidis, and H. Baird (1987). "On the Recognition of Printed Characters of Any Font and Size". *IEEE Transactions on Pattern Analysis and Machine Intelligence,* vol. 9, no. 2, pp. 274–288.

Kahneman, Daniel and Dale Miller (1986). "Norm Theory: Comparing Reality to Its Alternatives". *Psychological Review,* vol. 93, no. 2, pp. 136–153.

Kanerva, Pentti (1988). *Sparse Distributed Memory.* Cambridge, MA: Bradford Books/MIT Press.

Kedar-Cabelli, Smadar (1988a). "Towards a Computational Model of Purpose-Directed Analogy". In A. Prieditis (ed.), *Analogica.* Los Altos, CA: Morgan Kaufmann.

———— (1988b). "Analogy — from a Unified Perspective". In D. H. Helman (ed.), *Analogical Reasoning,* pp. 65–103. Dordrecht, Holland: Kluwer.

Kirkpatrick, S., C. D. Gelatt, Jr., and M. P. Vecchi (1983). "Optimization by Simulated Annealing". *Science,* vol. 220, pp. 671–680.

Knuth, Donald E. (1982). "The Concept of a Meta-Font". *Visible Language,* vol. 16, no. 1, pp. 3–27.

Kokinov, Boicho (1994a). "The DUAL Cognitive Architecture: A Hybrid Multi-Agent Approach". In *Proceedings of the Eleventh European Conference on Artificial Intelligence.* London: John Wiley.

———— (1994b). "A Hybrid Model of Reasoning by Analogy". In K. Holyoak and J. Barnden (eds.), *Advances in Connectionist and Neural Computation Theory, Vol. II: Analogical Connections,* pp. 247–318. Norwood, NJ: Ablex.

Kolodner, Janet (1993). *Case-Based Reasoning.* San Mateo, CA: Morgan Kaufmann.

Kuhn, Thomas (1970). *The Structure of Scientific Revolutions* (2nd ed.). Chicago: University of Chicago Press.

Kurzweil, Raymond (1990). *The Age of Intelligent Machines.* Cambridge, MA: MIT Press.

Laird, John, Paul Rosenbloom, and Allen Newell (1987). "Soar: An Architecture for General Intelligence". Technical Report #2, Cognitive Science and Machine Intelligence Laboratory, University of Michigan.

Lakoff, George (1987). *Women, Fire, and Dangerous Things: What Categories Reveal About the Mind.* Chicago: University of Chicago Press.

Langley, Patrick *et al.* (1987). *Scientific Discovery: Computational Explorations of the Creative Process.* Cambridge, MA: MIT Press.

Lea, W. A., ed. (1980). *Trends in Speech Recognition.* Englewood Cliffs, NJ: Prentice-Hall.

Lehninger, Albert (1975). *Biochemistry* (2nd ed.). New York: Worth Publishers.

Lenat, Douglas B. (1979). "On Automated Scientific Theory Formation: A Case Study Using the AM Program". In J. Hayes, D. Michie, and O. Mikulich (eds.), *Machine Intelligence 9,* pp. 251–283. Chichester, U.K.: Ellis Horwood.

————— (1982). "AM: Discovery in Mathematics as Heuristic Search". In R. Davis and D. Lenat (eds.), *Knowledge-Based Systems in Artificial Intelligence,* pp. 1–25. New York: McGraw-Hill.

————— (1983a). "The Role of Heuristics in Learning by Discovery: Three Case Studies". In Michalski, Carbonell, & Mitchell, 1983, pp. 243–306.

————— (1983b). "EURISKO: A Program that Learns New Heuristics and Domain Concepts". *Artificial Intelligence,* vol. 21, no. 1, 2, pp. 61–98.

————— (1983c). "Why AM and Eurisko Appear to Work". In *Proceedings of the American Association of Artificial Intelligence,* pp. 236–240.

Lucas, John R. (1961). "Minds, Machines, and Gödel". *Philosophy,* vol. 31, pp. 112–127.

Maier, N. R. F. (1931). "Reasoning in Humans, II. The Solution of a Problem and Its Appearance in Consciousness". *Journal of Comparative Psychology,* vol. 12, pp. 181–194.

Mantas, J. (1986). "An Overview of Character Recognition Methodologies". *Pattern Recognition,* vol. 19, no. 6, pp. 425–430.

Marr, David (1977). "Artificial Intelligence: A Personal View". *Artificial Intelligence,* vol. 9, pp. 37–48.

Marslen-Wilson, William *et al.* (1992). "Abstractness and Transparency in the Mental Lexicon". In *Proceedings of the Fourteenth Annual Conference of the Cognitive Science Society,* pp. 84–88. Hillsdale, NJ: Lawrence Erlbaum.

McCarthy, John and Patrick Hayes (1969). "Some Philosophical Problems from the Standpoint of Artificial Intelligence". In B. Meltzer and D. Michie (eds.), *Machine Intelligence 4.* Edinburgh, U.K.: Edinburgh University Press.

McClelland, James L. and David E. Rumelhart (1981). "An Interactive Activation Model of Context Effects in Letter Perception: Part I. An Account of Basic Findings". *Psychological Review,* vol. 88, pp. 375–407.

McClelland, James L., David E. Rumelhart, and the PDP Research Group (1986). *Parallel Distributed Processing: Explorations in the Microstructure of Cognition. Vol. II: Psychological and Biological Models.* Cambridge, MA: Bradford Books/MIT Press.

McClelland, James L., David E. Rumelhart, and Geoffrey E. Hinton (1986). "The Appeal of Parallel Distributed Processing". In Rumelhart, McClelland, & the PDP Research Group, 1986, pp. 3–44.

McCorduck, Pamela (1991). *Aaron's Code: Meta-Art, Artificial Intelligence, and the Work of Harold Cohen.* New York: Freeman.

McDermott, Drew (1976). "Artificial Intelligence Meets Natural Stupidity". *SIGART Newsletter,* no. 57, April 1976. Reprinted in J. Haugeland (ed.), *Mind Design.* Montgomery, VT: Bradford Books, 1981.

McGraw, Gary E. and Daniel Drasin (1993). "Recognition of Gridletters: Probing the Behavior of Three Competing Models". In *Proceedings of the Fifth Midwest AI and Cognitive Science Conference.* Carbondale, IL: Southern Illinois University Press.

McGraw, Gary, John Rehling, and Robert Goldstone (1994a). "Letter Perception: Toward a Conceptual Approach". In *Proceedings of the Sixteenth Annual Conference of the Cognitive Science Society,* pp. 613–618. Hillsdale, NJ: Lawrence Erlbaum.

——— (1994b). "Letter Perception: Human Data and Computer Models". Available as Publication #90, Center for Research on Concepts and Cognition, Indiana University, Bloomington. Also submitted for journal publication.

Meredith, Marsha J. (1986). *Seek-Whence: A Model of Pattern Perception.* Doctoral dissertation, Computer Science Department, Indiana University, Bloomington.

——— (1991). "Data Modeling: A Process for Pattern Induction". *Journal for Experimental and Theoretical Artificial Intelligence,* vol. 3, pp. 43–68.

Michalski, Ryszard S., Jaime G. Carbonell, and Thomas M. Mitchell, eds. (1983). *Machine Learning: An Artificial Intelligence Approach.* Palo Alto, CA: Tioga Press. Also reprinted by Morgan Kaufmann (Los Altos, CA).

——— (1986). *Machine Learning: An Artificial Intelligence Approach, Vol. II.* Los Altos, CA: Morgan Kaufmann.

Minsky, Marvin, L. (1985). *The Society of Mind.* New York: Simon & Schuster.

Mitchell, Melanie (1993). *Analogy-Making as Perception.* Cambridge, MA: Bradford Books/MIT Press.

Mitchell, Melanie and Douglas R. Hofstadter (1990a). "The Emergence of Understanding in a Computer Model of Analogy-making". *Physica D,* vol. 42, pp. 322–334.

——— (1990b). "The Right Concept at the Right Time: How Concepts Emerge as Relevant in Response to Context-dependent Pressures". In *Proceedings of the Twelfth Annual Conference of the Cognitive Science Society,* pp. 174–181. Hillsdale, NJ: Lawrence Erlbaum.

Moore, James and Allen Newell (1974). "How Can MERLIN Understand?" In L. W. Gregg (ed.), *Knowledge and Cognition.* Potomac, MD: Lawrence Erlbaum.

Moser, David J. (1991). "Sze-chuan Pepper and Coca-Cola: The Translation of *Gödel, Escher, Bach* into Chinese". *Babel,* vol. 37, no. 2, pp. 75–95.

Nanard, M. *et al.* (1989). "A Declarative Approach for Font Design by Incremental Learning". In J. André & R. Hirsch (eds.), *Raster Imaging and Digital Typography.* Cambridge, U.K.: Cambridge University Press.

Newell, Allen (1990). *Unified Theories of Cognition.* Cambridge, MA: Harvard University Press.

Newell, Allen and Herbert A. Simon (1976). "Computer Science as Empirical Inquiry: Symbols and Search". *Communications of the Association for Computing Machinery,* vol. 19, pp. 113–126. Reprinted in J. Haugeland (ed.), *Mind Design.* Montgomery, VT: Bradford Books, 1981.

Norman, Donald (1981). "Categorization of Action Slips". *Psychological Review,* vol. 88, pp. 1–15.

Novick, Laura R., and N. Coté (1992). "The Nature of Expertise in Anagram Solution". In *Proceedings of the Fourteenth Annual Conference of the Cognitive Science Society,* pp. 450–455. Hillsdale, NJ: Lawrence Erlbaum.

O'Hara, Scott (1992). "A Model of the 'Redescription' Process in the Context of Geometric Proportional Analogy Problems". In Klaus P. Jantke (ed.), *Analogical and Inductive Inference,* pp. 268–293. Berlin: Springer-Verlag.

——— (1994a). Personal communication.

———— (1994b). "A Blackboard Architecture for Case Re-interpretation", in *Proceedings of the Second European Workshop on Case-Based Reasoning.* Chantilly, France: Fondation Royaumont.

O'Hara, Scott and Bipin Indurkhya (1993). "Incorporating (Re-)Interpretation in Case-Based Reasoning". In Stefan Weiss, Klaus-Dieter Althoff, and Michael M. Richter (eds.), *Topics in Case-Based Reasoning, Selected Papers from the First European Workshop on Case-Based Reasoning,* pp. 246–260. Berlin: Springer-Verlag.

Palmer, S. (1977). "Hierarchical Structure in Perceptual Representation". *Cognitive Psychology,* vol. 9, pp. 441–474.

———— (1978). "Structural Aspects of Visual Similarity". *Memory and Cognition,* vol. 6, no. 2, pp. 91–97.

Persson, Staffan (1966). "Some Sequence Extrapolating Programs: A Study of Representation and Modeling in Inquiring Systems". Technical Report STAN-CS-66-050, Computer Science Department, Stanford University.

Pivar, M. and M. Finkelstein (1964). "Automation, Using LISP, of Inductive Inference on Sequences". In E. C. Berkeley and D. Bobrow (eds.), *The Programming Language LISP: Its Operation and Applications,* pp. 125–136. Cambridge, MA: Information International.

Pylyshyn, Zenon (1980). "Cognition and Computation". *Behavioral and Brain Sciences,* vol. 3, pp. 111–132.

Qin, Y. and Herbert A. Simon (1990). "Laboratory Replication of Scientific Discovery Processes". *Cognitive Science,* vol. 14, pp. 281–310.

Racter (1984). *The Policeman's Beard is Half Constructed.* New York: Warner Books.

Ray, Thomas (1992). "An Approach to the Synthesis of Life". In Christopher G. Langton *et al.* (eds.), *Artificial Life II,* pp. 371–408. Redwood City, CA: Addison-Wesley.

Reddy, D. Raj *et al.* (1976). "Working Papers in Speech Recognition, IV: The HEARSAY II System." Technical Report, Computer Science Department, Carnegie-Mellon University.

Reitman, Walter (1965). *Cognition and Thought: An Information-Processing Approach.* New York: John Wiley.

Riesbeck, Christopher K. and Roger C. Schank (1989). *Inside Case-Based Reasoning.* Hillsdale, NJ: Lawrence Erlbaum.

Ritchie, G. and F. Hanna (1990). "AM: A Case-Study in AI Methodology". In D. Partridge and Y. Wilks (eds.), *The Foundations of AI: A Sourcebook.* New York: Cambridge University Press.

Rowe, J. and Derek Partridge (1991). "Creativity: A Survey of AI Approaches". Technical Report #R-214, Computer Science Department, University of Exeter.

Rumelhart, David E., Geoffrey E. Hinton, and Ronald Williams (1986). "Learning Internal Representations by Error Propagation". In Rumelhart, McClelland, and the PDP Research Group, 1986, pp. 319–362.

David E. Rumelhart, James L. McClelland, & the PDP Research Group (1986). *Parallel Distributed Processing: Explorations in the Microstructure of Cognition. Vol. I: Foundations.* Cambridge, MA: Bradford Books/MIT Press.

Rumelhart, David E. and Donald Norman (1982). "Simulating a Skilled Typist: A Study of Skilled Cognitive–Motor Performance". *Cognitive Science,* vol. 6, no. 1, pp. 1–36.

Schank, Roger C. (1980). "Language and Memory". *Cognitive Science,* vol. 4, no. 3, pp. 243–284.

———— (1982). *Dynamic Memory*. New York: Cambridge University Press.

Searle, John (1980). "Minds, Brains, and Programs". *Behavioral and Brain Sciences,* vol. 3, pp. 417–458. Also reprinted in Hofstadter & Dennett (eds.), 1981.

Shrager, J. (1990). "Commonsense Perception and the Psychology of Theory Formation". In J. Shrager and P. Langley (eds.), *Computational Models of Scientific Discovery and Theory Formation*. Los Altos, CA: Morgan Kaufmann.

Simon, Herbert A. (1981). "1980 Procter Lecture: Studying Human Intelligence by Creating Artificial Intelligence". *American Scientist,* vol. 69, no. 3, pp. 300–309.

———— (1982). Personal communication, Oct. 21, 1982.

———— (1989). "The Scientist as Problem Solver". In David Klahr and Kenneth Kotovsky (eds.), *Complex Information Processing*. Hillsdale, NJ: Lawrence Erlbaum.

Simon, Herbert A. and Kenneth Kotovsky (1963). "Human Acquisition of Concepts for Sequential Patterns". *Psychological Review,* vol. 70, no. 6, pp. 534–546.

Skorstad, J., Brian Falkenhainer, and Dedre Gentner (1987). "Analogical Processing: A Simulation and Empirical Corroboration". In *Proceedings of the 1987 Conference of the American Association for Artificial Intelligence*. Los Altos, CA: Morgan Kaufmann.

Smith, Brian C. (1982). "Reflection and Semantics in a Procedural Language". Technical Report #272, Laboratory for Computer Science, Massachusetts Institute of Technology.

Smolensky, Paul (1983a). Personal communication.

———— (1983b). "Harmony Theory: A Mathematical Framework for Stochastic Parallel Processing". In *Proceedings of the 1983 Conference of the American Association of Artificial Intelligence*.

———— (1986). "Information Processing in Dynamical Systems: Foundations of Harmony Theory". In Rumelhart, McClelland, and the PDP Research Group, 1986, pp. 194–281.

———— (1988). "On the Proper Treatment of Connectionism". *Behavioral and Brain Sciences,* vol. 11, no. 1, pp. 1–74.

Treisman, Anne (1988). "Features and Objects: The Fourteenth Bartlett Memorial Lecture". *Quarterly Journal of Experimental Psychology,* vol. 40A, pp. 201–237.

Treisman, Anne and G. Gelade (1980). "A Feature-Integration Theory of Attention". *Cognitive Psychology,* vol. 12, no. 12, pp. 97–136.

Turing, Alan M. (1950). "Computing Machinery and Intelligence", *Mind,* vol. 59, no. 236. Reprinted in A. R. Anderson (ed.), *Minds and Machines*. Englewood Cliffs, NJ: Prentice-Hall, 1964.

Waldrop, M. Mitchell (1987). "Causality, Structure, and Common Sense". *Science,* vol. 237, pp. 1297–1299.

Waterman, D. A. and Frederick Hayes-Roth (1978). *Pattern-Directed Inference Systems*. New York: Academic Press.

Weizenbaum, Joseph (1976). *Computer Power and Human Reason: From Judgment to Calculation*. San Francisco: Freeman.

Winograd, Terry A. (1972). *Understanding Natural Language*. New York: Academic Press.

Winston, Patrick H. (1982). "Learning New Principles from Precedents and Exercises". *Artificial Intelligence,* vol. 19, pp. 321–350.

Zapf, Hermann (1970). *About Alphabets: Some Marginal Notes on Type Design*. Cambridge, MA: MIT Press.

Index